Springer Series in

MATERIALS SCIENCE 41

Springer-Verlag
Berlin Heidelberg GmbH

Physics and Astronomy ONLINE LIBRARY

http://www.springer.de/phys/

Springer Series in
MATERIALS SCIENCE

The Springer Series in Materials Science covers the complete spectrum of materials physics, including fundamental principles, physical properties, materials theory and design. Recognizing the increasing importance of materials science in future device technologies, the book titles in this series reflect the state-of-the-art in understanding and controlling the structure and properties of all important classes of materials.

27 **Physics of New Materials**
Editor: F. E. Fujita 2nd Edition

28 **Laser Ablation**
Principles and Applications
Editor: J. C. Miller

29 **Elements of Rapid Solidification**
Fundamentals and Applications
Editor: M. A. Otooni

30 **Process Technology for Semiconductor Lasers**
Crystal Growth and Microprocesses
By K. Iga and S. Kinoshita

31 **Nanostructures and Quantum Effects**
By H. Sakaki and H. Noge

32 **Nitride Semiconductors and Devices**
By H. Morkoç

33 **Supercarbon**
Synthesis, Properties and Applications
Editors: S. Yoshimura and R. P. H. Chang

34 **Computational Materials Design**
Editor: T. Saito

35 **Macromolecular Science and Engineering**
New Aspects
Editor: Y. Tanabe

36 **Ceramics**
Mechanical Properties, Failure Behaviour, Materials Selection
By D. Munz and T. Fett

37 **Technology and Applications of Amorphous Silicon**
Editor: R. A. Street

38 **Fullerene Polymers and Fullerene Polymer Composites**
Editors: P. C. Eklund and A. M. Rao

39 **Semiconducting Silicides**
Editor: V.E. Borisenko

40 **Reference Materials in Analytical Chemistry**
A Guide for Selection and Use
Editor: A. Zschunke

41 **Organic Electronic Materials**
Conjugated Polymers and Low Molecular Weight Organic Solids
Editors: R. Farchioni and G. Grosso

42 **Raman Scattering in Materials Science**
Editors: W. H. Weber and R. Merlin

43 **The Atomistic Nature of Crystal Growth**
By B. Mutaftschiev

44 **Thermodynamic Basis of Crystal Growth**
P–T–X Phase Equilibrium and Nonstoichiometry
By J.H. Greenberg

45 **Principles of Thermoelectrics**
Basics and New Materials Developments
By G.S. Nolas, J. Sharp, and H.J. Goldsmid

46 **Fundamental Aspects of Silicon Oxidation**
Editor: Y. J. Chabal

47 **Disorder and Order in Strongly Non-Stoichiometric Compounds**
Transition Metal Carbides, Nitrides and Oxides
By A.I. Gusev, A.A. Rempel, and A.J. Magerl

48 **The Glass Transition**
Relaxation Dynamics in Liquids and Disordered Materials
By E. Donth

Series homepage – http://www.springer.de/phys/books/ssms/

Volumes 1–26 are listed at the end of the book.

R. Farchioni G. Grosso (Eds.)

Organic Electronic Materials

Conjugated Polymers and Low Molecular Weight Organic Solids

With 200 Figures

Springer

Dr. Riccardo Farchioni
Professor Giuseppe Grosso
INFM and Dipartimento di Fisica,
Università di Pisa,
Piazza Torricelli 2,
56126 Pisa, Italy

Series Editors:

Prof. Alex Zunger
NREL
National Renewable Energy Laboratory
1617 Cole Boulevard
Golden Colorado 80401-3393, USA

Prof. Robert Hull
University of Virginia
Dept. of Materials Science and Engineering
Thornton Hall
Charlottesville, VA 22903-2442, USA

Prof. R. M. Osgood, Jr.
Microelectronics Science Laboratory
Department of Electrical Engineering
Columbia University
Seeley W. Mudd Building
New York, NY 10027, USA

Prof. H. Sakaki
Institute of Industrial Science
University of Tokyo
7-22-1 Roppongi, Minato-ku
Tokyo 106, Japan

ISBN 978-3-642-63085-9 ISBN 978-3-642-56425-3 (eBook)
DOI 10.1007/978-3-642-56425-3

Library of Congress Cataloging-in-Publication Data. Organic electronic materials: conjugated polymers and low molecular weight organic solids/R. Farchioni, G. Grosso (eds.) p. cm. – (Springer series in materials science, ISSN 0933-033X; 41) Includes bibliographical references and index. ISBN 3540667210 (alk. paper) 1. Conducting polymers. 2. Organic conductors. I. Farchioni, Riccardo. II. Grosso, G. (Giuseppe)., 1948– III. Springer series in materials science ; v. 41. QD382.C66 O74 2001 547'.70457–dc21 00-067932

http://www.springer.de

Originally published by Springer-Verlag Berlin Heidelberg New York in 2001
Softcover reprint of the hardcover 1st edition 2001

Data conversion: Frank Herweg, Leutershausen
Cover concept: eStudio Calamar Steinen
Cover production: *design & production* GmbH, Heidelberg

Printed on acid-free paper SPIN: 11375159 57/3111 5 4 3 2

Preface

This volume brings together selected reviews on basic ideas and phenomena relevant to the study of the electronic and optical properties of organic materials. The contributions focus on two branches, one dedicated to conjugated conducting polymers, the other to low molecular weight organic solids. For each branch a general introductory review provides information on the main physical aspects of the materials and a background knowledge of the language and of the points useful for the comprehension of the remaining parts of the book. Theoretical methods and numerical procedures for the evaluation of the electronic energy levels are then overviewed; this is useful also to provide support to the models often employed for the explanation of the phenomenology of these materials. The other chapters complete the understanding of the underlying physics and chemistry through the discussion of chemical structures, including electron–lattice interactions, surface effects, doping, photophysics, metallic conductivity, multilayer structures, optical and dielectric properties with inclusion of excitonic effects, and nonlinear spectroscopy, with reference to experimental research and applications.

The first part of the book is devoted to conjugated polymers. The introductory overview to electronic polymers by A.J. Epstein highlights the effect of the nanoscale structure on their semiconducting and metallic phenomena. The role of morphology and disorder in the photophysics, device physics, charge transport and optical properties of these materials is discussed.

Theoretical studies of materials properties by means of parameter-free Hamiltonians are presented in the review by M. Springborg, K. Schmidt, H. Meider and L. De Maria. Idealizations and models used to apply theoretical methods are discussed and it is shown that good agreement between experimental and theoretical results can be achieved. Response of the materials to external electromagnetic fields and effects due to structure modifications, composition and defects are also examined.

R. Farchioni, G. Grosso and P. Vignolo then show how to exploit real-space recursive procedures within the tight-binding method to construct effective one-dimensional chains representing the true polymer. By means of the recursion and the renormalization methods one can evaluate the Green's function of the polymer and from it obtain its electronic and transmission properties also in the presence of disorder and interchain interactions.

The photophysics of conjugated polymers is the subject of the review by E.M. Conwell. Possible excitations such as polarons, solitons, singlet and triplet excitons, biexcitons, bipolarons and excimers, created by photons, are examined, and optical absorption is discussed. Excitations involving interchain interactions, photogeneration of charge carriers and photoconductivity in conducting polymers are also considered.

The contribution of M. Fahlman and W.R. Salaneck provides an overview of the electronic structure of conjugated polymers and of interfaces between polymers and metals as obtained by means of photoelectric experiments interpreted with quantum chemistry calculations. This kind of spectroscopy for interfaces proves to be a precious tool in solving key questions encountered in the application of organic materials.

The second part of the volume deals with low molecular weight organic solids. The introduction to the subject by N. Karl elucidates the main differences of these materials with respect to typical inorganic materials. It also shows the importance of the comprehension of the nature of the excited states for understanding dynamic, optoelectronic and device applications of this class of organic materials.

First-principles approaches and basic notions concerning electronic states in organic molecular crystals in near-to-equilibrium situations are then reviewed by V. Čápek and I. Muzikante. Exciton states, charge-transfer states and the influence of phonons are also considered for ideal crystals and the role and detection of localized states created by structural defects or impurities are examined.

The fundamental concepts of charge carrier mobility and experiments for their detection are then overviewed by N. Karl; this contribution focuses on the role of intermolecular interactions on the local transfer of a charge carrier to a neighbouring molecule and the effect of impurities.

A description of elementary excitations in molecular crystals based on the dielectric approach is presented in the contribution by R.W. Munn and P. Petelenz. This approach is then applied to nonlinear optics and to the mixing of charge transfer and Frenkel excitons. Comparisons between Hamiltonian and dielectric approach descriptions stress the exciton role in these materials.

Layered structures made of heterojunctions between molecular solids and inorganic semiconductors or of donor–acceptor interfaces between crystalline organic materials are the subject of the review by V.M. Agranovich, G.C. La Rocca and F. Bassani. Theoretical analysis of peculiar effects such as resonances between Wannier excitons and Frenkel excitons, energy transfer mechanisms, resonant optical nonlinearities and photovoltaic effects are reported.

Fundamental experimental results on the behaviour of exciton processes in doped molecular organic films are then reviewed by V. Bulović, M.A. Baldo and S.R. Forrest. Energy transfer processes between dissimilar organic molecules and exciton transport are discussed and their contribution to the performance of light-emitting devices emphasized.

The style and the level of presentation is such as to render this book valuable to post graduate students and researchers working in solid-state physics and chemistry, material science, electronic materials and devices, and related fields. An effort has been made to obtain self-contained reviews with a coherent tutorial style whenever possible. Thus, the reader is brought to the study of these materials after having appreciated the development of appropriate theories and phenomenological models. The presentation of relevant experimental facts is always done with attention to the physical models and to the principles at the heart of their interpretation.

Pisa, January 2001

Riccardo Farchioni
Giuseppe Grosso

Contents

List of Contributors

V. M. Agranovich
Institute of Spectroscopy,
Russian Academy of Sciences,
142190 Troitsk, Russia
`agran@isan.troitsk.ru`

M. A. Baldo
POEM, Department of Electrical Engineering, and
the Princeton Materials Institute,
Princeton University,
Princeton, NJ 08544, U.S.A.
`baldo@ee.princeton.edu`

F. Bassani
INFM and Scuola Normale Superiore,
Piazza dei Cavalieri 7,
I-56126 Pisa, Italy
`bassani@sns.it`

V. Bulović
POEM, J-303 Engineering QUAD,
Princeton University,
Princeton, NJ 08544, U.S.A.,
and Universal Display Corporation,
Ewing, NJ 08618, U.S.A.
`vladimir@princeton.edu`

V. Čápek
Institute of Physics of Charles University, Faculty of Mathematics and Physics,
Ke Karlovu 5,
CZ-121 16 Prague 2, Czech Republic
`capek@karlov.mff.cuni.cz`

E. M. Conwell
Department of Chemistry,
University of Rochester,
200 Hutchison Hall,
Rochester, New York 14627 U.S.A.
`conwell@chem.chem.rochester.edu`

L. De Maria
Unité Conformation de
Macromolécules Biologiques,
Universite Libre de Bruxelles
Avenue Paul Higer,
CP-160/16,
B-1050 Bruxelles, Belgium
`leonardo@ucmb.ulb.ac.be`

A. J. Epstein
Department of Physics and
Department of Chemistry,
The Ohio State University,
174 West 18th Avenue, Columbus,
Ohio 43210-1106, U.S.A.
`epstein.2@osu.edu`

M. Fahlman
Department of Surface Physics
and Chemistry, IFM,
Linköping University,
S-58183 Linköping, Sweden
`mafah@ifm.liu.se`

R. Farchioni
INFM and Department of Physics,
University of Pisa
Piazza Torricelli 2, I-56126 Pisa,
Italy
farchion@df.unipi.it

S. R. Forrest
POEM, Department of Electrical
Engineering and
the Princeton Materials Institute,
EQUAD B-210,
Princeton University,
Princeton, NJ 08544, U.S.A.
forrest@ee.princeton.edu

G. Grosso
INFM and Department of Physics,
University of Pisa
Piazza Torricelli 2,
I-56126 Pisa, Italy
grosso@df.unipi.it

N. Karl
3. Institute of Physics,
University of Stuttgart,
70550 Stuttgart, Germany
n.karl@physik.uni-stuttgart.de

G. C. La Rocca
Department of Physics,
University of Salerno,
via S. Allende,
I-84081 Baronissi (Sa), Italy;
and
INFM, and Scuola Normale
Superiore,
Piazza dei Cavalieri 7,
I-56126 Pisa, Italy
larocca@sns.it

H. Meider
Department of Chemistry,
University of Konstanz,
Universitätsstrasse 10, Fach M722
78457 Konstanz, Germany
heiko.meider@uni-konstanz.de

R. W. Munn
Department of Chemistry, UMIST,
University of Manchester,
Manchester M60 1QD,
United Kingdom
r.w.munn@umist.ac.uk

I. Muzikante
Institute of Physical Energetics,
Academy of Science of Latvia,
Aizkraukles ielā 21,
LV-1006 Riga, Latvia
intam@edi.lv

P. Petelenz
Faculty of Chemistry,
ul. Ingardena 3,
30-060 Krakow, Poland
petelenz@chemia.uj.edu.pl

W. R. Salaneck
Department of Surface Physics and
Chemistry, IFM,
Linköping University,
S-581 83 Linköping, Sweden
bisal@ifm.liu.se

K. Schmidt
Department of Physics,
Tuluane University,
LA 70118 New Orleans, U.S.A.
kschmidt@mailhost.tcs.tuluane.edu

M. Springborg
Physical Chemistry,
University of Saarland,
D-66123 Saarbrücken, Germany
m.springborg@mx.uni-saarland.de

P. Vignolo
INFM and Scuola Normale
Superiore,
Piazza dei Cavalieri 7,
I-56126 Pisa, Italy
vignolo@cibs.sns.it

Part I

Conjugated Polymers

1 Introduction to Electronic Polymers: Influence of Nanostructure on Electronic Phenomena

A.J. Epstein

1.1 Introduction

While polymers have been known and utilized broadly for many decades, polymers with the intrinsic properties of semiconductors and metals are a relatively recent phenomenon. The report in 1977 about doping polyacetylene to achieve relatively high conductivity [1] opened up important new vistas for physics and chemistry, and for technology in general. Early studies recognized that a key feature of the electronic polymers and organic materials was a backbone consisting of alternating single and double bonds resulting in a "π-conjugated network". This in turn led to a relatively small energy gap, enabling the appearance of both semiconducting and metallic properties. The following two chapters by Springborg et al. and Grosso et al. present an introduction to the important concepts of conjugated polymer chains. A review of the literature of conjugated polymers is found in [2].

Initially these polymers were unstable in air and not readily processed. Over the past two decades major advances have occurred in the synthesis of new forms of conducting and semiconducting polymers that enable processing under a broad range of conditions including organic solvents, inorganic solvents, and aqueous media. There are even melt-processable versions of some of the electronic polymers.

A prime focus of the field has been the determination of the mechanisms for charge conduction and the intrinsic conductivity of these fascinating materials, especially doped polymers. In the past decade, interest also has increased in the semiconducting (generally undoped) forms of these polymers and organic materials, including their photophysics and their use in a wide variety of devices. The reports of light-emitting devices fabricated from molecular constituents, initially in the 1960s and subsequently in the mid-1980s [3], and from polymeric constituents in 1990 [4] stimulated interest in this area of research.

The ideal chemical structures of the parent polymers extensively studied in the conducting state are shown in Fig. 1.1, while many of the polymers studied to examine their semiconducting properties are shown in Fig. 1.2. The one-dimensional electronic structure of these polymers is discussed in detail in the following chapters. The roles of electron–phonon interaction, Peierls distortion, Coulomb interaction, etc., are well established. However, many

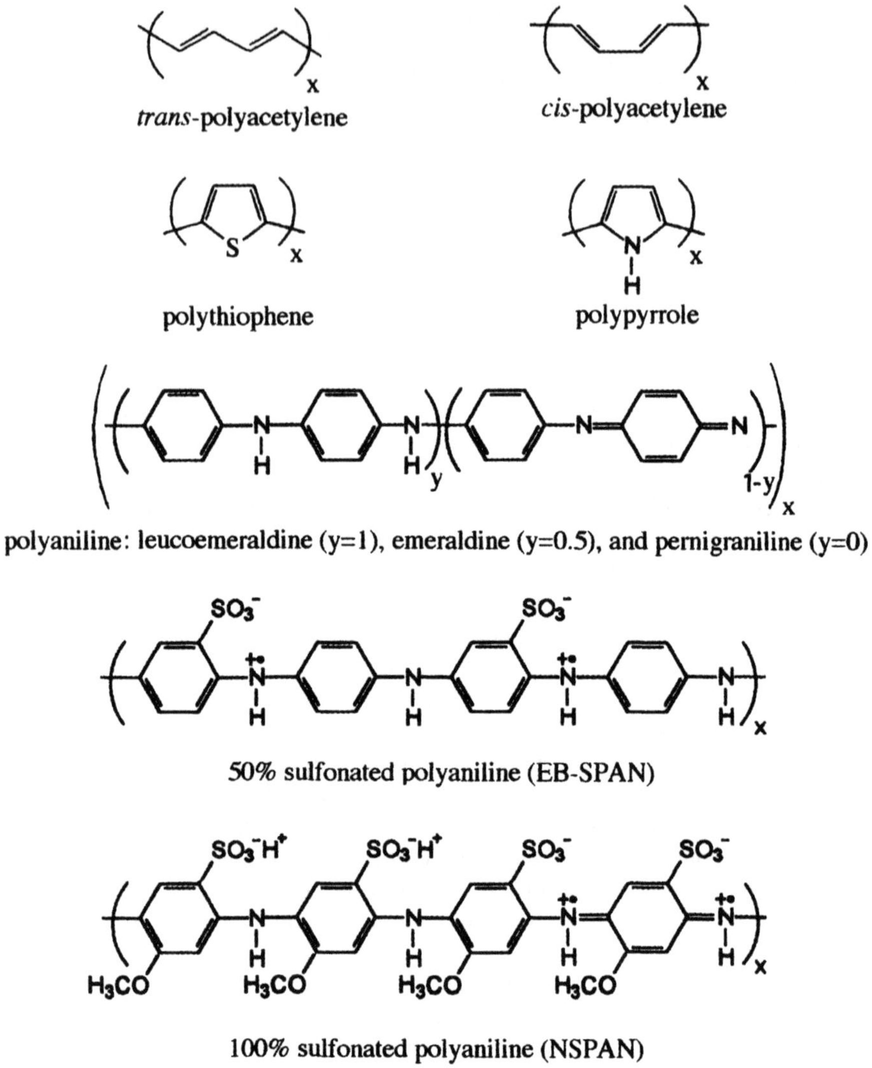

Fig. 1.1. Schematic structure of trans-polyacetylene, cis-polyacetylene, polythiophene, polypyrrole, and polyaniline (leucoemeraldine, emeraldine, and pernigraniline forms). These polymers and their derivatives, when doped, become conducting, with conductivities varying with dopant and processing. Also shown are the 50% and 100% sulfonated polyanilines (self-doped polymers)

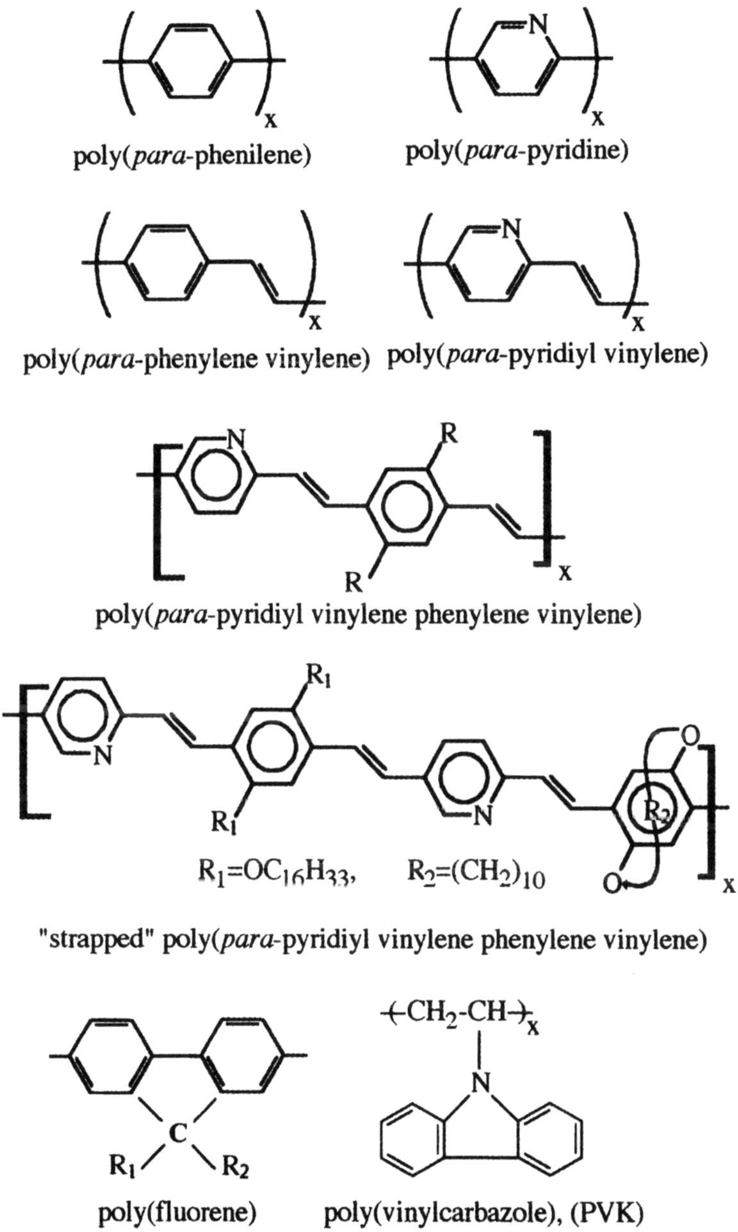

Fig. 1.2. Schematic structure of selected semiconducting polymers. Note that all but PVK are "conjugated" polymers

of the experimental data acquired for semiconducting and doped conducting forms of these polymers do not fit simple models developed for isolated one-dimensional chains. The following chapter by Salaneck on photoelectron spectroscopy studies of the electronic and chemical structures of conjugated polymers provides more insight into these issues.

One important issue is the morphology of individual chains (e.g. twisting and bending of the chain backbone and the variation of the radius of curvature used to describe these changes). Another key and nearly ubiquitous feature of electronic polymers is the important role of interchain interactions which strongly depend upon the variation on the nanoscale of local order in this system. Extensive structural studies of polymers have shown the development of inhomogeneous disorder leaving regions within polymer films where the chains are relatively well packed, and nearly crystalline and regions between these crystalline-like portions in which interchain correlations are very short range (Fig. 1.3a). Further, as indicated in Fig. 1.3b there are disordered regions which may be rod-like and disordered regions which may be coil-like. The local morphology is very sensitive to the preparation of materials and the means by which the films and powders are prepared. Hence materials of the same chemical composition may have very different electronic properties depending on the solvents used and the means of preparing the films and powders.

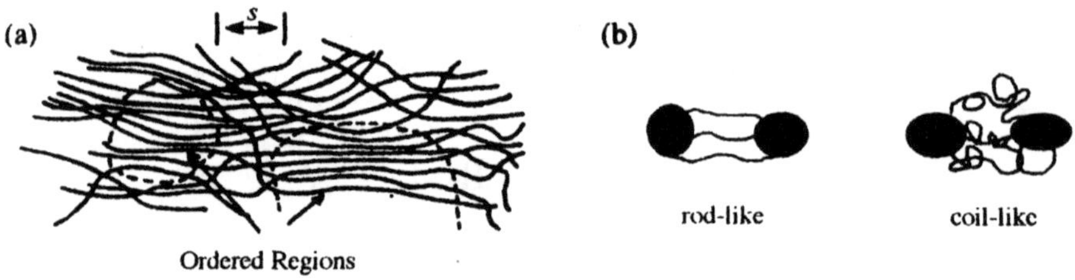

Fig. 1.3. (**a**) Schematic of the inhomogeneously disordered state of metallic and semiconducting polymers. (**b**) Schematic of rod-like and coil-like morphology of disordered regions

In this overview, the experimental detection of the inhomogeneous morphology and the implications for the control of the structural order on the photophysics and device physics of semiconducting light-emitting polymers and on the charge transport and optical properties of doped conducting polymers are discussed. These nanoscale variations in local order control much of the physical phenomena in these materials.

1.2 Inhomogeneous Structure of Electronic Polymers

Extensive X-ray diffraction studies as well as other probes of local order have been reported. To illustrate the richness in the structural order, selected results for polyacetylene, polypyrrole and polyaniline are presented. Experimental studies of other polymers including polythiophene and poly(phenylene vinylene)-based materials [5] lead to similar conclusions concerning the variability of the local order.

1.2.1 Pristine and Iodine-Doped trans-$(CH)_x$

Polyacetylene, $(CH)_x$, is usually prepared in the cis- form which can be converted into the thermodynamically more stable trans- form by thermal isomerization. Pristine $(CH)_x$ is found to be highly crystalline, with a fraction of crystallinity ($X_c \geq 90\%$) independent of the degree of isomerization [6]. In the crystalline regions cis- and trans-$(CH)_x$ adopt the same 2D *Pgg* hexagonal-like chain array with two chains per unit cell. Its lattice parameters evolve continuously with the degree of isomerization [6]. The 3D structure of cis-$(CH)_x$ is orthorhombic (*Pnam* space group [7a]) and a later refinement [7b] confirmed that the $P2_1/n^*$ monoclinic structure, in which the bond alternation is out of phase between the two chains of the unit cell, is adopted by trans-$(CH)_x$.

The 3D crystalline array is, however, spatially limited. In the crudest approximation, assuming that atomic position–position correlations are lost in narrow domain walls, crystalline domain sizes of the order of $L \sim 10$ nm, a value independent of the synthesis method within a factor of two [6,8,9], were determined both in interchain and intrachain directions. The loss of the crystalline order can be analyzed by paracrystalline models [10] for an improved description [11,12] where, for example, the broadening of the equatorial reflections is due to cumulative fluctuations of the interchain distances inside a crystalline domain.

Table 1.1 [11] compares the interchain domain size, $L_\perp$ and the percentage of paracrystallinity, σ_r/d, obtained from the analysis of the FWHM of the equatorial peaks of Tsukamoto polyacetylene, T-$(CH)_x$, with similar quantities deduced from the same analysis of the data of $(CH)_x$ prepared using the Shirakawa (S-$(CH)_x$ [12,13]), Akagi (A-$(CH)_x$ [14]) and Durham (D-$(CH)_x$ [15]) methods. With an average distance between first neighboring $(CH)_x$ chains of $d = 0.42$ nm, the first neighbor mean fluctuation σ_r amounts to between 0.014 and 0.026 nm. The interchain order is not improved in T-$(CH)_x$ [16] with respect to earlier methods of synthesis. The interchain crystalline domain size of T-$(CH)_x$ is comparable to that of Naarmann [9] and D-$(CH)_x$ [15] materials. It is even approximately a factor of two smaller than in S-$(CH)_x$ [12] and A-$(CH)_x$ [14]. The reduced interchain crystalline domain size for the undoped T-$(CH)_x$ may enable a more homogeneous doping and, hence, a more metallic doped material.

Table 1.1. Interchain crystalline domain size, $L_\perp$, and percentage of radial paracrystallinity σ_r/d, deduced from the analysis of the FWHM of the equatorial diffraction peaks of $(CH)_x$ prepared using different synthesis methods and different degrees of isomerization. (From [11])

Synthesis	Isomerization	$L_\perp$ (Å)	σ_r/d (%)	Ref.
T-$(CH)_x$	50% cis–50% trans	55	5±0.3	[11]
	trans	55	4±0.3	
S-$(CH)_x$	~ 90% cis	90	3.8	[12]
	trans	90	3.8	
A-$(CH)_x$	trans	90	3.3	[14]
D-$(CH)_x$	trans	45	6.3	[15]

Fig. 1.4 shows also that: (a) the interchain crystalline domain size does not vary significantly with the degree of isomerization; and (b) paracrystalline fluctuations are enhanced in the 50% cis–50% trans sample with respect to those of the nearly completely isomerized trans- sample. The first feature was previously noticed [6] in structural studies of partly isomerized S-$(CH)_x$. The second feature reflects the increase of lateral disorder expected in a sample containing about the same amount of cis and trans isomers.

The similarity in the intercalated iodine–$(CH)_x$ structures for heavily doped polyacetylene of different preparations suggests that the origins of their thousand-fold difference in conductivity (from about 50 to 50 000 S/cm) are subtle. In particular, the reduction of the I_3^- tilting angle for doped T-$(CH)_x$ suggests that small changes in the relative placement of the anion dopant and the host polymer (Table 1.2) [11,17–19] may make a very substantial difference in the interchain charge transfer and, hence, control the final metallic state.

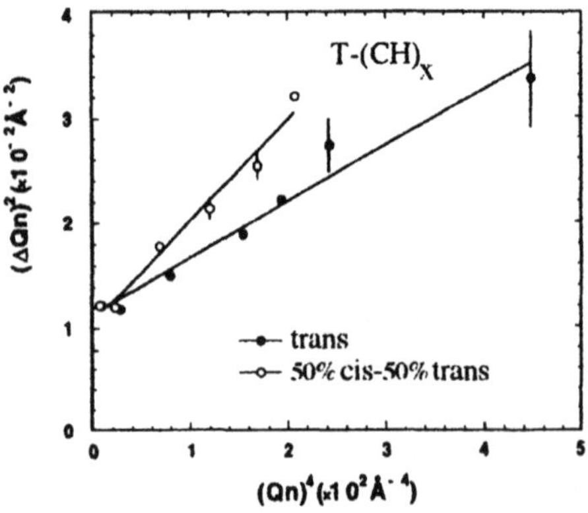

Fig. 1.4. The width of the nth diffraction peak, ΔQ_n, centered at the wavevector $Q_n = n(2\pi/d)$, squared (corrected for instrumental broadening) of the equatorial (hk0) reflections of the Debye–Scherrer spectrum otained from a nearly completely isomerized trans-$(CH)_x$ and from a 50%–cis- 50%–trans-$(CH)_x$ as a function of $(Q_n)^4$. (From [11])

Table 1.2. I_3^- repeat distance, d, I–I intra-unit distance, u, and amplitude of first neighbor fluctuations of distance, σ, in the chain direction for heavily iodine-doped A- and T-CH_x at room and low temperatures (RT and LT, respectively). These values are compared to those found in the organic crystal DIPS$\phi_4(I_3)_{0.76}$. (From [11])

	d (Å)	u (Å)	σ_{RT} (Å)	σ_{LT} (Å)	Ref.
T-$[CH(I_3)_y]_x$ $y \sim 0.08$	9.7±0.1	3.05±0.05	0.4	0.3 (25 K)	[11]
A-$[CH(I_3)_y]_x$ $y \sim 0.06$	9.35	2.92	0.34	0.23 (33 K)	[17]
DIPS$\phi_4(I_3)_{0.76}$	9.79	3.00	0.35	0.14 (50 K)	[19]

The coexistence of both metallic (finite conductivity at low temperatures and negative microwave frequency dielectric constant) and non-metallic states (conductivity vanishing at low temperatures and positive microwave frequency dielectric constant) even within different portions of samples of heavily iodine-doped stretched polyacetylene was reported [18]. A huge negative dielectric constant ε_{mw} was recorded for the center portions and a relatively small and positive ε_{mw} was measured for the neck and end portions of a six-fold stretched sample, demonstrating the sensitivity of the electronic structure and the associated phenomena to the local order.

1.2.2 Polyaniline

Polyaniline forms a rich family of electronic polymers. However, the local structure and degree of order depend crucially on the method of preparation. The electronic structure and associated phenomena are very sensitive to variations in the local order because of the steric repulsion between adjacent C_6H_4 rings of polyaniline and the ease with which disorder will induce their deviation from coplanarity with adjacent rings of the chain.

For powders of polyaniline, two classes of materials have clearly been distinguished [20]. Class I consists of materials prepared in the conducting doped emeraldine salt (ES) form, while class II consists of materials prepared in the insulating emeraldine base (EB) form. As-prepared hydrochloride ES-I powders are substantially "crystalline" (fraction of crystallinity $X_c \sim 50\%$), thereby introducing a nanoscale modulation of the structure. A description of the structure of the ES-I crystalline regions is given in [20]. Dedoping of ES-I leads to an essentially amorphous emeraldine base EB-I, the local structure of which was characterized [5,11,21,22] via the determination of the atomic radial distribution function (RDF). Redoped EB-I yields substantially "crystalline" ES-I. EB-II powders are also substantially "crystalline" (X_c up to 50%) with a local structure in accordance with the *Pbcn* space group [20] found for analogous polymers, such as poly(paraphenylene sulfide) (PPS) and poly(paraphenylene oxide) (PPO). Upon doping with HCl, EB-II powders are converted into substantially "crystalline" ES-II with a local structure in accordance with the *Pc2a* space group [20]. Dedoped ES-II leads to essentially amorphous EB-II. Its RDF, $W(r)$, has peaks observed at 0.14, 0.24 and 0.29 nm in both EB-I and EB-II, correspond to polymer interatomic distances. Those observed for $r > 0.034$ nm correspond mostly to interchain position–position correlations. Fig. 1.5 shows that these correlations slightly differ in EB-I and EB-II which had been doped with HCl then dedoped. However, the main difference concerns the amplitude of oscillations of $W(r)$ which remains sizeable for r values as large as 1 nm in EB-I, but which nearly vanishes for $r > 0.6$ nm in EB-II. These features show that the extent of the local order in amorphous polyaniline depends on the sample history through the number of doping–dedoping cycles performed during the preparation: EB-I powder studied is obtained after one dedoping process, while EB-II powder studied is obtained after a doping process followed by a dedoping process. Similar conclusions were also obtained [23] from the study of EB-II polyaniline films.

Unstretched acid-cast emeraldine films are also substantially crystalline ($X_c \sim 40\%$ when doped with a Cl/N ratio of about 0.5) with an ES-I local structure. Their physical properties are described in [24]. NMP cast films exhibit a larger variety of structural behavior. Most of the unstretched un-

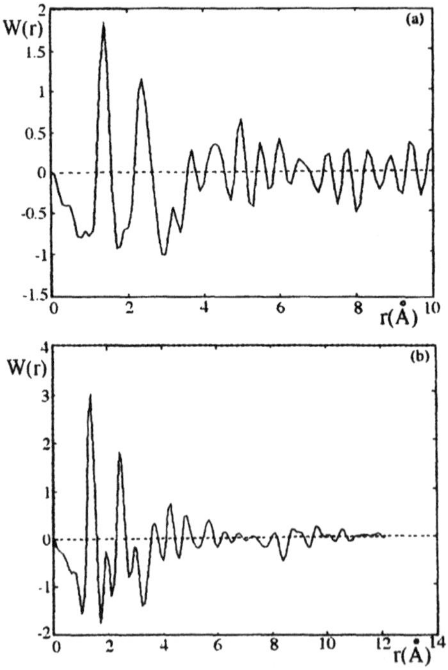

Fig. 1.5. Radial distribution function $W(r)$ of (**a**) EB-I powder (from [21]) and (**b**) EB-II, an EB-II powder doped with HCl, then dedoped (from [11])

doped (EB) and HCl-doped (ES) films are weakly crystalline. Non-crosslinked (NXL) films prepared using the procedure of [25] are always amorphous. Sometimes crosslinked (XL) films are found to have a fraction of crystallinity X_c up to about 20–30%. This amount could be controlled by the fraction of NMP remaining in the film which allows the formation of a free volume essential for a conformational reorganization of the polymeric chains. The RDF analysis of an amorphous NMP-cast EB film shows that the local order is similar to that of "crystalline" EB-II [23]. Either ES-I local order or ES-II local order has been observed in HCl-doped NMP-cast films.

A variation in crystallinity was observed from sample to sample. This could be due to aging effects associated with a slow evaporation of the solvent. In particular, a slight improvement of the crystalline order was observed over 8 months in our *m*-cresol PAN-CSA (CSA, camphor sulfonic acid) film. The different structures for the polyaniline chains of PAN-CSA could be related to the expected change from a "rod-like" conformation to a "coil-like" conformation of the polymeric chain when the material is processed with *m*-cresol and $CHCl_3$ solvents, respectively. Such a change could be due to a weaker interaction between the polymer chain and the solvent when *m*-cresol is replaced by $CHCl_3$ [26]. These different chain structures are important for understanding the dramatic drop in conductivity and the drastic change in the microwave frequency dielectric constant and electronic spectra from PAN-CSA (*m*-cresol) to PAN-CSA ($CHCl_3$) [26,27].

Some of the structural features that are important for understanding the electronic properties of the polyaniline family of electronic polymers are summarized here. First, polyaniline contains "crystalline" and amorphous regions that exhibit the same local order [23]. It is only the spatial extent of position–position correlations between polymeric chains which are smaller in the so-called amorphous regions. Thus, as schematically shown in Fig. 1.3a, the interface between crystalline (or, more likely, paracrystalline) regions and amorphous regions is very diffuse. As the extent of position–position correlations varies with the site taken as the origin for the diffraction, polyaniline material can be considered as a structurally "inhomogeneous" material. In this respect the charge delocalization will be better achieved in the crystalline or paracrystalline regions, where the polymer chains are better aligned than in the amorphous regions. In addition, the nature of the disorder of the "amorphous" links controls the dc charge transport properties between the 3D "crystalline" regions where the electron wavefunction is delocalized. Dielectric response measurements [27] show that the size of the 3D crystalline regions appear to control the microwave charge transport properties at low temperature for non-metallic samples, and that knowledge of the nature of the "amorphous" link between the crystalline domains is essential to understand the enhancement of the electronic delocalization observed when the temperature increases [28, 29]. The charge delocalization inside the "crystalline" regions could depend on the local structure adopted by the dopant and the polyaniline chains and on the degree of paracrystallinity. The details of the interaction of the chains may lead to a fractal motion of charge and spin [30]. In this respect, spin diffusion measurements [31] indicate that the transverse diffusion coefficient is much larger in the ES-I structure than in the ES-II structure. Such an influence of the chain packing could explain why unstretched PAN-CSA (*m*-cresol) exhibits a higher conductivity than stretched PAN-HCl. In contrast, the ring-substituted polyanilines doped with HCI show more localized conduction processes, even when the material exhibits a substantial crystallinity [32–36]. Such a difference could be attributed to the effect of bulky side-group substituents increasing the interchain separation and inducing intrachain conformational defects, together with an enhancement of the interchain disorder. Similarly, the steric and electronic interactions between adjacent rings and chains in sulfonated polyanilines leads to a loss of crystallinity and a conductivity of order 0.1 S/cm, depending upon the processing conditions and sulfonation level [37–39]. All these structural features seriously limit the intrachain electronic delocalization length and the interchain charge transfer.

1.2.3 PF_6^--Doped Polypyrrole

In contrast with polyaniline, the amorphous parts of PF_6^--doped polypyrrole, HPPy, are not substantially converted into crystalline ones upon stretching. The broad amorphous halo of HPPy occurs at a *d* spacing of about 0.45 nm [11], a typical value for the average packing distance of aromatic rings. An amorphous halo at about the same *d* spacing is also observed in EB-I polyaniline [20]. In HPPy this *d* spacing is, however, quite different from those of the reflections originating from the crystalline regions. This implies that the local order between polymeric chains is different in the amorphous and crystalline regions of HPPy, at variance with what is observed in polyaniline [23].

In the crystalline parts, the polymer chains are tightly packed in (b, c) planes separated along a^* by the PF_6^- anions. In these layers the polypyrrole chains are shifted with respect to each other in order to avoid face-to-face repulsion between adjacent aromatic rings. A tilt angle $(\pi/2 - \alpha)$ of about 25° between b^* and the polymer direction, b, is deduced from the data [11]. A similar tilted arrangement of polymer chains is observed in poly-(3-alkylthiophene) [40]. It could also occur in the ES-I structure of the polyaniline doped with HCl and CSA. The tilt angle of the polypyrrole chains is found to vary by several degrees with synthesis conditions and under stretching. This structural degree of freedom could control the interchain electronic transfer especially along the transverse direction where the aromatic rings have a close contact distance of about 0.34 nm (a value insensitive to the conditions of preparation and to the degree of stretching). Parameter a^* is also found to vary significantly with the conditions of preparation and under stretching. In particular, $d_{(100)}$ $(= 2\pi/a^*)$ is found to increase from 1.18 to

1.34 nm by stretching HPPy two-fold. Such a variation of more than 10% of the distance between two successive (b, c) layers of polypyrrole chains implies a considerable reorganization of the PF_6^- anions with respect to the polymeric chains. This could strongly influence the interlayer charge transport processes.

The achievement of very highly conducting doped polypyrrole despite the presence of different local orders in the amorphous and crystalline regions and of crystalline regions of very small size contrasts with the behavior of doped polyaniline for which substantial local order accompanies the occurrence of the metallic behavior.

It is interesting to speculate why doped polyaniline requires substantial order to obtain high conductivity while doped polypyrrole with little order can achieve similar high conductivity. The difference in behavior may be due to differences in the intrachain steric repulsion between adjacent rings. For both doped polypyrrole and doped polyaniline, the half-filled energy band resulting from the π–π overlap of the adjacent rings of the chain has its lowest energy with the adjacent rings coplanar. Although for polypyrrole the adjacent C_4NH rings have little steric repulsion, for polyaniline there is substantial in-chain steric repulsion of the C_6 rings [41,42]. Hence, the interchain disorder potentials may be sufficient (together with the in-chain steric repulsion energy) to force adjacent rings into varying torsion angles for polyaniline, while for polypyrrole the interchain disorder may not be sufficient to cause the adjacent rings to deviate substantially from coplanarity. The continued presence of near coplanarity may result in relatively little backscattering of the conduction electrons within disordered regions of doped polypyrrole, as compared to substantial backscattering of conduction electrons in the disordered regions of doped polyaniline yielding the observed higher conductivity of polypyrrole in the presence of substantial disorder. The coplanarity of the chains in the disordered regions facilitates both resonance tunneling [43] and hopping transport between the more ordered regions [29, 30].

1.3 Semiconducting Polymers

The role of interchain interactions in the photoluminescence (PL) properties of conjugated polymers is a topic of great interest, as well as controversy [44–69]. An important impetus for understanding these properties is that the maximum attainable efficiency of conjugated polymer light-emitting diodes is determined in part by the PL efficiency. With few exceptions [70–72], interchain interactions lead to reduced PL efficiency in these systems. Therefore, a complete understanding of the nature of interchain interactions in conjugated polymers, their role in determining PL efficiency, and means of avoiding the detrimental effects of these interactions is of scientific and technological importance.

A reasonably complete understanding of photophysical processes in conjugated polymers has evolved in recent years, although the details are still under debate. Rauscher et al. have suggested, based on morphological considerations, that poly (p-phenylene vinylene) (PPV) films should be viewed as a collection of oligomers of various length, leading to an inhomogeneously broadened density of states for the system [73,74]. Photoexcitation produces a *singlet exciton* of mixed Frenkel- and Wannier-like character that is localized to a conjugated segment. The exciton migrates to the lowest-energy (i.e. longest) segments of the sample, leading to redshifted emission. Evidence for such a model comes from measurements of "site-selective fluorescence", wherein the shape and position of the luminescence spectrum is monitored as a function of excitation energy. The work of Rauscher et al. demonstrates that while the apparent Stokes shift (redshift) between absorption and emission maxima is large in real samples due to inhomogeneous broadening, the *intrinsic* Stokes shift (i.e. that of ideal, well-extended PPV chains) is quite small, $< 100\ \mathrm{cm}^{-1}$. This small value is inconsistent with early electron–phonon based models, which predict a much larger Stokes shift due to geometric relaxation [75].

While most authors agree that the intrinsic photophysics of luminescent polymers is excitonic, the role of interchain interactions is a topic of widespread debate. The role of the excimer (short for "excited dimer") was suggested by Jenekhe and Osaheni [46, 70] for rigid-rod polymers such as poly (p-phenylene-2,6-benzobisoxazole). The excimer is an emissive excited-state complex which is delocalized over two molecular units [76]. The ground state of the excimer is dissociative; that is, the ground state of the dimer spontaneously dissociates into two ground-state molecules. Furthermore, the excimer cannot be directly excited optically. Photoexcitation in excimer-forming systems therefore leads to production of an intramolecular singlet exciton that later delocalizes over two molecules, forming the excimer. Excimer formation is accompanied by a strong geometric distortion along the intermolecular axis that, when combined with the dissociative nature of the ground state, leads to featureless, strongly Stokes-shifted emission in comparison to dilute solution [76]. A classical example of an excimer-forming system is pyrene in solution [76]. As the solution concentration is increased, the structured pyrene emission is gradually replaced by a featureless excimer band. The absorption spectrum, however, is found to be concentration independent.

A closely related concept is that of the "polaron pair" [44, 45, 77–79]. These entities are suggested to consist of an electron and hole residing on opposite chains bound by their mutual Coulomb attraction. (In comparison, excimers may contain contributions from neutral excited states in addition to charge-transfer states.) Yan et al. suggested that the majority of species created on photoexcitation of PPV are polaron pairs. They further suggest that these species are nonemissive, leading to reduced PL quantum efficiency in

PPV films. Based on their picosecond photoinduced absorption results, they further suggested that the polaron pair is created only from "hot" excitons and not from relaxed ones. Therefore, unlike in the case of the excimer, the branching ratio between emissive intrachain singlet excitons and interchain polaron pairs is determined on a femtosecond time–scale.

While excimers exist only in the excited state, interchain interaction may also lead to ground-state interactions, a manifestation of which is the formation of *aggregate states* [80]. These species have been suggested to occur in "ladder" polyphenylenes based on photoluminescence [47] and photocurrent [81] measurements. Upon aggregation, both the ground- and excited-state wavefunctions are delocalized over several polymer chains. The aggregate is therefore directly accessible spectroscopically. The delocalization leads to a redshift in both emission and absorption vs. solution. Unlike the excimer, which tends to form spontaneously, the aggregate usually results from a forced proximity of two or more molecules, an example of which is the anthracene sandwich dimer [80]. While the energy of the molecular ground state is lowered upon aggregate formation, the first excited state splits into several distinct levels. The allowedness of transitions to these levels is determined by the alignment of the individual dipoles of the molecules which form the aggregate. Parallel alignment typically leads to a less-allowed or forbidden lowest transition [80, 82]. Aggregation therefore leads to reduced PL efficiency in this case.

As an example of the roles of the substituent group, processing and sample preparation in determining the nanoscale structure, and hence the photophysical properties of semiconducting polymers, selected results [83–90] of one of the earliest and most thoroughly studied systems, pyridine-based polymers (Fig. 1.2), are presented here. These polymers are promising for light-emitting diode applications [91–93] due to their increased oxygen stability over that of PPV and their high electron affinity. The role of aggregates in the emission of film samples of the pyridine-based polymers was shown. In particular, film samples show strongly red-shifted luminescence with low quantum yield, which we demonstrated to result from the formation of aggregates. As-precipitated powder samples, on the other hand, behave like solution samples, indicating that aggregate formation is strongly morphology dependent and can be avoided. As such, the powder samples provide a convenient laboratory for the examination of solid-state effects in the absence of aggregate formation. We demonstrated the generality of these phenomena in the pyridine-based systems despite the differing properties of the repeat units.

All polymers [94–96] were precipitated from solution during or after polymerization and these as-precipitated solids are labeled as "powders" in this paper. The powder samples were mixed with KBr and pressed into pellets for study. Some of the powder sample were dissolved to form solution samples of varying concentration. The poly (p-pyridyl vinylene) (PPyV) and poly (p-pyridine) (PPy) samples are generally soluble in weak acids such as formic acid (HCOOH). The poly (p-pyridyl vinylene p-phenylene vinylene) (PPyVPV) copolymers are soluble in common organics by virtue of their hydrocarbon sidegroups. Here, results are shown for symmetrically substituted copolymers (see Fig. 1.2) with sidegroups R=$C_{12}H_{25}$, $OC_{16}H_{33}$ and $COOC_{12}H_{25}$ [89].

1.3.1 Fluorescence Properties of Solutions

Typical absorption (solid), PL (dotted) and PL excitation (PLE, dashed) spectra for representative pyridine-based polymers in dilute solution ($< 10^{-5}$ M) are shown in Fig. 1.6. Weak vibronic structure is observed in the PL spectra; however, no vibronic structure can be distinguished in the absorption spectra. Such diffuse spectra are characteristic of "floppy" systems which possess a ring-rotational degree of freedom, due to the generally weak steric hindrance towards ring rotations in these systems [97]. At room temperature, a distribution of ground states with differing ring-torsion angles is present, leading to the featureless absorption spectrum. The excited state of these systems is generally more planar due to aromatic-to-quinoid conversion, leading to the Stokes-shifted PL spectrum.

Unlike in the case of PPV [52], for all the pyridine-based polymers studied the PLE spectrum closely follows the absorption spectrum, in accordance with Vavilov's law [76]. Such a result indicates that internal conversion (phonon emission) to the first excited state is more efficient than other processes, such as radiative decay, in dissipating energy in the polymer. In MEH-PPV, a discrepancy between the PLE and absorption spectra of dilute solutions has been suggested to originate from a subpicosecond internal conversion to the ground state which becomes more likely under high-energy excitation. For the pyridine-based polymers, such a process apparently does not occur efficiently.

Figure 1.7 shows the PL spectrum (solid) of $OC_{16}H_{33}$ -derivatized PPyVPV in dilute solution. Unlike the spectra of Fig. 1.6, the PL spectrum of this copolymer shows more pronounced vibronic structure due to hydrogen bonding between the OC_{16} sidegroups and the vinylene bridge, which assists the planarity of the molecule in the ground and excited states. Assuming strong coupling to a single vibrational mode of wavenumber ν_0, the relative intensities $I(n)$ of the various vibronics n in the PL spectrum can be described by the Huang–Rhys parameter, S, given by

$$I(n) = \mathrm{e}^{-S} S^n/(n!).$$

The parameter S is proportional to the change in elastic energy upon geometric relaxation. The dotted line in Fig. 1.7 represents a fit to the equation above (assuming a Gaussian broadening), and the vertical lines in the figure represent the various vibronics. The fit corresponds to a Huang–Rhys

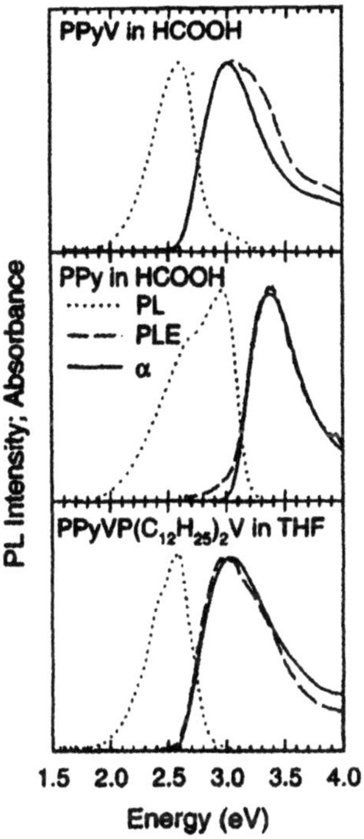

Fig. 1.6. Absorption (*solid*), photoluminescence (PL, *dotted*) and PL excitation (PLE, *dashed*) spectra for dilute solutions ($< 10^{-5}$ M) of PPyV in HCOOH (*upper*), PPy in HCOOH (*middle*), and PPyVPV in THF (*lower*). (From [89])

parameter of $S = 0.53$. For comparison, $S = 0.47$ has been suggested for gel-aligned MEH-PPV [98]. For PPyVPV, the strongly coupled vibration is found to be at $\nu_0 = 1302\ \text{cm}^{-1}$. A similar mode ($1300\ \text{cm}^{-1}$) provides the vibronic structure in the PL spectrum of distyryl benzene (the three-ring oligomer of PPV), where it has been assigned to the vinylene C=C stretch. The parameter S is related to the geometrical relaxation energy, E_{rel}, via the relation $S = E_{\text{rel}}/hc\nu_0$. For PPyVPV, we find $E_{\text{rel}} = 86$ meV, indicating relatively weak geometrical relaxation for the singlet exciton.

The PL efficiencies are generally high, especially for the PPyVPV copolymers. The PPy efficiency is lower, possibly due to the age and quality of the sample. All PL decays were found to be nearly ($> 90\%$) single-exponential. The PL decays were generally independent of wavelength, although the PPy sample shows a slight redshift to the PL with time. The single-exponential decay together with the close agreement between PLE and absorption spectra allows the computation of the radiative lifetime τ_{rad} from the quantum efficiency (QE) and the observed PL lifetime τ_{obs}: $\tau_{\text{rad}} = \tau_{\text{obs}}/\text{QE}$ [76]. The radiative lifetime is found to be $\sim$ 1–2 ns for all polymers studied, similar to results reported for substituted PPV in solution.

The PL spectrum is nearly independent of concentration, with the slight difference due to self-absorption at higher emission energies. This concentration independence suggests that the pyridine-based polymers do not have a

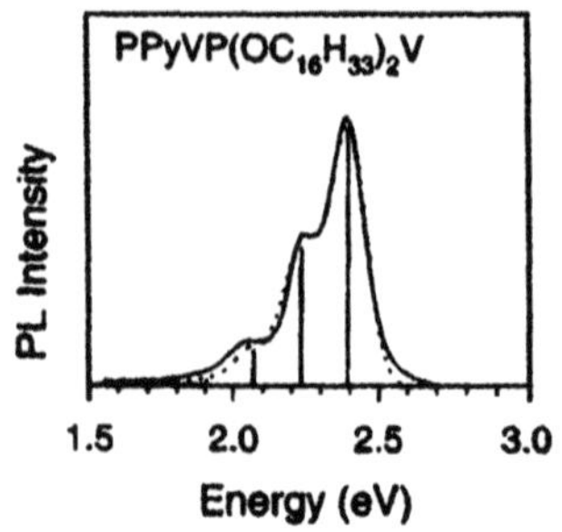

Fig. 1.7. PL of 10^{-5} M $PPyVP(OPC_{16}H_{33})_2V$ in THF (*solid line*). *Dotted line* is a fit to the Huang–Rhys expression (see text). (From [89])

strong tendency towards excimer formation. In contrast, in excimer-forming systems like pyrene [76], redshifted excimer-like emission replaces the single-molecule emission at relatively low concentrations ($< 10^{-3}$ M). The PL decay at 2.6 eV of the concentrated PPyVPV solution closely mirrors the single-exponential behavior of the dilute solution. The PL decay at 2.0 eV of the concentrated solution, however, shows a slight deviation from single-exponential behavior that becomes most evident after $\sim$ 2 ns. This additional long-lived component may be due to interchain association which is a precursor of aggregation in film samples. Similarly, in PPyV solutions, the emission and its time dynamics are concentration independent in the 10^{-6}–10^{-1} M range.

1.3.2 Fluorescence of Solid Samples: Aggregate Formation

While the PL of solution samples shows the high quantum efficiency characteristic of aromatic molecular systems, the PL of film samples does not share this property. In contrast to the solution results, the internal quantum efficiencies of the films are all quite low, typically 1/2 to 1/3 of the solution value. The internal quantum efficiency of the $PPyVP(OC_{16}H_{33})_2V$ film is particularly low, only about 0.02, due to formation of C=O sites due to oxygen exposure. These sites act as efficient luminescence quenchers. The susceptibility of the OC_{16} copolymer to oxygen attack is likely the result of the electron-donating character of the sidegroups. Significant C=O formation is not present in the other copolymers or in PPyV and PPy.

The origin of the reduced quantum efficiency of films becomes apparent when the cw spectroscopic properties of the film samples are compared with those of the solution. The PL spectrum of film (solid line) samples is compared with that of solution (dotted) and powder (dashed) samples for representative pyridine-based polymers in Fig. 1.8. (Again, the powder samples are distinguished from the film samples by the fact that powder samples are precipitated from solution during polymerization.) In all cases, the film spectrum is redshifted by $\sim$ 0.5 eV vs. solution and is featureless. In contrast, the powder spectrum more closely resembles the solution result in all cases, particularly in the case of PPyV (upper portion of the figure). Below, we demonstrate that the redshifted film spectra result from the formation of

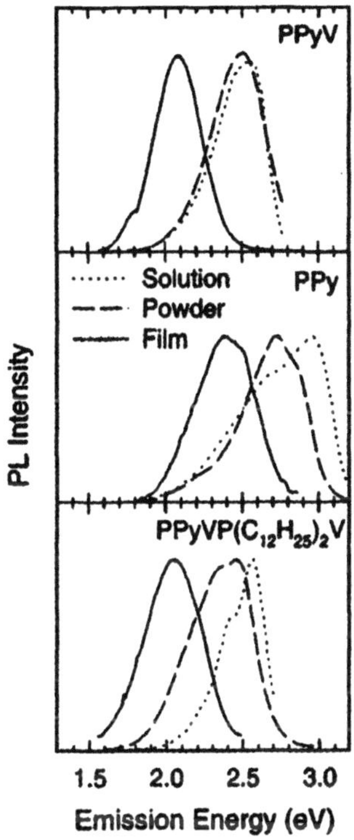

Fig. 1.8. PL spectrum of solution (*dotted*), powder (*dashed*) and film (*solid*) samples of PPyV (*upper*), PPy (*middle*) and PPyVP$(C_{12}H_{25})_2$V (*lower*). (From [89])

aggregate states in these samples. The fact that the film and powder spectra are different indicates that aggregate formation is *morphology dependent*. In addition to the difference between films and powders, films cast from different solvents generally display different PL spectra, again indicating a morphology dependence to the emission. Finally, films cast from the *same solvent* also have been seen to have different PL spectra depending on the preparation conditions. In particular, spin-cast films, for which the solvent evaporates rapidly, generally show less of a redshift than drop-cast films, for which the solvent evaporates slowly.

Several intrinsic and extrinsic mechanisms that might produce a redshifted film spectrum can be ruled out. First, a redshifted spectrum vs. solution may simply result from the fact that the index of refraction of the film sample is different from that of the solution. This and related "solid-state" phenomena are ruled out by the fact that the film and powder spectra are different, despite similar refractive indices. Furthermore, the effects of self-absorption can be ruled out, as the film and powder samples have similar optical densities in all cases. Oxygen contamination of film samples can also be ruled out as the source of the redshift from infrared-absorption studies. Indeed, the film samples were found to be extremely resistant to photooxidation (with the exception of the R=OC_{16} copolymer). Furthermore, attempts to deliberately photo-oxidize the powder samples did not lead to a redshifted

PL spectrum for these samples. Finally, deep traps due to residual solvent can be ruled out. The solvents used for the casting of the PPyVPV films (THF, chloroform) are quite volatile and are therefore not expected to remain in the film samples. Attempts to remove residual solvent by a combination of high vacuum and heating did not result in a reduced redshift for the film spectrum.

Although we have eliminated a number of factors which might produce a redshift in film samples, a distinction between aggregate formation and excimer formation cannot be made based on PL data alone. The concentration independence of the solution data presented above are already suggestive of aggregate formation rather than excimer formation, as excimers are expected to form spontaneously in solution, whereas aggregates usually result from "forced" interaction due to packing considerations. To fully rule out excimer formation, however, one must examine other spectroscopic properties. In particular, if excimer formation were responsible for the redshifted emission of films, one would expect the absorption and PLE spectra of film samples to be identical to those of solution. Furthermore, the Stokes shift of the emission should be large in the case of the excimer. On the other hand, since aggregates can be directly accessed optically, one expects redshifted absorption and PLE spectra and a relatively small Stokes shift for the emission in the aggregate case.

The direct absorption spectra of solutions (dotted), powders (dashed) and films (solid lines) are shown in Fig. 1.9 for the same polymers as in the previous figure. The film absorption spectra are redshifted from the solution results by a similar amount as in the case of the PL (~ 0.5 eV), showing additional low-energy oscillator strength below the solution absorption onset. While these spectra are not corrected for reflection, the reflectivity of the samples was measured and actually peaks well below the solution absorption edge. Therefore, a reflection correction would actually *enhance* the absorption spectrum below 2.5 eV in all cases. Additional complications arise from scattering losses, which likely contribute substantially to the tail of the absorption below 2.2 eV in the PPyV and PPyVPV spectra. Nevertheless, in the PPy spectrum, where such losses are minimal, it is clear that the film spectrum is substantially redshifted from the solution result, indicating the presence of additional, low-energy, directly accessible excitations in film samples. This result is in direct contrast with the expectations for excimer formation, as, in general, excimers cannot be directly excited [80]. As with the PL spectra, the powder absorption spectra closely follow the solution spectra, particularly in the case of PPyV. The PPy and PPyVPV powder absorption spectra deviate somewhat from the solution result, as with the PL spectra. Again, these results point to the morphology dependence of aggregate formation in these polymers.

Fig. 1.10 shows the PLE spectra for solutions, powders and films of the various polymers. Here, the powder PLE spectra are somewhat different from

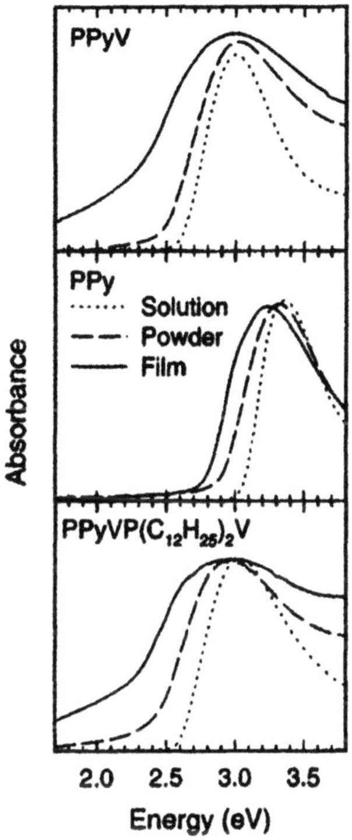

Fig. 1.9. Absorption spectrum of solution (*dotted*), powder (*dashed*) and film (*solid*) samples of PPyV (*upper*), PPy (*middle*) and PPyVP$(C_{12}H_{25})_2$V (*lower*). (From [89])

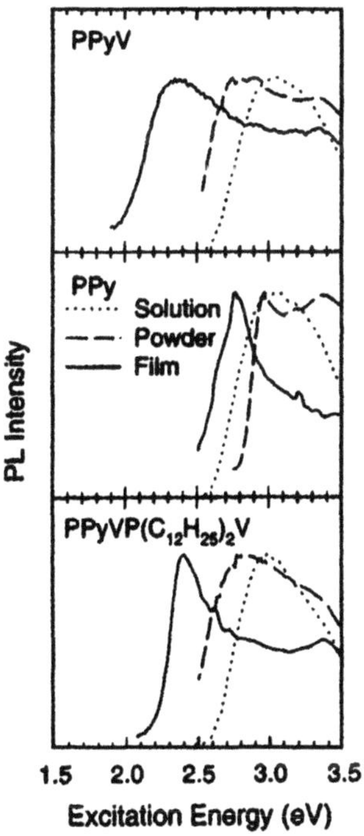

Fig. 1.10. PLE spectrum of solution (*dotted*), powder (*dashed*) and film (*solid*) samples of PPyV (*upper*), PPy (*middle*) and PPyVP$(C_{12}H_{25})_2$V (*lower*). (From [89])

the corresponding solution spectra. This deviation reflects the fact the powder samples are optically thick, whereas the solution samples are optically thin. Unlike the absorption spectrum, which is essentially the logarithm of the quantum efficiency for absorption, the PLE directly measures the quantum efficiency for emission (i.e. no logarithm involved), resulting in a "flattening" of the spectrum at high optical densities. The film PLE spectra are dramatically different from the solution spectra, peaking in a region where the solution PLE and absorption are essentially zero. The peak of the film PLE is close to the peak of the PL spectrum in all cases, suggesting a small intrinsic Stokes shift. Again, the redshifted PLE spectra of film samples suggests additional, low-energy, directly accessible states in film samples, inconsistent with excimer formation.

While the PLE spectra suggest a small Stokes shift for the emitting species, this value can be directly measured using the site-selective fluorescence technique. Ideally, one should use a tunable laser to measure such effects, as the spectral purity of such a source allows monitoring of emission wavelengths that are very close to the exciting line. An upper bound on the intrinsic Stokes shift can be obtained, however, by using a conventional PL spectrometer, which allows measurement of emission to within 5 nm of the exciting line. The PL spectra at varying excitation energies are shown for films of the pyridine-based polymers in Fig. 1.11. For the various spectra, the

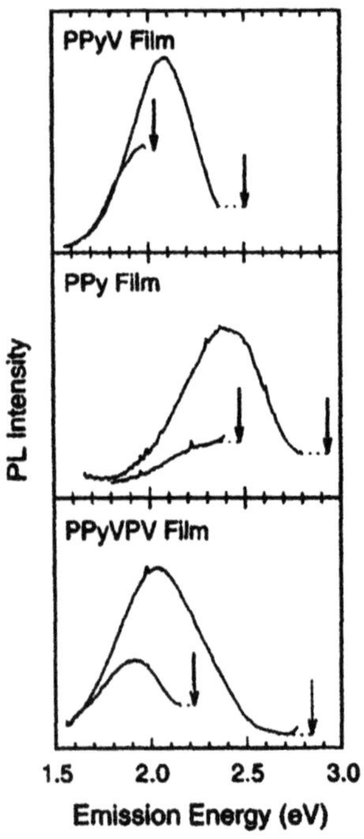

Fig. 1.11. PL spectra of film samples of PPyV (*upper*), PPy (*middle*) and PPyVPV (*lower*), demonstrating site-selective effects. Arrows indicate excitation energy for each spectrum. (From [89])

exciting wavelength is indicated by the downward arrow. In all cases, substantial luminescence can be excited with low-energy radiation. Furthermore, the peak of the emission under low-energy excitation is different from that under high-energy excitation, indicating site-selective effects (i.e. multiple emitting sites in the sample). An upper bound on the intrinsic Stokes shift can be estimated from the difference between the excitation energy and the peak of the resulting emission under low-energy excitation. For PPyV, the emission under 2.04 eV excitation peaks at 1.98 eV, indicating that the intrinsic Stokes shift of the low-energy emission is a very small 60 meV. This is again in contrast with expectation for excimer emission, for which the Stokes shift should be quite large ($\sim$ 0.5 eV).

An additional argument for the formation of aggregates comes from near-field scanning optical microscopy (NSOM), a technique which allows direct spatial imaging of emission and absorption in polymer films to a resolution of 100 nm. NSOM measurements of PPyV films indicate that emission is localized in partially aligned regions of the film of domain size $\sim$ 200 nm [87]. While this size is too large to correspond to a single aggregate, it does suggest that aggregates are most likely to form in aligned regions of the sample, as expected.

1.3.3 Time-Resolved Luminescence

As mentioned above, the solution PL decays are nearly single-exponential with a radiative lifetime of 1–2 ns, and the decays are independent of wavelength. In contrast, the film PL decays are highly non-exponential and show a strong wavelength dependence. Fig. 1.12 displays the time evolution of the PL spectrum of a PPyVPV film. For reference, the cw film and solution PL spectra are shown in the lower frame. In comparison with the cw spectra, the 0 ps spectrum displays both solution-like (2.6 eV) and aggregate-like (2.0 eV) components in roughly equal proportions. By 250 ps, the solution-like component is greatly diminished and by 1 ns has vanished. The aggregate-like component decays with little or no spectral diffusion. A complimentary picture, the evolution of the PL at a fixed wavelength vs. time delay, is shown for three representative polymers in Fig. 1.13.

In all cases, the decays are highly non-exponential at all wavelengths. The 2.6 eV solution-like decay is initially very fast, at the resolution of the experiment. (The instrument response is denoted IRF in the upper portion of the figure.) This fast decay suggests rapid diffusion of excitons to aggregate sites. The diffusion is likely facilitated by efficient Förster transfer due to the strong overlap between the intrachain solution-like emission and the aggregate absorption. On the other hand, the 2.0 eV aggregate-like emission is very long-lived. The PPyVP$(C_{12}H_{25})_2$V decay at this energy has a time constant of about 2.5 ns, greater than the predicted radiative lifetime of the solution emission (941 ps). Similar long-lived decays are seen for PPyV and PPy films, as seen in the figure. The very long-lived observed lifetime for the aggregate

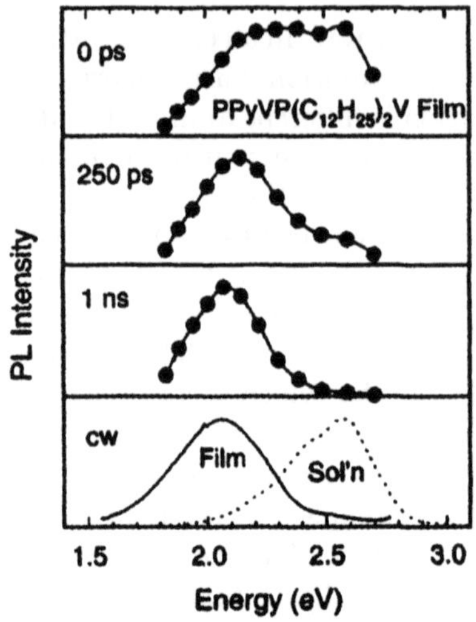

Fig. 1.12. PL spectrum of $PPyVP(C_{12}H_{25})_2V$ film at 0 ps, 250 ps and 1 ns delay following excitation at 2.8 eV. Lower frame shows cw PL spectra of $PPyVP(C_{12}H_{25})_2V$ solution (*dotted*) and film (*solid*) samples. (From [89])

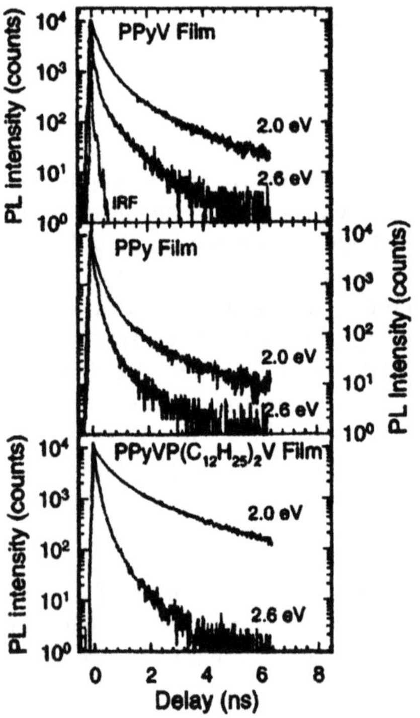

Fig. 1.13. PL decays of PPyV (*upper*), PPy (*middle*) and $PPyVP(C_{12}H_{25})_2V$ (*lower*) films excited at 2.8 eV. Decays are shown for 2.0 and 2.6 eV emission, as indicated. Instrument response function (IRF) is shown in the upper portion of the figure. (From [89])

emission, together with the low quantum yield of the film PL, suggests that the radiative lifetime of the aggregate is much longer than that of the single-chain emission. The $\sim$ 30% quantum yield of the PPyVP($C_{12}H_{25}$)$_2$V film emission places a lower bound of about 8 ns on the aggregate radiative lifetime. From the Einstein relations, a longer radiative lifetime is associated with a less-allowed transition. A less-allowed optical transition is expected for aggregates in which the individual transition dipoles are aligned parallel to each other, as in a card-pack arrangement. As the longer radiative lifetime indicates a lower probability for radiative decay, provided nonradiative decay pathways are similar for the aggregate and interchain emission, *a long radiative lifetime implies reduced quantum efficiency for the aggregate.*

Despite the presence of aggregates in the film samples, the film absorption spectrum is actually quite similar to that of the solution (e.g., the absorption maxima are at nearly the same energy), indicating that 2.8 eV excitation creates mainly intrachain excitons in the films which later migrate to the aggregate sites. Nevertheless, a rise in the aggregate emission could not be detected in any of the samples studied. The observation of such a rise is hindered by the very rapid initial migration to aggregate sites, which is likely faster than the experimental resolution, as well as the spectral overlap between the solution-like and aggregate emission spectra. Substantial sub-50-ps migration to aggregate sites is evidenced by the "0 ps" spectrum of the PPyVPV film, which suggests that at least half of the initial (< 50 ps) emission is from the aggregates. Since emission from aggregate excitons apparently occurs with a lower quantum yield than the intrachain emission, it can be concluded that the initial density of excited aggregates (i.e. the number of aggregates excited within the 50 ps resolution) is actually greater than the *density* of the intrachain species. It is therefore not surprising that a rise in the aggregate emission could not be detected.

The time evolution of the PL spectrum of powder samples is considered in Fig. 1.14. Here we only consider PPyV powders, which, from the cw PL data of Fig. 1.8, are the least affected by aggregate formation. The single-wavelength decays of the powder samples retain much ($\sim$ 90%) of the single-exponential character of the solution; however, a spectral redshift is seen with time, again suggestive of exciton migration. We note that the spectral diffusion in the powder samples is much less than that of the film samples: at 2 ns, the 2.0 eV powder emission is roughly twice the emission at 2.6 eV, assuming both decays are normalized at $t = 0$. In contrast, for film samples, at 2 ns the 2.0 eV emission is over one order of magnitude greater than the 2.6 eV emission. Furthermore, unlike in the case of the film samples, for the powder samples the total (energy-integrated) PL decay closely follows that of the solution, as is shown in the lower portion of Fig. 1.14. The small discrepancy between the solution and powder total PL decays is likely due to enhanced triplet production in the powder samples. These results suggest that the spectral diffusion observed in powder samples is an artifact of exciton

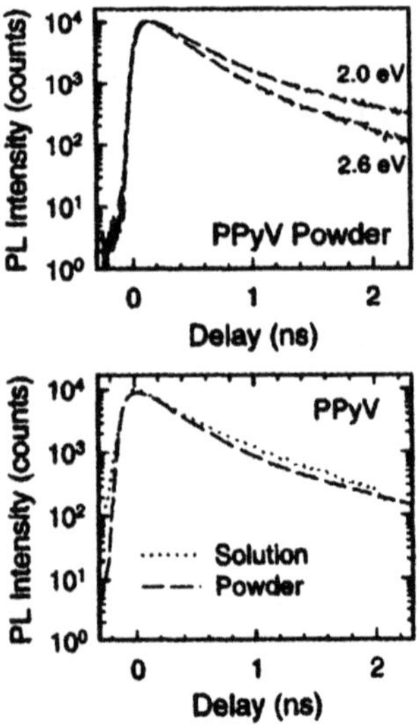

Fig. 1.14. PL decays at 2.0 and 2.6 eV (*upper*) and energy-integrated PL decay (*lower*) for PPyV powder (*dashed lines*) excited at 2.8 eV. Solution PL decay (2.6 eV) is shown for comparison (*dotted line*) in the lower portion of the figure. (From [89])

migration in an inhomogeneously broadened density of states, and *not* due to the presence of multiple emission components, as would result from aggregation. Therefore, the time-resolved data are consistent with the observation that substantial aggregation does not occur in PPyV powder samples.

1.3.4 Morphology

Although the PPyV powder and film samples show large differences in their photophysical properties, their morphological properties are actually quite similar. Both the powder and film samples lack substantial structural order. On the other hand, the NSOM measurements discussed briefly above indicate large regions ($\sim$ 200 nm) of gross alignment for the PPyV film samples. We therefore suggest that the formation of aggregates in the pyridine-based polymers is more closely related to liquid-crystalline order than crystalline order. A possible origin for the morphology dependence of aggregate formation in this system is that powder samples are precipitated from a solvent in which the polymer is very poorly soluble, leading to a disordered, coil-like morphology. On the other hand, film samples are prepared from solvents in which the polymer dissolves well, leading to liquid-crystalline order in concentrated solutions. As the solvent evaporates upon film formation, the correlations between neighboring chains persist, leading to grossly aligned regions in the film. Aggregate sites are likely to form in these regions.

1.4 Conducting Polymers

1.4.1 Models for Electrical Conductivity

Much work has focused on the nature of the charge carriers in the highly doped metallic state. Even though there is a high density of conduction electrons at the Fermi level (chemical potential) for the highly doped state, the carriers may be spatially localized due to the nanoscale disorder so the carriers cannot participate in transport except through hopping [2, 99–102]. The prime source of localization which has been studied is structural disorder in the polymers. X-ray diffraction studies of these systems show that they are generally of modest crystallinity, with regions of the material which are more ordered while other regions are more disordered (see Section 1.2). Also, the fibrillar nature of many of the conducting polymers may lead to localization by reducing the effective dimensionality of the electrons delocalized in a bundle of polymer chains. Fig. 1.3a presents a schematic view of the inhomogeneous disorder in these doped polymers, with individual polymer chains passing through both ordered regions (typically 3–10 nm across) and disordered regions. The percent "crystallinity" may vary from near zero to 50 or 60% for polypyrroles and polyanilines, respectively, to greater then 80% for polyacetylenes. It is noted that the chains in the disordered regions may be either relatively straight, tightly coiled, or intermediate in disorder.

In a perfect crystal with periodic potentials, electron wavefunctions form delocalized Bloch waves. Impurities and lattice defects in disordered systems introduce backward scattering of these electron waves with resulting "Anderson localization" [103]. The electronic structure of the system strongly depends on the degree of disorder. The energy fluctuation in the random potentials broadens the bandwidth and creates smooth "band tails". Due to these band tails, the original bandgap between the conduction and the valence bands of a semiconductor may be closed. The ramifications, including a finite density of states $N(E_F)$ produced at the Fermi level E_F between mobility edges, were discussed by Sir Neville Mott [104]. When the Fermi level lies in the localized region, the conductivity at zero temperature is zero even for a system with a finite density of states. The Mott variable range hopping (VRH) model is applicable to systems with strong disorder such that the disorder energy is much greater than the bandwidth. The general form of the temperature-dependent conductivity σ in Mott's model is described as

$$\sigma = \sigma_0 \exp\left[-\left(T_0/T\right)^{1/(d+1)}\right] \tag{1.1}$$

where d is the dimensionality and, for three-dimensional systems,

$$\sigma = \sigma_0^{3d} \exp\left[-\left(T_0^{3d}/T\right)^{1/4}\right] \quad \text{and} \quad T_0^{3d} = c/(k_B N(E_F) L^3) \tag{1.2}$$

where c is the proportionality constant, k_B is the Boltzmann constant, L is the localization length, and σ_0 is a temperature (T) independent or weakly-temperature dependent prefactor. Equation (1.1) for $d = 1$ corresponds to the behavior of quasi-one-dimensional variable range hopping (Q1D-VRH) for the circumstance where the charge carriers primarily hop along a "one-dimensional" chain and occasionally hop to a nearest-neighbor chain thereby avoiding energetically large barriers. Thus

$$\sigma^{\mathrm{Q1D-VRH}} = \sigma_0^{\mathrm{Q1D-VRH}} \exp\left[-(T_0^{\mathrm{Q1D-VRH}}/T)^{1/2}\right], \tag{1.3}$$

with a corresponding

$$T_0^{\mathrm{Q1D-VRH}} = 16/\left[k_B N(E_F) L z\right]. \tag{1.4}$$

Here L is the one-dimensional localization length and z is the number of nearest neighbor chains. For an isolated one-dimensional chain (1.1) does not apply, and the temperature dependence of the conductivity is thermally activated reflecting the energy ΔE_B for excitation over the largest barrier encountered during charge transport along the chain,

$$\sigma^{1D} = \sigma_0^{1D} \exp\left[-(\Delta E_B/T)\right]. \tag{1.5}$$

If the Fermi level is at an energy such that the electronic states are extended, then finite conductivity at zero temperature is expected. This model assumes that the substantial disorder is homogeneous through the isotropic three-dimensional sample. Other external parameters such as magnetic field or pressure can affect the localization/delocalization transition and the localization lengths. This model has received much experimental attention for doped and ion-implanted polymers, although more recent studies of ion-implanted rigid rod and ladder polymers reveals a three-dimensional semi-metallic conductor with weak localization effects [105].

For an isolated one-dimensional metallic chain localization of charge carriers arises for even weak disorder because of quantum interference due to static backscattering of electrons. This contrasts to the strong disorder required for localization in three-dimensional systems. The localization effects in the nanoscale inhomogeneously disordered (partially crystalline) conducting polymers are proposed to originate from one-dimensional localization in the disordered regions. The inhomogeneous disorder model [106, 107] represents the doped polymer as relatively ordered regions (or "crystalline islands") interconnected through polymer chains traversing disordered regions (Fig. 1.3). The individual polymer chains often are longer then the island and inter-island length scales. Within this model, conduction electrons are three-dimensionally delocalized in the "crystalline" ordered regions (though the effects of paracrystalline disorder may limit delocalization within these regions). In order to transit between ordered regions, the conduction electrons must diffuse along electronically isolated chains through the disordered regions where the electrons readily become localized. The localization length of these electrons depends on the details of the disorder (e.g. electrons traveling along tightly coiled chains are expected to have much shorter localization lengths than electrons traveling along expanded coils or relatively straight chains). Phonon-induced increases in the localization length increase the conductivity with increasing temperature. This model accounts for localized behavior at low temperature despite conductivities at room temperature in excess of the Mott minimum conductivity.

Three-dimensional crystalline order facilitates delocalization. It has been shown that nematic-like order can also increase delocalization, though less effectively. If the localization length for some conduction electrons exceed the separation between ordered regions then the total delocalization will be substantially enhanced. Other approaches have emphasized [108] the "homogeneous" disorder limit, assuming that the localization length is larger than the structural coherence length and that the electrons respond to an average structure. For the fully doped conducting polymers which are disordered and poorly conducting, the conduction path may be treated [30] as charge motion on an anisotropic fractal network whose dimensionality is close to one. For this case the dc and ac conductivities are determined to follow the expectations of the quasi-one-dimensional Mott variable range hopping law for fractal systems of dimensionality close to one. As the dimensionality increases, it is predicted that both the ac conductivity and the dielectric constant would have stronger temperature dependence than for the quasi-one-dimensional or three-dimensional variable range hopping cases. Charge motion between nanoscale three-dimensionally ordered regions through disordered polymer chain has also been described in terms of resonant quantum tunneling. This model can account for the anomalously long transport times determined in microwave and optical frequency studies of highly conducting polymers [109].

Other researchers have considered the behavior of models of metallic strands interrupted by barriers [110]. A.B. Kaiser has applied a modification of this model including charge carriers either being thermally activated over thin barriers (dominating at high temperatures) or tunneling through the barriers (dominating at low temperatures) [111]. In this case the conductivity may be represented by

$$\sigma^{\mathrm{B}} = \sigma_0^{\mathrm{B}} \exp\left[-\left(T_0/(T+T_1)\right)\right]. \tag{1.6}$$

Here σ_0^{B}, T_0 and T_1 are constants.

For conventional metals, many of the electrical transport properties can be described by the Drude model within which electrons are treated as free particles in a gas with a single scattering time τ. Despite its simplified assumptions, the Drude model explains high and frequency-independent conductivity of traditional metals from dc to the microwave ($\sim 10^{10}$ Hz) frequency range, and a real part of the dielectric constant (ε_r) which is negative below the screened plasma frequency ($\omega_p^2 = 4\pi ne^2/m^*\varepsilon_b$; n is the density of carriers, m^* is the carrier effective mass, and ε_b is the background dielectric constant). Within the low-frequency Drude limit ($\omega\tau \ll 1$), the Drude response can be deduced as $\varepsilon_r = -\omega_p^2\tau^2$ and $\varepsilon_i = \omega_p^2\tau/\omega$, where ε_i is the imaginary part of the dielectric constant and ω is the frequency. The frequency-dependent response of the conductivity and dielectric constant of an inhomogeneously disordered polymer with crystalline islands has been modeled in terms of a chain-linked network of quantum dots [109]. The model leads to multiple zero crossings of the dielectric function as a function of frequency and an anomalously long scattering time.

1.4.2 Electrical Conductivity of Conducting Polymers

The temperature-dependent conductivity $\sigma(T)$ of heavily iodine-doped $(CH)_x$ and hexafluorophosphate (PF_6)-doped PPy down to the mK range vary as a function of aging (disorder) [112]. The highest σ_{dc} at room temperature reported in this study is $\sim 5\times10^4$ S/cm for I_3 doped T-$(CH)_x$ and $\sim 10^3$ S/cm for the highest conducting PPy(PF_6). For both of these materials, the conductivity decreases with decreasing temperature to a minimum at $T_m \sim 10$ K. Below T_m, σ increases by $\sim 20\%$ and then is constant to 1 mK. Some highly conducting preparations of PAN-CSA show similar behavior. Samples of the same chemical composition but prepared from different solvents may have different local order on the nanoscale, and thus very different conductivities. Less highly conducting samples of doped polyacetylene, doped polyaniline, and doped polypyrrole become insulating at low temperatures.

Hydrochloric acid as well as camphor sulfonic acid doped polyaniline prepared in chloroform often have [99, 100] $\log\sigma$ proportional to $T^{-1/2}$, as expected for quasi-one-dimensional variable range hopping [see (1.3) and (1.4)]. Generally, the higher conductivity samples have a weaker temperature dependence at low temperatures ($T_0 \sim$ 700-1000 K for $T < 80$ K), and lower conductivity samples a stronger temperature dependence ($T_0 \sim 4000$ K). The smaller T_0 for the more highly conducting samples has been associated with weaker localization due to improved nanoscale intrachain and interchain order.

The microwave frequency dielectric constant provides a measure of the charge delocalization in individual samples. The low-temperature dielectric constant, ε_{mw}, for a series of emeraldine hydrochloride samples is proportional to the square of the crystalline coherence length, ξ^2, independent of the direction of orientation of the sample with regard to the microwave frequency electric field [29]. This demonstrates that the charge is delocalized three-dimensionally within the crystalline regions of these samples. Using a simple metallic box model, $\varepsilon = \varepsilon_{\text{infinity}} + (2^{9/2}/\pi^3)e^2N(E_F)L^2$, and for the low-temperature localization length taking the X-ray crystalline correlation length determined by X-ray diffraction, $N(E_F)$ = 1.23 state/(eV 2-rings) (0.088 state/eV-(C+N)) for PAN-HCl. This compares very favorably with the value obtained from magnetic susceptibility experiments.

The sign, magnitude and temperature dependence of the 6.5×10^9 Hz dielectric constant for very highly conducting T-$[CH(I_3)_y]_x$, PPy-PF_6, and *m*-cresol-prepared PAN-CSA are quite striking [99,107]. For example, PAN-CSA (*m*-cresol) has a metallic negative dielectric constant (-4×10^4) and features a maximum in microwave frequency conductivity at 180 K. A similar large and negative value of $\varepsilon_{\mathrm{mw}}$ and a temperature dependence of $\varepsilon_{\mathrm{mw}}$ were determined for heavily iodine-doped stretched Tsukamoto polyacetylene (-1.5×10^6) and PF_6-doped polypyrrole [107] (-1×10^5). Using the Drude model for low frequencies ($\omega\tau < 1$), a plasma frequency of ω_p=0.015 eV (120 cm^{-1}) and a room-temperature scattering time of $\sim 1.2\times10^{-11}$ s were calculated for the PAN-CSA (*m*-cresol) system, though the exact values correlate with the sample preparation conditions. Similar values are obtained for heavily iodine-doped stretched Tsukamoto polyacetylene and PF_6-doped polypyrrole. These values of ω_p are much smaller than one expects from the usual Drude model, suggesting that only a small fraction of the conduction band electrons participate in this low-frequency plasma response. Similarly, the value of τ is two orders of magnitude larger than usual for an alkali, noble or transition metal. The origin of the anomalously large scattering time was suggested to reflect the effects of the time for the conduction of electrons to transit between metallic quantum dots [109].

For the conducting forms of doped polyacetylene and other conducting polymers, there are zero, two, three or one zero crossings of the real part of the dielectric function (ε_1) as the frequency is decreased [99, 100]. For the least-conducting materials, ε_1 remains positive for the entire optical frequency range (50–50000 cm^{-1}), reaching values of several hundred at microwave frequencies. For higher-conductivity materials, ε_1 crosses zero between 1 and 3 eV (the all-conduction-electron plasma response) and then becomes positive again below 1000 cm^{-1}, reaching values in excess of 10^4 at microwave

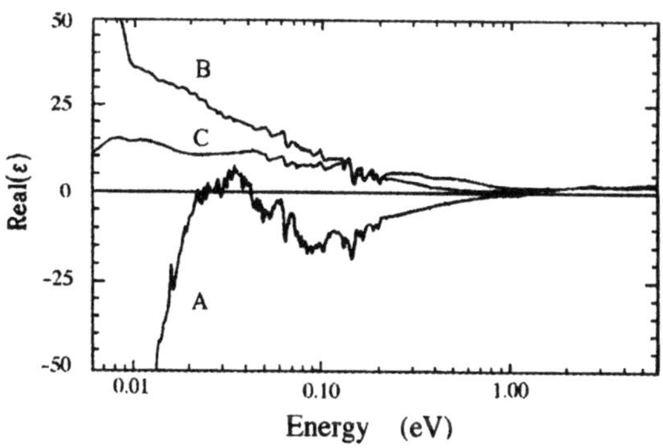

Fig. 1.15. Real part of room-temperature dielectric response vs. frequency for (A) PAN-CSA (*m*-cresol), (B) PAN-CSA (chloroform/*m*-cresol), and (C) PAN-HCl ("intermediate cross linked"). (From [99, 100])

frequencies. For the most metallic samples, two behaviors have been reported dependent upon the system. For doped PAN and PPy with $\sigma_{dc} \sim 400$ S/cm, ε_1 demonstrates the previous two zero crossings, and a third zero crossing occurs to negative values at a "delocalized conduction electron plasma frequency" of several hundred wavenumbers. Fig. 1.15 illustrates this range of behaviors for three differently doped samples of polyaniline. For very highly conducting doped polyacetylene, ε_1 crosses zero at the all-conduction-electron plasma frequency and remains negative down to the lowest measured optical frequencies. These multiple zero crossings of the dielectric are in accordance with expectations for the response of chain-linked quantum dots (see Fig. 1.16 and [109]).

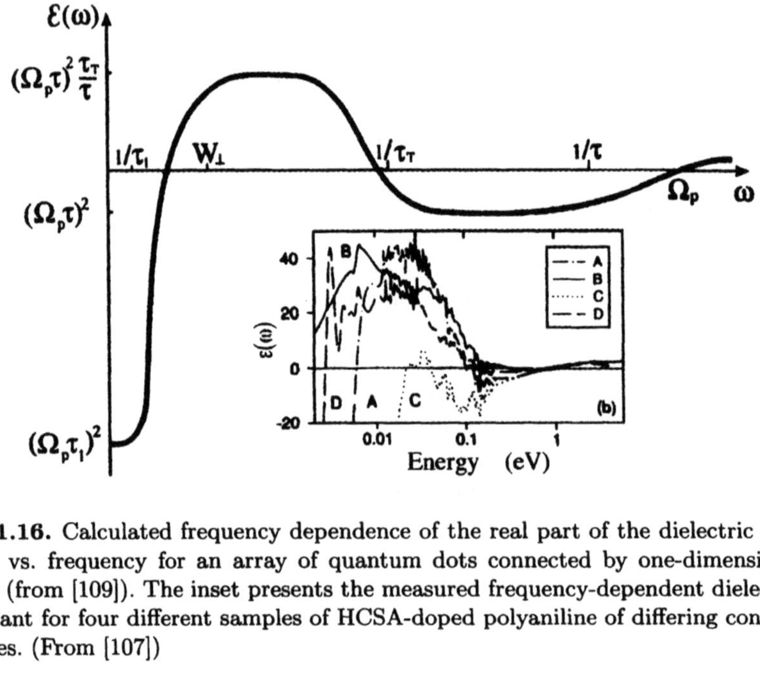

Fig. 1.16. Calculated frequency dependence of the real part of the dielectric constant vs. frequency for an array of quantum dots connected by one-dimensional wires (from [109]). The inset presents the measured frequency-dependent dielectric constant for four different samples of HCSA-doped polyaniline of differing conductivities. (From [107])

1.4.3 Summary

Electronic polymer behavior is strongly influenced by nanoscale structure. Small variation in local order have a dramatic effect on optical response and charge transport in these materials. Control of *intra*chain and *inter*chain order is key to controlling the electronic properties of these systems.

Acknowledgments. Stimulating collaborations with many scientists have contributed to the perspective presented here. I am deeply indebted to A.G.

MacDiarmid, J.P. Pouget, V. Prigodin, G. Ihas, T. Ishiguro, Y. Min, D. Tanner and L. Zuppiroli for extensive and stimulating discussions and collaborations. Contributions by those at The Ohio State University, especially J. Blatchford, T. Gustafson, J. Joo, R. Kohlman, Y.Z. Wang and Z.H. Wang are particularly important. This work was supported in part by ONR and NSF DMR-9508723.

References

1. C.K. Chiang, C.R. Fincher, Y.W. Park, A.J. Heeger, H. Shirakawa, E.J. Louis, S.C. Gua, A.G. MacDiarmid, Phys. Rev. Lett. **39**, 1098 (1977)
2. J.W. Blatchford, A.J. Epstein, Am. J. Phys. **64**, 120 (1996)
3. M. Pope, H.P. Kallmann, P. Magnante, J. Chem. Phys. **38**, 2042 (1962); W. Helfrich, W.G. Schneider, Phys. Rev. Lett. **14**, 229 (1965); C.W. Tang, S.A. Van Slyke, Appl. Phys. Lett. **51**, 913 (1987)
4. J.H. Burroughes, D.D.C. Bradley, A.R. Brown, R.N. Marks, K. Mackay, R.H. Friend, P.L. Burn, A.B. Holmes, Nature **347**, 539 (1990)
5. M.J. Winokur, In: *Handbook of Conducting Polymers*, 2nd ed., ed. by T.A. Skotheim, R.L. Elsenbaumer, J.R. Reynolds (Marcel Dekker, New York 1998) p. 707
6. P. Robin, J.P. Pouget, R. Comes, H.W. Gibson, A.J. Epstein, Phys. Rev. B **27**, 3938 (1983); P. Robin, J.P. Pouget, R. Comes, H.W. Gibson, A.J. Epstein, J. Phys. (Paris) Colloq. **44**, C3-87 (1983); the values of L given in these references must be divided by two
7. (a) R.H. Baughman, S.L. Hsu, G.P. Pez, A.J. Signorelli, J. Chem. Phys. **68**, 5405 (1978); (b) Q. Zhu, J.E. Fischer, R. Zuzok, S. Roth, Solid State Commun. **83**, 179 (1992)
8. J. Tsukamoto, Adv. Phys. **41**, 509 (1992)
9. D. Djurado, J. Ma, N. Theophilou, J.E. Fischer, Synth. Met. **30**, 395 (1989)
10. M. Kakudo, N. Kasai, *X-ray Diffraction by Polymer* (Kodansha, Tokyo/Elsevier, Amsterdam, 1972)
11. J.P. Pouget, Z. Oblakowski, Y. Nogami, P.A. Albouy, M. Laridjani, E.J. Oh, Y. Min, A.G. MacDiarmid, J. Tsukamoto, T. Ishiguro, A.J. Epstein, Synth. Met. **65**, 131 (1994)
12. J.P. Pouget, In: *Electronic Propenies of Polymers and Related Compounds, Springer Series in Solid State Sciences*, Vol. 63 (Springer, Berlin 1985) p. 26
13. T. Ito, H. Shirakawa, S. Ikeda, J. Polym. Sci. Polym. Chem. Ed. **12**, 11 (1974)
14. F. Saldi, M. Lelaurain, D. Billaud, J. Chem. Phys. **89**, 997 (1992)
15. M.M. Sokolowski, E.A. Marseglia, R.H. Friend, Polymer **27**, 1714 (1987)
16. T. Ishiguro, H. Kaneko, Y. Nogami, H. Ishimoto, H. Nigshiyama, J. Tsukamoto, A. Takahashi, M. Yamaura, T. Hagiwara, K. Sato, Phys. Rev. Lett. **69**, 660 (1992)
17. P.A. Albouy, J.P. Pouget, J. Halim, V. Enkelmann, G. Wegner, Makromol. Chem. **193**, 853 (1992)
18. J. Joo, G. Du, A.J. Epstein, V.N. Prigodin, J. Tsukamoto, Phys. Rev. B **52**, 8060 (1995); J. Joo, G. Du, J. Tsukamoto, A.J. Epstein, Synth. Met. **88**, 1 (1997)
19. P.A. Albouy, J.P. Pouget, H. Strzelecka, Phys. Rev. B **35**, 173 (1987)

20. J.P. Pouget, M.E. Jozefowicz, A.J. Epstein, X. Tang, A.G. MacDiarmid, Macromolecules **24**, 779 (1991)
21. M. Laridjani, J.P. Pouget, A.G. MacDiarmid, A.J. Epstein, J. Phys. I (France) **2**, 1003 (1992)
22. M. Laridjani, A.J. Epstein, Eur. Phys. J. B **7**, 585 (1999)
23. M. Laridjani, J.P. Pouget, E.M. Scherr, A.G. MacDiarmid, M.E. Jozefowicz, A.J. Epstein, Macromolecules **25**, 4106 (1992)
24. H.H.S. Javadi, K.R. Cromack, A.G. MacDiarmid, A.J. Epstein, Phys. Rev. B **39**, 3579 (1989) and earlier refs. therein
25. A.G. MacDiarmid, Y. Min, J.M. Wiesinger, E.J. Oh, E.M. Scherr, A.J. Epstein, Synth. Met. **55–57**, 753 (1993); E.J. Oh, Y. Min, J.M. Wiesinger, S.K. Manohar, E.M. Scherr, P.J. Prest, A.G. MacDiarmid, A.J. Epstein, Synth. Met. **55–57**, 977 (1993)
26. A.G. MacDiarmid, A.J. Epstein, *Conducting Polymers: Science and Technology 2nd Brazilian Polymer Conf.* Oct. 5–8, 1993 (Plenum, New York) p. 554; A.G. MacDiarmid, A.J. Epstein, Synth. Met. **65**, 103 (1994)
27. J. Joo, Z. Oblakowski, G. Du, J.P. Pouget, E.J. Oh, J.M. Wiesinger, Y. Min, A.G. MacDiarmid, A.J. Epstein, Phys. Rev. B **49**, 2977 (1994)
28. J. Joo, A.J. Epstein, V.N. Prigodin, Y. Min, A.G. MacDiarmid, Phys. Rev. B **50**, 12226 (1994)
29. J. Joo, S.M. Long, J.P. Pouget, E.J. Oh, A.G. MacDiarmid, A.J. Epstein, Phys. Rev. B **57**, 9567 (1998)
30. A.N. Samukhin, V.N. Prigodin, L. Jastrabík, A.J. Epstein, Phys. Rev. B **58**, 11354 (1998)
31. J.P. Travers, P. Le Guyadec, P.N. Adams, P.J. Laughlin, A.P. Monkman, Synth. Met. **65**, 159 (1994); M. Nechtschein, In: *Handbook of Conducting Polymers*, 2nd ed., ed. by T.A. Skotheim, R.L. Elsenbaumer, J.R. Reynolds (Marcel Dekker, New York 1998) p. 141
32. J.P. Pouget, S.L. Zhao, Z.H. Wang, Z. Oblakoswski, A.J. Epstein, S.K. Manohar, J.M. Wiesinger, A.G. MacDiarmid, C.H. Hsu, Synth. Met. **55–57**, 341 (1993)
33. J.P. Travers, P. Le Guyadec, P.N. Adams, P.J. Laughlin, A.P. Monkman, Synth. Met. **65**, 159 (1994)
34. M.E. Jozefowicz, A.J. Epstein, J.P. Pouget, J.G. Masters, A. Ray, A.G. MacDiarmid, Macromolecules **24**, 5863 (1991)
35. Z.H. Wang, A. Ray, A.G. MacDiarmid, A.J. Epstein, Phys. Rev. B **43**, 4373 (1991)
36. J.P. Pouget, M. Laridjani, M.E. Jozefowicz, A.J. Epstein, E.M. Scherr, A.G. MacDiarmid, Mater. Res. Soc. Symp. Proc. **247**, 589 (1992)
37. J. Yue, A.J. Epstein, J. Am. Chem. Soc. **112**, 2800 (1990); J. Yue, Z.H. Wang, K.R. Cromack, A.J. Epstein, A.G. MacDiarmid, J. Am. Chem. Soc. **113**, 1665 (1991)
38. X.-L. Wei, Y.Z. Wang, S.M. Long, C. Bobeczko, A.J. Epstein, J. Am. Chem. Soc. **118**, 2545 (1996)
39. W. Lee, G. Du, S.M. Long, S. Shimizu, T. Saitoh, M. Uzawa, A.J. Epstein, Synth. Met. **84**, 807 (1997)
40. J. Mardalen, E.J. Samuelsen, O.G. Gautun, P.M. Carlsen, Synth. Met. **48**, 363 (1992)
41. J.M. Ginder, A.J. Epstein, Phys. Rev. B **41**, 10674 (1990)

42. J.G. Masters, J.M. Ginder, A.G. MacDiarmid, A.J. Epstein, J. Chem. Phys. **96**, 4768 (1992)
43. V.N. Prigodin, A.J. Epstein, to be published
44. M. Yan, L.J. Rothberg, F. Papadimitrakopoulos, M.E. Galvin, T.M. Miller, Phys. Rev. Lett. **73**, 744 (1994)
45. M. Yan, L.J. Rothberg, E. W. Kwock, T. M. Miller, Phys. Rev. Lett. **75**, 1992 (1995)
46. S.A. Jenekhe, J.A. Osaheni, Science **265**, 739 (1994)
47. U. Lemmer, S. Heun, R.F. Mahrt, U. Scherf, M. Hopmeier, U. Siegner, E.O. Gobel, K. Mullen, H. Bassler, Chem. Phys. Lett. **240**, 373 (1995)
48. I.D.W. Samuel, G. Rumbles, C.J. Collison, Phys. Rev. B **52**, R11573 (1995)
49. R. Jakubiak, L.J. Rothberg, W. Wan, B.R. Hsieh, Synth. Met. **101**, 230 (1999)
50. R. Jakubiak, C.J. Collison, W.C. Wan, L.J. Rothberg, B.R. Hsieh, J. Phys. Chem. A **103**, 2394 (1999)
51. L.J. Rothberg, M. Yan, A.W.P. Fung, T.M. Jedju, E.W. Kwock, M.E. Galvin, Synth. Met. **84**, 537 (1997)
52. L.J. Rothberg, M. Yan, F. Papadimitrakopoulos, M.E. Galvin, E.W. Kwock, T.M. Miller, Synth. Met. **80**, 41 (1996)
53. M. Yan, L.J. Rothberg, E.W. Kwock, T.M. Miller, Phys. Rev. Lett. **75**, 1992 (1995)
54. T.Q. Nguyen, I.B. Martini, J. Liu, B.J. Schwartz, J. Phys. Chem. B **104**, 237 (2000)
55. A.K. Sheridan, I.D.W. Samuel, A. Bleyer, D.D.C. Brandley, Synth. Met. **101**, 259 (1999)
56. T.Q. Nguyen, V. Doan, B.J. Schwartz, J. Chem. Phys. **110**, 4068 (1999)
57. U. Stalmach, H. Detert, H. Meier, V. Gebhardt, D. Haarer, A. Bacher, H.W. Schmidt, Optical Mats. **9**, 77 (1998)
58. P.F. van Hutten, H.J. Brouwer, V.V. Krasnikov, L. Quali, U. Stalmach, G. IIadziioannou, Synth. Met. **102**, 1443 (1999)
59. M.W. Wu, E.M. Conwell, Phys. Rev. B **56**, 10060 (1997)
60. E. Conwell, Trends Polymer Sci. **5**, 218 (1997)
61. E.M. Conwell, Macromolecular Symp. **116**, 39 (1997)
62. E.M. Conwell, J. Perlstein, S. Shaik, Phys. Rev. B **54**, R2308 (1996)
63. H. Bassler, B. Schweitzer, Acc. Chem. Res. **32**, 173 (1999)
64. B. Schweitzer, H. Bassler, Synth. Met. **109**, 1 (2000)
65. M. Wohlgenannt, W. Graupner, G. Leising, Z.V. Vardeny, Phys. Rev. B **60**, 5321 (1999)
66. C.W. Lee, K.S. Wong, J.D. Huang, S.V. Frolov, Z.V. Vardeny, Chem. Phys. Lett. **314**, 564 (1999)
67. R. Osterbacka, M. Wohlgenannt, D. Chinn, Z.V. Vardeny, Phys. Rev. B **60**, R11253 (1999)
68. Z.V. Vardeny, X. Wei, In: *Handbook of Conducting Polymers*, 2nd ed., ed. by T.A. Skotheim, R.L. Elsenbaumer, J.R. Reynolds (Marcel Dekker, New York 1998) p. 22
69. A. Dogariu, A.J. Heeger, H.L. Wang, Phys. Rev. B **61**, 1635 (2000)
70. J.A. Osaheni, S.A. Jenekhe, Macromolecules **27**, 739 (1994)
71. D.D. Gebler, Y.Z. Wang, J.W. Blatchford, S.W. Jessen, D.-K. Fu, T.M. Swager, A.G. MacDiarmid, A.J. Epstein, Appl. Phys. Lett. **70**, 1644 (1997)
72. D.D. Gebler, Y.Z. Wang, D.-K. Fu, T.M. Swager, A.J. Epstein, J. Chem. Phys. **108**, 7842 (1998)

73. U. Rauscher, L. Schutz, A. Greiner, H. Bassler, J. Phys. Condens. Matt. **1**, 9751 (1989)
74. U. Rauscher, H. Bassler, D.D.C. Bradley, M. Hennecke, Phys. Rev. B **42**, 9830 (1990)
75. F.X. Bronold, A. Saxena, A.R. Bishop, Phys. Rev. B **48**, 13162 (1993)
76. J.B. Birks, *Photophysics of Aromatic Molecules* (Wiley Interscience, London 1970)
77. E.L. Frankevich, A.A. Lymarev, I. Sokolik, F.E. Karasz, S. Blumstengel, R.H. Baughmann, H.H. Horhold, Phys. Rev. B **46**, 9320 (1992)
78. J.W.P. Hsu, M. Yan, T.M. Jedju, L.J. Rothberg, B.R. Hseih, Phys. Rev. B **49**, 712 (1994)
79. H.A. Mizes, E.M. Conwell, Phys. Rev. B **50**, 11243 (1994)
80. M. Pope, C.E. Swenberg, *Electronic Processes in Organic Crystals* (Oxford University Press, New York 1982; 2nd edition, 1999)
81. A. Kohler, J. Gruner, R.H. Friend, K. Mullen, U. Scherf, Chem. Phys. Lett. **243**, 456 (1995)
82. R.S. Becker, *Theory and Interpretation of Fluorescence and Phosphorescence* (Wiley Interscience, New York 1969)
83. S.W. Jessen, D.D. Gebler, Y.Z. Wang, J.W. Blatchford, L.B. Lin, T.L. Gustafson, H.L. Wang, T.M. Swager, A.G. MacDiarmid, A.J. Epstein, Polym. Material Sci. Eng. **72**, 573 (1995)
84. J.W. Blatchford, S.W. Jessen, L.B. Lin, J.J. Lih, T.L. Gustafson, A.J. Epstein, D.-K. Fu, M.J. Marsella, T.M. Swager, A.G. MacDiarmid, S. Yamaguchi, H. Hamaguchi, Phys. Rev. Lett. **76**, 1513 (1996)
85. J.W. Blatchford, S.W. Jessen, L.B. Lin, T.L. Gustafson, A.J. Epstein, D.K. Fu, H.L. Wang, T.M. Swager, A.G. MacDiarmid, Materials Res. Soc. Symp. Proc. **413**, 671 (1996)
86. A.J. Epstein, J.W. Blatchford, Y.Z. Wang, S.W. Jessen, D.D. Gebler, L.B. Lin, T.L. Gustafson, H.-L. Wang, Y.W. Park, T.M. Swager, A.G. MacDiarmid, Synth. Met. **78**, 253 (1996)
87. J.W. Blatchford, T.L. Gustafson, A.J. Epstein, D.A. Vanden Bout, J. Kerimo, D.A. Higgins, P.F. Barbara, D.-K. Fu, T.M. Swager, A.G. MacDiarmid, Phys. Rev. B **54**, 3683 (1996)
88. J.W. Blatchford, T.L. Gustafson, A.J. Epstein, J. Chem. Phys. **105**, 9214 (1996)
89. J.W. Blatchford, S.W. Jessen, L.-B. Lin, T.L. Gustafson, D.-K. Fu, H.-L. Wang, T.M. Swager, A.G. MacDiarmid, A.J. Epstein, Phys. Rev. B **54**, 9180 (1996)
90. S.W. Jessen, J.W. Blatchford, L.-B. Lin, T.L. Gustafson, J. Partee, J. Shinar, D.-K. Fu, M.J. Marsella, T.M. Swager, A.G. MacDiarmid, A.J. Epstein, Synth. Met. **84**, 501 (1997)
91. D.D. Gebler, Y.Z. Wang, J.W. Blatchford, S.W. Jessen, L.B. Lin, T.L. Gustafson, H.L. Wang, T.M. Swager, A.G. MacDiarmid, A.J. Epstein, J. Appl. Phys. **78**, 4264 (1995)
92. Y.Z. Wang, D.D. Gebler, L.B. Lin, J.W. Blatchford, S.W. Jessen, H.L. Wang, A.J. Epstein, Appl. Phys. Lett. **68**, 894 (1996)
93. Y.Z. Wang, A.J. Epstein, Acc. Chem. Res. **32**, 217 (1999)
94. M.J. Marsella, D.-K. Fu, T.M. Swager, Adv. Mater. Res. **7**, 145 (1995)

95. T. Yamamoto, T. Maruyama, Z. Zhou, T. Ito, T. Fukuda, Y. Yoneda, F. Begum, T. Ikeda, S. Sasaki, H. Takezoe, A. Fukuda, K. Kubota, J. Am. Chem. Soc. **116**, 4832 (1994)
96. D.K. Fu, M.J. Marsella, T.M. Swager, Polym. Prepr. Am. Chem. Soc. Div. Polym. **36**, 585 (1995)
97. O. Lhost, J.L. Brédas, J. Chem. Phys. **96**, 5279 (1992)
98. T.W. Hagler, K. Pakbaz, K.F. Voss, A.J. Heeger, Phys. Rev. B **44**, 8652 (1991)
99. R.S. Kohlman, J. Joo, A.J. Epstein, In: *Physical Properties of Polymers Handbook*, ed. by J.E. Mark (AIP Press 1996) Chap. 34, p. 453
100. R.S. Kohlman, A.J. Epstein, In: *Handbook of Conducting Polymers*, ed. by T.A. Skotheim, R.L. Elsenbaumer, J.R. Reynolds (Marcel Dekker, New York 1998) p. 85
101. D. Baeriswyl, D.K. Campbell, S. Mazumdar, In: *Conjugated Conducting Polymers*, ed. by H.G. Kiess (Springer-Verlag, Berlin 1992), p. 7
102. E.M. Conwell, IEEE Transactions on Electrical Insulation EI-22, 591 (1987)
103. P.W. Anderson, Phys. Rev. **109**, 1492 (1958)
104. N.F. Mott, E. Davis, In: *Electronic Processes in Non-Crystalline Materials* (Clarendon Press, Oxford 1979) p. 6
105. G. Du, V.N. Prigodin, A. Burns, J. Joo, C.S. Wang, A.J. Epstein, Phys. Rev. B **58**, 4485 (1998); Phys. Rev. B **61**, 10142 (2000)
106. J.M. Ginder, A.F. Richter, A.G. MacDiarmid, A.J. Epstein, Solid State Commun. **63**, 97 (1987); Z.H. Wang, A. Ray, A.G. MacDiarmid, A.J. Epstein, Phys. Rev. B **43**, 4373 (1991); J. Joo, V.N. Prigodin, Y.G. Min, A.G. MacDiarmid, A.J. Epstein, Phys. Rev. B **50** 12226 (1994)
107. R.S. Kohlman, J.Joo, Y.Z. Wang, J.P. Pouget, H. Kaneko, T. Ishiguro, A.J. Epstein, Phys. Rev. Lett. **74**, 773 (1995); R.S. Kohlman, J. Joo, Y.G. Min, A.G. MacDiarmid, A.J. Epstein, Phys. Rev. Lett. **77**, 2766 (1996); R.S. Kohlman, A. Zibold, D.B. Tanner, G.G. Ihas, T. Ishiguro, Y.G. Min, A.G. MacDiarmid, A.J. Epstein, Phys. Rev. Lett. **78**, 3915 (1997)
108. Reghu Menon, C.O. Yoon, D. Moses, A.J. Heeger, In: *Handbook of Conducting Polymers*, ed. by T.A. Skotheim, R.L. Elsenbaumer, J.R. Reynolds (Marcel Dekker, New York 1998) p. 27
109. C.J. Bolton-Heaton, C.J. Lambert, V.I. Falko, V.N. Prigodin, A.J. Epstein, Phys. Rev. B 60, 10569 (1999); V.N. Prigodin, A.J. Epstein, to be published
110. M.J. Rice, J. Bernasconi, J. Phys. F Metal Phys. **2**, 905(1972)
111. A.B. Kaiser, Phys. Rev. B **40**, 2806 (1989)
112. T. Ishiguro, H. Kaneko, Y. Nogami, H. Nishiyama, J. Tsukamoto, A. Takahashi, M. Yamaura, J. Sato, Phys. Rev. Lett. **69**, 660(1992); H. Kaneko, T. Ishiguro, J. Tsukamoto, A. Takahashi, Solid State Commun. **90**, 83 (1994)

2 Theoretical Studies of Electronic Properties of Conjugated Polymers

M. Springborg, K. Schmidt, H. Meider and L. De Maria

2.1 Introduction

In our everyday life we continuously make use of different materials with different properties. We thereby exploit that through appropriate choices we can obtain materials with predefined material properties. Thus, for many applications the mechanical properties are most important, whereas for others (e.g., in the semiconductor industry) the electronic properties are of interest. For some applications not only one but the combination of a number of properties is of importance. Conjugated polymers are a class of materials with an unusual combination of materials properties, which makes them interesting in technological applications for certain special purposes.

Polymers are most well known for their mechanical properties; they form the basis for the majority of plastics. A typical example is polyethylene, shown in Fig. 2.1. Polyethylene consists of very long chains of covalently bonded, approximately tetrahedrally coordinated carbon atoms. A single chain is fairly flexible, and since the interactions between the chains are only weak, a macroscopic sample of polyethylene possesses the typical mechanical properties of plastics that we know for, e.g., household plastic films. Many other plastics are based on related macromolecules, so their overall mechanical properties are very similar.

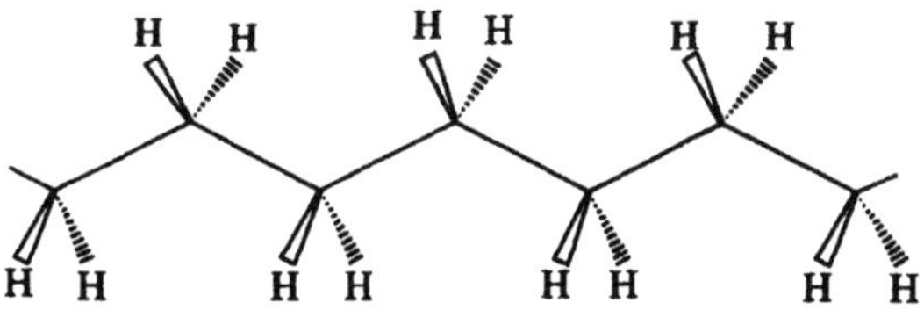

Fig. 2.1. Structure of a single polyethylene chain, $(CH_2)_x$, as well as that of a polysilane chain, $(SiH_2)_x$

The chemical bonds between the atoms of polyethylene can be interpreted as being due to sp^3 hybrids on the carbon atoms and $1s$ functions on the hydrogen atoms. These functions form strong, directed, covalent bonds between nearest neighbors. In the simplest approximation, electronic excitations involve exciting an electron from one of these bonding orbitals to the equivalent antibonding orbital, which requires a large amount of energy. Therefore, the electronic band structures for these materials are characterized by a large energy gap between occupied and unoccupied bands (see, e.g., Fig. 2.2), and the material is transparent.

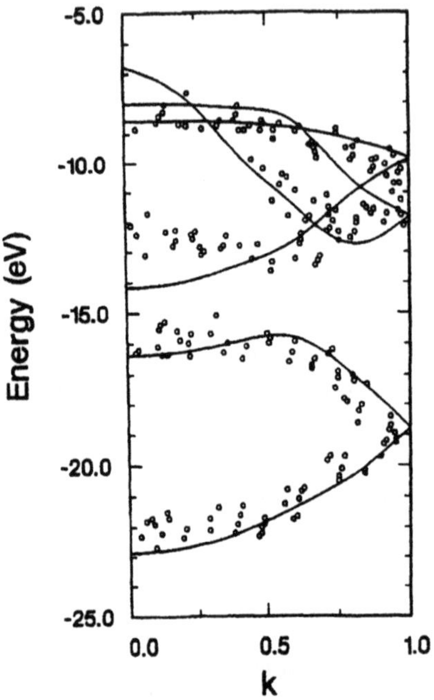

Fig. 2.2. Band structures of a single, infinite, periodic polyethylene chain from theory (*solid lines*) and experiment (*circles*). The lowest unoccupied orbitals appear at positive band energies. (From [7])

Conjugated polymers differ from these polymers in their electronic properties. The carbon atoms are not fourfold, but threefold coordinated. Figure 2.3 shows the prototype of conjugate polymers, polyacetylene. In this case, the chemical bonds are formed partly by sp^2 hybrids on the carbon atoms and $1s$ functions on the hydrogen atoms, which lead to strong σ bonds (symmetric orbitals with respect to reflection in the plane of the nuclei) between the atoms, and partly by those carbon p functions that are anti-symmetric with respect to the reflection. The latter lead to π bonds between the carbon atoms, and since these bonds are weaker, the orbitals appear at higher energies. Moreover, the energy required to excite an electron from a bonding π orbital to an anti-bonding one is relatively low, and, therefore, the energy gap separating unoccupied and occupied orbitals is much smaller than for, e.g., polyethylene (see, e.g., Fig. 2.4). For conjugated polymers this energy gap is typically in the range 1–3 eV, and is thus comparable with typical values for conventional, crystalline semiconductors. Conjugated polymers are accordingly semiconductors with the same mechanical properties as plastics.

One of the important differences between conjugated polymers and non-conjugated polymers is the presence of the extended π-electron systems in the

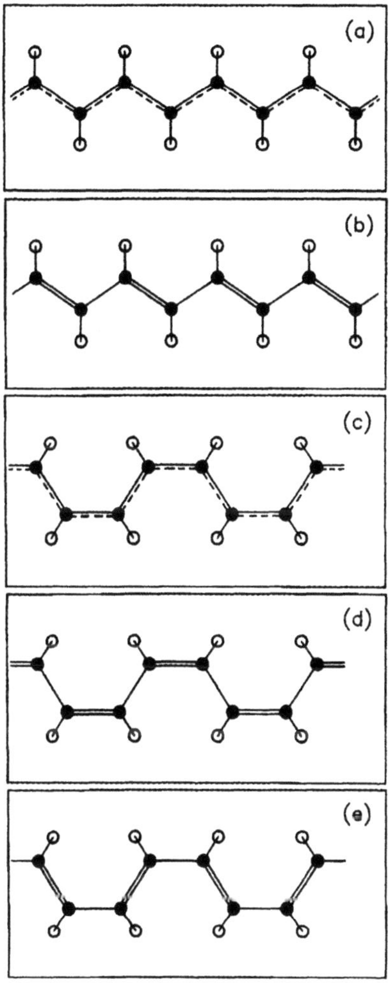

Fig. 2.3. Structure of (**a**,**b**) trans-polyacetylene and (**c**,**d**,**e**) cis-polyacetylene with either (**b**,**d**,**e**) alternating C–C bond lengths or (**a**,**c**) non-alternating C–C bond lengths. *Black* and *white circles* represent carbon and hydrogen atoms, respectively

former. There is, however, another important difference. The band structures for the trans-polyacetylene structure of Fig. 2.3a (i.e., Fig. 2.4a) show that this material has a vanishing gap. Focusing only on the π electrons, one may consider this material to be a quasi-one-dimensional system with one (π) electron per repeated (CH) unit, and as such a metal with one exactly half-filled band. As pointed out by Peierls [1], such a material is unstable against a doubling of the unit cell which will lead to alternating bond lengths as well as to a gap at the Fermi level (see Fig. 2.4b). Therefore, the bond lengths of trans-polyacetylene alternate as in Fig. 2.3b.

For the cis-isomers of Fig. 2.3c–e these arguments do not apply directly, but, nevertheless, also for this material, as well as for most other conju-

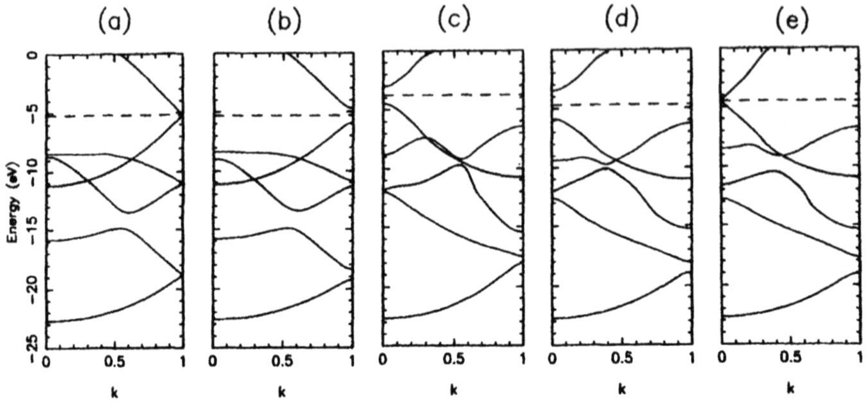

Fig. 2.4a–e. The band structures for the five structures shown in Fig. 2.3. The *horizontal dashed lines* mark the Fermi level; $k = 0$ and $k = 1$ represent the center and the edge of the first Brillouin zone, respectively

gated polymers, the C–C bond lengths along the backbone alternate between slightly different values.

For trans-polyacetylene it may now be observed that the structure of Fig. 2.3b as well as the one obtained by interchanging the short and the long bonds is energetically degenerate. Therefore, structural defects that are interfaces between these two structures may occur, as shown in Fig. 2.5a.

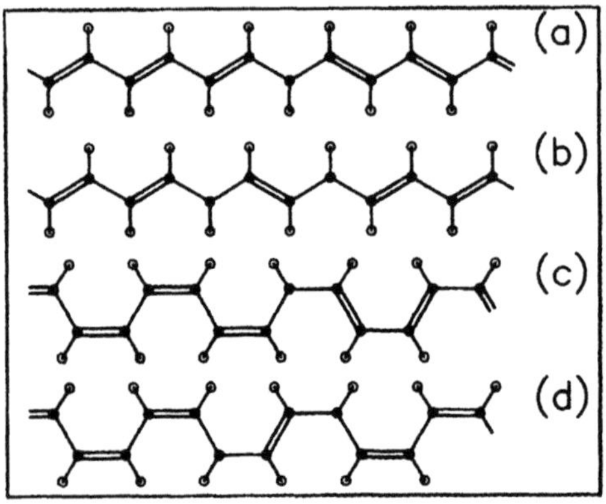

Fig. 2.5. Structure of (**a**,**c**) a soliton and (**b**,**d**) a polaron in (**a**,**b**) trans- and (**c**,**d**) cis-polyacetylene. The presentation is as in Fig. 2.3. It should be stressed that the real defects are much more extended than sketched in the figure

For an infinite, isolated system, these may move freely along the chain. For a neutral system, they will contain one unpaired electron and therefore carry spin. However, charge added to the system may localize at such defects, whereby the defect possesses no spin, but charge. The defect, the so-called soliton, can accordingly be held responsible for transport of spin and charge through the chains; therefore, much effort is devoted to understanding and exploring its properties. It turns out that in order to obtain a proper description of it, it is important to take into account that the properties of the π orbitals depend sensitively on the bond lengths, i.e., the materials possess strong electron–phonon couplings.

For cis polyacetylene, the two isomers of Fig. 2.3d,e are not energetically degenerate, and, therefore, a hypothetical soliton (Fig. 2.5c) will move only in that direction that leads to the largest segment with the low-energy structure. However, local distortions of the bond-length alternation (Fig. 2.5d) may exist and carry spin or charge throughout the system. This defect is the so-called polaron, and, as for the soliton, its existence may lead to extra states in the gap between occupied and unoccupied orbitals.

Although we have so far only discussed a single conjugated polymer, polyacetylene, it should be stressed that there are very many others. Most of those are obtained by adding heteroatoms to the backbone, by substituting the sidegroups with others, or through synthesis of a slightly more complicated backbone. Nevertheless, they all possess the same characteristic properties: they have the mechanical properties of conventional plastics but the electronic properties of semiconductors; they consist of only weakly interacting, long chains with threefold coordinated carbon atoms; the orbitals closest to the Fermi level are formed by carbon-centered π functions; the π orbitals have large electron–phonon couplings; upon charging and/or exciting the material, structural distortions such as solitons and polarons may occur; these are extended defects that can transport charge and spin, and that lead to extra levels in the optical gap.

In order to study these materials theoretically, one encounters various problems. First, they are approximately infinite in one dimension and (when neglecting the weak interchain interactions) finite in the other two. Furthermore, for the structural defects the periodicity is destroyed. For parameter-free studies these problems represent serious bottlenecks. Nevertheless, there has been a large number of such studies on these materials. Alternatively, one may apply simpler approaches based on model Hamiltonians whereby more complicated systems and properties can be treated. These depend, however, sensitively on the accuracy with which the parameter values entering the models can be determined, and since the conjugated polymers often are relatively ill-defined materials, this may be a critical issue.

It is the purpose of this chapter to show how parameter-free electronic-structure calculations can be employed in studying the properties of the conjugated polymers. Such methods have the advantage that any specific material, in principle, can be studied, but the disadvantage that the calculations easily become computationally involved. Therefore, the systems and properties that can be studied are limited. Nevertheless, we shall see here how conjugated polymers can be treated with such methods. Emphasis will be placed on the applications of one specific electronic-structure method, but, whenever relevant, other approaches will be mentioned.

This chapter is organized as follows. In Sect. 2.2 we briefly describe the various parameter-free electronic-structure methods that may be applied. Sect. 2.3 is devoted to applications on infinite, periodic systems, and in Sect. 2.4 we study effects of symmetry distortions as they occur when solitons and polarons exist, when the material is exposed to external electromagnetic fields, or when other perturbations due to extra charges or substitutions are present. Finally, Sect. 2.5 contains the conclusions.

2.2 Methods of Calculation

Experimental samples of conjugated polymers are often ill-defined. This means that the materials consist of more or less parallel chains each of which is either long or short. The interactions between the chains are much weaker than those along the chains. In total, therefore, when attempting to calculate the electronic properties of such materials with parameter-free methods, one must first address two fundamental questions: Which kind of method shall be applied? and How should the structure of the material be approximated? We shall address these questions in this section.

2.2.1 Hartree–Fock- or Density-Functional-Based Methods

Assuming that the Born–Oppenheimer approximation is valid, the problem of calculating the (static) electronic properties of interest becomes that of solving the (time-independent) Schrödinger equation

$$\hat{H}_e \Psi_e = E_e \Psi_e \,, \tag{2.1}$$

where Ψ_e is the electronic wavefunction that depends on the position and spin coordinates of all (N) electrons. The N-body Hamilton operator $\hat{H}_e$ contains the kinetic energy of the electrons, the potential energy from the (M) nuclei, and the electron–electron interactions,

$$\hat{H}_e = -\frac{1}{2}\sum_{i=1}^{N} \nabla_i^2 - \sum_{k=1}^{M}\sum_{i=1}^{N} \frac{Z_k}{|\boldsymbol{R}_k - \boldsymbol{r}_i|} + \frac{1}{2}\sum_{i\neq j=1}^{N} \frac{1}{|\boldsymbol{r}_i - \boldsymbol{r}_j|} \,. \tag{2.2}$$

We have used Hartree atomic units here ($m_e = |e| = \hbar = 4\pi\epsilon_0 \equiv 1$) and denoted the charge and position of the kth nucleus Z_k and $\boldsymbol{R}_k$, respectively. Furthermore, $\boldsymbol{r}_i$ is the position of the ith electron.

From a given set of nuclear positions $\{\boldsymbol{R}_k\}$ the total energy is calculated from the electronic energy E_e as follows:

$$E_{tot} = E_e + \frac{1}{2}\sum_{k\neq l=1}^{M} \frac{Z_k Z_l}{|\boldsymbol{R}_k - \boldsymbol{R}_l|} \,, \tag{2.3}$$

and by varying these one may determine the structure with the lowest total energy.

Unfortunately, (2.1) cannot be solved exactly for systems containing more than two particles and one has to resort to approximations. A common approximation is that of assuming that the ground-state eigenfunction Ψ_e can be written as a single Slater determinant, which leads to the well-known Hartree–Fock approximation as well as the Hartree–Fock single-particle equations,

$$\hat{F}\psi_i(\boldsymbol{r},\sigma) = \epsilon_i\psi_i(\boldsymbol{r},\sigma) \tag{2.4}$$

(σ being the spin). Here, the Fock operator, $\hat{F}$, contains a kinetic-energy term, one from the Coulomb potential of the nuclei, one from that of the electrons, and the exchange term,

$$\hat{F} = -\frac{1}{2}\nabla^2 - \sum_{k=1}^{M} \frac{Z_k}{|\boldsymbol{r} - \boldsymbol{R}_k|} + \sum_{j=1}^{N} \hat{J}_j - \sum_{j=1}^{N} \hat{K}_j \,. \tag{2.5}$$

The Coulomb and exchange operator of the jth orbital ($\hat{J}_j$ and $\hat{K}_j$, respectively) depend on the jth eigenfunction ψ_j.

The total electronic energy E_e is written as follows within the Hartree–Fock approximation:

$$E_e = -\frac{1}{2}\sum_{i=1}^{N}\langle\psi_i|\nabla^2|\psi_i\rangle - \sum_{i=1}^{N}\sum_{k=1}^{M}\left\langle\psi_i\left|\frac{Z_k}{|\boldsymbol{r} - \boldsymbol{R}_k|}\right|\psi_i\right\rangle + \frac{1}{2}\sum_{i,j=1}^{N}\langle\psi_i|\hat{J}_j - \hat{K}_j|\psi_i\rangle \,, \tag{2.6}$$

or, by assuming that (2.5) is solved exactly, as

$$E_e = \sum_{i=1}^{N}\epsilon_i - \frac{1}{2}\sum_{i,j=1}^{N}\langle\psi_i|\hat{J}_j - \hat{K}_j|\psi_i\rangle \,. \tag{2.7}$$

Correlation effects are by definition not included in this approach. In order to include these, one may follow one out of two different approaches. One possibility is to write the many-electron wavefunction Ψ_e explicitly as a linear combination of more (many) Slater determinants and through variation determine the expansion coefficients. This approach has, however, the disadvantage that it becomes prohibitively involved for larger systems, and, furthermore, it may suffer from the size-consistency problem, i.e., that a doubling of the system will not necessarily lead to a doubling of the total energy when the two halves are non-interacting. Alternatively, one may include the correlation effects via a perturbation theory such as that of Møller and Plesset [2], which does not suffer from the size-consistency problem.

As an alternative to these (so-called wavefunction-based) methods, one may apply (so-called density-based) methods whose origin is the density-functional theory of Kohn and coworkers. Hohenberg and Kohn [3] showed that the electronic energy E_e in (2.1) for the ground state is a unique functional of the electron density, although it was not possible to give its precise form. Kohn and Sham [4] reformulated the problem of calculating E_e as that of solving a set of single-particle equations

$$\hat{h}_{\text{eff}}\psi_i(\boldsymbol{r},\sigma) = \epsilon_i\psi_i(\boldsymbol{r},\sigma) \ . \tag{2.8}$$

Here, the Kohn–Sham operator is

$$\hat{h}_{\text{eff}} = -\frac{1}{2}\nabla^2 - \sum_{k=1}^{M} \frac{Z_k}{|\boldsymbol{r}-\boldsymbol{R}_k|} + V_{\text{C}}(\boldsymbol{r}) + V_{\text{xc}}(\boldsymbol{r}) \ , \tag{2.9}$$

and the total electronic energy E_e becomes

$$\begin{aligned} E_e = &-\frac{1}{2}\sum_{i=1}^{N}\langle\psi_i|\nabla^2|\psi_i\rangle - \sum_{i=1}^{N}\sum_{k=1}^{M}\left\langle\psi_i\left|\frac{Z_k}{|\boldsymbol{r}-\boldsymbol{R}_k|}\right|\psi_i\right\rangle \\ &+ \int\rho(\boldsymbol{r})\left[\frac{1}{2}V_{\text{C}}(\boldsymbol{r}) + \epsilon_{\text{xc}}(\boldsymbol{r})\right]d\boldsymbol{r} \ . \end{aligned} \tag{2.10}$$

$V_{\text{C}} = \sum_{j=1}^{N}\hat{J}_j$ is the Coulomb potential of the electron density,

$$\rho(\boldsymbol{r}) = \sum_{i=1}^{N/2}\sum_{\sigma}|\psi_i(\boldsymbol{r},\sigma)|^2 \ , \tag{2.11}$$

and V_{xc} and $\rho\cdot\epsilon_{\text{xc}}$ are the so-called exchange-correlation potential and energy density, respectively. They are related through

$$V_{\text{xc}}(\boldsymbol{r}) \equiv \frac{\delta}{\delta\rho(\boldsymbol{r})}E_{\text{xc}} \equiv \frac{\delta}{\delta\rho(\boldsymbol{r})}\int\epsilon_{\text{xc}}(\boldsymbol{r}')\rho(\boldsymbol{r}')d\boldsymbol{r}' \ , \tag{2.12}$$

with E_{xc} being the total exchange-correlation energy and when neglecting spin-polarization effects. Unfortunately, the precise form of this as a functional of ρ is not known, and one applies therefore various approximations. The most common one is the so-called local-density approximation (LDA) according to which V_{xc} and ϵ_{xc} at the point $\boldsymbol{r}$ are functions of the electron density at that point; however during the last decade also so-called generalized gradient approximations (GGAs), where also dependences on the gradient of the electron density at the point of interest are incorporated, have become important.

As above, one may assume that the Kohn–Sham equations (2.8) are solved exactly and then write E_e as

$$E_e = \sum_{i=1}^{N}\epsilon_i + \int\rho(\boldsymbol{r})\left[-\frac{1}{2}V_{\text{C}}(\boldsymbol{r}) + \epsilon_{\text{xc}}(\boldsymbol{r}) - V_{\text{xc}}(\boldsymbol{r})\right]d\boldsymbol{r} \ . \tag{2.13}$$

According to Koopmans' theorem [5], the eigenvalues of the Hartree–Fock equations (2.4) provide good approximations to the electronic excitation energies. However, for the single-particle eigenvalues of the Kohn–Sham equations (2.8) it is only known [6] that the highest energy of the occupied orbitals should equal the first ionization potential. Experience has shown that, for larger systems, the energy regions spanned by the Hartree–Fock eigenvalues are usually too wide, and often these energies are rescaled by a factor of 0.8–0.9 in order to improve the agreement with experimental information. In contrast, those of density-functional calculations are usually (although there are exceptions) in better agreement with experiments, but rigidly shifted towards higher energies. An example of this can be seen in Fig. 2.2, where we observe good agreement between the calculated band structures [7] obtained with a density-functional method and the experimentally observed bands [8], which was obtained by shifting the latter rigidly 2 eV up in energy.

For the unoccupied orbitals it has been found that the energy gap separating occupied and unoccupied orbitals is significantly overestimated by Hartree–Fock calculations (for semiconductors, this overestimate may amount to more than 5 eV). Density-functional calculations, on the other hand, lead to band gaps that are too small (typically 0.5–1 eV too small for semiconductors).

Independent of whether one chooses to use a Hartree–Fock or a density-functional method, one has to solve a set of single-particle equations,

$$\hat{h}\psi_i(\boldsymbol{r}) = \epsilon_i\psi_i(\boldsymbol{r}) \ . \tag{2.14}$$

Most often this is done by expanding the eigenfunctions in some set of basis functions,

$$\psi_i(\boldsymbol{r}) = \sum_{j=1}^{P} \chi_j(\boldsymbol{r})c_{ji} \ . \tag{2.15}$$

Gaussians and plane waves represent two choices of basis functions that have the advantage that the necessary matrix elements $\langle\chi_k|\chi_l\rangle$ and $\langle\chi_k|\hat{h}|\chi_l\rangle$ can be calculated easily, but the disadvantage that relatively many basis functions are required in order to obtain accurate results. Alternatively, one may use basis sets that are better adapted to the problem. These include augmented waves where one uses either plane waves or spherical waves in the so-called interstitial region (the region outside all atom-centered, non-overlapping spheres) that are matched continuously and differentiably onto numerical basis functions inside the spheres, Slater-type orbitals, and completely numerical basis functions.

In the studies we report below, we have used a specific density-functional method that is based on LMTOs (linearized muffin-tin orbitals – these are augmented spherical waves) as basis functions [9, 10].

2.2.2 Infinite or Finite Systems

Calculating the electronic structure of a macroscopic sample containing a more or less random distribution of short and long chains that are partly parallel and partly randomly oriented with respect to each other is an impossible task. Instead, one has to make approximations in the description of the structure.

We consider the example of the polyacetylene in Fig. 2.3. A finite oligomer of this may contain n repeated C_2H_2 units, and at each end one may saturate the dangling bonds to the absent neighboring C atoms with H atoms giving CH_2 units at the end. Thus, a model system for a long polyacetylene chain is $CH_2(C_2H_2)_nCH_2$. Most properties (e.g., the total energy per CH unit, bond lengths, electron density, etc.) converge relatively rapidly to their values for $n \to \infty$ (for some properties like the bond lengths it is then useful to concentrate on the central units). Therefore, since the interchain interactions are considerably weaker than the intrachain interactions, it is often a good approximation to calculate the electronic properties of the conjugated polymers by studying various finite oligomers and from the information on the properties as functions of n extrapolate to the $n \to \infty$ limit. An improved convergence as a function of n is obtained by studying the changes of the property of interest as a function of n. For example, for the total energy $E_{\text{tot}}(n)$, one may study either the relative total energy

$$E_{\text{rel}}(n) = E_{\text{tot}}(n)/n \tag{2.16}$$

or the quantity

$$\Delta E_{\text{tot}}(n) = E_{\text{tot}}(n) - E_{\text{tot}}(n-1) \; . \tag{2.17}$$

In the limit $n \to \infty$, they should converge to the same value, but it turns out that ΔE_{tot} converges faster than E_{rel} – this can be understood as being due to an optimal cancellation of end effects.

In addition, other quantities that, roughly, are proportional to the size of the system can be examined through equivalent approaches. There are, however, cases where such quantities converge only slowly as a function of n. To these belong the non-linear responses to external dc or ac fields, which shall be discussed in Sects. 2.4.5 and 2.4.6. The advantage with this approach is that commonly used quantum-chemistry programs for the calculation of molecular properties can, in principle, be directly applied to these materials.

As an alternative to the study of finite systems, one may directly study infinite systems. Thus, the boundary effects are avoided, but, on the other hand, the calculations become possible only if it can be assumed that the system is periodic. We will thus assume that the polymer of interest contains an infinite number of repeated, identical units placed along the (straight) polymer axis. Actually, there is no formal problem in generalizing to the case that the units are placed periodically in a helical way along a straight helical polymer axis. The symmetry is utilized in forming symmetry-adapted basis functions (Bloch waves). Thus, from the jth basis function of the qth unit cell χ_{jq} we construct

$$\chi_j^k = \lim_{Q\to\infty} \frac{1}{\sqrt{2Q+1}} \sum_{q=-Q}^{Q} e^{ikq\pi} \chi_{jq} \, . \tag{2.18}$$

Here, k is proportional to the (one-dimensional) Bloch wavenumber, and the first Brillouin zone is defined by $k \in]-1;1]$.

Using the symmetry properties of the identity operator and of the single-particle operator $\hat{h}$ of (2.14) (independent of whether this is the Fock or the Kohn–Sham operator), one obtains

$$\langle \chi_{j_1}^{k_1} | \chi_{j_2}^{k_2} \rangle = \delta_{k_1,k_2} \lim_{Q\to\infty} \sum_{q=-Q}^{Q} e^{ik_2 q\pi} \langle \chi_{j_1 0} | \chi_{j_2 q} \rangle \tag{2.19}$$

and

$$\langle \chi_{j_1}^{k_1} | \hat{h} | \chi_{j_2}^{k_2} \rangle = \delta_{k_1,k_2} \lim_{Q\to\infty} \sum_{q=-Q}^{Q} e^{ik_2 q\pi} \langle \chi_{j_1 0} | \hat{h} | \chi_{j_2 q} \rangle \, . \tag{2.20}$$

This approach has the advantage that instead of having to treat an infinitely large system directly, one needs only to consider problems whose size is defined by the number of basis functions per unit cell. The disadvantage is that the method is restricted to periodic systems. Thus, the solitons and polarons of Fig. 2.5 are not directly treatable with this method. In order to circumvent this problem, one may artificially repeat these structural distortions periodically, which, however, may lead to very large repeated units, or, alternatively, one may apply Green's function techniques explicitly developed for the study of defects. However, such calculations may also become very complicated.

In our applications we use a method that assumes the system to be infinite and periodic. Actually, the method has been explicitly designed for studies on systems that have a general helical screw-axis symmetry [9] for which the primitive symmetry operation is a combined translation of a and rotation of v.

2.2.3 Crystals or Chains

By just treating one isolated chain (assuming that it is periodic and infinite), it is utilized that the interchain interactions are so much weaker than the intrachain interactions that they to a first approximation can be neglected. For the general helical symmetry this is a clear advantage, since one may in such a case treat any helical symmetry without having to require that v is commensurate with 2π. On the other hand, special computer codes devoted to such systems are required and, most likely for that reason, only relatively few such programs exist based either on the Hartree–Fock (see, e.g., [11,12]) or on the density-functional approach [9,13].

Alternatively, one may still assume that the system is periodic, but then, in addition, also assume periodicity in the two directions perpendicular to the polymer axis. This amounts to repeating the single polymer chain periodically so that an idealized crystalline compound results for which it still is assumed that the distances between the chains are so large that they interact only weakly. With this construction traditional band-structure programs that have been developed for crystalline materials can be applied. Furthermore, (augmented) plane waves can also be used as basis functions, which not is the case when considering a single, isolated chain.

Due to the convenience related to being able to apply already existing computer programs, some studies on the properties of conjugated polymers obtained with such programs have been carried out. We shall discuss some of them below, but shall otherwise concentrate on applications of our own single-chain code.

2.3 Periodic Systems

After the general description given above of the various methods that are used in calculating the electronic properties of conjugated polymers, we shall now turn to representative examples of applications. We should stress at this point that this set of examples is far from complete but is instead chosen in accord with the authors' subjective interest.

We shall in this section study infinite, periodic conjugated polymers, starting with the prototype of those, polyacetylene, which was described in Sect. 2.1. Subsequently, we shall show how the general properties prevail – but also how they are modified – when the composition and structure are modified. Finally, we shall study systems that to a lesser extent belong to the class of conjugated polymers, although they show, as we shall demonstrate, many similarities with them.

2.3.1 Trans- and Cis-Polyacetylene: Structure and Band Structures

In Fig. 2.3 we showed the structures of trans- and cis-polyacetylene, both when it was assumed that all C–C bonds have the same length and when they were allowed to alternate. The band structures (Fig. 2.4) showed that for trans-polyacetylene the bond-length alternation was accompanied by the opening up of a band gap at the Fermi level, which was identified as a physical realization of a Peierls transition. The existence of the bond-length alternation is of ultimate importance for the possibility of solitons (Fig. 2.5; also called the mid-gap states), and, therefore, we optimized the structural parameters for an isolated, infinite, periodic trans-polyacetylene chain [10]. This led to C–C bond lengths of alternating 2.58 and 2.76 a.u., C–H bond lengths of 2.08 a.u., and C–C–C bond angles of 128°. Experimental studies [14, 15] have yielded 2.57 and 2.72 a.u. for the C–C bond lengths, 2.05–2.10 a.u. for the C–H bond lengths, and 123° for the C–C–C bond angles, which is in good agreement with our results.

In order to obtain more information about the bond-length alternation we kept the unit-cell length constant but reduced the C–C bond-length alternation gradually. Separating the total energy into two terms, one being the sum of the energies of the occupied π-electron levels and the other being the remainder [see (2.7) and (2.13), and Sects. 2.4.1 and 2.4.2], we obtained the curves shown in Fig. 2.6.

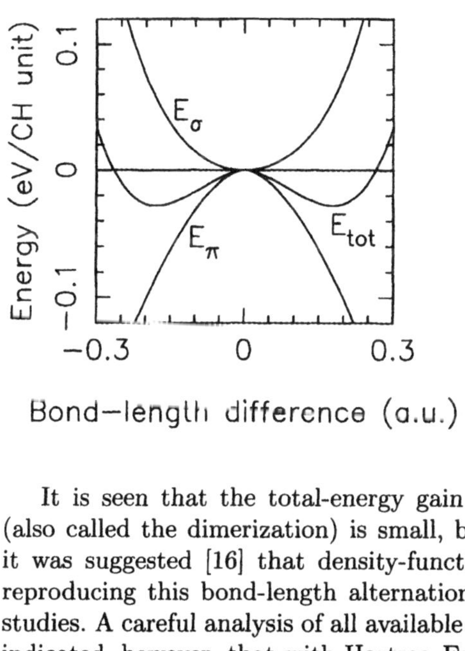

Fig. 2.6. Total energy, E_{tot}, the sum of the single-particle energies of the occupied π orbitals, E_π, and the remaining part of the total energy, E_σ, for trans-polyacetylene as functions of the C–C bond-length alternation

It is seen that the total-energy gain upon the bond-length alternation (also called the dimerization) is small, but non-vanishing. Some years ago, it was suggested [16] that density-functional calculations are incapable of reproducing this bond-length alternation, in contrast to the results of our studies. A careful analysis of all available theoretical studies of this issue [17] indicated, however, that with Hartree–Fock-based methods the bond-length alternation was always found. Also, when correlation effects where included, it was non-vanishing although reduced. For density-functional methods there was a tendency to underestimate it, although only when an attempt was made to model the material in a crystalline form, was the bond-length alternation markedly underestimated, if not completely lacking. This suggests that the crystal approximation, and not the density-functional formalism, is responsible for the problems in describing the bond-length alternation. In particular, the LDA may lead to overestimates in the (weak) interactions between the chains, thereby possibly obscuring the intrachain interactions. More recently,

Suhai [18] optimized the structure of an infinite, periodic trans-polyacetylene chain using both density-functional methods and Hartree–Fock methods, and for the latter both without and with the inclusion of correlation effects. He found that the bond-length alternation was found by all computational methods, although its size as well as the accompanying energy gain depends on the method.

Not only the structural properties but also the electronic ones are important for conjugated polymers. In Fig. 2.2 we showed that the experimental band structures of polyethylene are reproduced well by the density-functional calculations on a single, infinite, periodic chain. Unfortunately, equivalent, detailed, experimental information on conjugated polymers is lacking. Therefore, we show in Fig. 2.7 the density of states of trans-polyacetylene together with experimental photoelectron spectra recorded at various photon energies [19].

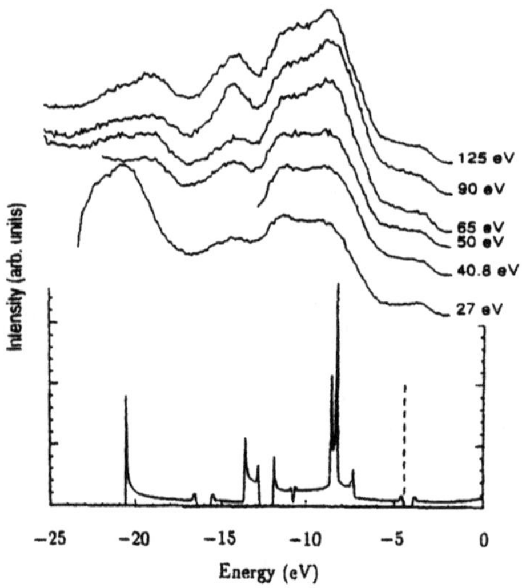

Fig. 2.7. Experimental photoelectron spectra recorded for the photon energies given on the curves (*upper curves*; from [19]) and theoretical valence-band density of states (*bottom curve*) for trans-polyacetylene

Except for a rigid shift of the experimental spectra of about 1.5 eV, no adjustment was made, and, accordingly, very good agreement is observed.

In Figs. 2.4 and 2.6 we saw how the bond-length alternation affects the band structures and the total energy, respectively. However, the interatomic electronic interactions are also modified through this, after all, relatively small structural change. To see this, the COOP (for a review, see, e.g., [20]) and COHP [21] are excellent tools. The crystal-orbital overlap population (COOP) is based on the overlap matrix elements, whereas the crystal-orbital Hamilton population COHP is based on the single-particle eigenvalues. For the COOP we use the fact that any eigenfunction is normalized, and from (2.19) we obtain

$$1 = \sum_n \sum_{p_1,p_2} \left[\sum_{\alpha_1,\alpha_2} (c^k_{j,p_1\alpha_1})^* c^k_{j,p_2\alpha_2} e^{ik\pi n} \langle \chi_{p_1\alpha_1,0} | \chi_{p_2\alpha_2,n} \rangle \right] , \tag{2.21}$$

where $\chi_{p\alpha,n}$ is the αth basis function centered at the pth atom in the nth unit cell, and $c^k_{j,p\alpha}$ is the coefficient of this basis function. Moreover, j and k are band indices and k number, respectively. Pairs of atoms for which the expression in the square bracket for that particular orbital is positive have bonding interactions. By modulating the density of states with these terms one obtains the COOP that provides a qualitative picture of the bonding interactions as a function of orbital energy.

Alternatively, the COHP is based on the fact that energies are the important ingredient in determining stability. Thus, focusing on the first part of the right-hand side of (2.7) and (2.13) one finds

$$\epsilon^k_j = \sum_n \sum_{p_1,p_2} \left[\sum_{\alpha_1,\alpha_2} (c^k_{j,p_1\alpha_1})^* c^k_{j,p_2\alpha_2} e^{ik\pi n} \langle \chi_{p_1\alpha_1,0} | \hat{h}_{\text{eff}} | \chi_{p_2\alpha_2,n} \rangle \right] . \tag{2.22}$$

Also here is the density of states modulated by the expression in the square brackets, but here negative contributions are interpreted as indicating bonding interactions.

In Fig. 2.8 we show the COOP and COHP curves for the interactions between the nearest carbon neighbors of trans-polyacetylene. It can be seen that the bond-length alternation (dimerization) leads to a significant change in the populations along these bonds. For example, whereas in the energy region closest to the Fermi energy they appear as monotonous functions for the undimerized structure, with a tendency towards indicating weaker bonding at higher energies, they show much more structure for the dimerized structure. Here, the interactions along the shorter (double) bonds become much more highly bonded up to the Fermi level, above which significant anti-bonding interactions are recognized. On the other hand, for the longer (single) bonds such anti-bonding interactions are found just below the Fermi level. This means that these populations clearly demonstrate how the bonds change character upon the bond-length alternation.

In Fig. 2.3 we also showed the structures of the cis isomer of polyacetylene. Although the structure without bond-length alternation (Fig. 2.3c) has a non-vanishing band gap at the Fermi energy (see Fig. 2.4), the experimentally observed structure has a bond-length alternation and corresponds, in fact, to the so-called cis-trans isomer of Fig. 2.3d. As seen in Fig. 2.4, this structural change leads to an increase in the band gap at the Fermi level. Furthermore, the trans-cis isomer of Fig. 2.3e has a smaller band gap than the structure without bond-length alternation. This may be the explanation for the recent finding by Hirata, Torii, and Tasumi [22] from density-functional calculations on finite oligomers that the trans-cis isomer is at most metastable, if it exists at all.

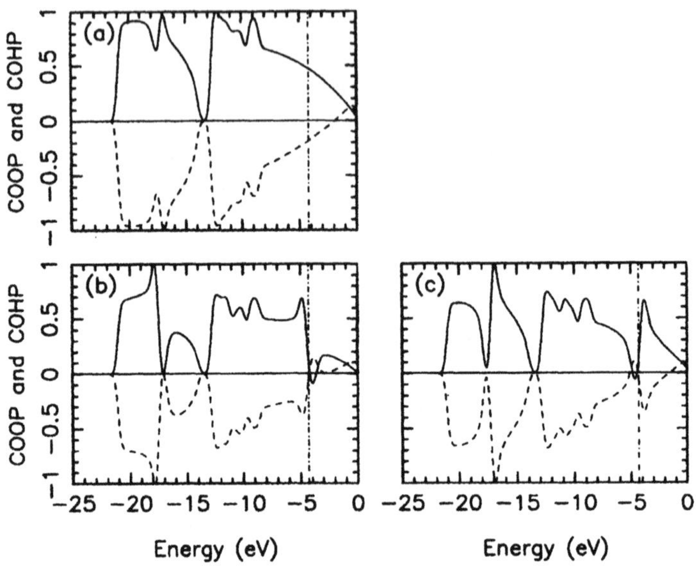

Fig. 2.8. The crystal-orbital populations COOP (*solid curves*) and COHP (*dashed curves*) for the interactions between nearest carbon neighbors for trans-polyacetylene for (**a**) the undimerized structure and (**b,c**) the dimerized structure along (**b**) the double bonds and (**c**) the single bonds. The panels have only approximately the same ordinate scales, and the *vertical dash-dotted lines* mark the Fermi level. The curves have been broadened with Gaussians

2.3.2 Conjugated Polymers Based on C, H, O, N, and S Atoms

One of the advantages of conjugated polymers is that, to some extent, it is possible to control and vary their properties through appropriate modifications of their structure. One can either modify the local arrangement of the carbon atoms of the backbone, incorporate heteroatoms into this, or replace the hydrogen atoms with other sidegroups. Here, we shall briefly discuss some of the consequences of such modifications. Polycarbonitrile (Fig. 2.9) represents one of the simplest possibilities for the incorporation of heteroatoms into the backbone. Starting with the trans-isomer of polyacetylene, polycarbonitrile can be obtained by replacing every second CH group with an (isoelectronic) N atom. In the previous subsection as well as in Sect. 2.1,

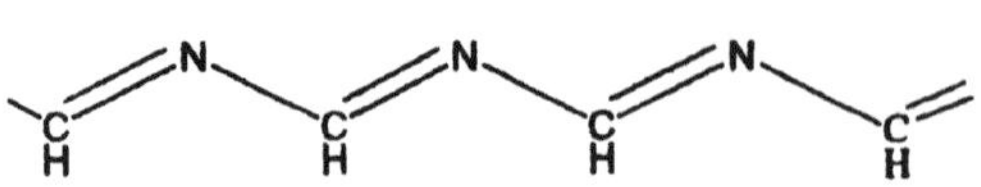

Fig. 2.9. Structure of polycarbonitrile

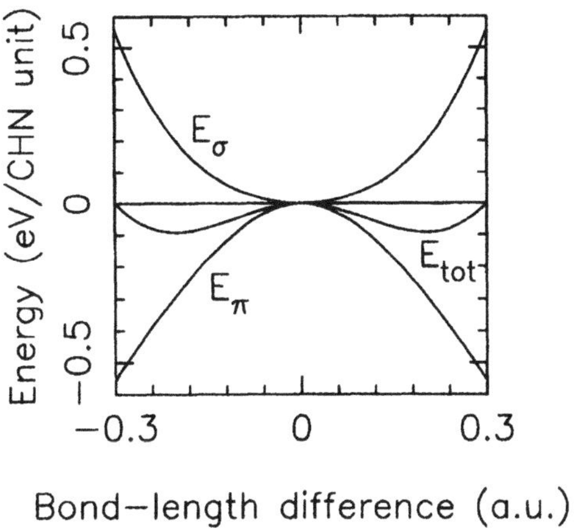

Fig. 2.10. As Fig. 2.6, but for polycarbonitrile and as functions of the C–N bond-length alternation

we demonstrated that the occurrence of a bond-length alternation for trans-polyacetylene could be considered as a Peierls distortion, i.e., a doubling of the unit cell that accompanies the creation of a gap at the Fermi level. For polycarbonitrile, the presence of the N atoms leads by itself to a doubling of the unit cell and, therefore, one might not directly expect a C–N bond-length alternation. Nevertheless, Fig. 2.10 shows that the total energy is lowered upon such an alternation [23]. As seen in Fig. 2.11, the structure without bond-length alternation has a non-vanishing gap at the Fermi level, as expected from the discussion above, but the bond-length alternation leads to an increase of its size. Furthermore, this polymer has the unusual property of having a σ band very close to the Fermi level. This (occupied) band is formed by N lone-pair orbitals (n orbitals). Finally, the optimized structural parameters were found [23] to be 2.54 and 2.67 a.u. for the C–N bond lengths, 2.05 a.u. for the C–H bond lengths, and 122° for the C–N–C and N–C–N bond angles.

Currently, poly(paraphenylene vinylene) (PPV) is one of the most studied conjugated polymers (see, e.g., [24]) and was the first one that was used in the construction of a light-emitting diode based on conjugated polymers [25]. As Fig. 2.12 shows, it is, in its simplest form, based solely on carbon atoms along the backbone and can in that respect be considered closely related to polyacetylene. In order to optimize the materials properties for applications as semiconductors, it is important to be able to control the frontier orbitals. Therefore, we calculated the band structures for the pure, infinite chain of Fig. 2.12 [26]; they are shown in Fig. 2.13a. We did not optimize the structural parameters with the density-functional method but instead used values obtained from an optimization on a finite oligomer by employing the semi-empirical AM1 method (this approach was also used for the polymers discussed in the remaining part of this subsection). Due to the larger unit cell compared with polyacetylene, there are many more bands. Therefore, the

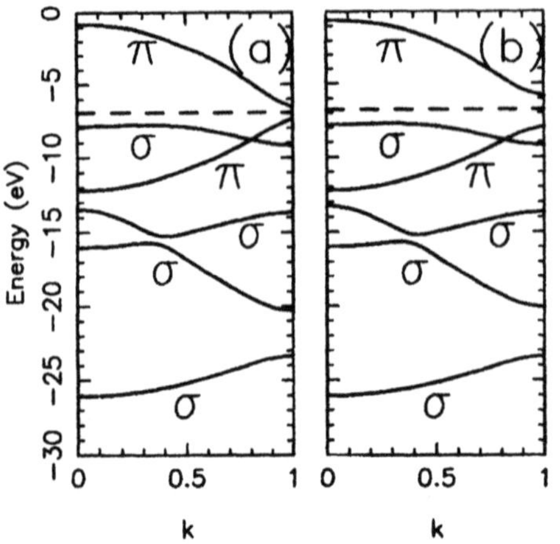

Fig. 2.11. As Figs. 2.4a and 2.4b, but for polycarbonitrile (**a**) without and (**b**) with a C–N bond-length alternation

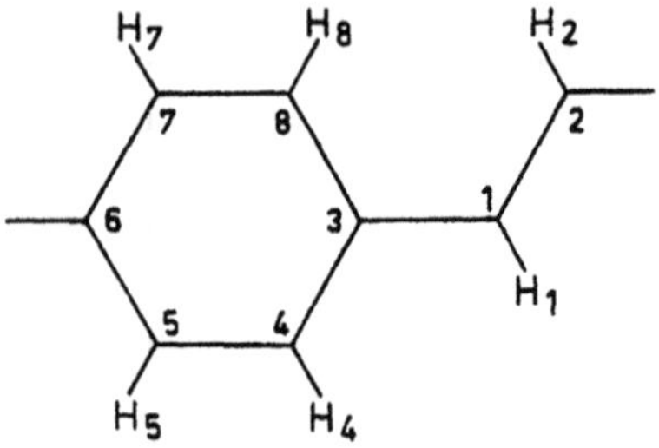

Fig. 2.12. Structure of poly(para-phenylene vinylene) (PPV)

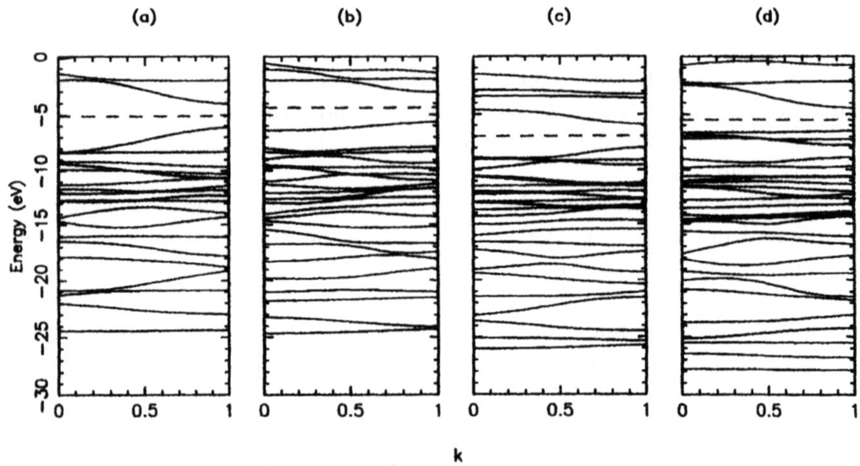

Fig. 2.13. Band structures of (**a**) pure PPV and (**b**)–(**d**) PPV substituted with (**b**) one NH_2 group, (**c**) one CN group, and (**d**) one NH_2 and one CN group per vinylene linkage. The *dashed lines* mark the Fermi level; $k = 0$ and $k = 1$ represent the center and edge of the first Brillouin zone, respectively

total width of the valence bands is somewhat larger than for polyacetylene, but the position of the Fermi level hardly changes. Here, however, the most important result is that both the highest occupied and the lowest unoccupied band are, as expected, of π symmetry and, in addition, have their major components on the vinylene linkages between the phenylene rings. Therefore, when attempting to modify the electronic properties, one would hope to achieve this best by replacing the H_1 and H_2 atoms of Fig. 2.12 by other sidegroups. We examined this by replacing one of these hydrogen atoms with either a NH_2 group or a CN group and show in Fig. 2.13b and c the resulting band structures. Finally, Fig. 2.13d shows the bands when both substituents are incorporated simultaneously. It is seen that the electron donor NH_2 leads to an overall upward shift of the bands in energy, whereas the electron acceptor CN produces an overall downward shift. However, a more careful study shows that there are also more subtle effects that are specific to the substituents, and that, furthermore, the effects of the double substitution are not simply a superposition of the individual substitutions.

Instead of replacing the hydrogen atoms with other sidegroups, one may replace some of the backbone atoms with others in order to modify the electronic properties, equivalent to passing from polyacetylene to polycarbonitrile. Thus, by replacing one of the CH groups of the phenylene rings of PPV with an N atom, one arrives at poly(2,5-pyridine vinylene), PPyV, which is shown in Fig. 2.14a. Conversely, replacing the vinylene linkages with either single N atoms or single NH groups results in polyaniline, Fig. 2.14b. Polyaniline is another of the presently very popular conjugated polymers; it differs in a couple of respect from most others: Due to steric hindrances the backbone of this polymer is not planar, but the polymer instead forms helices. Therefore, the separation into σ and π orbitals is only approximate. The lack of strict planarity has led to the proposal that local distortions, so-called ring-twist distortions, may exist. For these, the angles between neighboring phenylene rings are locally distorted [27]. One may also vary the material

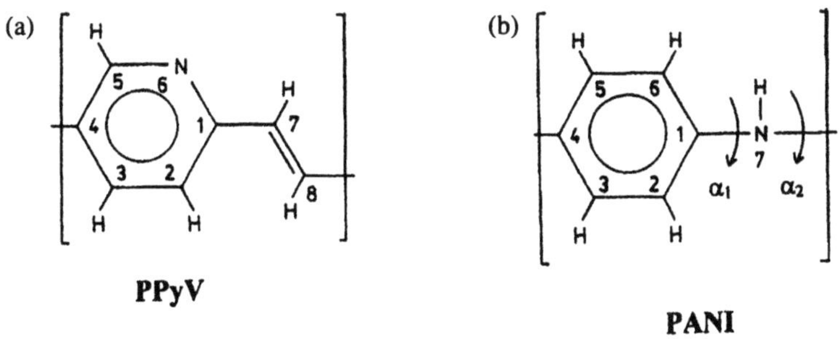

Fig. 2.14. Structure of (**a**) poly(2,5-pyridine vinylene) (PPyV) and (**b**) polyaniline (PANI)

properties by having all, half, or none of the links as NH groups, whereas the rest are single N atoms.

The band structures for the two systems in Fig. 2.14 are shown in Fig. 2.15 [26]. Compared with the bands for PPV (see Fig. 2.13a), those of PPyV show many similarities, but the Fermi level is pushed towards higher energies, as also was the case when passing from polyacetylene to polycarbonitrile.

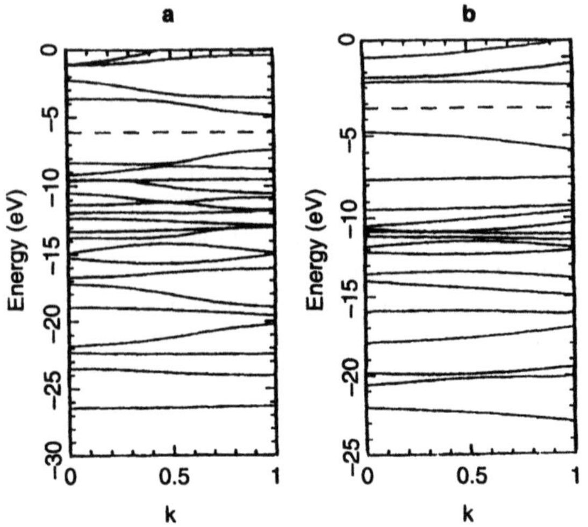

Fig. 2.15. Band structures of (**a**) PPyV and (**b**) polyaniline. The *dashed lines* mark the Fermi level; $k = 0$ and $k = 1$ represent the center and edge of the first Brillouin zone, respectively. (From [26])

A further similarity between these two systems is the occurrence of extra (σ) bands at the top of the valence-band region. In contrast, the bands of polyaniline in Fig. 2.15b are markedly different from those of PPV, and the fact that they are so narrow indicates that the electronic interactions for this material are reduced. This is partly due to the non-planarity of the backbone.

Compared with polyacetylene and polycarbonitrile, the PPV- and polyaniline-based polymers contain more or less aromatic rings as well as, in some cases, heteroatoms. A simpler system, that in addition is also one of the intensively studied conjugated polymers, is polythiophene shown in Fig. 2.16. In its simplest form it consists of a planar zigzag sequence of C_4H_2S units, and of the two forms in Fig. 2.16, the aromatic one (Fig. 2.16a) has a lower total energy than the quinoid one (Fig. 2.16b). As Fig. 2.17 shows, the aromatic structure has simultaneously a larger band gap at the Fermi level than the quinoid structure, but both structures have π bands close to the Fermi level. For the sake of the discussion in Sect. 2.4.1 we show in Fig. 2.18 the densities of the two orbitals closest to the Fermi level. Fig. 2.18 shows that the highest occupied orbital has strong bonding components along the C–C double bonds and anti-bonding components along the C–C single bonds. For the lowest unoccupied orbital the situation is reversed, with the additional aspect that there are also significant contributions from the S atoms.

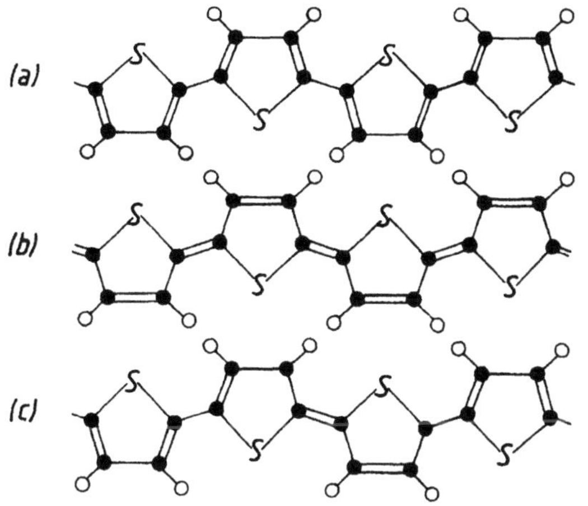

Fig. 2.16. Structure of (**a**) aromatic and (**b**) quinoid polythiophene. (**c**) A polaron-containing chain

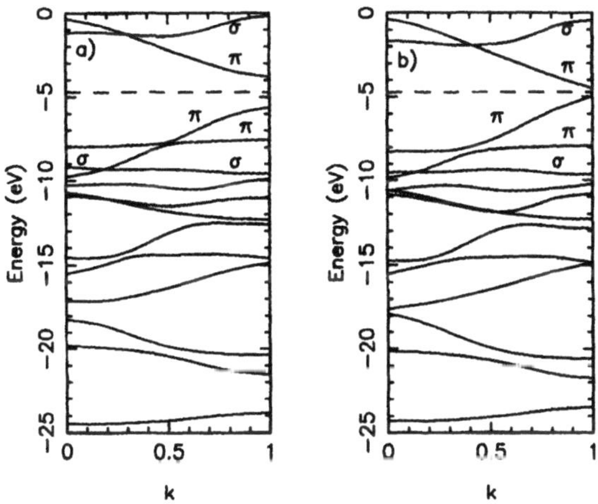

Fig. 2.17. Band structures of (**a**) aromatic and (**b**) quinoid polythiophene. The *dashed lines* mark the Fermi level

As we discussed above, one way of modifying the electronic properties is to replace the hydrogen atoms with other side groups, e.g., alkyl groups. The simplest case leads to poly-3-methylthiophene, shown in Fig. 2.19, for which we in Fig. 2.20 show the band structures.

All structural coordinates were kept as for the unsubstituted compound, which makes it possible to identify the effects due to the substitution and separate them from those due to structural relaxations. Compared with the band structures for the unsubstituted compound (Fig. 2.17a) we observe only modest changes. Due to the larger number of electrons and orbitals, extra bands occur and band-pushing effects can be held responsible for the slightly larger width of the valence bands in Fig. 2.20. Finally, the frontier orbitals (not shown) turn out to be very similar for the two systems. These results suggest that the effects of alkyl-substitution on the electronic properties are

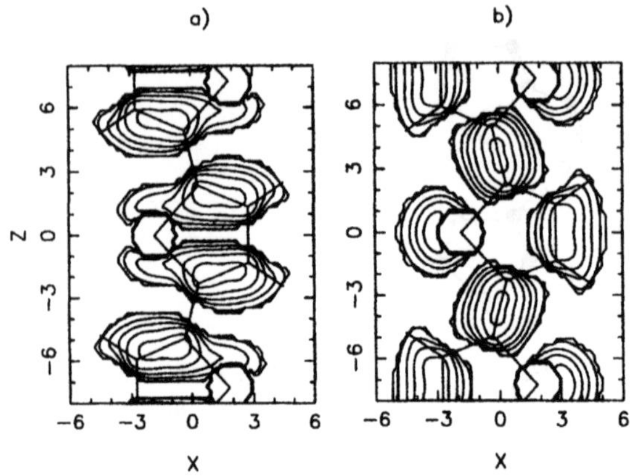

Fig. 2.18. Contour plots of (a) the highest occupied and (b) the lowest unoccupied orbital of aromatic polythiophene shown in a plane parallel to that of the nuclear backbone (also shown)

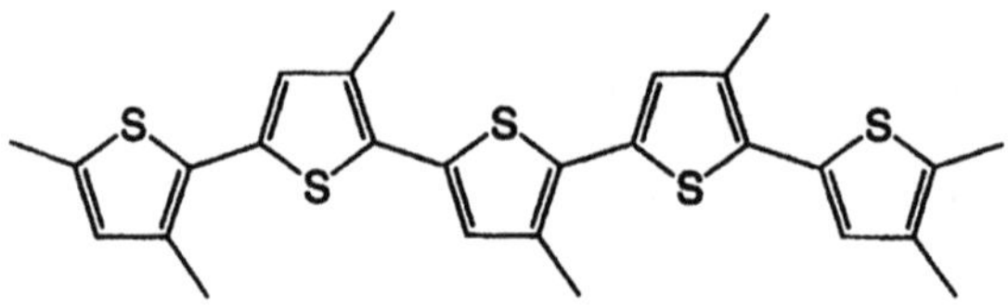

Fig. 2.19. Structure of poly-3-methylthiophene

only indirectly due to the alkyl groups themselves; instead, their presence may lead to structural changes in the backbone of the polymer. Thus, for larger alkyl groups, steric hindrances may force the backbone to have a non-planar structure, whereby the changes in the structure of the backbone lead to changes in the electronic properties. In order to study the latter, we considered the simplest polythiophene, shown in Fig. 2.16, and introduced two parameters describing its structure. One of those, γ, is defined as the dihedral angle between neighboring thiophene units and is 180° for the structures in Fig. 2.16 but reduced for general helical structures. The other, δ, describes whether the system is aromatic as in Fig. 2.16a or quinoid as in Fig. 2.16b. $\delta = -1$ and $\delta = +1$ are defined as corresponding to the aromatic and quinoid structures, respectively, and all other structural parameters are assumed to vary linearly as functions of δ.

In Fig. 2.21 we show the variation of the band gap and the first ionization potential for three values of γ (180°, 165°, and 150°) as functions of δ. It is seen that leaving the planar structure can have profound effects on the electronic properties, first of all since then interactions between the π orbitals of neighboring thiophene rings become weaker. However, in all cases, the aromatic compound has a larger ionization potential and band gap than the quinoid compound.

By separating the thiophene rings with vinylene bridges, one arrives at polythiophenylene vinylene, shown in Fig. 2.22, for which we in Fig. 2.23 show the band structures.

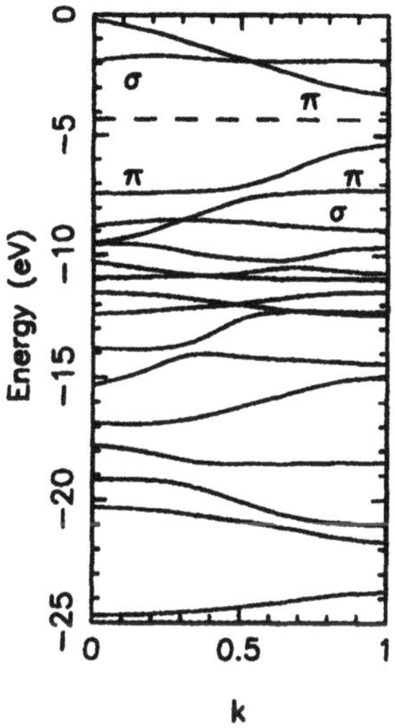

Fig. 2.20. Band structures of poly-3-methylthiophene

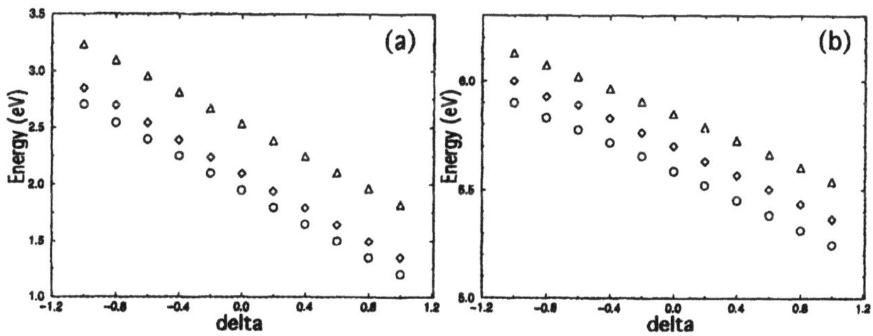

Fig. 2.21. Variation in (**a**) the band gap and (**b**) the first ionization potential for polythiophene as a function of the structural parameter δ (describing the change from aromatic to quinoid structure) for a dihedral angle, γ, of 180° (*circles*), 165° (*diamonds*), and 150° (*triangles*)

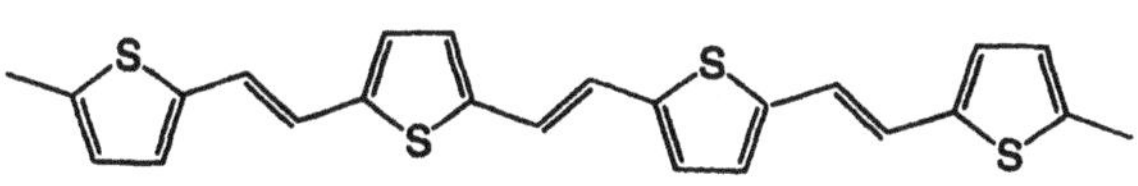

Fig. 2.22. Polythiophenylene vinylene

This compound has a significantly smaller band gap than polythiophene, which can be understood from the orbital pictures of Fig. 2.24. It is seen that, whereas the highest occupied orbital is confined to the thiophene rings, the lowest unoccupied one has large additional contributions from the vinylene groups. Finally, we studied the effects of replacing parts of the CH groups of polythiophene with N atoms. We thereby considered the two compounds shown in Fig. 2.25, polythiazole and polythiodiazole. Fig. 2.26 shows the band

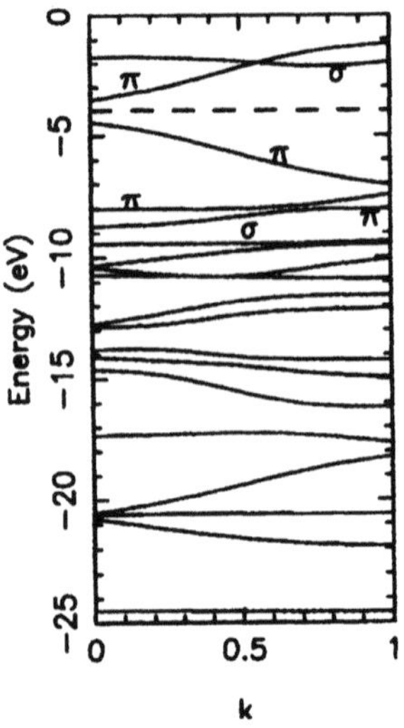

Fig. 2.23. Band structures of polythiophenylene vinylene

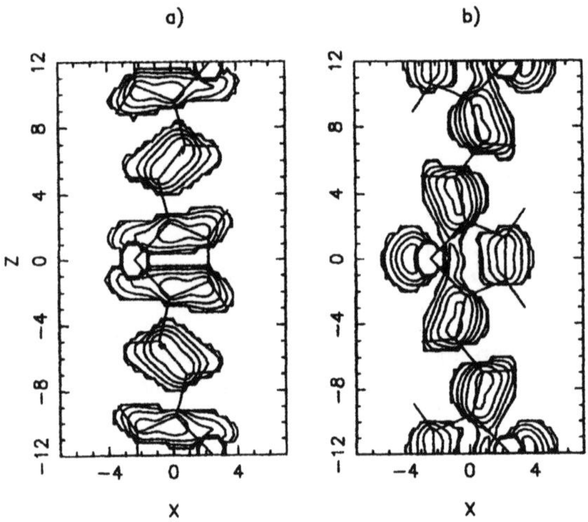

Fig. 2.24. Contour plots of (**a**) the highest occupied and (**b**) the lowest unoccupied orbital for polythiophenylene vinylene shown in a plane parallel to that of the nuclear backbone (also shown)

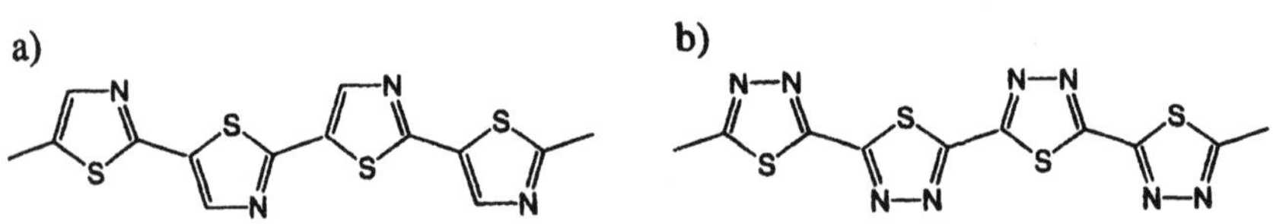

Fig. 2.25. (**a**) Polythiazole and (**b**) polythiodiazole

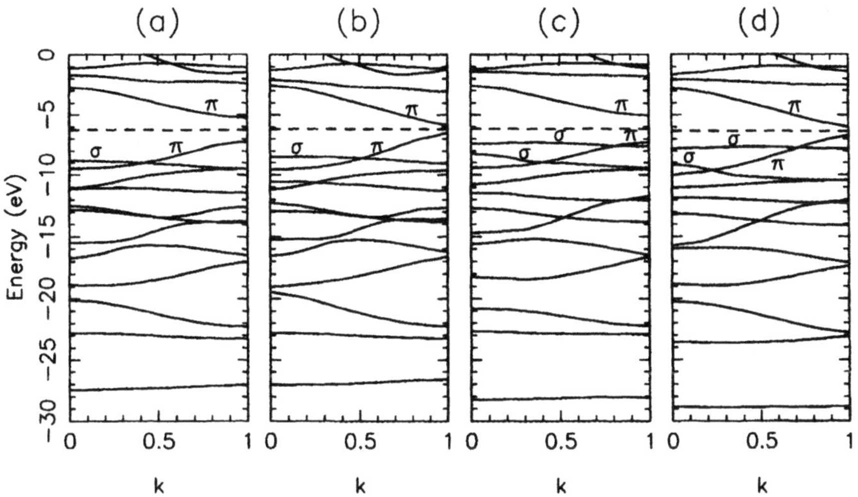

Fig. 2.26. Band structures of (**a**) aromatic and (**b**) quinoid polythiazole, and (**c**) aromatic and (**d**) quinoid polythiodiazole

structures for these. In all cases we recognize high-lying σ valence bands that are all due to nitrogen lone-pair orbitals, just as was the case for polycarbonitrile. Furthermore, Fig. 2.26 shows that the band gap is also in this case larger for the aromatic than for the quinoid structure. However, although the band structures for the different compounds of Fig. 2.25 do show some differences, the largest differences are observed when they are compared with the band structures of the unsubstituted polythiophene i.e., the Fermi energy is lowered by more than 1 eV upon introduction of the nitrogen atoms. A similar effect is also seen when the band structures of polyacetylene (Fig. 2.4) are compared with those of polycarbonitrile (Fig. 2.11).

2.3.3 Incorporating Metal Atoms

In the previous subsection we discussed how the properties of the conjugated polymers can be varied by replacing some of the carbon atoms with other atoms, most notably nitrogen or sulphur atoms. Since these atoms have *s* and *p* valence electrons, as does carbon, the chemical bonds are expected to be relatively similar. This situation may change when, instead, metal atoms with *d* valence electrons are incorporated into the backbone. Wittmann et al. [28] have reported experimental results for conjugated polymers containing Pd or Pt atoms. As a simplification of these systems, we considered the one shown in Fig. 2.27 [29].

For the experimental system, the middle C_2 pairs of the C_6 segments are replaced by phenylene rings, and the hydrogen atoms of the PH_3 groups by larger alkyl groups; however, for the purpose of analyzing how the presence

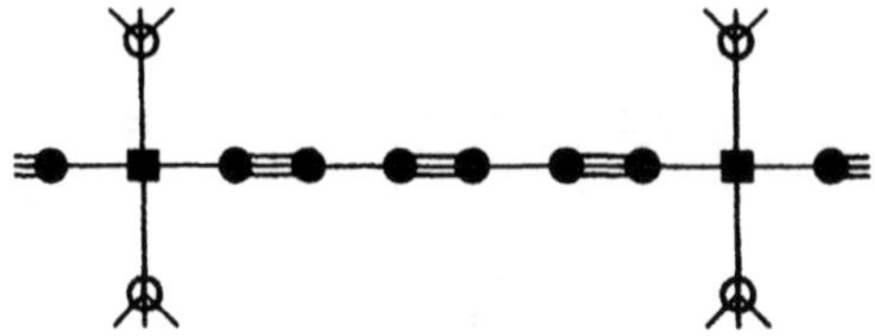

Fig. 2.27. A Pt-containing polyyne. The *solid squares*, *solid circles*, and *open circles* represent Pt atoms, C atoms, and PH_3 groups, respectively

of the metal atoms modifies the electronic properties of the conjugated polymers, these differences are less important. In Fig. 2.28 we show the band structures for this polymer, together with those of a linear carbon chain as well as the energies of a finite C_6 chain, a Pt atom, and a PH_3 group. Most important is that close to the Fermi level all parts of the system contribute to the orbitals. Thus, the presence of the metal atoms does not destroy the conjugation along the backbone, which actually can be seen by analyzing the orbitals individually (not shown). Finally, the inclusion of relativistic effects (which are important for Pt) does not change this conclusion significantly.

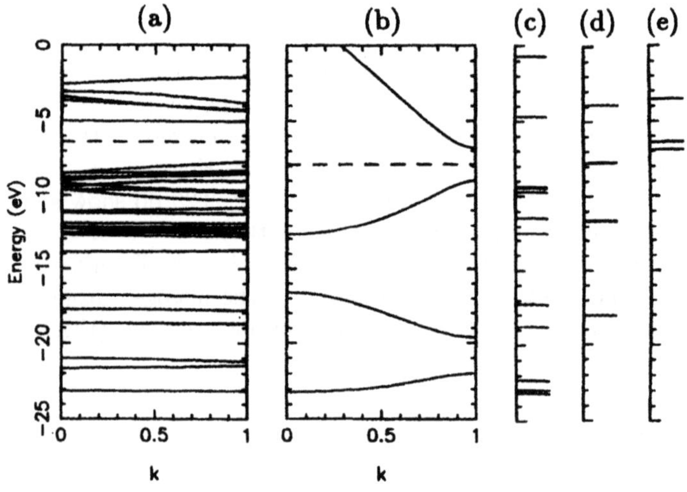

Fig. 2.28. Band structures of (**a**) the Pt-containing polyyne shown in Fig. 2.27 and of (**b**) a linear carbon chain with alternating bond lengths, together with the orbital energies of (**c**) a finite C_6 chain, (**d**) a PH_3 group, and (**e**) a Pt atom. The *dashed lines* separate occupied and unoccupied orbitals. (From [29])

2.3.4 Polysilanes

The most important properties of conjugated polymers are that they are semiconducting and quasi-one-dimensional and that they possess large electron–phonon couplings which lead to the possibility of local structural distortions

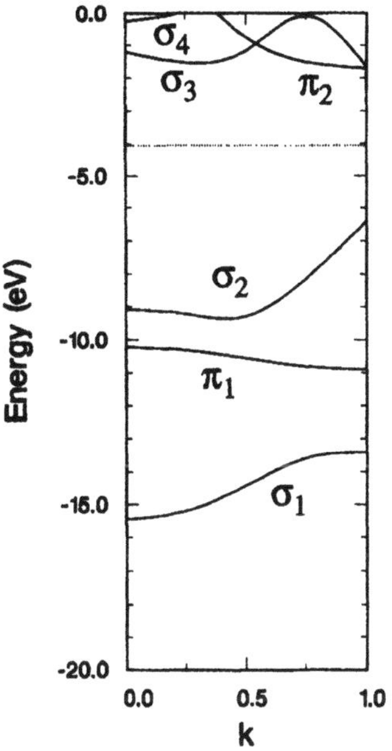

Fig. 2.29. Band structures of polysilane, $(SiH_2)_x$. (From [31])

when the material is excited or charged. There are, however, other classes of materials that have similar properties, and in this and the following subsection we shall briefly discuss two of these.

Whereas diamond is an insulator, crystalline silicon is a semiconductor. Similarly, replacing all the carbon atoms of polyethylene (Fig. 2.1) with silicon atoms leads to a polymer that has a markedly smaller band gap than polyethylene. Thus, although the chemical bonds of this material are based on sp^3 and not sp^2 hybrids (as for the carbon-based conjugated polymers), it has similarities with the materials we have discussed above.

Moreover, the properties of polysilane may be varied by replacing some of the hydrogen atoms with other (often larger organic) sidegroups, in which case a helical structure often results. Finally, whereas structural distortions such as solitons and polarons for conjugated polymers (see Fig. 2.5) involve local displacements of the carbon atoms essentially perpendicular to the bond-angle bisectors, it has been suggested that polarons with local distortions along the bond-angle bisectors result when polysilanes are charged or excited [30]. It should be added that a further difference to the organic conjugated polymers is that polysilanes do not possess any bond-length alternation along the backbone.

We studied the simplest of these materials, i.e., a planar polysilane, as shown in Fig. 2.1, with hydrogen atoms as the sidegroups [31].

Fig. 2.29 shows the band structures, and compared with polyethylene we observe a significant reduction of the band gap, although it is still fairly large compared with that of the conjugated carbon-based polymers discussed so far.

More important is that using the results of the density-functional calculations in analyzing whether polarons as those described above may occur led to the conclusion that such polarons are not stable. Thus, despite the similarities with the carbon-based conjugated polymers, we have to conclude that the polysilanes, after all, are quite different materials.

2.3.5 MX-Chain Compounds

At a first sight, the material shown in Fig. 2.30 seems to be very different from those we have discussed above. It is one example of the so-called MX-chain compounds, which all share the common characteristic that, although they are crystalline, they consist of chains with a backbone of alternating metal (M, in our case Pt) and halogen (X, here Br) atoms. Furthermore, the backbone possesses an alternation, i.e., the X atoms are not placed symmetrically between neighboring M atoms but are displaced so that XMX and M units alternate along the chain. This alternation can be locally disrupted, leading to the occurrence of solitons and polarons that are conceptually similar to those discussed for the carbon-based conjugated polymers.

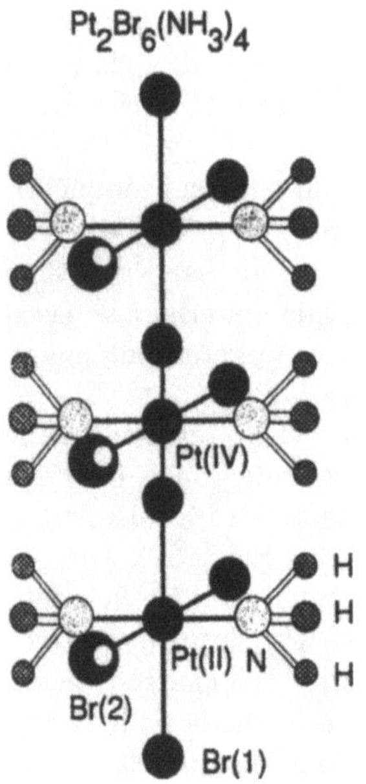

Fig. 2.30. Structure of a single chain of $Pt_2Br_6(NH_3)_4$

We studied the material in Fig. 2.30 [32]. Special emphasis was put on the question of whether the NH_3 groups (ammonia) were important in the description of the physical properties of the system. It turned out (see Fig. 2.31) that without them the calculations predicted that the structure with the lowest total energy has the Br atoms sitting symmetrically between the neighboring Pt atoms, and only upon inclusion of the NH_3 groups were the correct displacements obtained. Since the NH_3 groups have closed shells, this result is mainly due to electronic polarizations.

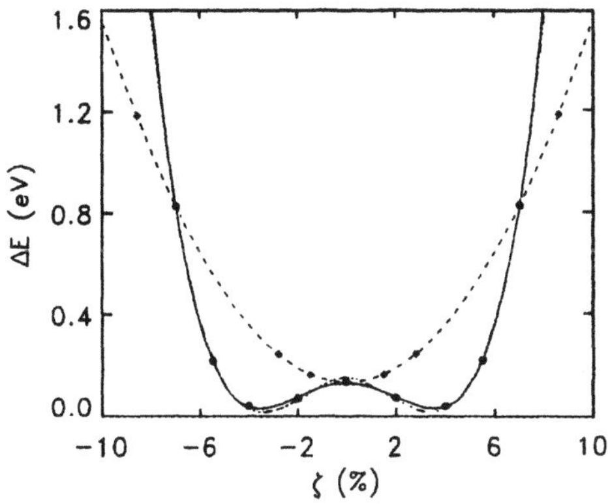

Fig. 2.31. The total energy per formula unit for the system Fig. 2.30 as a function of the displacement ζ of the Br atoms away from the symmetric position. ζ is measured relative to the Pt–Pt distance, and the *dashed* and *full curves* give the results without and with inclusion of the NH_3 groups, respectively. (From [32])

Finally, as Fig. 2.32 shows, the material is a semiconductor with a band gap that is very similar to that of the organic conjugated polymers. In this case, however, the frontier orbitals are formed by d_{z^2} functions on the metal atoms and p_z functions on the halogen atoms, if it is assumed that the chain is parallel to the z axis.

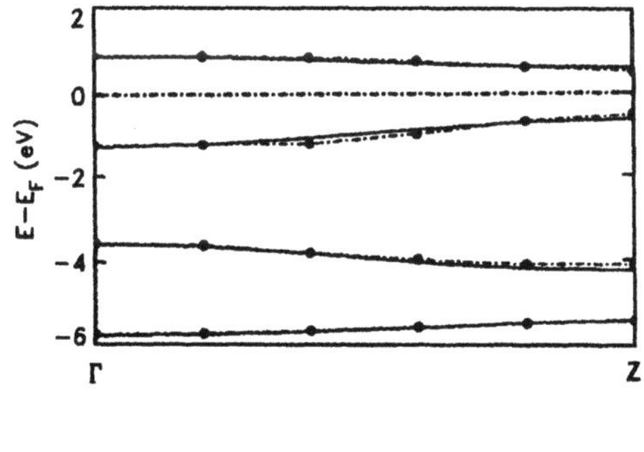

Fig. 2.32. The σ bands closest to the Fermi level (which is set equal to 0) for the system shown in Fig. 2.30. These bands are formed by Pt d_{z^2} and Br p_z functions, with the z axis parallel to the chain axis. The *dash-dotted curves* show a tight-binding fit. (From [32])

2.4 Breaking the Symmetry

In the previous section we discussed the basic properties of different conjugated polymers and related materials under the assumption that the chains are long and periodic. Many properties depend, however, on how the material of interest responds to distortions and external interactions, and a few examples of those shall be addressed here. To these belong the properties of the solitons and polarons that we discussed in Sect. 2.1. The existence of these is closely related to the strong coupling between the phonons and the electronic frontier orbitals, but they destroy the periodicity of the system and are simultaneously extended. Thus, they are highly difficult to access with parameter-free electronic-structure methods. Instead, one may attempt to study their properties using parametrized models. We shall in the first subsection here discuss the results of such studies as well as present a general scheme that can be used in giving some insight into these defects without actually having to study them explicitly. Thereby, the parameter-free electronic-structure methods can be employed. Subsequently, we shall present a more general method that can be used in extracting model parameters from the parameter-free methods. Then, we shall discuss some consequences of extra charges and of isolated substitutions. Finally, we shall discuss two issues that currently are of great interest: the problem of calculating properties of the materials when they are exposed to external electrostatic fields, and that of calculating the linear and non-linear responses to external time-dependent electromagnetic fields.

2.4.1 Solitons and Polarons

Let us as a first example consider the soliton shown in Fig. 2.5a. For the neutral chain, each carbon atom possesses one π electron and these occupy bonding orbitals essentially centered along the C–C double bonds. Thus, they will have lower energies than the isolated π orbital centered at the soliton, which, in turn, will have a lower energy than the unoccupied anti-bonding orbitals formed by the π functions. Thus, one may expect that the π orbital of the soliton will have a single-particle energy in the gap between occupied and unoccupied orbitals, which is exactly the case. Since the π orbitals, in the neutral case, are half-filled, it is energetically favorable for the system to place added charge in that orbital. Therefore, these structural defects are believed to be generated when charging the system, and since they are fairly delocalized (over more than 10 CH units), they can move relatively easily up and down the chain and eventually carry the charge with them.

For the polaron of Fig. 2.5b the situation is very similar. One may consider the polaron to be formed by two solitons; the two orbitals at the solitons will then form an energetically low bonding combination and an energetically high anti-bonding combination. Both may lie in the optical gap and, once again, be responsible for the transport of charge or spin throughout the chain.

These defect-induced gap states are accordingly important for transport properties. Identifying them in, for example an optical experiment provides a means for proving their existence. Therefore, a lot of experimental and theoretical effort is devoted to studying these gap states. Here, we shall assume that a single-particle description is appropriate; in the next subsection we shall briefly discuss many-body effects. We shall accordingly here assume that the electronic properties of interest can be described with the Hamiltonian

$$\hat{H}_{\text{sp}} = \sum_n \sum_j \sum_\sigma \epsilon_{jn} \hat{c}^\dagger_{\sigma jn} \hat{c}_{\sigma jn} + \sum_n \sum_m \sum_j \sum_k \sum_\sigma t_{nm,jk} \hat{c}^\dagger_{\sigma jn} \hat{c}_{\sigma km} \ . \tag{2.23}$$

Here, $\hat{c}_{\sigma jn}$ and $\hat{c}^\dagger_{\sigma jn}$ are the annihilation and creation operator, respectively, for an electron in the jth orbital of the nth unit cell having spin σ (i.e., of the Wannier function $\chi_{\sigma jn}$). ϵ and t are on-site energies and hopping integrals, respectively, and the second sum is supposed to exclude terms with $(j, n) = (k, m)$. For conjugated polymers each unit may be one atom of the backbone and the orbitals the corresponding π functions, whereas for the MX chains each Br and Pt atom of the backbone defines one site and the functions are the d_{z^2} functions of Pt and the p_z functions of Br.

The structure of the material is described through a set of so-called configuration coordinates, $\{y_p\}$. y_n describes the structure of the nth unit and it is assumed that all other structural degrees of freedom are varied simultaneously with y_n.

For trans-polyacetylene, these parameters could be the u_n shown in Fig. 2.33, which originally were introduced by Su et al. [33] in their famous work, where they qualitatively and quantitatively discussed the role of solitons in trans-polyacetylene. For periodic, undistorted trans-polyacetylene, we have

$$u_n = (-1)^n u_0 \ , \tag{2.24}$$

i.e., $(-1)^n u_n$ is constant, whereas it has a node for the soliton and shows local variation for the polaron.

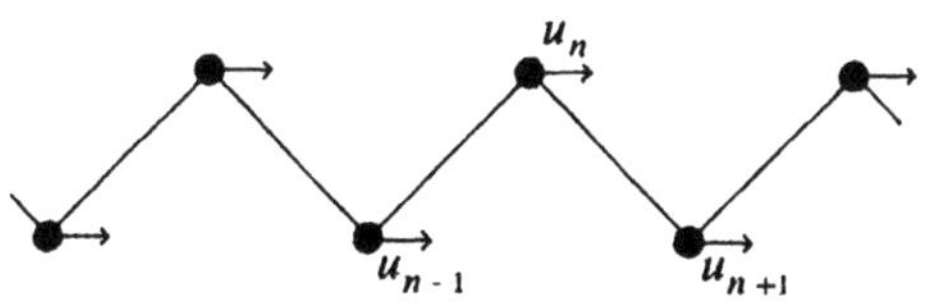

Fig. 2.33. The configuration coordinates u_n of trans-polyacetylene as introduced by Su et al. [33]

The on-site energies and hopping integrals entering (2.23) will depend on the structure, i.e., on all $\{y_p\}$. One may hope that this dependence is simple and that it can be determined through calculations on periodic structures. Thus, Su et al. assumed that the hopping integrals depend linearly on the corresponding bond lengths, i.e.,

$$t_{n,n+1} = t_0 - \alpha(u_{n+1} - u_n) \ , \tag{2.25}$$

where, in addition, the j and k indices can be dropped, since there only is one orbital per unit, and it is assumed that only nearest-neighbor hopping integrals are non-zero. In (2.25) the constants t_0 and α can, for example, be determined from parameter-free calculations on the periodic structures obeying (2.24).

Finally, that part of the total energy that is not contained within the single-particle Hamiltonian of (2.23) may be approximated through an analytical function of the configuration coordinates $\{y_p\}$,

$$H_{\rm r} = f(\{y_p\}) \; . \tag{2.26}$$

In the model of Su et al., it was assumed that $H_{\rm r}$ could be written as a harmonic function of nearest-neighbor bond lengths,

$$H_{\rm r} = \frac{K}{2} \sum_n (u_{n+1} - u_n)^2 \; . \tag{2.27}$$

The constant K can also be determined from calculations on the periodic structures.

Such a separation of the total energy was demonstrated already in Figs. 2.6 and 2.10 for trans-polyacetylene and polycarbonitrile, respectively. Its formal justification can be found in (2.7) and (2.13).

Assuming that the single-particle model of (2.23) and (2.26) is applicable both when the chain is periodic and when it contains structural defects such as solitons and polarons, and, moreover, that the same parameters and parameter values can be used in both cases, the latter can be determined from, for example, density-functional or Hartree–Fock calculations on the periodic structures and subsequently be applied in model calculations on the defect-containing structures. Here, the parameters of the single-particle Hamiltonian of (2.23) can be determined from the calculated single-particle energies as a function of structure, and once the function of (2.26) has been specified, the parameters entering it can be determined from the variation of the total energy as a function of periodic distortions. Such methods have been applied on a number of different polymers (see, for instance, [34]).

By parameterizing the results of calculations on infinite, periodic structures and subsequently using these in the study of the defect-containing system, it should be obvious that the properties of the defect-free system indirectly determine those of the defect-containing one. It might therefore be hoped that, at least in certain cases, one can avoid the steps of extracting the model parameters from the parameter-free calculations and subsequently perform model calculations and instead directly use the information on the defect-free system in predicting properties of the defects. It turns out that when the defect is extended in space and one is interested in only the single-particle energies of the orbitals induced by the defect, the simplification proposed above is possible. These energies form an important part of the stability of the defects, although, by only studying the single-particle energies, no final information about the stability of the defects is obtained.

The mathematical foundations for the scheme have been described in detail elsewhere [35–38]. The scheme is based on the transformation of the single-particle Schrödinger equation

$$\hat{H}_{\mathrm{sp}}\psi = \epsilon\psi \ , \tag{2.28}$$

with

$$\psi = \sum_{jn} \chi_{\sigma jn} c_{jn} \ , \tag{2.29}$$

into the analysis of a set of transfer-matrix equations,

$$\underline{c}_{n+1} = \underline{\underline{T}}_n \cdot \underline{c}_n \ , \tag{2.30}$$

where $\underline{c}_n$ contains the coefficients c_{jn} for the Wannier functions of the nth unit and $\underline{\underline{T}}_n$ depends on the (unknown) eigenvalue ϵ and the on-site energies and hopping integrals of $\hat{H}_{\mathrm{sp}}$. We shall not discuss the foundations for the method further but instead show how it works with some examples. As the first example we consider the soliton of Fig. 2.5d which is also shown in Fig. 2.34a. Locally, the soliton involves different structures (i.e., bond-length alternations) similar to those that are repeated periodically for the structures shown in Fig. 2.34b–h. Instead of studying the structure of Fig. 2.34a we shall consider the whole class of periodic structures, such as those shown in Fig. 2.34b–h. As long as the defect is extended (which is the case for solitons and polarons in conjugated polymers) it can be shown [35–38] that this is a good approximation. For each of the periodic structures we can perform electronic-structure calculations, e.g., with the density-functional method. For each structure we will have energy bands at some energies and at other energies there will be gaps. In Fig. 2.4a and b we showed the band structures for the structures of Fig. 2.34b and h and for that of Fig. 2.34e, i.e., for the structures with and without a bond-length alternation. Focusing on the π bands of Fig. 2.4, their energies will change as shown in the left part of Fig. 2.35 when changing the structure from that of Fig. 2.34b (corresponding to the vertical dashed line on the left in the left part of Fig. 2.35), via that of Fig. 2.34e (the middle of the left part of Fig. 2.35), to that of Fig. 2.34h (the vertical dashed line on the right). Outside the defect the electronic bands will have energies as shown by the two vertical dashed lines in the left part of Fig. 2.35, which gives the major parts of the density of states shown in the right part of Fig. 2.35. Inside the defect the structure changes from that of the one dashed line to that of the other. Of the energies for the periodic structures we chose that or those which are the farthest away (in energy) from the energies of the orbitals outside the defect (for the mathematical arguments, see [35–38]). This gives, in the present case, the orbital in the middle of the gap between occupied and unoccupied π bands as shown in the right part of Fig. 2.35. That the soliton gives this state has been known since

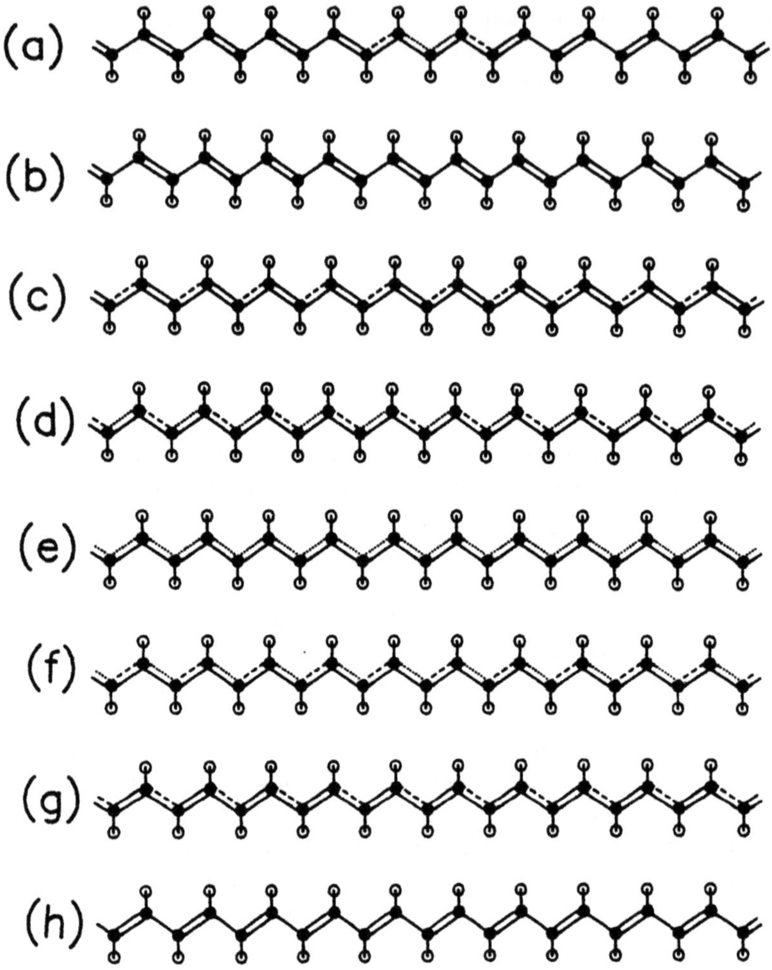

Fig. 2.34. (**a**) A soliton in trans-polyacetylene, and (**b**)–(**h**) various periodic structures. The presentation is as in Fig. 2.5

the pioneering work of Su et al. [33] about 20 years ago, but is here obtained in quite a different – and more general – way.

Replacing every second CH group in trans-polyacetylene with an N atom leads to polycarbonitrile, the structure of which is shown in Fig. 2.9. This material already has a non-zero band gap for the structure without a bond-length alternation, but the band gap increases upon a bond-length alternation (see Fig. 2.11), and simultaneously the total energy is reduced (see Fig. 2.10). In this case a soliton will thus lead to two gap levels, as shown in Fig. 2.36.

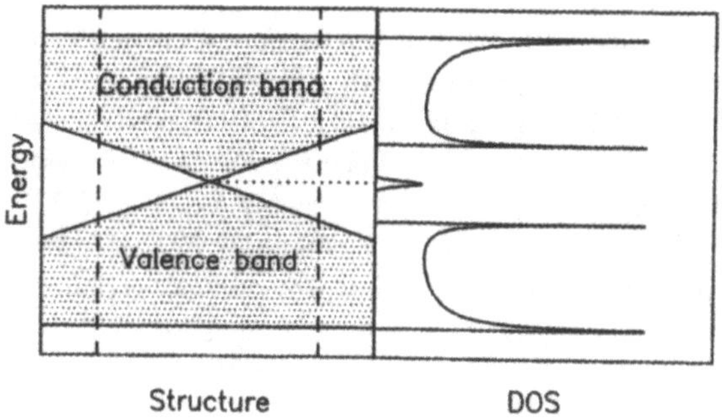

Fig. 2.35. On the left, the variation in the energy ranges spanned by the π bands for trans-polyacetylene as a function of bond-length alternation is shown. On the right, the density of states for the soliton-containing chain of Fig. 2.34a is shown. For further details, see the text

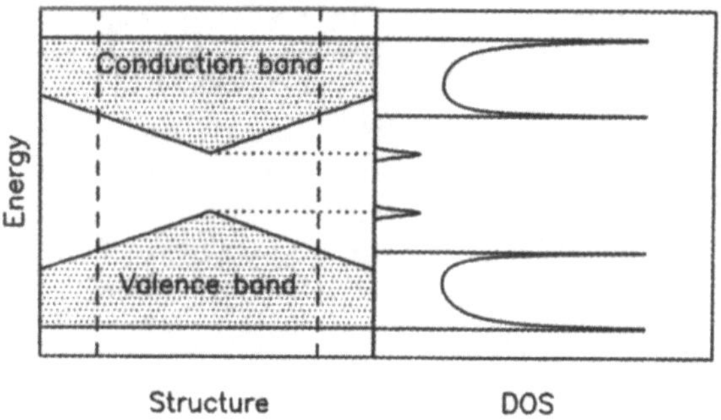

Fig. 2.36. As Fig. 2.35, but for polycarbonitrile

The fact that the soliton-induced levels appear at the energies of the band edges for the structure without a bond-length alternation was found earlier in more precise model calculations [23,39], but appears here as a trivial outcome of the more general scheme.

The case for polythiophene (Fig. 2.16) is slightly more complicated. In this case, the two structures differing in the C–C bond-length alternation along the backbone are no longer energetically degenerate, and the aromatic structure shown in Fig. 2.16a has a lower total energy than the quinoid structure shown in Fig. 2.16b. Polarons (Fig. 2.16c) can be generated upon excitation or doping, and recently it was found that a polaron leads to two levels in the gap and that these are placed asymmetric with respect to the band edges [40]. In contrast to the proposal in the experimental paper [40], we shall show that this finding can be explained with band-structure effects without invoking many-body or correlation effects.

In Fig. 2.18 we showed the highest occupied and the lowest unoccupied (π) orbital in the neutral case, i.e., the HOMO and the LUMO. Fig. 2.18a shows that when passing from the aromatic to the quinoid structure (as locally is the case for a polaron) the anti-bonding (bonding) interactions of the HOMO are made stronger (weaker), so that the energy of this orbital increases. Equivalently, Fig. 2.18b implies that the energy of the LUMO is reduced for this distortion, but since this orbital has large components on the sulphur atoms, these changes are less pronounced. Therefore, we end up with a diagram such as that shown in Fig. 2.37, where the left part shows the changes in the band structures closest to the Fermi level as a function of the bond-length alternation (i.e., essentially the parameter δ of Fig. 2.21), whereas the right part shows the density of states for the polaron-containing chain. Thus, the appearance of asymmetric polaron states can be explained as due to single-particle effects.

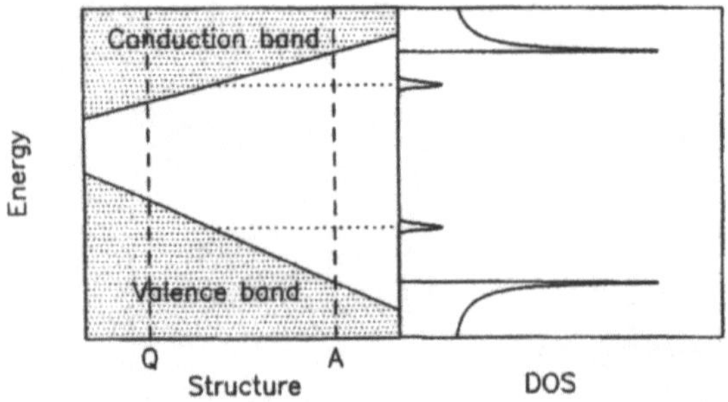

Fig. 2.37. As Fig. 2.35, but for a polaron in polythiophene. A and Q mark the aromatic and quinoid structures, respectively, and a polaron corresponds to a local change from A to Q and back again

The system of Fig. 2.27 is considered to be an idealization of the metal-containing polyynes synthesized and studied by Wittmann et al. [28]. By analyzing the frontier orbitals of this system (not shown) we found [29] that an increase in the C–C triple-bond lengths and a decrease in the C–C single-bond lengths (as is relevant when generating a polaron) led to an increase in the energies of the highest occupied orbitals, and a (smaller) reduction in the energies of the lowest unoccupied ones. Thus, in this case we also arrive at a situation such as that shown in Fig. 2.37, with a polaron giving two asymmetric gap states. This prediction has not yet been confirmed experimentally.

We have here presented a few examples of how parameter-free calculations on periodic structures can be used to give information on the energies of localized defect-induced orbitals. We close this subsection by stressing that the approach is not restricted to either a specific parameter-free approach or the examples shown here. Thus, the approach can also be applied, for example, to the ring-torsion defects in polyaniline as well as the polarons in polysilanes and the solitons and polarons in the MX-chain compounds, as well as others. But we shall not discuss these here.

2.4.2 Model Parameters

In the preceding subsection we assumed that the properties of the polymers could be described using a single-particle model for the relevant orbitals (i.e., the π orbitals). However, in some cases many-body effects need to be included.

For instance, when exciting or adding charge to trans-polyacetylene, solitons will be generated. According to the discussion above, a soliton will lead to a state very close to the midgap position independent of the charge and spin of the system. However, experimental studies show that the energy of this state does depend on charge and spin, which can only be explained by many-body effects (for a detailed discussion, see [7]). This means that the single-particle Hamiltonian of (2.23) has to be extended with many-body terms. Considering the π electrons of conjugated polymers, a Hubbard-like Hamiltonian is often useful. The electronic part of such a model is

$$\hat{H}_{\text{el}} = \left(\sum_{n,\sigma} \epsilon_n \hat{c}^{\dagger}_{n,\sigma} \hat{c}_{n,\sigma} - \sum_{n \neq m,\sigma} t_{n,m} (\hat{c}^{\dagger}_{n,\sigma} \hat{c}_{m,\sigma} + \hat{c}^{\dagger}_{m,\sigma} \hat{c}_{n,\sigma}) \right) + \left(\sum_n U_n \hat{\rho}_{n,\uparrow} \hat{\rho}_{n,\downarrow} + \sum_n V_{n,n+1} \hat{\rho}_n \hat{\rho}_{n+1} \right) \equiv \hat{h}_1 + \hat{h}_2 \ , \tag{2.31}$$

with

$$\hat{\rho}_{n,\sigma} = \hat{c}^{\dagger}_{n,\sigma} \hat{c}_{n,\sigma} \tag{2.32}$$

and

$$\hat{\rho}_n = \hat{\rho}_{n,\uparrow} + \hat{\rho}_{n,\downarrow} \ . \tag{2.33}$$

The operators $\hat{c}^{\dagger}_{n,\sigma}$ and $\hat{c}_{n,\sigma}$ are the creation and annihilation operators, respectively, for a π function centered at the nth C atom of the backbone with spin σ. Hence, it is indirectly assumed that the π electrons can be described using a set of orthonormal, atom-centered Wannier functions $\{\chi_{\sigma n}\}$, although their precise form not is specified. The parameters ϵ_n, $t_{n,m}$, U_n, and $V_{n,n+1}$ are often determined from experimental information on macroscopic samples, which, however, makes a direct comparison with the model difficult.

As we have seen, the single-particle energies of the density-functional calculation provide accurate information on electronic single-particle excitations, and, therefore, these can be used in the determination of the parameters for the single-particle part of $\hat{H}_{\text{el}}$, i.e., of $\hat{h}_1$. However, the parameters of $\hat{h}_2$ can be determined only by knowing how the total energy or the band structures change when the number of electrons ascribed to one or more atoms is varied. Here, the problem arises that the density-functional or Hartree–Fock calculations employ a set of basis functions that are neither orthonormal nor minimal, in contrast to the indirect assumptions in $\hat{H}_{\text{el}}$. In order to circumvent this problem we have recently [42] developed a method based on constrained density-functional calculations [43]. This method allows for controlled modifications of the number of electrons that can be ascribed to the individual atoms and can therefore be used in forming the link to the model Hamiltonian.

We have applied our method on trans-polyacetylene. By assuming that the nearest-neighbor hopping integrals take two alternatingly values (t_- and

t_+), that the next-nearest-neighbor hopping integrals (t_2) also may be non-vanishing, that only on-site ($U \equiv U_n$) and that nearest-neighbor ($V \equiv V_{n,n+1}$) Hubbard parameters are non-zero, we found [44] $t_- = 2.88$ eV, $t_+ = 3.35$ eV, $t_2 = -0.25$ eV, $U = 10.7$ eV, and $V = 0.3$ eV. These values were found under the assumption that the Hartree–Fock approximation could be applied for the model Hamiltonian. Whereas the single-particle parameters and U compare well with the values found in other studies, V is found to be significantly smaller than what often is assumed, when it is included at all. Finally, when repeating the study for the cis isomers of polyacetylene, some changes in the nearest-neighbor hopping integrals were found (actually, the two different kinds of bonds have to be described by two different types of hopping integrals), but the values of the many-body parameters change only little [44]. An exception is the trans-cis isomer, for which $V \simeq 1$ eV.

2.4.3 Charged Chains

As mentioned in the preceding subsection, many-body effects show up in soliton-induced gap states that are placed asymmetric in the band gap, i.e., neutral but spin-containing solitons lead to energies different to those for charged but spin-free solitons. The many-body effects are most important when studying perturbations in the otherwise periodic electron distribution as they occur, e.g., for the charged chains or for the excited (but neutral) chains. Above, we discussed the case in which the chains are neutral but experience a redistribution of the electrons. Here we shall discuss the other case, i.e., when charge is added to the chain.

When approximating a conjugated polymer as a finite chain, there is, in principle, no fundamental problem in treating a charged chain. However, when treating an infinite, charged chain one encounters a problem. Since the density-functional or Hartree–Fock calculations become possible only if periodicity is assumed, an extra charge has to be added periodically. This will, however, lead to a diverging Coulomb potential which can only be avoided when adding a compensating charge distribution. There is a significant arbitrariness in how this compensating distribution is chosen, but here we have chosen to add it as constant densities inside all so-called muffin-tin spheres (i.e., atom-centered, non-overlapping spheres) [45]. It can be shown [45] that this, in fact, leads to a realistic extra potential from the added charge.

In Sect. 2.3.1 we studied the Peierls distortion for trans-polyacetylene, i.e., the occurrence of a bond-length alternation accompanied by a gap at the Fermi level. For charged chains, the bond-length alternation will not lead to a gap at the Fermi level, and one may suggest that the Peierls distortion is then suppressed. That this in fact is the case is shown in Fig. 2.38, obtained from density-functional calculations on charged chains [45]. Most remarkable is that the system reacts differently with extra and less electrons per unit cell, i.e., these results clearly show the lack of electron–hole symmetry. This

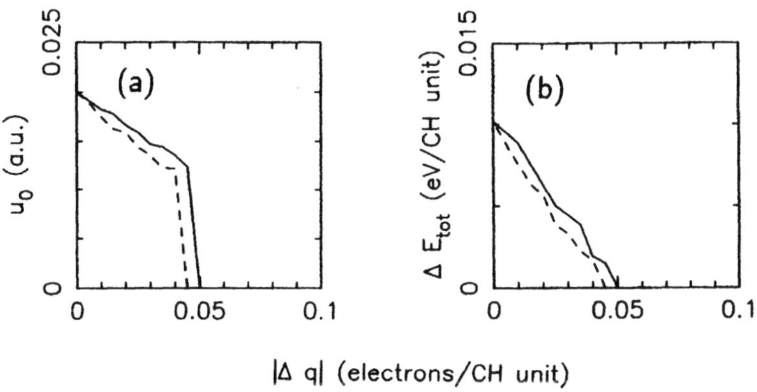

Fig. 2.38. Results from calculations on charged chains of trans-polyacetylene. (**a**) the size of the bond-length alternation [given as u_0 of (2.24)]; (**b**) the accompanying energy gain upon the alternation. *Dashed* and *solid curves* correspond to less and extra electrons per unit, respectively

asymmetry is partly a single-particle effect, i.e., the fact that next-nearest-neighbor hopping integrals need to be included in the single-particle Hamiltonian of (2.23), but also many-body effects lead to this asymmetry [45].

Finally, we add that the results of Fig. 2.38 are somewhat unrealistic. Since periodicity is assumed, also the added charge is periodically distributed. Therefore, the system is given only a minimal possibility to screen the extra charge and, accordingly, the energy costs upon adding the extra charge are overestimated.

2.4.4 Substitutions: Periodic and Isolated

Only in the rarest cases does an experimentally produced sample of well-ordered chains consist of a well-defined (finite or infinite) length. Instead, short and long chains that interact more or less strongly with other chains that may or may not be parallel are found. Moreover, the individual chains may not be completely regular, but disruptions or impurities may occur. These can, for example, be CH_2 groups instead of CH groups or cross-linkings between the chains. Also, other substitutions of parts of the backbone or of the sidegroups may exist. In all cases, these modifications may influence the mechanical and electronic properties of the material.

In Sect. 2.3.2 we saw how one can make use of the fact that periodic substitutions lead to modifications in the electronic properties in producing polymers with different properties. Moreover, in Sect. 2.4.1 we described a simple method that allowed us to obtain accurate information on chains containing structural distortions. The question is whether single, isolated substitutions can be treated using a similar method. There is a number of differences between the two situations.

First, the structural defects treated in Sect. 2.4.1 (most notably, solitons and polarons) are highly mobile and are important in transport properties. In contrast, the isolated substitutions are not mobile and instead act as traps for the transport of charge or spin. Second, the scheme we discussed in Sect. 2.4.1 relies heavily on the assumption that the spatial extent of the defect is large, i.e., that the hopping integrals and on-site energies of the single-particle Hamiltonian of (2.23) vary only slowly across the defect. For the substitutional defects, these matrix elements change abruptly, and there may even be a different number of relevant orbitals for the units around the substitutions than for those far away from the substitutions. Finally, the scheme of Sect. 2.4.1 is based on a graphical representation in which one gradually turns on the defect, whereas this is conceptually not possible for the substitutions (How does one gradually change a C atom into a N atom?).

Since it is desirable to be able to study the effects of single substitutions, we are currently exploring whether one may apply the results of calculations on various periodic structures in the exploration of the properties of the single substitutions, despite the problems above. Since the work not is completed yet, we shall give only a brief description of the preliminary results.

We consider an infinite periodic chain $(A)_x$, where A is a single atom or a group of atoms (e.g., a CH group). Subsequently, we substitute one of the A groups with a B group, leading to a structure like $\cdots$ A·A·A·A·B·A·A·A·A$\cdots$. Considering a simple single-particle model such as that of (2.23) with one orbital per (A or B) unit, and assuming that the substitution changes only the corresponding on-site energy, it turns out [46] that results for the periodic $(AA)_x$ and $(AB)_x$ structures completely specify the energies of the extra orbitals induced by the B impurity in the otherwise periodic A chain.

Currently, this proposal is being explored further via density-functional calculations on periodic structures of real systems, but no final results have been obtained yet.

2.4.5 dc Fields: Polarizabilities and Hyperpolarizabilities

As was described in Sect. 2.2.1, the interesting semiconducting properties of conjugated polymers can be ascribed to the π electrons. These are intermediate between being strongly bonded between nearest-neighbor atoms and being free-electron-like. Therefore, they respond relatively easily to external perturbations, for instance, due to structural or compositional changes. Another area where these properties have become of great interest is the field of non-linear optics (see, e.g., [47–49]). Any material that is exposed to an external electromagnetic field will respond to it. One may quantify the response through either the dipole moment or the total energy. For the former one obtains

$$\mu_i = \mu_i^{(0)} + \sum_j \alpha_{ij} E_j + \sum_{jk} \beta_{ijk} E_j E_k + \gamma_{ijkl} E_j E_k E_l + \cdots \ , \qquad (2.34)$$

where μ_i is the ith component of the dipole moment, E_i that of the electromagnetic field, $\mu_i^{(0)}$ is the static dipole moment, α is the polarizability, and β, γ, $\cdots$ are the first, second, $\cdots$ hyperpolarizabilities.

For the total energy one obtains

$$E_{\mathrm{tot}} = E_{\mathrm{tot}}^{(0)} - \sum_i \mu_i E_i \ , \tag{2.35}$$

with $E_{\mathrm{tot}}^{(0)}$ being the total energy in the absence of the external field.

For free electrons, only α will be non-zero, whereas for strongly bonded electrons, all polarizabilities and hyperpolarizabilities will be very small. The intermediate characteristics of the π electrons of conjugated polymers make both the polarizabilities and hyperpolarizabilities large compared with the values of other materials. Another advantage for technological applications is the short response times.

Interest in these properties stems from the fact that when the external fields are dynamic with frequencies ω_2, ω_3, $\cdots$ for E_j, E_k, $\cdots$, the responses will also be frequency dependent,

$$\begin{aligned} \alpha_{ij} &= \alpha_{ij}(\omega_1, \omega_2) \ , \\ \beta_{ijk} &= \beta_{ijk}(\omega_1, \omega_2, \omega_3) \ , \\ \gamma_{ijkl} &= \gamma_{ijkl}(\omega_1, \omega_2, \omega_3, \omega_4) \ , \end{aligned} \tag{2.36}$$

etc., with ω_1 being the frequency of the response. For any of those, the sum of all arguments has to vanish, e.g., for the second hyperpolarizability γ,

$$\omega_1 \pm \omega_2 \pm \omega_3 \pm \omega_4 = 0 \ . \tag{2.37}$$

Thus, responses may be generated with frequencies other than those of the external fields. Non-vanishing β and γ lead to effects like the dc-Kerr effect, second and third-harmonic generation, etc., which makes it possible to manipulate light with light.

Of course, it would be advantageous to be able to calculate these quantities using parameter-free electronic-structure methods. In this subsection we shall discuss the static case, i.e., all $\omega_p = 0$. One may then either attempt to include the effects of the external field directly in the calculations or, alternatively, to apply a perturbation-theory approach. The first approach corresponds to adding the term

$$\sum_{i=1}^{N} \boldsymbol{E} \cdot \boldsymbol{r}_i \tag{2.38}$$

to the electronic Hamilton operator of (2.2) and the term

$$\sum_{k=1}^{M} -Z_k \boldsymbol{E} \cdot \boldsymbol{R}_k \tag{2.39}$$

to the total energy of (2.3).

When the field has a non-vanishing component along the polymer axis, it is not directly possible to use this approach when assuming that the polymer is infinite. Firstly, the Hamilton operator of (2.38) diverges at both ends. Secondly, the presence of the field destroys the symmetry, so symmetry-adapted Bloch waves as described in Sect. 2.2.2 cannot be used directly. Therefore, the secular equation would be as large as the system, i.e., of infinite dimension.

Instead, one may study finite systems, and from the properties of these, as a function of their size, extrapolate to the infinite system in the spirit of (2.16) and (2.17). It turns out, however, that in order to obtain converged properties one needs rather large systems, as has been shown experimentally [50] and theoretically [51, 52]. As one example we mention the work of Champagne et al. [51], who considered finite oligomers of trans-polyacetylene $CH_2(C_2H_2)_NCH_2$. Assuming that these are placed in the (x, z) plane with the z axis parallel to the chain axis, they found for the polarizabilities α_{ii} that the quantity

$$\Delta\alpha_{ii}(N) = \alpha_{ii}(N) - \alpha_{ii}(N-1) \tag{2.40}$$

for the $i = z$ component for $N = 11$ was still 5–10% different from its extrapolated $N \to \infty$ value. The other components, i.e., $i = x$ and $i = y$, converge more rapidly, but are, on the other hand, roughly one order of magnitude smaller than the $i = z$ component. The latter should not be surprising taking into account that the electrons can be polarized most easily along the chain direction. An even slower convergence as a function of N was found by Champagne et al. [52] for the first hyperpolarizabilities β_{ijk} for polycarbonitrile [52]. This study also showed that the components along the chain axis are markedly larger than the others.

The approaches based on perturbation theory have the advantage that they also can be applied for dynamic (ac) fields (treated in Sect. 2.4.6). Here, we discuss two more aspects related to the theoretical treatment of the static polarizabilities and hyperpolarizabilities. First, as seen in (2.39), the field also affects the nuclei, and, therefore, the structure may change when the system is exposed to an external electrostatic field. In fact, Champagne et al. [51, 53] have shown that this coupling of the external field to the phonons of the system may provide up to ~50% of the total polarizabilities. Second, since the perturbation-theory approaches require fairly large basis sets, some proposals exist to avoid the problems described above when treating infinite chains. When the field has no components along the polymer axis, the terms of (2.38) and (2.39) can be included relatively easily. Doing so has the advantage that the effects of the external fields and the accuracy of the approach can be estimated.

As an example of such studies we show in Fig. 2.39 results for a polycarbonitrile chain that is assumed to be placed in the (x, z) plane with the z axis parallel to the polymer axis and that is exposed to an electrostatic field

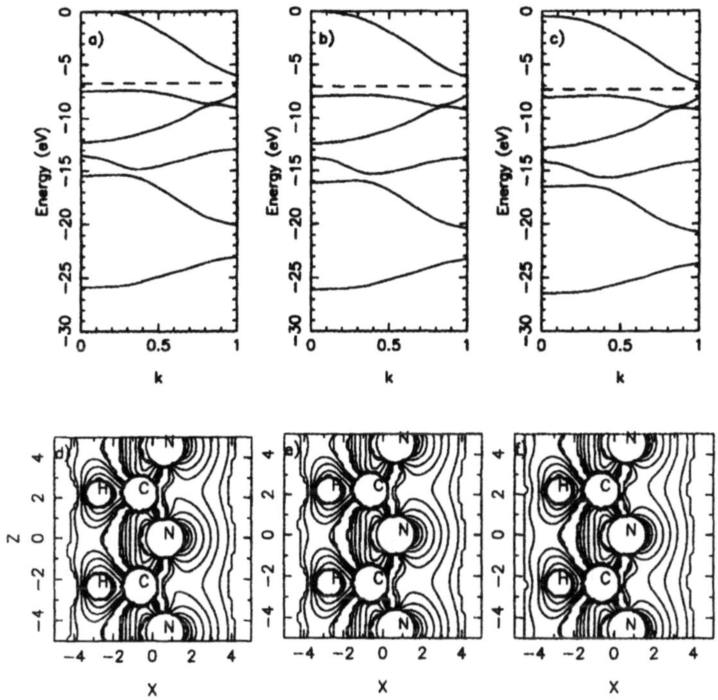

Fig. 2.39. (**a–c**) Band structures and (**d–f**) the density of the highest occupied σ orbital for polycarbonitrile when exposed to an external dc field of strengths (a,d) −0.06 a.u., (b,c) 0 a.u., and (c,f) +0.06 a.u. perpendicular to the chain direction but in its plane. (From [54])

along the x axis [54]. The field was given three different strengths, −0.06 a.u., 0 a.u., and +0.06 a.u., which are slightly larger than what is currently produced in experiments where non-linear optical effects are measured. The band structures are only marginally affected, although some overall shifts can be recognized. However, the changes are significantly less pronounced than those we have found when modifying the structure or the composition of the polymers. Furthermore, the highest occupied σ orbital (which is due to nitrogen lone-pairs and which, accordingly, should response particularly strongly to the external field) shows only modest changes when the external field is turned on.

By calculating various key quantities as functions of the strengths and direction of the external dc field, they can subsequently be fitted with series such as those of (2.34) and (2.35). In Fig. 2.40 we show results of such calculations both for trans-polyacetylene and for polycarbonitrile. It can be shown that due to the presence of an inversion symmetry for the former, all odd hyperpolarizabilities (e.g., all β) vanish for it, whereas they may be

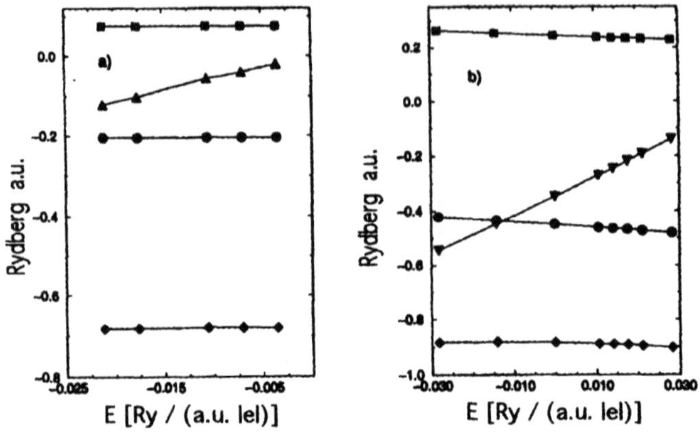

Fig. 2.40. Various key quantities for (**a**) trans-polyacetylene and (**b**) polycarbonitrile as functions of the strength of an external dc field. The chains are assumed to be lying in the (x, z) plane with the z axis parallel to the chain axis, and the dc field is assumed to be parallel to the x axis. *Diamonds*, *circles*, *squares*, and *triangles* represent relative total energy per C_2H_2 and CHN unit, the Fermi energy, the band gap, and the x component of the dipole moment in Rydberg a.u., respectively

non-zero for the latter. From the results of Fig. 2.40, μ, α, β, γ, ... can be extracted [54].

When the electrostatic field has components parallel to the chain axis, it is not possible to include its effects directly in the Schrödinger (or Hartree–Fock or Kohn–Sham) equations, when the chain is assumed being infinite. There are different ways to circumvent this problem. The simplest approach is by assuming that the field leads to a redistribution of the electrons, but that this is periodic like the underlying lattice, and, therefore, that the total potential felt by the electrons also has this symmetry. In one dimension and for a single electron, the dc field gives a potential of the form

$$\Delta V = E x \, . \tag{2.41}$$

Assuming that the potential felt by the electron in the absence of the external field is periodic, with the period a, one may replace the potential of (2.41) with

$$\Delta V' = \sum_n \Delta V'_n \tag{2.42}$$

with

$$\Delta V'_n = \begin{cases} E \cdot (x - na) & \text{for } x \in]na; (n+1)a] \\ 0 & \text{otherwise .} \end{cases} \tag{2.43}$$

This results in a piecewise linear potential, and it can be included in the calculations directly. This approach has the disadvantage that it is not obvious how it can be improved systematically and the approximations may become uncontrolled.

As an alternative one may exploit the fact that a typical calculation for an infinite periodic system involves solving the single-particle equations for a discrete set of k points. Denoting the number of equidistant k points N_k, such a calculation actually corresponds to assuming that the electronic orbitals are periodic with the Born von Kármán zone, i.e., with the period $N_k \cdot a$. Therefore, an improvement of (2.42) and (2.43) corresponds to [54]

$$\Delta V = \sum_n \Delta V_n'' \tag{2.44}$$

with

$$\Delta V_n'' = \begin{cases} E \cdot (x - nN_k a) & \text{for } x \in]nN_k a; (n+1)N_k a] \\ 0 & \text{otherwise} . \end{cases} \tag{2.45}$$

This approach has the advantage that it can systematically be made more accurate by simply increasing the number of points k. As a further improvement, it turns out that in order to avoid having to deal with diverging matrix elements between Bloch functions that are spread out over the whole infinite system, it is very useful instead to deal with localized Wannier functions constructed from the Bloch functions. Thereby, the size of the basis set can be reduced considerably without losing significant accuracy. We are currently implementing this method, but have no final results yet [54].

2.4.6 ac Fields: Non-linear Optics

In the treatment of the effects of non-static external fields with theoretical methods, to our knowledge, only results based on perturbation theory exist. The fundamentals of these methods were originally presented by Genkin and Mednis [55] and Orr and Ward [56]. We assume that we have solved the Schrödinger equation for the unperturbed system exactly and know all eigenfunctions and eigenvalues. In our case, this corresponds to assuming that the Hartree–Fock or Kohn–Sham equations provide the exact solutions to the Schrödinger equation, and that only single-particle excitations are relevant for the excited states. Moreover, due to finite basis sets, we will not have *all* the excited states at our disposal. Using g as a label for the ground state and letting

$$E_{ng} = E_n - E_g - i\Gamma_{ng} \tag{2.46}$$

be the energy difference between the ground state and the nth excited state (modified by an imaginary lifetime broadening $i\Gamma_{ng}$), the general expressions for the polarizabilities and hyperpolarizabilities become

$$\alpha_{ij}(\omega_1, \omega_2) = K_\alpha(\omega_1, \omega_2) \sum_p \left[\sum_n \left(\frac{\langle g|\hat{\mu}_i|n\rangle\langle n|\hat{\mu}_j|g\rangle}{(E_{ng} - \omega_2)} + \frac{\langle g|\hat{\mu}_j|n\rangle\langle n|\hat{\mu}_i|g\rangle}{(E_{ng}^* + \omega_2)} \right) \right]$$

$$\beta_{ijk}(\omega_1,\omega_2,\omega_3) = K_\beta(\omega_1,\omega_2,\omega_3)\sum_p\left[\sum_{m,n}\left(\frac{\langle g|\hat{\mu}_i|m\rangle\langle m|\hat{\mu}_k|n\rangle\langle n|\hat{\mu}_j|g\rangle}{(E_{mg}-\omega_1)(E_{ng}-\omega_2)}\right.\right.$$
$$\left.\left.+\frac{\langle g|\hat{\mu}_k|m\rangle\langle m|\hat{\mu}_j|n\rangle\langle n|\hat{\mu}_i|g\rangle}{(E^*_{mg}+\omega_3)(E^*_{ng}+\omega_1)}+\frac{\langle g|\hat{\mu}_k|m\rangle\langle m|\hat{\mu}_i|n\rangle\langle n|\hat{\mu}_j|g\rangle}{(E^*_{mg}+\omega_3)(E_{ng}-\omega_2)}\right)\right]$$

$$\gamma_{ijkl}(\omega_1,\omega_2,\omega_3,\omega_4) = K_\gamma(\omega_1,\omega_2,\omega_3,\omega_4)$$
$$\times\sum_p\left[\sum_{l,m,n}\left(\frac{\langle g|\hat{\mu}_i|l\rangle\langle l|\hat{\mu}_l|m\rangle\langle m|\hat{\mu}_k|n\rangle\langle n|\hat{\mu}_j|g\rangle}{(E_{lg}-\omega_1)(E_{mg}-\omega_2-\omega_3)(E_{ng}-\omega_2)}\right.\right.$$
$$+\frac{\langle g|\hat{\mu}_j|l\rangle\langle l|\hat{\mu}_k|m\rangle\langle m|\hat{\mu}_i|n\rangle\langle n|\hat{\mu}_l|g\rangle}{(E^*_{lg}+\omega_2)(E^*_{mg}+\omega_2+\omega_3)(E_{ng}-\omega_4)}$$
$$+\frac{\langle g|\hat{\mu}_l|l\rangle\langle l|\hat{\mu}_i|m\rangle\langle m|\hat{\mu}_k|n\rangle\langle n|\hat{\mu}_j|g\rangle}{(E^*_{lg}+\omega_4)(E_{mg}-\omega_2-\omega_3)(E_{ng}-\omega_2)}$$
$$\left.+\frac{\langle g|\hat{\mu}_j|l\rangle\langle l|\hat{\mu}_k|m\rangle\langle m|\hat{\mu}_l|n\rangle\langle n|\hat{\mu}_i|g\rangle}{(E^*_{lg}+\omega_2)(E^*_{mg}+\omega_2+\omega_3)(E^*_{ng}+\omega_1)}\right)$$
$$-\sum_{m,n}\left(\frac{\langle g|\hat{\mu}_i|m\rangle\langle m|\hat{\mu}_l|g\rangle\langle g|\hat{\mu}_k|n\rangle\langle n|\hat{\mu}_j|g\rangle}{(E_{mg}-\omega_1)(E_{ng}-\omega_4)(E_{ng}-\omega_2)}\right.$$
$$+\frac{\langle g|\hat{\mu}_i|m\rangle\langle m|\hat{\mu}_l|g\rangle\langle g|\hat{\mu}_k|n\rangle\langle n|\hat{\mu}_j|g\rangle}{(E_{mg}-\omega_4)(E^*_{ng}+\omega_3)(E_{ng}-\omega_2)}$$
$$+\frac{\langle g|\hat{\mu}_l|m\rangle\langle m|\hat{\mu}_i|g\rangle\langle g|\hat{\mu}_j|n\rangle\langle n|\hat{\mu}_k|g\rangle}{(E^*_{mg}+\omega_1)(E^*_{ng}+\omega_4)(E^*_{ng}+\omega_2)}$$
$$\left.\left.+\frac{\langle g|\hat{\mu}_l|m\rangle\langle m|\hat{\mu}_i|g\rangle\langle g|\hat{\mu}_j|n\rangle\langle n|\hat{\mu}_k|g\rangle}{(E^*_{mg}+\omega_4)(E_{ng}-\omega_3)(E^*_{ng}+\omega_2)}\right)\right]. \qquad (2.47)$$

Here, K are arithmetic constants. Furthermore, the p summation includes all permutations of the frequencies $\omega_2, \omega_3, \cdots$. Finally, $\hat{\mu}$ are the dipole-moment operators.

This approach can be used for both static and dynamical fields. When using single-particle methods like Hartree–Fock or density-functional methods, the matrix elements are single-particle matrix elements of the dipole-moment operators. For the infinite system, these matrix elements will diverge, unless approaches such as those described in Sect. 2.4.5 are applied. Alternatively, one may also consider finite systems and attempt to estimate the limiting values for the size of the system going to infinity. However, we stress that these limits are often first reached for quite large systems for which the necessary truncations in the sizes of the basis sets lead to other errors.

Finally, for sufficiently high frequencies ω_i the electromagnetic fields do not couple with the phonons so that only the electronic degrees of freedoms have to be considered. For lower frequencies the nuclear degrees of freedom should, in principle, also be included in the calculations, as we saw in Sect. 2.4.5.

2.5 Conclusions

This chapter has been devoted to theoretical studies of material properties of conjugated polymers when using parameter-free electronic-structure methods. We have thereby focused on the results of applications of our own method, but stress that this does not mean that there are no other methods for such studies, nor does it imply that no other method can provide similar (or equally accurate) information.

Conjugated polymers are often ill-defined materials and when treating them with theoretical methods one has to make idealizations, as we have discussed. It is hoped that these do not obscure the final results of the calculations in any uncontrollable way. On the other hand, it also means that it is of ultimate importance to compare the results with experimental information whenever possible, remembering that these may also suffer from unknown effects due to material imperfections. In the present chapter we have seen that in most cases good agreement between experimental and theoretical results could be achieved.

Having stressed the importance of the interplay between theory and experiment, it should also be pointed out that the theoretical studies often can provide information that can only can be obtained with difficulty (if at all) by experimentation. For example, theoretical studies are able to establish relations between material structure and composition and material properties. Here, we have demonstrated this both through the modification of the electronic properties of periodic conjugated polymers when modifying their structure or composition and through the relations between the properties of the periodic chains and those of defect-containing chains.

Current parameter-free electronic structure methods are constructed mainly for studies of ground-state properties of materials that either are relatively small or are infinite and periodic and that are subject to no external forces. However, for conjugated polymers some of the technologically most interesting applications are within non-linear optics and semiconductor devices. In both cases, external forces act on the materials, and they are not isolated but closely connected with other materials. We have given a few examples on how at present such systems are addressed with electronic-structure calculations. However, these methods and their applications are still in their infancy, and it is an area that will definitely (have to) develop considerably in the not too distant future.

Acknowledgments. One of the authors (MS) is grateful to Fonds der Chemischen Industrie for very generous support. This work was supported by the Deutsche Forschungsgemeinschaft (DFG) through projects no. Sp439/2 and Sp439/6 and by the Optikzentrum Konstanz. Some of the calculations were carried out at the Höchstleistungsrechenzentrum Jülich.

References

1. R. Peierls, *Quantum Theory of Solids* (Clarendon, Oxford 1955)
2. C. Møller, M.S. Plesset, Phys. Rev. **46**, 618 (1934)
3. P. Hohenberg, W. Kohn, Phys. Rev. **136**, B864 (1964)
4. W. Kohn, L.J. Sham, Phys. Rev. **140**, A1133 (1965)
5. T. Koopmans, Physica **1**, 104 (1933)
6. J.F. Janak, Phys. Rev. B **18**, 7165 (1978)
7. M. Springborg, M. Lev, Phys. Rev. B **40**, 3333 (1989)
8. K. Seki, N. Ueno, U.O. Karlsson, R. Engelhardt, E.-E. Koch, Chem. Phys. **105**, 247 (1986)
9. M. Springborg, O.K. Andersen, J. Chem. Phys. **87**, 7125 (1987)
10. M. Springborg, J.-L. Calais, O. Goscinski, L.A. Eriksson, Phys. Rev. B **44**, 12713 (1991)
11. J.-M. André, J. Delhalle, J.-L. Brédas, *Quantum Chemistry Aided Design of Organic Polymers. An Introduction to the Quantum Chemistry of Polymers and its Applications* (World Scientific, Singapore 1991)
12. J.J. Ladik, *Quantum Theory of Polymers as Solids* (Plenum, New York 1988)
13. J.W. Mintmire, C.T. White, Phys. Rev. Lett. **50**, 101 (1983)
14. C.S. Yannoni, T.C. Clarke, Phys. Rev. Lett. **51**, 1191 (1983)
15. C.R. Fincher, Jr., C.-E. Chen, A.J. Heeger, A.G. MacDiarmid, J.B. Hastings, Phys. Rev. Lett. **48**, 100 (1982)
16. J. Ashkenazi, W.E. Pickett, H. Krakauer, C.S. Wang, B.M. Klein, S.R. Chubb, Phys. Rev. Lett. **62**, 2016 (1989)
17. M. Springborg, S.-L. Drechsler, J. Málek, Phys. Rev. B **41**, 11954 (1990)
18. S. Suhai, Phys. Rev. B **51**, 16553 (1995)
19. J. Rasmusson, S. Stafström, M. Lögdlund, W.R. Salaneck, U. Karlsson, D.B. Swanson, A.G. MacDiarmid, G.A. Arbuckle, Synth. Met. **41–43**, 1365 (1991)
20. R. Hoffmann, *Solids and Surfaces: A Chemist's View of Bonding in Extended Structures*, (VCH, Weinheim 1988)
21. R. Dronskowski, P.E. Blöchl, J. Phys. Chem. **97**, 8617 (1993)
22. S. Hirata, H. Torii, M. Tasumi, Phys. Rev. B **57**, 11994 (1998)
23. M. Springborg, Ber. Bunsenges. Phys. Chem. **95**, 1238 (1991)
24. *Proceedings of the International Conference on Science and Technology of Synthetic Metals 1996*, Synth. Met. **84–86** (1997)
25. J.H. Borroughes, D.D.C. Bradley, A.R. Brown, R.N. Marks, K. MacKay, R.H. Friend, P.L. Burn, A.B. Holmes, Nature **347**, 539 (1990)
26. M.E. Vaschetto, M. Springborg, J. Mol. Struct. (Theochem.) **468**, 181 (1999)
27. J.M. Ginder, A.J. Epstein, Phys. Rev. B **41**, 10674 (1990)
28. H.F. Wittmann, R.H. Friend, M.S. Khan, J. Lewis, J. Chem. Phys. **101**, 2693 (1994)
29. M. Springborg, R.C. Albers, Phys. Rev. B **53**, 10626 (1996)

30. M.J. Rice, S.R. Phillpot, Phys. Rev. Lett. **58**, 937 (1987)
31. M. Springborg, Phys. Rev. B **40**, 7839 (1990)
32. M. Alouani, R.C. Albers, J.M. Wills, M. Springborg, Phys. Rev. Lett. **69**, 3104 (1992)
33. W.P. Su, J.R. Schrieffer, A.J. Heeger, Phys. Rev. Lett. **42**, 1698 (1979)
34. M. Springborg, Int. Rev. Phys. Chem. **12**, 241 (1993)
35. M. Springborg, Solid State Commun. **89**, 665 (1994)
36. M. Springborg, Z. Phys. B **95**, 363 (1994)
37. M. Springborg, Int. J. Quant. Chem. **58**, 717 (1996)
38. M. Springborg, Synth. Met. **85**, 1037 (1997)
39. M.J. Rice, E.J. Mele, Phys. Rev. Lett. **49**, 1455 (1982)
40. J.C. Gustafsson, Q. Pei, O. Inganäs, Solid State Commun. **87**, 265 (1993)
41. D. Baeriswyl, D.K. Campbell, S. Mazumdar, In: *Conjugated Conducting Polymers*, ed. by H. Kiess (Springer Verlag, Berlin 1992) pp. 7
42. H. Meider, M. Springborg, J. Phys. Condens. Matt. **10**, 6953 (1998)
43. P.H. Dederichs, S. Blügel, R. Zeller, H. Akai, Phys. Rev. Lett. **53**, 2512 (1984)
44. H. Meider, M. Springborg, Chem. Phys. Lett. **300**, 339 (1999)
45. M. Springborg, C. Arcangeli, J. Mol. Struct. (Theochem) **531**, 211 (2000)
46. L. De Maria, M. Springborg (unpublished)
47. J. Messier, F. Kajzar, P. Prasad, D. Ulrich (eds.) *Nonlinear Optical Effects in Organic Polymers* (Kluwer, Dordrecht 1989)
48. J.L. Brédas, R.R. Chance (eds.) *Conjugated Polymeric Materials: Opportunities in Electronics, and Molecular Electronics* (Kluwer, Dordrecht 1990)
49. J. Messier, F. Kajzar, P. Prasad (eds.) *Organic Molecules for Nonlinear Optics and Photonics* (Kluwer, Dordrecht 1991)
50. I.D.W. Samuel, I. Ledoux, C. Dhenaut, J. Zyss, H.H. Fox, R.R. Schrock, R.J. Silbey, Science **265**, 1070 (1994)
51. B. Champagne, E.A. Perpète, J.-M. André, J. Chem. Phys. **101**, 10796 (1994)
52. B. Champagne, D. Jacquemin, J.-M. André, B. Kirtman, J. Phys. Chem. **101**, 3158 (1997)
53. B. Champagne, H. Vanderhoeven, É.A. Perpète, J. M. André, Chem. Phys. Lett. **248**, 301 (1996)
54. K. Schmidt, M. Springborg, Phys. Chem. Chem. Phys. **1**, 1743 (1999)
55. V.N. Genkin, P.M. Mednis, Sov. Phys. JETP **27**, 609 (1968)
56. B.J. Orr, J.F. Ward, Mol. Phys. **20**, 513 (1971)

3 Recursive Algorithms for Polymeric Chains

R. Farchioni, G. Grosso and P. Vignolo

3.1 Introduction

The implementation of real-space recursive methods for the evaluation of the electronic structure and transport properties of selected representative conjugated polymers is the main topic of this chapter. For this it is convenient to evaluate the matrix elements of the resolvent operator of the polymer Hamiltonian. In fact, by knowing these matrix elements we can infer information on the energy spectrum, on the density of states, and on the transmittivity of finite and infinite strands of polymer. The evaluation of the Green's function matrix becomes computationally simpler when the corresponding Hamiltonian is represented on a local basis; moreover, the transport properties can also be easier deduced if this kind of basis is assumed. A very attractive point is that the Green's function matrix elements can be represented by iterative procedures as continued fractions [1–4] which can be calculated without the explicit knowledge of the eigenvalues and eigenvectors of the Hamiltonian: this is very convenient for large-scale eigenvalue problems in particular for systems without periodicity or with complex unit cells.

In the first part of the chapter we briefly outline some general features of iterative methods and present illustrative examples with reference to polymeric structures. In Sect. 3.2 we recall the essential terminology and quantities of the tight-binding formalism which is adopted for the description of the Hamiltonian of π-conjugated polymers. In Sect. 3.3 and 3.4 we introduce the renormalization and the recursion methods, while Sect. 3.5 recalls some relations concerning the Green's function matrix elements, important for numerical computations. In the second part of the chapter, with reference to the case of polythiophene and leucoemeraldine, we show (Sect. 3.6) how the band structure of a conjugated polymer can be obtained through the generation of an effective Hamiltonian for a strictly one-dimensional chain with a single site per primitive cell [5]. The linear chain model is also convenient for the determination of the transmittivity [6] of the polymer as discussed in Sect. 3.7. The procedure outlined for perfect polymers, is extended to handle disordered structures also in the presence of interchain coupling (Sect. 3.8). Sect. 3.9 contains the conclusions.

3.2 The Tight-Binding Method

For the evaluation of the electronic structure of a polynuclear system, we have to solve self-consistently the one-electron Schrödinger equation in a given fixed nuclear configuration:

$$\left[\frac{p^2}{2m} + V(\boldsymbol{r})\right] \psi(\boldsymbol{r}) = E\psi(\boldsymbol{r}) \,. \tag{3.1}$$

Here the potential $V(\boldsymbol{r})$ contains the Coulomb interaction with the nuclei and the electron–electron interaction made by Coulomb, exchange and correlation contributions as indicated, for instance, in Chap. 2 and [7]. In the tight-binding (or LCAO) method the one-electron wavefunction $\psi(\boldsymbol{r})$ is expressed in terms of an appropriate set of orthonormal localized functions $\{\phi_{m\mu}\}$ built from the μ orbitals centered at the atomic positions $\boldsymbol{r}_m$ of the system:

$$\psi(\boldsymbol{r}) = \sum_{m,\mu} c_{m\mu}\phi_{m\mu}(\boldsymbol{r}) \,. \tag{3.2}$$

Projected on this manifold, the Hamiltonian of (3.1) takes the tight-binding form:

$$\begin{aligned} H &= \sum_{m\mu} |\,\phi_{m\mu}\rangle\langle\phi_{m\mu}\,| \left(\frac{p^2}{2m} + V(\boldsymbol{r})\right) \sum_{n\nu} |\,\phi_{n\nu}\rangle\langle\phi_{n\nu}\,| \\ &= \sum_{m\mu} E_{m\mu} |\,\phi_{m\mu}\rangle\langle\phi_{m\mu}\,| + \sum_{m\mu\neq n\nu} t_{m\mu,n\nu} |\,\phi_{m\mu}\rangle\langle\phi_{n\nu}\,| \,, \end{aligned} \tag{3.3}$$

where $E_{m\mu}$ are the "site energies" corresponding to the orbital $\phi_{m\mu}$ and $t_{m\mu,n\nu}$ the "hopping-interaction energies" between pairs of orbitals. Combining the Schrödinger equation (3.1) with (3.2) leads to the secular equation

$$\|\, H_{m\mu,n\nu} - E\delta_{m\mu,n\nu} \,\| = 0 \,, \tag{3.4}$$

where

$$H_{m\mu,n\nu} = \langle\phi_{m\mu}|H|\phi_{n\nu}\rangle \quad \text{and} \quad \delta_{m\mu,n\nu} = \langle\phi_{m\mu}|\phi_{n\nu}\rangle \,. \tag{3.5}$$

In the case of periodic structures, the crystal wave function $\psi(\boldsymbol{k})$ has the typical Bloch form and a convenient basis set for its expansion is given in terms of Bloch sums built from the orbitals in the primitive cell of the crystal. In this case the matrix elements $H_{m\mu,n\nu}$ are functions of the vector $\boldsymbol{k}$ belonging to the first Brillouin zone, and (3.4) provides the band structure $E(\boldsymbol{k})$. Often, as in the case of π-conjugated polymers, a single effective orbital per site is considered; in this case the Hamiltonian is written as

$$H = \sum_{m} E_m |\,\phi_m\rangle\langle\phi_m\,| + \sum_{m\neq n} t_{m,n} |\,\phi_m\rangle\langle\phi_n\,| \,. \tag{3.6}$$

The simple formal expression (3.3) has several assumptions and simplifications associated with it that are worth noting; they are as follows:

- The orthonormality condition of the orbitals $\{\phi_{m\mu}\}$ is generally introduced to simplify the expressions of $E_{m\mu}$ and $t_{m\mu,n\nu}$; it can be overcome by considering the matrix overlap $S_{m\mu,n\nu} = \langle\phi_{m\mu} \mid \phi_{n\nu}\rangle$ and the projection operator $P = \sum_{m\mu} \sum_{n\nu} \mid \phi_{m\mu}\rangle S^{-1}_{m\mu,n\nu} \langle\phi_{n\nu} \mid$ in (3.3).
- The interactions $t_{m\mu,n\nu}$ are generally limited to a small number of neighboring shells; an often-adopted choice involves only nearest neighbors.
- The explicit determination of the basis functions, and the consequent evaluation of mono- and multi-center matrix elements, can be avoided if these elements are considered as fitting parameters used to reproduce experimental results (band gaps, optical transitions, band widths, etc.) or more sophisticated electronic-structure calculations.

A lot of mathematical and numerical results are known for the spectral properties of Hamiltonians of the type given in (3.6) with different forms of potential, in particular for one-dimensional systems. This has allowed the determination of the electronic properties of several classes of materials. We shall see in the next sections how it is possible to exactly map the Hamiltonian of a real conjugated polymer into an effective Hamiltonian, of the form of (3.6), describing a fictitious system equivalent to the polymer, i.e., with the same electronic structure and transmittivity.

3.3 Decimation Technique and Renormalization Method

Let us consider an Hamiltonian H represented on a given orthonormal basis set, as, for instance, the tight-binding Hamiltonian given in (3.3). We can show that by a suitable partition of the space S, in which H is defined, we can focus on a reduced subsystem, with renormalized interactions, described in terms of the Hamiltonian matrix elements of the original system. This is done by means of the Dyson equation, which allows the Green's function of the reduced subsystem to be obtained and thus its renormalized Hamiltonian to be defined. We present here only few lines of the procedure; the complete treatment can be found in [2–4].

The space S can be arbitrarily partitioned into two complementary parts, S_{A} and S_{B}, where S_{A} is the subspace of interest and S_{B} the space to be decimated. By means of the projection operators

$$P_{\mathrm{A}} = \sum_{i \epsilon \mathrm{A}} \mid \phi_i\rangle\langle\phi_i \mid \quad \text{and} \quad P_{\mathrm{B}} = 1 - P_{\mathrm{A}} \ ,$$

the Hamiltonian can be written as $H = H^0 + W$, where $H^0 = H_{\mathrm{AA}} + H_{\mathrm{BB}}$ does not mix the subspaces S_{A} and S_{B}, and $W = H_{\mathrm{AB}} + H_{\mathrm{BA}}$ represents the interaction between them. Let $G^0 = 1/(E - H^0)$ be the resolvent of Hamiltonian H^0; from the Dyson equation we have for the resolvent of the complete system: $G = G^0 + G^0WG$. Projecting also the operator G onto the subspaces S_A and S_B, after some manipulations one obtains for the Green's function on the subspace of interest, S_A:

$$G_{S_A} = \frac{1}{E - H_{S_A}} , \tag{3.7}$$

where

$$H_{S_A} = H_{AA} + H_{AB} \frac{1}{E - H_{BB}} H_{BA} . \tag{3.8}$$

Thus, after decimation of the subspace S_B, we can describe the complementary system A by means of the effective energy-dependent Hamiltonian H_{S_A}. Equation (3.8) is a very important result for practical applications, and it is at the heart of the renormalization method and of the recursion method.

For instance, let us consider a chain with site energies E_n and (real) nearest-neighbor hopping $t_{n,n+1}$ as shown in Fig. 3.1.

Fig. 3.1. Linear chain with site energies E_n and nearest-neighbor hopping $t_{n,n+1}$

Using (3.8) the decimation of site N (system B) leads to the following renormalized energies for the sites N − 1 and N + 1:

$$\tilde{E}_{N-1} = E_{N-1} + \frac{t^2_{N-1,N}}{E - E_N} \quad \text{and} \quad \tilde{E}_{N+1} = E_{N+1} + \frac{t^2_{N,N+1}}{E - E_N} ,$$

the interaction between them takes the following form:

$$\tilde{t}_{N-1,N+1} = \frac{t_{N-1,N} t_{N,N+1}}{E - E_N} .$$

As we shall show in the illustrative examples below, this partition technique is particularly useful in mapping the polymeric structure into a strictly one-dimensional chain. Once a linear chain is obtained, successive decimations can safely be performed iteratively because at any step, after a decimation, the surviving subsystem has the same linear structure as the original one.

3.3.1 Illustrative Examples of the Renormalization Method

As the first example of the application of the renormalization procedure to a polymeric chain, we consider an infinite crystal made by planar arrangement of pentagons, as in the schematic structure shown in Fig. 3.2a, which represents the polypirrole (polythiophene) polymer if the lowest X vertex of the pentagons are occupied by nitrogen (sulfur) atoms.

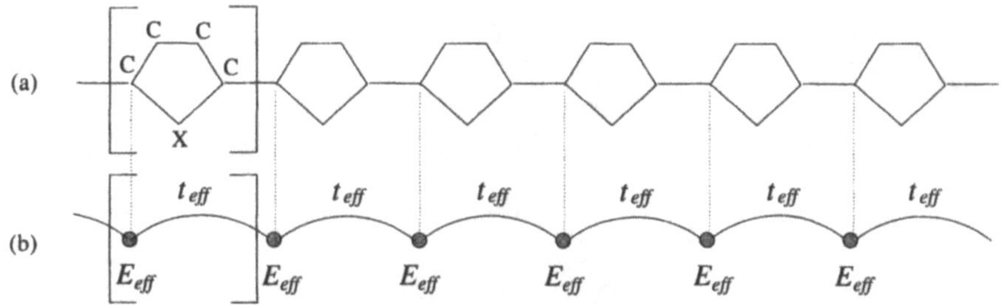

Fig. 3.2. (**a**) Schematic structure of polypirrole (X = N) or polythiophene (X = S) crystals; the primitive unit cell is indicated by large square brackets. (**b**) Equivalent renormalized crystal

Let us consider the situation in which there is a single effective p_z orbital per site; due to the symmetry of the pentagons we consider as equivalent the carbon orbitals at the sites 1 and 4 (thus, we fix $E_1 = E_4$) and the carbon orbitals at the sites 2 and 3 (thus, $E_2 = E_3$) (see Fig. 3.3a).

Moreover, for the hopping interactions we have $t_{1,2} = t_{3,4} = t$, $t_{1,5} = t_{4,5} = t'$, and $t_{2,3} = t''$, and for the interaction between carbon 4 and carbon 1 belonging to the adjacent ring we have $t_{4,1} = \tau$. If we consider the space spanned by the orbital at site 1 and the orbital at site 4 as the subspace of interest S_{A}, then, after the decimation of site 5, by means of (3.8), we obtain

$$E_1^{(1)} = E_1 + \frac{t'^2}{E - E_5}, \quad E_4^{(1)} = E_4 + \frac{t'^2}{E - E_5} \quad \text{and} \quad t_{1,4} = \frac{t'^2}{E - E_5} \, .$$

Superscripted indices in parenthesis indicate the step of renormalization. The renormalized primitive cell so obtained is shown in Fig. 3.3b. Then, decimation of site 2 leads to the cell shown in Fig. 3.3c with renormalized energies

$$E_1^{(2)} = E_1^{(1)} + \frac{t^2}{E - E_2}, \quad E_3^{(1)} = E_3 + \frac{t''^2}{E - E_2} \quad \text{and} \quad t_{1,3} = \frac{t \cdot t''}{E - E_2} \, .$$

A further decimation of site 3 leads to the cell in Fig. 3.3d with renormalized energies

$$E_1^{(3)} = E_1^{(2)} + \frac{t_{1,3}^2}{E - E_3^{(1)}}, \quad E_4^{(2)} = E_4^{(1)} + \frac{t^2}{E - E_3^{(1)}} \quad \text{and} \quad t_{1,4}^{(1)} = t_{1,4} + \frac{t_{1,3} t}{E - E_3^{(1)}} \, .$$

Finally, the decimation of site 4 leads to a lattice with one site per cell (see Fig. 3.3e and 3.2b), with effective site energies E_{eff} and hopping interactions t_{eff} given by

$$E_{\mathrm{eff}}(E) = E_1^{(3)} + \frac{\left[t_{1,4}^{(1)}\right]^2}{E - E_4^{(2)}} \qquad \text{and} \qquad t_{\mathrm{eff}}(E) = \frac{t_{1,4}^{(1)} \cdot \tau}{E - E_4^{(2)}} \, . \tag{3.9}$$

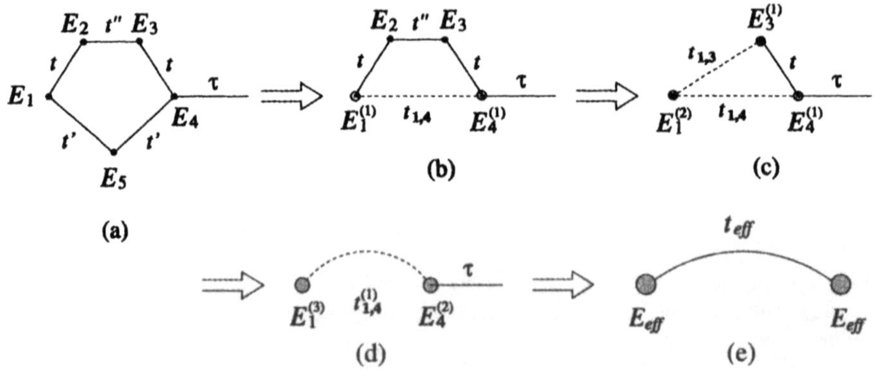

Fig. 3.3. Renormalization steps to reduce the unit cell of polypirrole or polythiophene to the unit cell of a linear chain as in Fig. 3.2b

The Green's function at site 1 is then given by

$$G_{1,1}(E) = \frac{1}{E - E_{\text{eff}}(E)} \,. \tag{3.10}$$

For the evaluation of the total density of states, knowledge of the Green's function at each site of the primitive cell of the original crystal is necessary. For this, different strategies can be followed. For instance, let us deduce the expression of the Green's function projected on the "internal" sites of the pentagon (the equivalent carbons at sites 2 and 3 and the nitrogen at site 5). In this case it is convenient to map the overall original periodic structure (see Fig. 3.4a) into the structure obtained by renormalizing all the sites to the left of carbon 1 and to the right of carbon 4 following the procedure outlined in the example shown above (Fig. 3.4b). A further renormalization of the semi-infinite linear chains leads to the structure shown in Fig. 3.4c, where the self-energies $\Sigma(E)$ summarize all the left and right semi-infinite crystals adjacent to the reference central pentagon. One can now iterate the same renormalization procedure on each of the remaining five sites of the pentagon and evaluate the Green's function matrix element required. For instance, after having reduced the polymeric chain of Fig. 3.4a to the equivalent renormalized molecule of Fig. 3.4c, by further application of (3.8) we obtain (see the results for the constant semi-infinite chain in Sect. 3.4.1)

$$\Sigma(E) = E_1 + \frac{1}{2}\left((E - E_{\text{eff}}) - \sqrt{(E - E_{\text{eff}})^2 - 4t_{\text{eff}}^2}\right) \,, \tag{3.11}$$

$$G_{2,2}(E) = G_{3,3}(E) = \frac{1}{E - \tilde{E}_2} \quad \text{and} \quad G_{5,5}(E) = \frac{1}{E - \tilde{E}_5} \,, \tag{3.12}$$

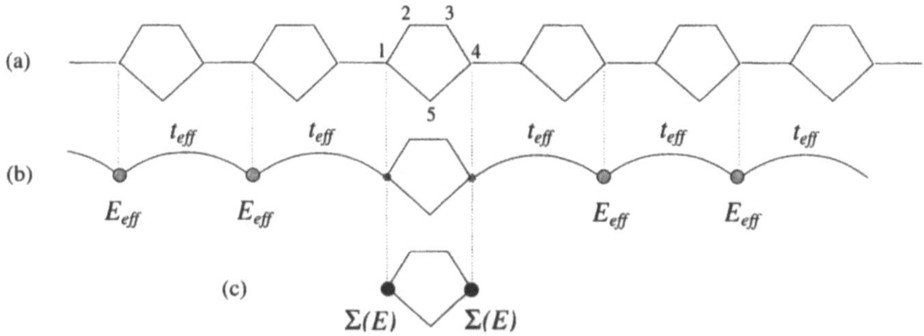

Fig. 3.4. (**a**) Schematic structure of polypirrole or polythiophene crystal. (**b**) Equivalent renormalized crystal. (**c**) Effective molecule for the evaluation of bands and density of states

where $\tilde{E}_2$ ($\tilde{E}_5$) is the value of the effective energy at site 2 (5) when all the other sites are decimated. $\tilde{E}_2$ is given by

$$\tilde{E}_2 = E_2 + at^2 + b + (t'' + b)^2 (E - E_2 - at^2 - b)^{-1} \ , \tag{3.13}$$

where $a = a(E) = \dfrac{1}{E - \Sigma(E)}$ and $b = b(E) = \dfrac{(att')^2}{E - E_5 - at'^2}$.
Similarly, for $\tilde{E}_5$ we have

$$\tilde{E}_5 = E_5 + at'^2 + b' + (att' + \frac{b'}{att'} t'')^2 (E - E_5 - at'^2 - b')^{-1} \ , \tag{3.14}$$

where $b' = b'(E) = \dfrac{(att')^2}{E - E_2 - at^2}$.
From (3.10) and (3.12) the partial and total density of states of the polymer can be easily evaluated by means of (3.22) and (3.23), respectively, which are given in Sect. 3.5.

As a further example of the renormalization procedure we consider the case of the leucoemeraldine base polymer, which is made of aromatic rings connected by nitrogen atoms (see Fig. 3.5a). Sites 1 to 6 are occupied by carbons, and site 0 is occupied by a nitrogen atom. Due to the symmetry of each cell, we have $t_{1,2} = t_{1,6} = t_{3,4} = t_{4,5} = t$, and $t_{2,3} = t_{5,6} = t'$, $t_{4,0} = \tau$ (interaction between sites 4 and 0 on adjacent rings).

After the decimation of sites 1 to 6, we obtain the single site per cell structure shown in Fig. 3.5b with energies [8]

$$E_{\text{eff}} = E_0 + \frac{t_{0,4}}{E - \tilde{E}_4} \frac{t_{0,1}^2 + \tilde{t}_{1,0}^2}{E - \tilde{E}_1 - \dfrac{\tilde{t}_{1,4}^2}{E - \tilde{E}_4}} \quad \text{and} \quad t_{\text{eff}} = \frac{t_{0,1}\tilde{t}_{1,0}}{E - \tilde{E}_1 - \dfrac{\tilde{t}_{1,4}^2}{E - \tilde{E}_4}} \ , \tag{3.15}$$

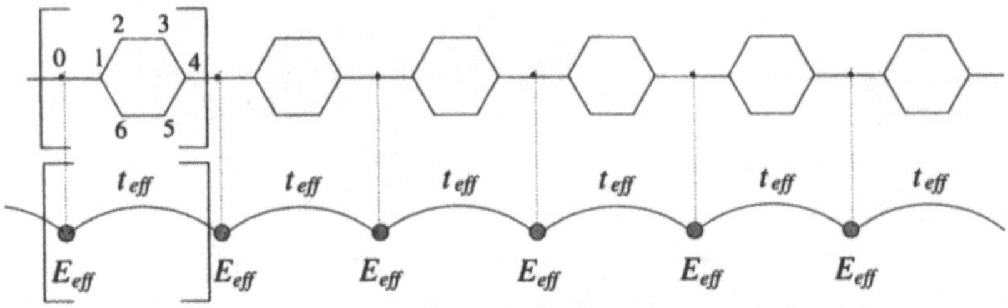

Fig. 3.5. (**a**) Schematic structure of leucoemeraldine polymer. (**b**) Renormalized effective crystal

where

$$\tilde{E}_1 = E_1 + \frac{t_{1,2}^2}{E - E_2} + \frac{t_{1,6}^2}{E - E_6} + \left(\frac{t_{1,2}t_{2,3}}{E - E_2}\right)^2 \frac{1}{E - E_3 - \dfrac{t_{2,3}^2}{E - E_2}}$$

$$+ \left(\frac{t_{1,6}t_{5,6}}{E - E_6}\right)^2 \frac{1}{E - E_5 - \dfrac{t_{5,6}^2}{E - E_6}} ,$$

$$\tilde{E}_4 = E_4 + \frac{t_{3,4}^2}{E - E_3 - \dfrac{t_{2,3}^2}{E - E_2}} + \frac{t_{4,5}^2}{E - E_5 - \dfrac{t_{5,6}^2}{E - E_6}} ,$$

$$\tilde{t}_{1,4} = \left(\frac{t_{1,2}t_{2,3}}{E - E_2}\right) \frac{t_{3,4}}{E - E_3 - \dfrac{t_{2,3}^2}{E - E_2}} + \left(\frac{t_{1,6}t_{5,6}}{E - E_6}\right) \frac{t_{4,5}}{E - E_5 - \dfrac{t_{5,6}^2}{E - E_6}} ,$$

$$\tilde{t}_{1,0} = \frac{\tilde{t}_{1,4}t_{0,4}}{E - \tilde{E}_4} .$$

Analytic but cumbersome expressions are obtained if we let internal sites 1 to 6 survive and have the Green's function projected onto them. For these cases we prefer to use the alternative procedure outlined below following the recursion method.

3.4 The Recursion Method and the Continued Fraction Representation of the Green's Function

In the recursion method [1] one starts from a given normalized state $| f_0\rangle$ belonging to the space S spanned by the orthonormal basis set $\{|\phi_i\rangle\}$, where the Hamiltonian matrix elements $H_{ij} = \langle\phi_i |H| \phi_j\rangle$ are known. By iterative application of the operator H, the recursion method allows the generation of a new orthonormal basis set $\{|\, f_n\rangle\}$ on which the Hamiltonian H is represented by a tridiagonal matrix. Therefore, this procedure automatically maps the original system into a linear semi-infinite chain. The vectors $\{|\, f_n\rangle\}$ are constructed as follows. Given the normalized state $|\, f_0\rangle$ and the projection operator $P_0 = |\, f_0\rangle\langle f_0\,|$, we can define

$$|\, F_1\rangle = (1 - P_0)H|\, f_0\rangle = H|\, f_0\rangle - a_0|\, f_0\rangle \; , \tag{3.16}$$

where $a_0 = \langle f_0\,|H|\, f_0\rangle$. Let us indicate the normalization of $|\, F_1\rangle$ by b_1^2, i.e., $b_1^2 = \langle F_1\,|\, F_1\rangle$; then the normalized vector orthogonal to $|\, f_0\rangle$ is $|\, f_1\rangle \equiv |\, F_1\rangle / b_1$.

For $n > 1$, the vectors $|\, f_n\rangle$ are obtained from the three-term recursive relation:

$$|\, F_{n+1}\rangle = (H - a_n)|\, f_n\rangle - b_n|\, f_{n-1}\rangle \; , \tag{3.17a}$$

$$\text{where} \quad a_{n+1} = \langle f_{n+1}\,|H|\, f_{n+1}\rangle \tag{3.17b}$$

$$\text{and} \quad (b_{n+1})^2 = \langle F_{n+1}\,|\, F_{n+1}\rangle \; . \tag{3.17c}$$

A noticeable aspect of the procedure is that the orthogonalization of $|\, f_{n+1}\rangle$ to the two predecessors $|\, f_n\rangle$ and $|\, f_{n-1}\rangle$, as required by (3.17), automatically guarantees its orthogonality to all the other previous vectors $|\, f_0\rangle, |\, f_1\rangle \ldots |\, f_{n-2}\rangle$. This feature is very convenient for computer storage capabilities when one considers matrices of very large rank and of sparse character. On the basis $|\, f_n\rangle$ the operator H is represented by the tridiagonal matrix:

$$H = \begin{pmatrix} a_0 & b_1 & & & \\ b_1 & a_1 & b_2 & & \\ & b_2 & a_2 & b_3 & \\ & & \ddots & \ddots & \ddots \end{pmatrix} , \tag{3.18}$$

which clearly emphasizes the chain aspect assumed by the original system. From (3.18) one can easily deduce the matrix for the operator $E - H$, and from the upper left element of its inverse, the Green's function of the seed state $|\, f_0\rangle$ is obtained in the form of an infinite continued fraction:

$$G_{0,0}(E) = \langle f_0\,|\frac{1}{E-H}|\, f_0\rangle = \cfrac{1}{E - a_0 - \cfrac{b_1^2}{E - a_1 - \cfrac{b_2^2}{\cdots}}} \; . \tag{3.19}$$

The above result allows the evaluation of the diagonal Green's function matrix element $G_{0,0}(E)$, for any energy, by simply summing the continued fraction. In (3.19) the energy dependence is factorized from the definition of a_n and b_n, which only depend on the Hamiltonian H and the seed state $|\, f_0\rangle$.

Off-diagonal matrix elements of the type $G_{0,n}$ between $|\, f_0\rangle$ and $|\, f_n\rangle$ can always be evaluated by the recursion method, using a combination of diagonal matrix elements as follows:

$$
\begin{aligned}
G_{0,n} &= \langle f_0 |(E-H)^{-1}| f_n\rangle \\
&= \frac{1}{4}\langle f_0{+}f_n |(E-H)^{-1}| f_0{+}f_n\rangle - \frac{1}{4}\langle f_0{-}f_n |(E-H)^{-1}| f_0{-}f_n\rangle \\
&\quad -\frac{i}{4}\langle f_0{+}if_n |(E-H)^{-1}| f_0{+}if_n\rangle + \frac{i}{4}\langle f_0{-}if_n |(E-H)^{-1}| f_0{-}if_n\rangle \, .
\end{aligned}
$$

We do not dwell here on technical details of how to control and reproduce the asymptotic behavior of the tail of (3.19), mainly necessary in the presence of critical points and gaps in the energy spectrum (for this we suggest [9] and references therein).

3.4.1 Illustrative Examples of the Recursion Method

As a first application of the recursion method, we show how to obtain an effective linear chain for the polyacene polymer both in the trans- and cis-configurations (Fig. 3.6a and b, respectively).

(a)

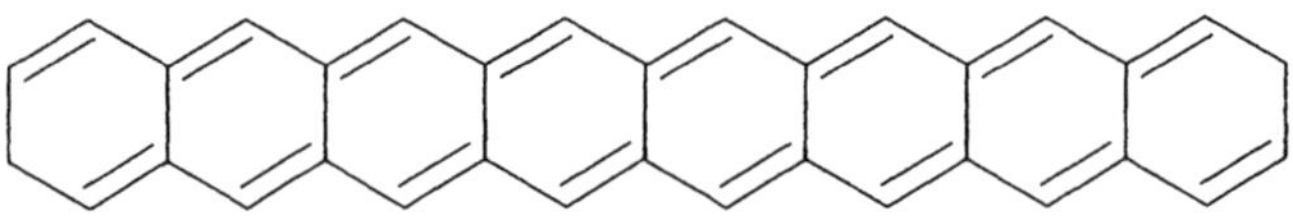

(b)

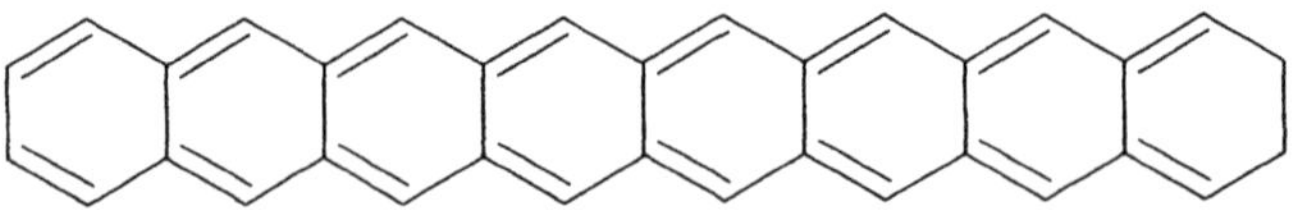

Fig. 3.6. Polyacene polymer: (**a**) trans- and (**b**) cis- configurations

Let us describe the details of the procedure for the trans- configuration (Fig. 3.7). Let E_0, E_1 and $E_{\tilde{0}}, E_{\tilde{1}}$ be the site energies of the p_z orbitals in the primitive cell of the polymer, at the positions 0 and 1 in the lower and upper chains respectively (see Fig. 3.7a). Let $t_{i,j}$ be the nearest-neighbor interactions on the same chain and $t_{i,\tilde{i}}$ the interaction energy between adjacent sites on different chains. Due to the symmetry of the problem, we can consider sites 0 and $\tilde{0}$ equivalent and set $E_{\tilde{0}} = E_0$, similarly for sites 1 and $\tilde{1}$ we set $E_{\tilde{1}} = E_1$. Moreover, we have $t_{0,1} = t_{\tilde{1},\tilde{2}} \equiv t'$, $t_{\tilde{0},\tilde{1}} = t_{1,2} \equiv t''$ and $t_{0,\tilde{0}} \equiv \tau$.

To generate an *infinite* periodic linear chain corresponding to the structure in Fig. 3.7a, we start writing the polymer Hamiltonian on the basis of the localized orbitals $|\, n\rangle$ and $|\, \tilde{n}\rangle$, belonging to the lower and the upper chains respectively, in the form $H = H_{\rm R} + H_{\rm L} + H_{\rm C}$, where we have separated the Hamiltonians at the left and at the right of the "central" states $|\, 0\rangle$ and $|\, \tilde{0}\rangle$

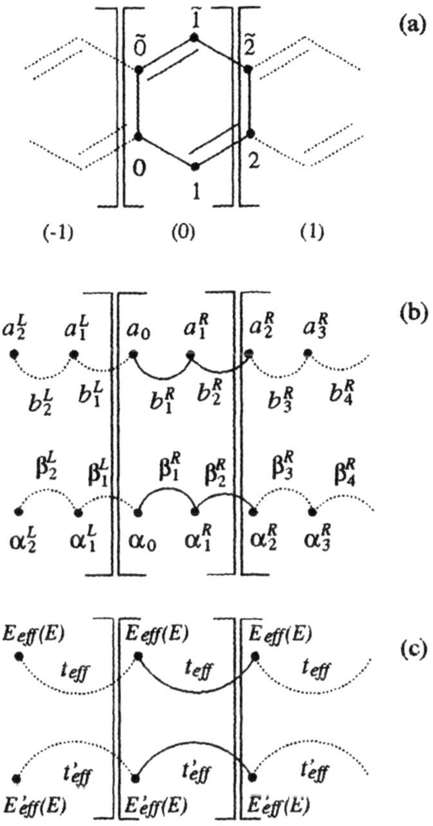

Fig. 3.7. (a) Polymeric chain for polyacene in the trans- configuration. A single unit cell is shown in detail. (b) Exact mapping of the crystal into two non-interacting equivalent linear chains. (c) Exact mapping of the two linear chains of the system in (b) into two equivalent chains with a single atom per cell

as follows:

$$H_{\rm C} = \langle 0\,|H|\,0\rangle|\,0\rangle\langle 0\,| + \langle\tilde{0}\,|H|\,\tilde{0}\rangle|\,\tilde{0}\rangle\langle\tilde{0}\,| + \langle 0\,|H|\,\tilde{0}\rangle|\,\tilde{0}\rangle\langle 0\,| + \langle\tilde{0}\,|H|\,0\rangle|\,0\rangle\langle\tilde{0}\,| \,,$$

$$H_{\rm L} = \sum_{n,m\le 0} \langle n\,|H|\,m\rangle|\,m\rangle\langle n\,| + \langle\tilde{n}\,|H|\,\tilde{m}\rangle|\,\tilde{m}\rangle\langle\tilde{n}\,| + \langle\tilde{n}\,|H|\,n\rangle|\,n\rangle\langle\tilde{n}\,| + \langle n\,|H|\,\tilde{n}\rangle|\,\tilde{n}\rangle\langle n\,| - H_{\rm C} \,,$$

$$H_{\rm R} = \sum_{n,m\ge 0} \langle n\,|H|\,m\rangle|\,m\rangle\langle n\,| + \langle\tilde{n}\,|H|\,\tilde{m}\rangle|\,\tilde{m}\rangle\langle\tilde{n}\,| + \langle\tilde{n}\,|H|\,n\rangle|\,n\rangle\langle\tilde{n}\,| + \langle n\,|H|\,\tilde{n}\rangle|\,\tilde{n}\rangle\langle n\,| - H_{\rm C} \,.$$

We now perform separate iterations of (3.17) operating with the Hamiltonian H on the states at the left and at the right of the "symmetric" and "antisymmetric" seed states $|\,0^{\rm S}\rangle = \frac{1}{\sqrt{2}}(|\,0\rangle + |\,\tilde{0}\rangle)$ and $|\,0^{\rm A}\rangle = \frac{1}{\sqrt{2}}(|\,0\rangle - |\,\tilde{0}\rangle)$. The result of this procedure is to map the original crystal into two *non-interacting* linear chains, as shown in Fig. 3.7b.

Starting from the state $|0^S\rangle$ and the Hamiltonian H, we can now generate a new orthonormal basis $\{\phi_n^L, \phi_0, \phi_n^R\}$ from the three-term recursion relation:

$$|\Phi_n^{L,R}\rangle = b_n^{L,R}|\phi_n^{L,R}\rangle$$
$$= H_{L,R}|\phi_{n-1}^{L,R}\rangle - a_{n-1}^{L,R}|\phi_{n-1}^{L,R}\rangle - b_{n-1}^{L,R}|\phi_{n-2}^{L,R}\rangle \quad n>0\,, \tag{3.20}$$

with

$$|\phi_0^L\rangle = |\phi_0^R\rangle = |\phi_0\rangle = |0^S\rangle \qquad \text{and} \qquad |\phi_{-1}^L\rangle = |\phi_{-1}^R\rangle = 0\,.$$

The same reasoning can be used in starting from $|0^A\rangle$ and generating the corresponding basis $\{\chi_n^L, \chi_0, \chi_n^R\}$.

The coefficients $a_n^{L,R}$ and $b_n^{L,R}$ can be expressed in terms of the energies E_n, $t_{i,j}$ and $t_{i,\tilde{i}}$. Equivalent expressions are written for the states χ_n and for the corresponding recursion coefficients $\alpha_n^{L,R}$ and $\beta_n^{L,R}$. By construction we have

$$a_0^{L,R} = 0, \quad b_0^{L,R} = 1 \qquad \text{and} \qquad \alpha_0^{L,R} = 0, \quad \beta_0^{L,R} = 1\,;$$

moreover, we define $a_0 = \langle 0^S|H_C|0^S\rangle$ and $\alpha_0 = \langle 0^A|H_C|0^A\rangle$. For $n \geq 1$, we exploit (3.17) and obtain the explicit expression of the coefficients a_n, b_n and α_n, β_n reported in Table 3.1.

It can be verified that in the case of the cis- configuration, setting $t_{0,1} = t_{\tilde{0},\tilde{1}} \equiv t'$, $t_{\tilde{1},\tilde{2}} = t_{1,2} \equiv t''$ and $t_{0,\tilde{0}} \equiv \tau$, following the same procedure as before, we map the polymer into two non-interacting linear chains, as shown in Fig. 3.7b. In this case the coefficients a_n, b_n and α_n, β_n are as reported in Table 3.2.

Table 3.1. Site and hopping energies for the two decoupled chains representing the trans-polyacene polymer

$a_0 = E_0 + \tau$		$\alpha_0 = E_0 - \tau$	
$a_1^{L,R} = E_1$	$b_1^{L,R} = \sqrt{\dfrac{t'^2 + t''^2}{2}}$	$\alpha_1^{L,R} = E_1$	$\beta_1^{L,R} = \sqrt{\dfrac{t'^2 + t''^2}{2}}$
$a_2^{L,R} = E_0 + \tau$	$b_2^{L,R} = \dfrac{\sqrt{2}\,t't''}{\sqrt{t'^2 + t''^2}}$	$\alpha_2^{L,R} = E_0 - \tau$	$\beta_2^{L,R} = \dfrac{\sqrt{2}\,t't''}{\sqrt{t'^2 + t''^2}}$
$a_3^{L,R} = E_1$	$b_3^{L,R} = \sqrt{\dfrac{t'^2 + t''}{2}}$	$\alpha_3^{L,R} = E_1$	$\beta_3^{L,R} = \sqrt{\dfrac{t'^2 + t''^2}{2}}$
$\vdots$	$\vdots$	$\vdots$	$\vdots$

Table 3.2. Site and hopping energies for the two decoupled chains representing the cis-polyacene polymer

$a_0 = E_0 + \tau$		$\alpha_0 = E_0 - \tau$	
$a_1^{\mathrm{L,R}} = E_1$	$b_1^{\mathrm{L,R}} = t''$	$\alpha_1^{\mathrm{L,R}} = E_1$	$\beta_1^{\mathrm{L,R}} = t''$
$a_2^{\mathrm{L,R}} = E_0 + \tau$	$b_2^{\mathrm{L,R}} = t'$	$\alpha_2^{\mathrm{L,R}} = E_0 - \tau$	$\beta_2^{\mathrm{L,R}} = t$
$a_3^{\mathrm{L,R}} = E_1$	$b_3^{\mathrm{L,R}} = t''$	$\alpha_3^{\mathrm{L,R}} = E_1$	$\beta_3^{\mathrm{L,R}} = t''$
$\vdots$	$\vdots$	$\vdots$	$\vdots$

We can then simplify the two orbitals per cell independent linear chains shown in Fig. 3.7b by decimating one of the two sites in each cell. With the structures so obtained (one effective orbital per cell; see Fig. 3.7c), we can calculate the band structure of polyacene. One can see also that E_{eff}, E'_{eff}, t_{eff} and t'_{eff} have the same analytical expressions both in the trans- and in the cis-configurations, that is, when only nearest-neighbor interactions are included, the trans- and the cis-polyacene polymers are energetically equivalent.

As a further useful example we can apply the recursion method to the case of the leucoemeraldine base polymer (see Fig. 3.8a) and reduce it to a linear infinite chain with a single site plus a non-interacting dimer per cell.

Let $E_0 \dots E_6$ be the site energies of the p_z orbitals in the positions $0 \dots 6$ (see Fig. 3.8a) and $t_{i,j}$ the nearest-neighbor interactions. Due to the symmetry of the aromatic rings, we can consider sites 1 and 4 to be equivalent (paracarbons), and sites 2, 3, 5 and 6 to be equivalent (orthocarbons). Site 0 is occupied by the nitrogen p_z orbital.

To generate an infinite periodic linear chain corresponding to the structure represented in Fig. 3.8a, we chose a seed state $|\,0\rangle$ (for instance, on the site occupied by the nitrogen orbital) and separated the Hamiltonian at the left and at the right of the "central" state $|\,0\rangle$, as has been shown in the case of polyacene.

The result of this procedure is to map the original crystal into a structure made by a linear chain and a couple of orbitals (generated by p_z orthocarbons) non-interacting with the chain, as shown in Fig. 3.8b. The details of the procedure are reported in [5].

The linear cells shown in Figs. 3.7b and 3.8b are particularly convenient for the determination of the density of states projected on each site of the cell (see Sect. 3.5). In fact, one can now choose any "reference site" of the cell and decimate all the other sites of the same cell, so to arrive at a single effective orbital per cell Hamiltonian as in Figs. 3.7c and 3.8c. The final expressions for the effective site energies $E_{\text{eff}}(E)$ and $E'_{\text{eff}}(E)$ and the effective hopping interactions $t_{\text{eff}}(E)$ and $t'_{\text{eff}}(E)$ are analytical and are given in terms of the original energies E_i and $t_{i,j}$.

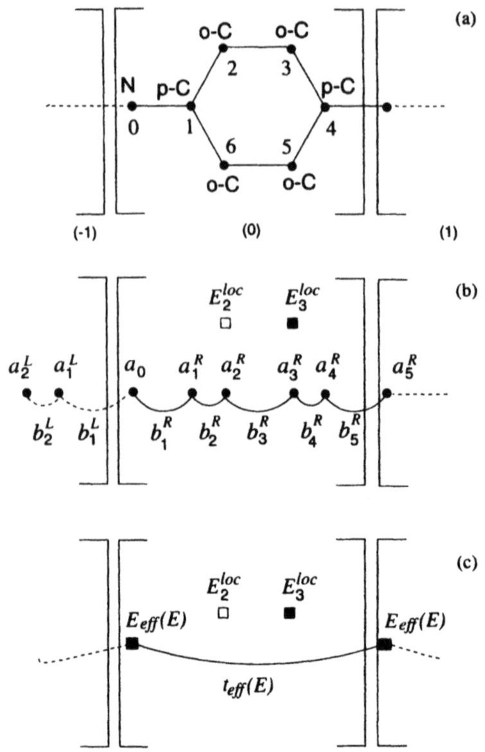

Fig. 3.8. (**a**) Polymeric chain for the leucoemeraldine base polymer. A single unit cell is shown in detail. (**b**) Exact mapping of the crystal into an equivalent linear chain and a dimer in every cell. (**c**) Exact mapping of the linear chain of the system in (**b**) into an equivalent single atom per cell chain

3.5 Green's Function and Density of States

The general definition of the Green's function diagonal matrix element $G_{0,0} = \langle f_0 | G | f_0 \rangle$, in terms of the eigenstates ψ_α and eigenvalues E_α of the Hamiltonian H, is (see, for example, [10])

$$G_{0,0}(E + i\varepsilon) = \sum_\alpha \frac{1}{E + i\varepsilon - E_\alpha} |\langle \psi_\alpha \, | \, f_0 \rangle|^2 \ , \tag{3.21}$$

where a positive, infinitesimal imaginary quantity $i\varepsilon$ is added to the energy E so that $G_{0,0}(E)$ is analytic and $\mathrm{Im}G_{0,0}(E) < 0$. If we consider the imaginary part of (3.21), keep the limit $\varepsilon \to 0^+$ and remember that

$$\lim_{\varepsilon \to 0^+} \frac{1}{\pi} \frac{\varepsilon}{(E - E_\alpha)^2 + \varepsilon^2} = \delta(E - E_\alpha) \ ,$$

we obtain the relation

$$n_0(E) = -\frac{1}{\pi} \lim_{\varepsilon \to 0^+} \mathrm{Im}G_{0,0}(E + i\varepsilon) = \sum_\alpha |\langle \psi_\alpha \, | \, f_0 \rangle|^2 \delta(E - E_\alpha) \ . \tag{3.22}$$

The quantity $n_0(E)$ is the density of states *projected* on the seed state $|f_0\rangle$; its meaning is evident from (3.22): it counts eigenstates of the system per unit energy, at the energy E, having non-vanishing overlap with the state $|f_0\rangle$. If we consider the contribution from all the states of the system (in a periodic structure it suffices to consider only the contribution from the states in the primitive cell), the total density of states $N(E)$ is recovered:

$$N(E) = \sum_i n_i(E) = -\frac{1}{\pi} \lim_{\varepsilon \to 0^+} \operatorname{Im} \operatorname{Tr} G(E + i\varepsilon) \ . \tag{3.23}$$

The advantage of the continued fraction expression (3.19) with respect to the form of (3.21) is evident, in fact, for any given energy E, in (3.19) we only need to know the coefficients a_n and b_n to sum the continued fraction. For infinite continued fractions this can be done by suitable extrapolation of the asymptotic coefficients and by introducing appropriate terminators, or, if a truncated fraction is adopted, by generating a Lorentzian broadening of each delta contribution, assigning a positive imaginary contribution to the energy E [2,3,11,12].

As in the case of polymers, one often has to evaluate the total density of states of strands of chains represented by an Hamiltonian matrix of the form

$$H = \begin{pmatrix} E_1 & t_{1,2} & & & \\ t_{2,1} & E_2 & t_{2,3} & & \\ & t_{3,2} & E_3 & t_{3,4} & \\ & & \ddots & \ddots & \ddots \\ & & & t_{N,N-1} & E_N \end{pmatrix} .$$

In this case one can see that

$$\begin{aligned} G_{1,N}(E) &= \langle 1 | \frac{1}{E-H} | N \rangle = (-1)^{N+1} \frac{\det(E-H)_{N,1}}{\det(E-H)} \\ &= \frac{\prod_{i=1}^{N-1} t_{i,i+1}}{\prod_{i=1}^{N} (E-\lambda_i)} , \end{aligned} \tag{3.24}$$

where λ_i are the eigenvalues of the Hamiltonian. If we consider that

$$\frac{\partial \ln G_{1,N}(E)}{\partial E} = \sum_i \frac{-1}{E-\lambda_i} = -\operatorname{Tr} G(E) ,$$

from (3.23) one thus obtains that $N(E)$ can be expressed also in terms of the single matrix element $G_{1,N}(E)$ by the relation [13]

$$N(E) = \frac{1}{\pi} \operatorname{Im} \frac{\partial}{\partial E} \ln G_{1,N}(E) . \tag{3.25}$$

The above expression can be generalized to obtain the density of states of systems represented by several interacting chains [14]. In the case of m chains of length N, the density of states $N(E)$ is given by

$$N(E) = \frac{1}{\pi} \operatorname{Im} \left(\frac{\partial}{\partial E} \ln \det [G_{1,N}(E)] \right) , \tag{3.26}$$

where the $m \times m$ matrix $G_{1N}(E)$ is defined only in terms of the first and the last sites of the m chains. Equation (3.26) can be extended to deal with systems described by energy-dependent effective Hamiltonians, provided the following auxiliary Green's function is defined:

$$G(\delta, E) = \frac{1}{E + \delta - H_{\text{eff}}(E)}$$

where δ is a real, positive quantity. In this case, in fact, one can obtain [15]

$$N(E) = \frac{1}{\pi} \operatorname{Im} \left(\frac{\partial}{\partial \delta} \ln \det [G_{1,N}(\delta, E)] \right)_{\delta=0} . \tag{3.27}$$

Finally, it is worth noting that (3.25), (3.26) and (3.27) for the density of states can be also written in the space domain to obtain the space density $n(x) = \sum_i \delta(x - x_i)$, whenever an m-diagonal matrix representation of the position operator can be given [16]; a peculiar example is that of the harmonic oscillator, where the position operator has a tridiagonal matrix representation on the eigenfunctions of the Hamiltonian.

3.5.1 Illustrative Example: Green's Function for the Infinite, Constant Linear Chain

The effective Hamiltonians represented in Figs. 3.2b, 3.5b, 3.7c and 3.8c are examples of periodic, infinite, constant chains. For their treatment it is useful to have explicit expressions of the matrix elements of the Green's function. We start with the semi-infinite chain and then pass to the infinite one.

Semi-infinite, Constant, Linear Chain

Consider an Hamiltonian of the form given in (3.6), for a tight-binding crystal with one orbital per site, with constant site energies $E_m = E_0$, and with constant nearest-neighbor interactions t as shown in Fig. 3.9.

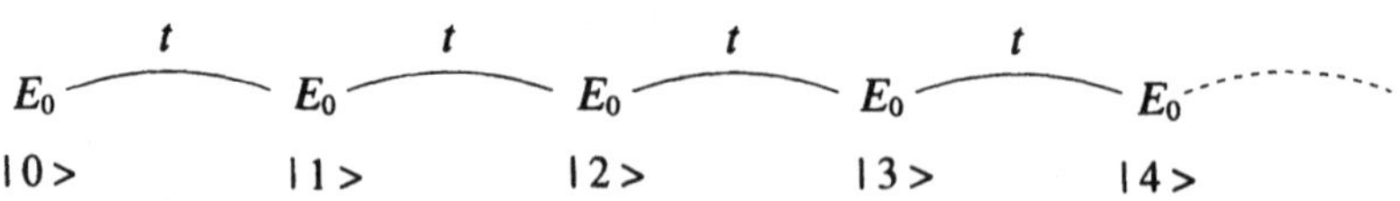

Fig. 3.9. Semi-infinite, constant, linear chain

The above system is represented by a tridiagonal Hamiltonian; thus, for the Green's function $g(E)$ of the semi-infinite chain evaluated at the site at the origin, it holds that

$$g_{0,0}(E) = \langle 0 | g(E) | 0 \rangle = \cfrac{1}{E - E_0 - \cfrac{t^2}{E - E_0 - \cfrac{t^2}{\cdots}}} = \frac{1}{E - E_0 - T(E)} ,$$

where the "terminator" $T(E) = t^2/[E - E_0 - T(E)]$ exactly sums the asymptotic tail of the infinite continued fraction.

To guarantee that $\mathrm{Im} g_{0,0}(E + i\varepsilon) < 0$ for $\varepsilon \to 0^+$, we have to choose for the terminator the expression

$$T(E) = \frac{1}{2}\left((E - E_0) - \sqrt{(E - E_0)^2 - 4t^2} \right) . \tag{3.28}$$

By introducing the variable $\tilde{E} = \dfrac{E - E_0}{2|t|}$, we obtain

$$g_{0,0}(E) = \frac{1}{|t|}\left(\tilde{E} - \sqrt{\tilde{E}^2 - 1}\right) = \begin{cases} \frac{1}{|t|}\left(\tilde{E} - \sqrt{|\tilde{E}^2 - 1|}\right) & \text{if } \tilde{E} > 1 \\ \frac{1}{|t|}\left(\tilde{E} - i\sqrt{|\tilde{E}^2 - 1|}\right) & \text{if } -1 < \tilde{E} < 1 \\ \frac{1}{|t|}\left(\tilde{E} + \sqrt{|\tilde{E}^2 - 1|}\right) & \text{if } \tilde{E} < -1 . \end{cases} \tag{3.29}$$

Infinite, Constant Linear Chain

Let us consider the chain in Fig. 3.10. By the recursion method one can show that for the above system the diagonal elements of the Green's function are given by

$$G_{n,n}(E) = G_{0,0}(E) = \frac{1}{E - E_0 - 2t^2 g_{0,0}(E)} , \tag{3.30}$$

where $g_{0,0}(E)$ is the Green's function of the semi-infinite chain. From (3.29) and (3.30) we obtain

$$G_{0,0}(E) = \frac{1}{2|t|\sqrt{\tilde{E}^2 - 1}} , \tag{3.31}$$

and by exploiting (3.22) we obtain the following for the local density of states:

$$n_0(E) = \frac{1}{\pi}\frac{\Theta(1 - \tilde{E}^2)}{2|t|\sqrt{\tilde{E}^2 - 1}} , \tag{3.32}$$

where $\Theta(x)$ is the step function $[\Theta(x) = 1$ for $x \geq 0$, $\Theta(x) = 0$ for $x < 0]$. To obtain the off-diagonal matrix elements of the Green's function, it is convenient to write the Hamiltonian as $H = H^0 + W$, where $H^0 = H_{\rm L} + H_{\rm R}$ is made by the left and right semi-infinite chains, as shown in Fig. 3.10, and $W = t(|\,0\rangle\langle 1\,| + |\,1\rangle\langle 0\,|)$ is their connection.

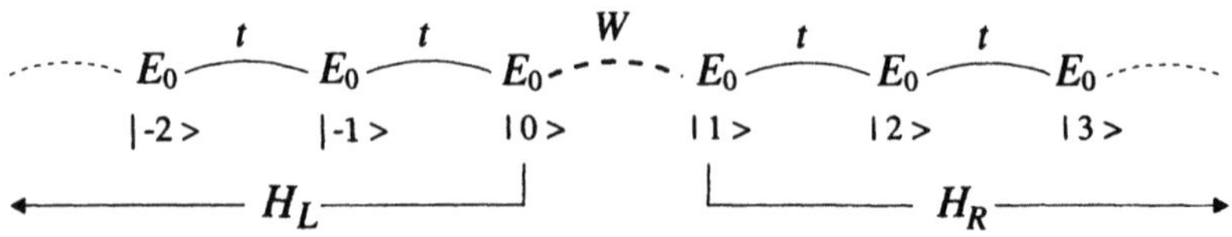

Fig. 3.10. Infinite chain seen as two semi-infinite chains, left ($H_{\rm L}$) and right ($H_{\rm R}$), connected by the interaction $W = t$

Let G and G^0 be the Green's functions relative to H and H^0 respectively. We can write G^0 in terms of the left and right semi-infinite chains Green's functions: $G^0 = g^{(\rm L)} + g^{(\rm R)}$. Then, we exploit the Dyson equation $G = G^0 + G^0WG$ and evaluate the term $G_{0,n}(E) = \langle 0\,|G(E)|\,n\rangle$; considering that by translation symmetry $G_{1,n}(E) = G_{0,n-1}(E)$, one obtains

$$G_{0,n}(E) = [g_{0,0}(E)t]^{|n|}G_{0,0}(E) = \frac{1}{2|t|}\left(\frac{t}{|t|}\right)^n \frac{\left(\tilde{E} - \sqrt{\tilde{E}^2 - 1}\right)^{|n|}}{\sqrt{\tilde{E}^2 - 1}} \,. \quad (3.33)$$

Relations (3.32) and (3.33) will prove useful in the treatment of the transmission properties of the polymeric effective chains.

3.6 Effective Tight-Binding Hamiltonians and Electronic States for Conjugated Polymers

Effective Hamiltonians such as those corresponding to the structures in Figs. 3.2b, 3.5b, 3.7c and 3.8c are diagonalized in a Bloch representation, and the eigenvalues are given by

$$E(k) = E_{\rm eff}(E) + 2t_{\rm eff}(E)\cos ka \;, \quad (3.34)$$

where a is the lattice parameter of the linear chain, and $k \in [-\pi/a, \pi/a)$. In spite of its simplicity, by varying the wavenumber k, (3.34) provides the band structure of the polymer; at a generic (low-symmetry) k-point we expect to obtain band structures with a number of bands equal to the number of orbitals in the original primitive cells represented in Figs. 3.2a, 3.5a, 3.7a and 3.8a respectively. By counting the electrons per orbital in the cells, we can thus define the number of valence and conduction bands (at temperature $T = 0$), filling each band at a given k with two electrons of opposite spin.

These one-dimensional effective chains can be exploited to fit electronic structures obtained by more sophisticated calculations and experimental results with the original three-dimensional systems. In fact, the definitions of the quantities $E_{\text{eff}}(E)$ and $t_{\text{eff}}(E)$ [see for instance (3.9)] are given in terms of the site energies E_n and hopping interactions $t_{n,m}$ of the original crystal. Interchain interactions and disorder effects can be treated separately.

Once the quantities E_{eff} and t_{eff} for the polymer under study are determined, capturing the essential physical aspects of the electronic structure of the conjugated polymer, we feel confident that they can also be used to evaluate its transmission properties.

We have now a simple and efficient way to generate band structures from the original parameters E_n and $t_{n,m}$, which enter into the definition of the effective quantities of (3.34). In particular, we have precise control over the role of each parameter in the final band structure.

To illustrate a product of this procedure, we show in Fig. 3.11 the band structure generated from π orbitals of carbon and sulfur atoms in polythiophene (PT).

By varying the parameters E_1, E_2, E_5, t, t', t'' and τ independently (see Fig. 3.3), one can follow the modifications of the band structure. For instance, the fundamental gap E_G decreases as $|E_5|$, $|t''|$ and $|\tau|$ increase, while it increases as $|t|$ and $|t'|$ increase; the gap between the bands v_1 and v_2 (see Fig. 3.11) increases as E_1 and $|t''|$ increase; and the gap between the bands c_1 and c_2 increases as E_1 increases and $|t''|$ decreases. The global width of the valence bands v_2 and v_3 increases as $|\tau|$ increases, and it is essentially insensitive to the variation of the other parameters, which themselves control the separation between v_2 and v_3, i.e., the position of the flat section of v_2 and v_3. To reproduce the large dispersion of the bands v_2 and v_3 in PT a suitable balance of $|E_5|$, $|t''|$ and E_1 has to be found. In fact, as these parameters increase, the bandwidth of v_2 decreases, while the bandwidth of v_3 increases. The opposite effect on the bandwidth of v_1 and v_2 is given by $|t|$. At the end of this refinement process, a convenient choice of parameters is the one reported in Table 3.3; these parameters generate the band structures shown in Fig. 3.11.

Table 3.3. Tight-binding parameters of the polythiophene Hamiltonian

E_1	E_2	E_5	t	t'	t''	τ
−1 eV	0 eV	−3.8 eV	−3.5 eV	−2.5 eV	−2.8 eV	−3.2 eV

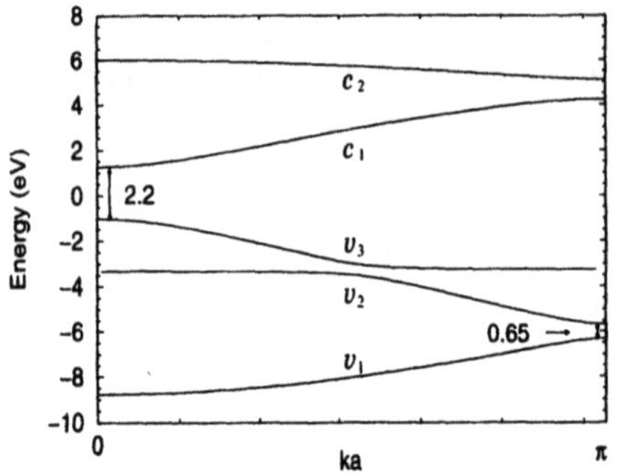

Fig. 3.11. Electronic π-band structure of polythiophene

Table 3.4 shows a numerical comparison between the band structure reported in Fig. 3.11, the bands reported in the literature [17] and the experimental value of the fundamental gap [18].

The orbital character of the states at different energies, obtained from the projected density of states, is shown in Fig. 3.12. We see, for instance, that the bands around the gap are largely made by the p_z carbon orbitals (Fig. 3.12a and 3.12b); the strong peak at about 4 eV below the Fermi level is made by p_z orbitals of the carbons at sites 2 and 3 (Fig. 3.12b) and the p_z sulfur orbitals (Fig. 3.12c); moreover, the high-dispersion parts of the second and the third valence bands correspond to the wavefunctions delocalized along the carbon backbone only, with essentially no weight on the sulfur atoms.

The above results agree with the results of Wu *et al.* [17], who reported the wavefunctions of polythiophene corresponding to the second and the third valence bands (v_2 and v_3), and confirm the nature and the delocalization of these wavefunctions.

Table 3.4. Comparison of (**a**) the band structure of polythiophene obtained in this work with (**b**) the valence band structure of [17] and ($\star$) the experimental optical transition ($v_3 \rightarrow c_1$) of ref. [18]

	Bandwidths					Transitions		
	v_1	v_2	v_3	c_1	c_2	$v_1 \rightarrow v_2$	$v_3 \rightarrow c_1$	$c_1 \rightarrow c_2$
(a)	2.4 eV	2.4 eV	2.2 eV	2.0 eV	1.0 eV	0.65 eV	2.2 eV	0.9 eV
(b)	2.6 eV	2.4 eV	2.1 eV			0.55 eV	2.1 eV*	

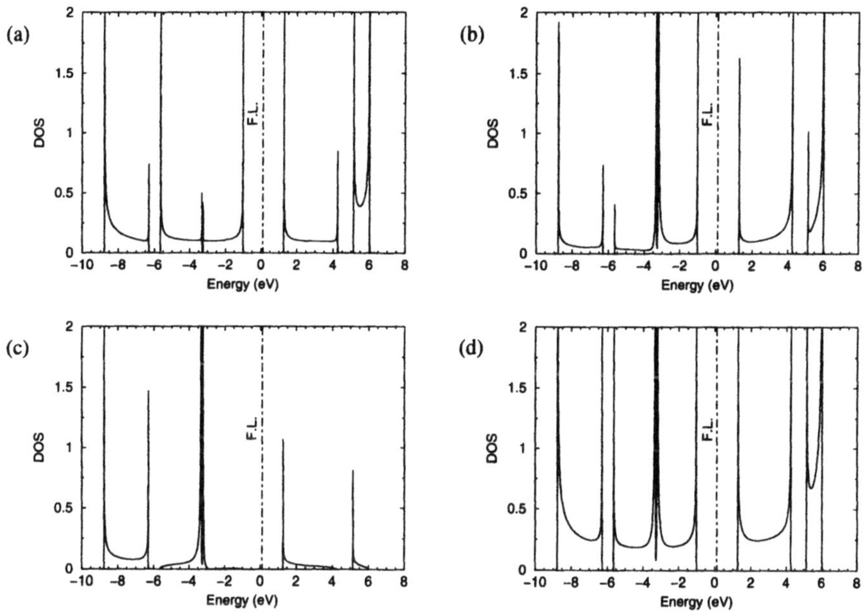

Fig. 3.12. Densities of states projected on different sites of polythiophene: (**a**) on the para-carbon atoms, (**b**) on the ortho-carbon atoms, (**c**) on the sulfur atom and (**d**) on the total cell. The Fermi level (F.L.) position is also shown

We shall see in the following sections that a reliable parameterization of the polymer Hamiltonian is very important for the study of its transport properties. For instance, to realize a good parameterization of emeraldine salt [5] in both the polaronic and bipolaronic configurations, one can first reproduce the band structure of the leucoemeraldine base, to obtain the parameterization of the benzenoid rings, then the band structure of the pernigraniline base (made up of 50% quinoid and 50% benzenoid rings), for the parameterization of the quinoid rings (see Fig. 3.13).

With the set of parameters given in Table 3.5, one can obtain the band structure of the leucoemeraldine base presented in Fig. 3.14, which reproduces the main theoretical [19–22] and experimental [23, 24] results.

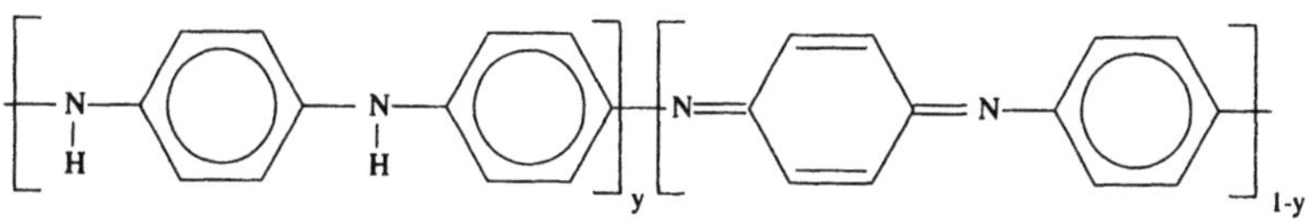

Fig. 3.13. Schematic of polyaniline base forms: $y = 1$, leucoemeraldine; $y = 0.5$, emeraldine; $y = 0$, pernigraniline

Table 3.5. Tight-binding parameters of leucoemeraldine base [5]

E_0	E_1	E_2	t'	t''	τ
−4.2 eV	1.1 eV	0 eV	−3.8 eV	−3.35 eV	−3.4 eV

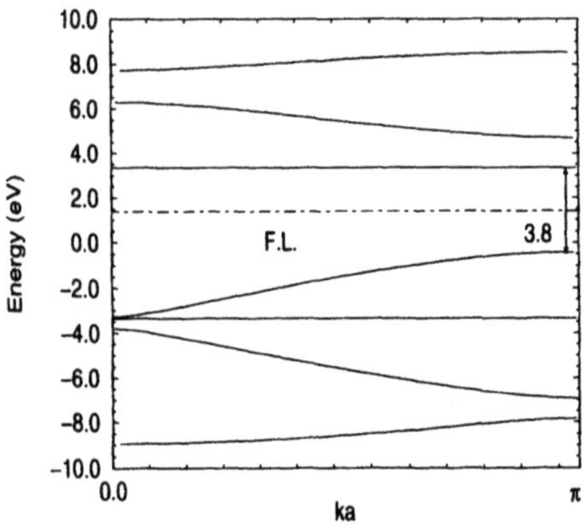

Fig. 3.14. Band structure of the leucoemeraldine base obtained with the set of parameters given in Table 3.5

The theoretical work on polyaniline since its early stages [25–27] has shown that with an oxidation process leucoemeraldine can be reduced to emeraldine base 25% of which is quinoid rings (see Fig. 3.13). By protonation of the imine atoms (the nitrogen atoms around the quinoid rings), bipolaron defects are produced, and an emeraldine salt in the bipolaronic lattice configuration is generated (Fig. 3.15a); the breaking of bipolaron defects leads to the polaronic lattice configuration (Fig. 3.15b). The polaronic lattice has an odd number of π electrons in the fundamental cell; thus, it is a metal with an half-occupied band (Fig. 3.16b). Details of the search for

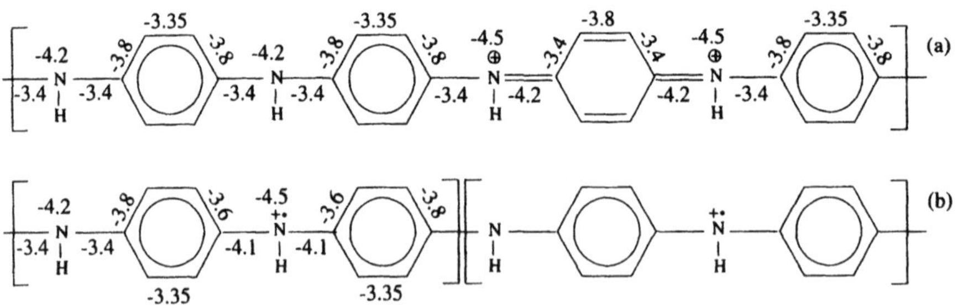

Fig. 3.15. Schematic of the unit cells and of the parameterization (in eV) of (**a**) emeraldine salt, bipolaronic lattice, and (**b**) emeraldine salt, polaronic lattice. The site energies of the orthocarbons are chosen as the reference zero of energy, and the site energies of the paracarbons are fixed at 1.1 eV

an optimal set of parameters, such as the ones reported in Fig. 3.15, giving the band structures of Fig. 3.16, can be found in [5]. Here we stress that for the bipolaronic lattice the band structure in Fig. 3.16a reproduces the known theoretical values [21, 28] of the fundamental gap ($E_G \approx 0.6\,\text{eV}$), of the gap between the second-to-last valence band and the conduction band ($\approx 2\,\text{eV}$) and of the gap between the third-to-last valence band and the conduction band ($\approx 2.5\,\text{eV}$).

The band structure in Fig. 3.16b for the polaronic lattice reproduces the experimental optical transitions at 1.5 eV and 2.9 eV well [23, 29–31] and is in general agreement with other theoretical calculations [21, 28].

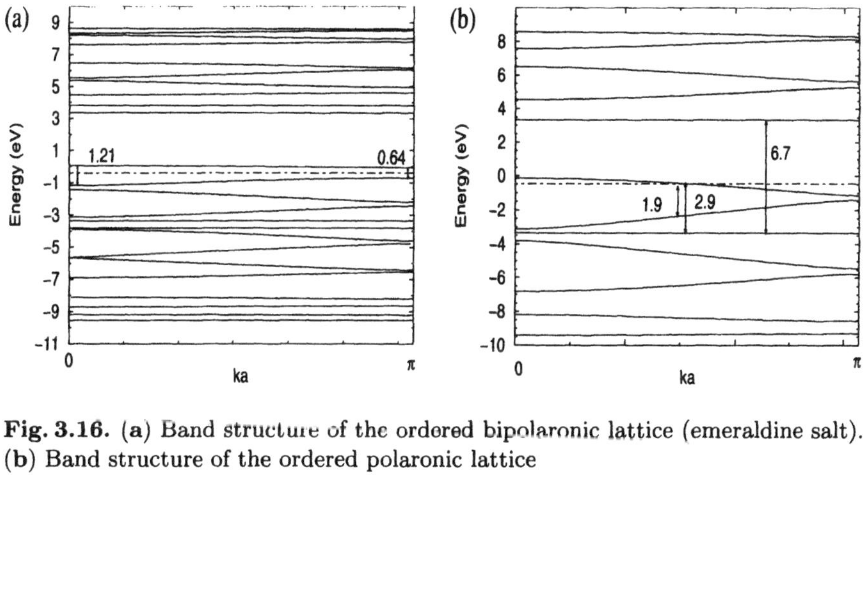

Fig. 3.16. (**a**) Band structure of the ordered bipolaronic lattice (emeraldine salt). (**b**) Band structure of the ordered polaronic lattice

3.7 Transmittivity of Polymeric Chains

Using basic elements of scattering theory, we present in this section how to calculate the transmission coefficient of single and double interacting chains without and with defects [6].

The scheme worked out here, together with the renormalization formalism, allows us to study realistic situations such as those that occur with the electronic transport in polymers.

3.7.1 Scattering by a Single Impurity Embedded in a Perfect Chain

Let us consider an infinite constant chain with all site energies $E_n = E_0$, except for the energy of the site at the origin, which has value $E_0 + \varepsilon$; let t be the constant nearest-neighbor interaction (Fig. 3.17).

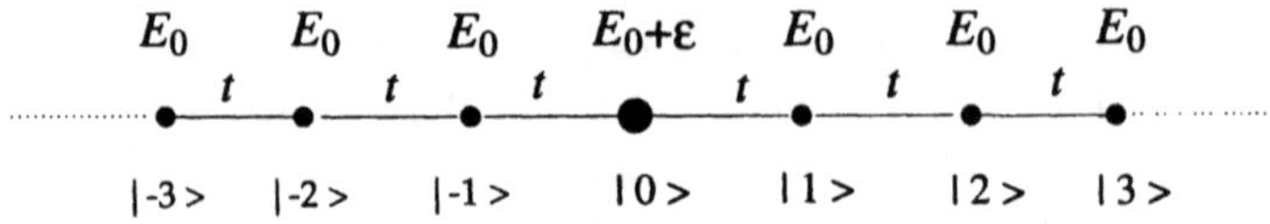

Fig. 3.17. Infinite linear chain with an impurity of energy $E_0+\varepsilon$ at the site $n=0$

The Hamiltonian H of this system can be written as

$$H = H_0 + H_{\mathrm{I}} \ ,$$

with H_0 being the unperturbed Hamiltonian of the crystal,

$$H_0 = \sum_m |\, m\rangle E_0 \langle m\,| + t \sum_m (|\, m\rangle\langle m+1\,| + |\, m+1\rangle\langle m|) \ ,$$

and H_{I} the impurity Hamiltonian,

$$H_{\mathrm{I}} = |\, 0\rangle\, \varepsilon \langle 0\,| \ .$$

The eigenfunctions $|\,\phi\rangle$ of H can be written in terms of the eigenfunctions $|\,k\rangle$ of H_0 as follows [10]:

$$|\,\phi\rangle = |\,k\rangle + G^0 H_{\mathrm{I}} \frac{1}{1 - G^0 H_{\mathrm{I}}} |\,k\rangle \ , \tag{3.35}$$

where G^0 is the Green's function of H_0. In the case of a single impurity at the site $n=0$, the general expression (3.35) projected on the generic site n becomes [10]

$$\langle n\,|\,\phi\rangle = \langle n\,|\,k\rangle + G^0_{n,0}\, \varepsilon \frac{1}{1 - G^0_{0,0}\, \varepsilon} \langle 0\,|\,k\rangle \ ,$$

where $G^0_{p,q} = \langle p\,|G^0(E)|\,q\rangle$ and $\langle m\,|\,k\rangle = \mathrm{e}^{ikma}$.

From (3.33), it is easy to demonstrate that, inside the energy band of the infinite perfect chain, $G^0_{n,0}$ can be written as follows

$$G^0_{n,0} = \begin{cases} G^0_{0,0}\, \mathrm{e}^{-ik|n|a} & \text{if } t > 0 \\ G^0_{0,0}\, \mathrm{e}^{ik|n|a} & \text{if } t < 0 \ . \end{cases} \tag{3.36}$$

Therefore, for $t<0$, we obtain

$$\langle n\,|\,\phi\rangle = \mathrm{e}^{ikna} + \frac{G^0_{0,0}\,\varepsilon}{1 - G^0_{0,0}\,\varepsilon}\, \mathrm{e}^{ik|n|a} \ ;$$

comparison with the expression defining the transmittance τ and the reflectance ρ,

$$\langle n\,|\,\phi\rangle = \begin{cases} \tau e^{ikna} & \text{if } n > 1 \\ e^{ikna} + \rho e^{-ikna} & \text{if } n < 0 \ , \end{cases}$$

gives

$$\tau = 1 + \frac{G^0_{0,0}\,\varepsilon}{1 - G^0_{0,0}\,\varepsilon} = \frac{1}{1 - G^0_{0,0}\,\varepsilon} \qquad \text{and} \qquad \rho = \frac{G^0_{0,0}\,\varepsilon}{1 - G^0_{0,0}\,\varepsilon}\,. \tag{3.37}$$

The transmittivity $\mathcal{T}$, in the case of a single chain, can thus be written [10] as

$$\mathcal{T} = |\,\tau\,|^2 = \left|\frac{1}{1 - G^0_{0,0}\,\varepsilon}\right|^2 ; \tag{3.38}$$

taking into account the explicit expression of $G^0_{0,0}$ (3.31), inside the band we have

$$\mathcal{T} = \frac{4t^2 - (E - E_0)^2}{\varepsilon^2 + (4t^2 - (E - E_0)^2)}\,, \tag{3.39}$$

which shows that, inside the band, the transmittivity is always less than 1.

3.7.2 Scattering by a Dimer Impurity Embedded in a Perfect Chain

Let us consider a couple of impurities with different energies (asymmetric dimer) embedded in a perfect periodic chain, as shown in Fig. 3.18. Let $E_0 + \varepsilon_0$ and $E_0 + \varepsilon_1$ be the dimer site energies and $t + \Delta t$ the interaction between the atoms of the dimer. E_0 defines the barycenter of the energy band and t the corresponding bandwidth of the leads connected to the dimer.

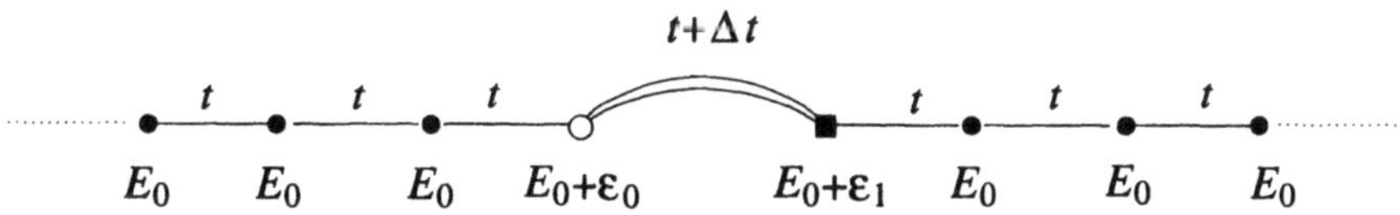

Fig. 3.18. Linear chain with a dimer impurity

The study of the scattering by a dimer impurity is particularly interesting because, by the renormalization procedure, a single strand of ordered or of disordered polymer can be always reduced to a dimer; in this case ε_0, ε_1 and Δt are functions of energy.

The system of Fig. 3.18 is described by the Hamiltonian

$$H = H_0 + H_{\mathrm{I}}\,,$$

where

$$H_0 = E_0 \sum_n |\,n\rangle\langle n\,| + t \sum_n (|\,n\rangle\langle n+1\,| + |\,n+1\rangle\langle n\,|)$$

and

$$H_{\mathrm{I}} = \varepsilon_0(|\,0\rangle\langle 0\,|) + \varepsilon_1(|\,1\rangle\langle 1\,|) + \Delta t(|\,0\rangle\langle 1\,| + |\,1\rangle\langle 0\,|)\,.$$

In this case the operator $T^+ = H_{\mathrm{I}} \cdot (1 - G^0 H_{\mathrm{I}})^{-1}$ and H_{I} are represented by 2×2 matrices which work in the space $\{|\,0\rangle, |\,1\rangle\}$. Projecting $|\,\phi\rangle$ on the generic site n, exploiting (3.36), gives

$$\langle n\,|\,\phi\rangle = \mathrm{e}^{ikna} + (G^0_{0,0}T^+_{0,0} + G^0_{1,0}T^+_{0,1} + G^0_{0,1}T^+_{1,0}\mathrm{e}^{-2ika} + G^0_{1,1}T^+_{1,1})\mathrm{e}^{ik|n|a}\,, \quad (3.40)$$

from which the following expressions for the transmittance and the reflectance are obtained:

$$\tau = 1 + G^0_{0,0}T^+_{0,0} + G^0_{1,0}T^+_{0,1} + G^0_{0,1}T^+_{1,0}\mathrm{e}^{-2ika} + G^0_{1,1}T^+_{1,1}\,, \quad (3.41)$$

and

$$\rho = G^0_{0,0}T^+_{0,0} + G^0_{1,0}T^+_{0,1} + G^0_{0,1}T^+_{1,0}\mathrm{e}^{-2ika} + G^0_{1,1}T^+_{1,1}\,. \quad (3.42)$$

(3.41) and (3.42) are consistent with the relations obtained with a different procedure in [32].

In the simple case of a symmetric dimer ($\varepsilon_1 = \varepsilon_0$) with the choice $\Delta t = 0$ and $E_0 = 0$, the transmission coefficient has the following expression:

$$\mathcal{T} = \frac{t^2(4t^2 - E^2)}{\varepsilon_0^2(E - \varepsilon_0)^2 + t^2(4t^2 - E^2)}\,. \quad (3.43)$$

From (3.43) we can see that $\mathcal{T}$ is equal to 1 for $E = \varepsilon_0$ if $|\,\varepsilon_0\,| < |\,2t\,|$.

3.7.3 Transmission Through Two Interacting Chains

Bands of a Couple of Interacting Chains

Let us consider the system represented in Fig. 3.19, made of two chains with site energies equal to E_0, hopping interaction t and interchain interaction $t_\perp$.

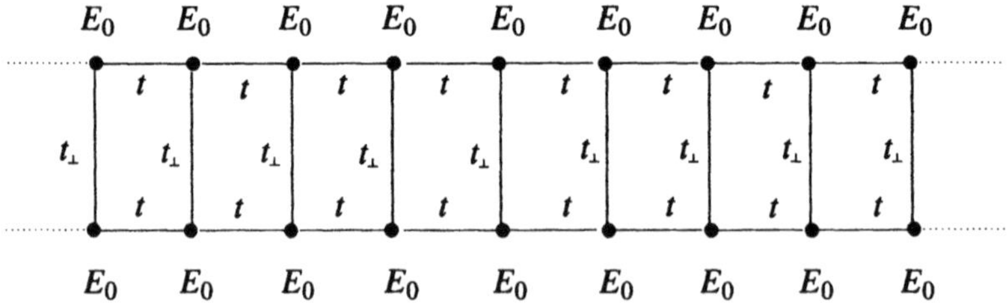

Fig. 3.19. Two identical linear chains; the interchain interaction is $t_\perp$, the intrachain interaction is t

Let $|n\rangle$ and $|\tilde{n}\rangle$ be the localized basis states of the lower and the upper chains respectively; on this basis the Hamiltonian of the system has the following form:

$$H_0 = \sum_n [E_0 \,|n\rangle\langle n| + t\,(|n\rangle\langle n+1| + |n+1\rangle\langle n|)] + \sum_{\tilde{n}} \Big[E_0 \,|\tilde{n}\rangle\langle\tilde{n}| \\ + t\left(|\tilde{n}\rangle\langle\widetilde{n+1}| + |\widetilde{n+1}\rangle\langle\tilde{n}|\right)\Big] + \sum_{n=\tilde{n}} t_\perp \,(|n\rangle\langle\tilde{n}| + |\tilde{n}\rangle\langle n|)\,. \qquad (3.44)$$

The two interacting chains can be decoupled if we choose the basis:

$$|n^{\mathrm{S}}\rangle = \frac{1}{\sqrt{2}}(|n\rangle + |\tilde{n}\rangle) \quad \text{and} \quad |n^{\mathrm{A}}\rangle = \frac{1}{\sqrt{2}}(|n\rangle - |\tilde{n}\rangle)\,. \qquad (3.45)$$

As shown in Fig. 3.20, on the basis $\{|n^{\mathrm{S}}\rangle, |n^{\mathrm{A}}\rangle\}$ the Hamiltonian H represents two new *non*-interacting linear chains; in fact, we have:

$$H_0 = H^{\mathrm{S}} + H^{\mathrm{A}},$$

where

$$H^{\mathrm{S}} = \sum_{n^{\mathrm{S}}} (E_0 + t_\perp)\,(|n^{\mathrm{S}}\rangle\langle n^{\mathrm{S}}|) + t\,(|n^{\mathrm{S}}\rangle\langle n+1^{\mathrm{S}}| + |n+1^{\mathrm{S}}\rangle\langle n^{\mathrm{S}}|)$$

and

$$H^{\mathrm{A}} = \sum_{n^{\mathrm{A}}} (E_0 - t_\perp)\,(|n^{\mathrm{A}}\rangle\langle n^{\mathrm{A}}|) + t\,(|n^{\mathrm{A}}\rangle\langle n+1^{\mathrm{A}}| + |n+1^{\mathrm{A}}\rangle\langle n^{\mathrm{A}}|)\ ;$$

H^{S} is the Hamiltonian of a single chain with site energies $E_0 + t_\perp$ and hopping interaction t; H^{A} is the Hamiltonian of a single chain with site energies $E_0 - t_\perp$ and hopping energies t. The dispersion relations for the two chains are

$$E = E_0 + t_\perp + 2t\cos k^{\mathrm{S}}a \quad \text{and} \quad E = E_0 - t_\perp + 2t\cos k^{\mathrm{A}}a$$

respectively (see Fig. 3.21), where a is the length of the unit cell.

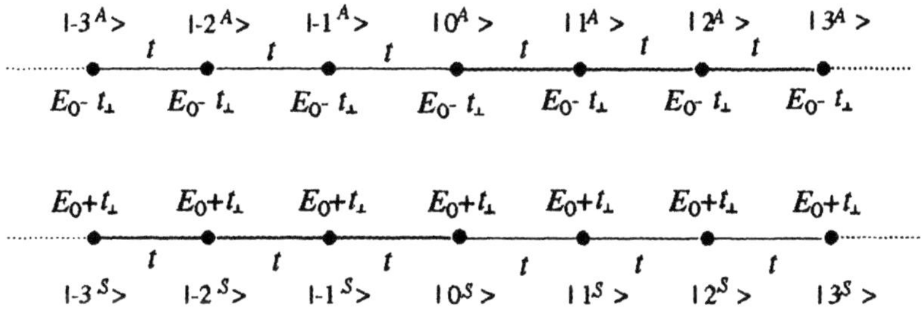

Fig. 3.20. On the "symmetric–antisymmetric" basis, the Hamiltonian H describes two non-interacting linear chains

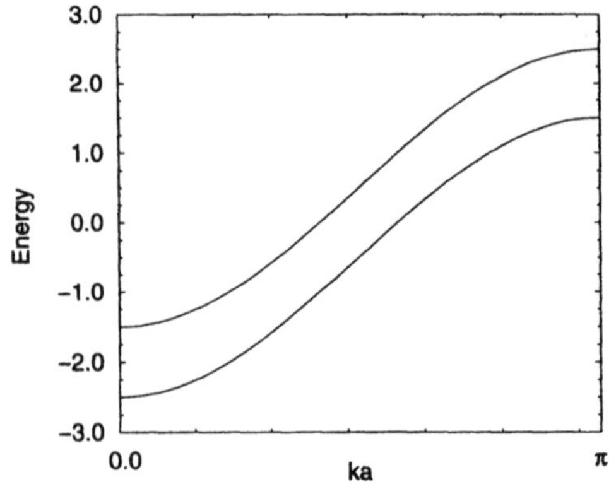

Fig. 3.21. Band structure for the system of Fig. 3.19, with $t = -1$, $t_\perp = -0.5$ and $E_0 = 0$

Electronic transport can take place through the "symmetric" chain for energy values inside the "symmetric" band, i.e.,

$$E_0 + t_\perp - 2|t| < E < E_0 + t_\perp + 2|t| \ ,$$

and also through the "antisymmetric" chain for energy values inside the "antisymmetric" band, i.e.,

$$E_0 - t_\perp - 2|t| < E < E_0 - t_\perp + 2|t| \ .$$

Scattering by a Single Impurity Embedded in Two Interacting Chains

Let us consider the system shown in Fig. 3.19 described by the Hamiltonian (3.44) and add an impurity term at the site $n = 0$ of the lowest chain:

$$H_\text{I} = \varepsilon | 0\rangle\langle 0 | \ .$$

Making the change of basis (3.45), one can realize that the symmetric and the antisymmetric chains interact only via the states $|0^\text{S}\rangle$ and $|0^\text{A}\rangle$, as shown in Fig. 3.22.

On the new basis, the Hamiltonian H_I can be written as

$$H_\text{I} = \frac{\varepsilon}{2}(|0^\text{S}\rangle\langle 0^\text{S}|) + \frac{\varepsilon}{2}(|0^\text{A}\rangle\langle 0^\text{A}|) + \frac{\varepsilon}{2}(|0^\text{S}\rangle\langle 0^\text{A}| + |0^\text{A}\rangle\langle 0^\text{S}|) \ .$$

Following the procedure of Sect. 3.7.1 and 3.7.2, one can calculate the projection of the outcoming wave on the symmetric and on the antisymmetric chains, for incoming waves $|k^\text{S}\rangle$ or $|k^\text{A}\rangle$.

So we obtain the following four terms for the transmittance:

$$\tau^\text{SS} = 1 + G^0_{0^\text{S},0^\text{S}} T^+_{0^\text{S},0^\text{S}} \ ,$$

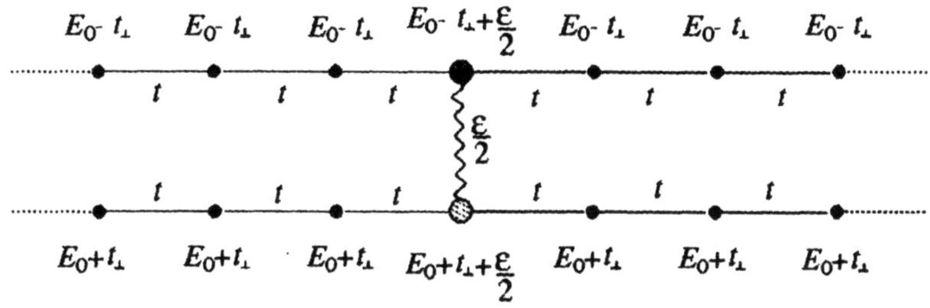

Fig. 3.22. On the "symmetric–antisymmetric" basis, the two chains interact at the sites where the impurity ε has been introduced

which is the transmittance of the incoming symmetric wave $|\,k^{\mathrm{S}}\rangle$ through the symmetric channel;

$$\tau^{\mathrm{SA}} = G^0_{0^{\mathrm{S}},0^{\mathrm{S}}} T^+_{0^{\mathrm{S}},0^{\mathrm{A}}} \ ,$$

which is the transmittance of the incoming symmetric wave $|\,k^{\mathrm{S}}\rangle$ through the antisymmetric channel;

$$\tau^{\mathrm{AS}} = G^0_{0^{\mathrm{A}},0^{\mathrm{A}}} T^+_{0^{\mathrm{A}},0^{\mathrm{S}}} \ ,$$

which is the transmittance of the incoming antisymmetric wave $|\,k^{\mathrm{A}}\rangle$ through the symmetric channel; and

$$\tau^{\mathrm{AA}} = 1 + G^0_{0^{\mathrm{A}},0^{\mathrm{A}}} T^+_{0^{\mathrm{A}},0^{\mathrm{A}}} \ ,$$

which is the transmittance of the incoming antisymmetric wave $|\,k^{\mathrm{A}}\rangle$ through the antisymmetric channel.

Finally, one has for the total transmissions of a symmetric ($\mathcal{T}^{\mathrm{S}}$) and of an antisymmetric ($\mathcal{T}^{\mathrm{A}}$) wave the expressions

$$\mathcal{T}^{\mathrm{S}} = \frac{v^{\mathrm{S}}|\,\tau^{\mathrm{SS}}\,|^2 + v^{\mathrm{A}}|\,\tau^{\mathrm{SA}}\,|^2}{v^{\mathrm{S}}} \quad \text{and} \quad \mathcal{T}^{\mathrm{A}} = \frac{v^{\mathrm{A}}|\,\tau^{\mathrm{AA}}\,|^2 + v^{\mathrm{S}}|\,\tau^{\mathrm{AS}}\,|^2}{v^{\mathrm{A}}} \ , \qquad (3.46)$$

where v^{S} and v^{A} are the electron velocities in the symmetric and antisymmetric chains.

As a test of the above procedure we reproduce the transmission property of a model system studied by Mizes and Conwell [33], made by a double chain with $t = t_\perp = -2.5$ eV and site energies equal to zero everywhere except for one site with energy ε. In Fig. 3.23 the total transmission of an incoming symmetric wave of the unperturbed system for different values of ε is reported. It can be seen that perfect transmission exists for $E = -2.5$ eV, for each value of ε, and that for negative values of ε there are energy values for which the transmission is zero, that is, the incoming wave is totally reflected.

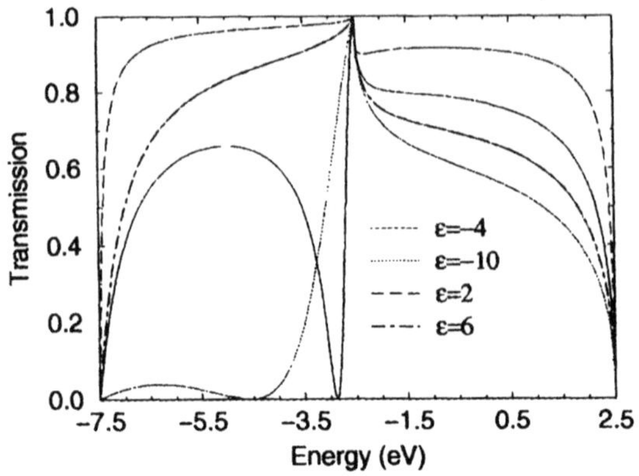

Fig. 3.23. Transmission from a single impurity in a double interacting chain

Scattering by a Couple of Dimers on Adjacent Interacting Chains

We focus now on a model system made of two different mutually interacting dimers on adjacent interacting chains (ladder polymer), as shown in Fig. 3.24. By suitable renormalization these polymers can be exactly reduced to a couple of interacting dimers. It can be verified that, in terms of the symmetric and antisymmetric states given in (3.45), the unperturbed part of the Hamiltonian is separated into two non-interacting linear chains, and that the Hamiltonian of the impurities in general cannot be decoupled, thus making the two linear chains interact, as shown in Fig. 3.25. One thus calculates [6] the transmittivity as in the case of a single defect, taking into account that in this case H_{I} and T^{+} are 4×4 matrices. As a simple test one can calculate the transmission probability through the symmetric channel for a double chain containing two identical perturbations; this system was studied in [34] using transfer matrix formalism. The unperturbed system is characterized by $E_0 = 0$ and $t = t_{\perp} = -1$; one chain contains two identical adjacent defect atoms with site energies equal to -1.2, the other chain contains, at the corresponding sites, two identical adjacent defects with site energies equal to -0.8. The dependence of $|\tau^{\mathrm{SS}}|$ on the energy, in units of $|\,t\,|$, is shown in Fig. 3.26; an identical curve was obtained in [34].

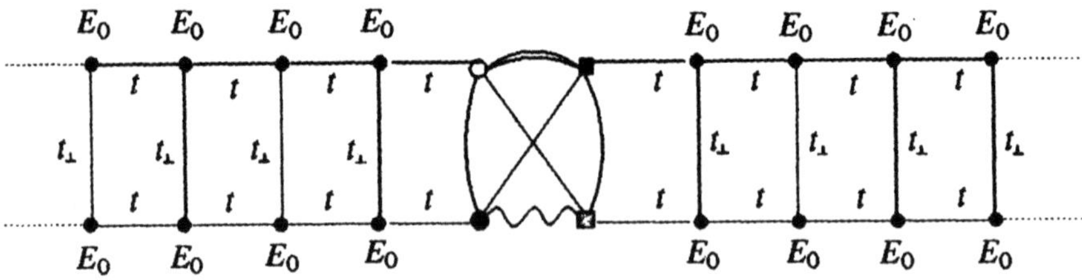

Fig. 3.24. A couple of interacting dimer impurities on adjacent interacting chains

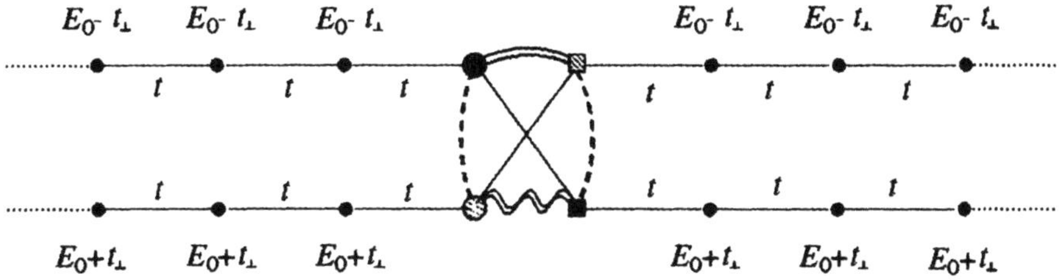

Fig. 3.25. The same system as in Fig. 3.24 with leads represented on the symmetric and antisymmetric states $|n^{\mathrm{S}}\rangle$ and $|n^{\mathrm{A}}\rangle$

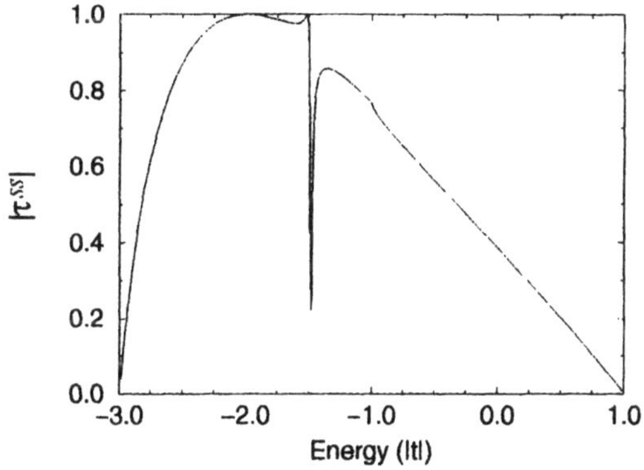

Fig. 3.26. Transmission probability in the symmetric channel for a double chain with $E_0 = 0$ and $t = t_\perp = -1$, containing two symmetric dimer impurities. One chain has identical adjacent defect atoms with site energies equal to -1.2, the other has the same, at corresponding sites, with site energies equal to -0.8

As an example of the application of the procedure outlined in this chapter in more realistic cases, we report in Figs. 3.27 and 3.28 the transmittivity of a single strand of emeraldine salt in ordered bipolaron and polaron structures respectively. The single polymeric strands have been reduced to single dimers by the renormalization procedure, and the 2×2 matrix representing the Hamiltonian H_{I} has energy-dependent elements.

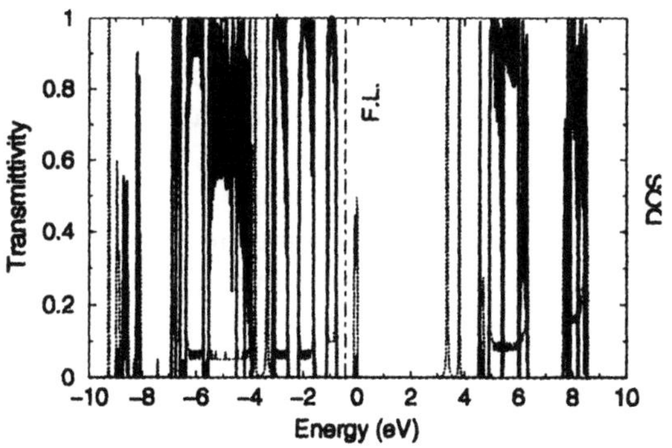

Fig. 3.27. Transmittivity (*solid line*) and DOS (*dotted line*, arbitrary units) of the ordered bipolaronic lattice

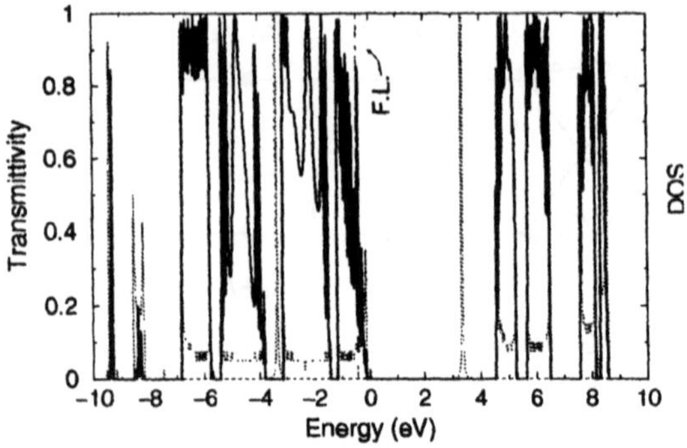

Fig. 3.28. Transmittivity (*solid line*) and DOS (*dotted line*, arbitrary units) of the ordered polaronic lattice. (From [6])

It is evident from Figs. 3.27 and 3.28, and also from the band structures in Fig. 3.16, that the ordered bipolaronic lattice is a semiconductor, while the ordered polaronic lattice is a metal. In fact, only in the polaronic lattice does the gap between valence and conduction bands close and the Fermi level penetrate into a region of highly transmitting states. The next step is to verify whether extra states induced by disorder and the effect of interchain interactions modify the above picture.

3.8 Effects of Disorder on the Electronic States

We have so far shown that the decimation–renormalization and recursion methods are efficient tools to describe band structures, densities of states and transmission of ideal, periodic polymeric chains. However, real polymers are often far from regular structures. Not only do the chains have finite length, but they are also affected by several kinds of disorder due to random distribution of the chain elements, impurities, chemical treatments, effect of environment, etc., so that they may often resemble amorphous systems. For this, models appropriate to amorphous semiconductors have often been introduced to explain the transport properties of polymers.

The interpretation of the nature of the metallic state of conjugated conducting polymers has focused on controversial issues. In fact, almost all of the systems have a Mott-type form for the conductivity, and debate has grown over their quasi-one-dimensional or three-dimensional nature. This is all connected with the complex morphology of conducting polymers.

A first step toward the interpretation of this phenomenon, is the study of the effect of the disorder which is realized by a random distribution of rings, or of defect atoms. Also in the presence of disorder an evaluation of the electronic states of the polymer can be made by appropriate extension of the recursive relations already presented for the periodic structures. In this case the accurate control of the extended and localized natures of the states and an evaluation of their localization length and transmittivity around the Fermi energy become essential in assessing their contribution to transport.

For this, one first maps the original random polymer into an effective one-dimensional lattice composed of randomly distributed units. Each of them is characterized by a typical internal structure which represents specific rings or atoms and interactions of the original polymer.

This stochastic distribution of well-defined units cannot be described as an ordinary one-dimensional random lattice; in fact, as shown below, it represents a disordered lattice with a short-range order. A well-known example of such lattices is the Random Dimer Model [35–37], which has been used to account for the high transmittivity of some families of conducting polymers [36].

To show how recursive methods can be employed to handle structurally disordered polymers, let us consider the case of emeraldine salt (Fig. 3.15) in the bipolaronic lattice (BL) and polaronic lattice (PL) configurations.

In the random BL, disorder is generated by the random distribution of protonated quinoid units, $N^+ = Q = N^+$, separated by benzenoid rings (the occurrence of adjacent protonated quinoid units is energetically unfavoured).

In the random PL, disorder is generated by the random distribution of protonated nitrogen atoms (with the constraint that configurations with two consecutive protonated nitrogens cannot occur).

From Fig. 3.15 one can see that the scheme shown in Fig. 3.29a is a convenient choice for the two fundamental tiles to generate ordered or disordered maps representing emeraldine salts.

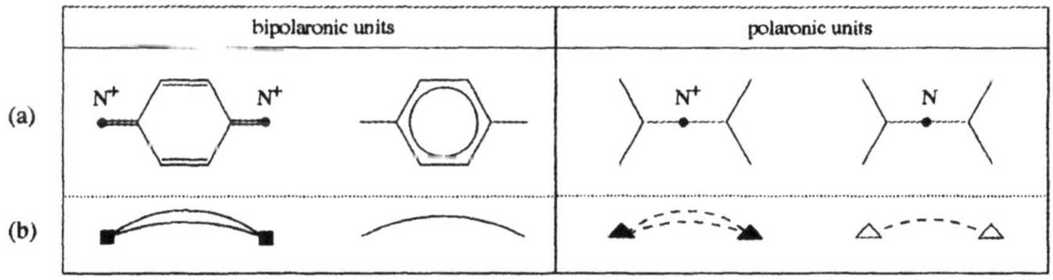

Fig. 3.29. (**a**) Elementary units of the bipolaronic and polaronic disordered lattices; (**b**) equivalent renormalized units

In the case of the BL, to reduce the two tiles shown in Fig. 3.29a to a minimal description, we apply the renormalization procedure with respect to the nitrogen atoms. The decimation of the paracarbons and the orthocarbons of the quinoid unit gives rise to a symmetric dimer, while the decimation of a complete benzenoid ring gives rise to an effective interaction (see Fig. 3.29b); thus, in correspondence with M consecutive nitrogen-connected benzenoid rings, we obtain $M - 1$ monomers.

In the case of the PL, we can first reduce the polymeric chain into a linear chain by the recursion method as shown in Section 3.4; then we can preserve the bonding combination of the orthocarbon orbitals of the two tiles and decimate the internal sites. We can thus see that the two tiles are equivalent to two symmetric dimers (Fig. 3.29b); moreover, two consecutive dimers cannot occur but they have to be intercalated by the orthocarbon–orthocarbon bonding.

From the form of the renormalized units (Fig. 3.29b), it is evident that the disordered BL and PL are exactly mapped into the random dimer model (RDM) and the random dimer alloy model (RDAM) respectively as shown in Fig. 3.30a and b (for the definition of these lattice models, see [38]).

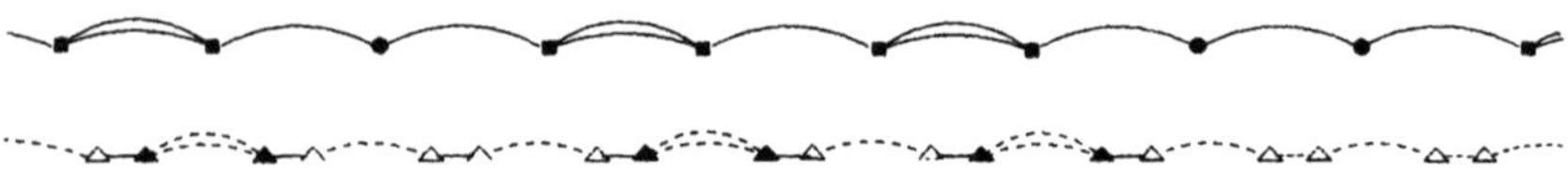

Fig. 3.30. (**a**) The random dimer model equivalent to the random bipolaronic lattice; (**b**) the random alloy dimer model equivalent to the random polaronic lattice

To add the effect of interchain interaction as well, we consider the map of two adjacent disordered fully protonated BLs mutually interacting when two imine–amine nitrogens have the same site index on the respective chains. This kind of interaction is supported by experimental considerations [39]. Fig. 3.31 shows the random-ladder polymer obtained by facing two single strands of disordered BLs together. The renormalized tiles of Fig. 3.29b preserve the nitrogen atoms and are thus ready to support the interaction between amine and imine nitrogens belonging to different chains. Other kinds of interactions between other elements of the chains can be introduced in the same way.

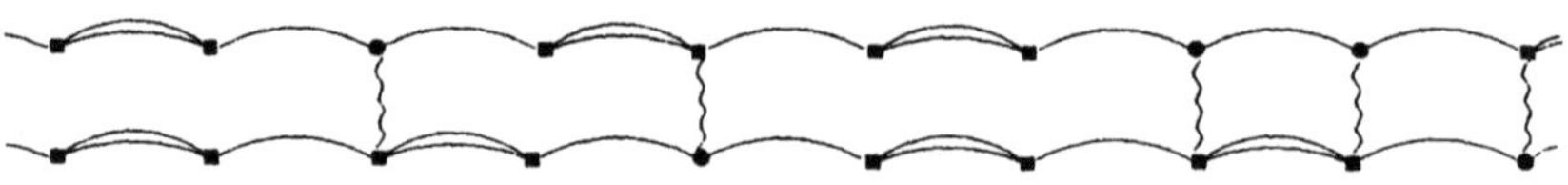

Fig. 3.31. The map of two interacting disordered fully protonated bipolaronic lattices

We can describe the systems in Fig. 3.31 by the site and hopping energies, which allow us to reproduce the correct band structures of the corresponding ordered crystals (see Sect. 3.6). In fact, one can once again exploit the renormalization procedure with respect to the surface sites, and thereby reduce the polymeric chain (or the two interacting polymeric chains) to a single dimer (or to a couple of interacting dimers) and evaluate the density of states and the transmittivity with the methods described respectively in Sect. 3.5 (and in [14]) and in Sect. 3.7 (and in [6]).

In Figs. 3.32 and 3.33 we show the transmittivity of a single strand respectively and of two interacting strands of random emeraldine BLs; we considered the mean value of 100 configurations of 200 ring-long strands, in fact, the experimentally estimated size of highly conducting emeraldine polymers

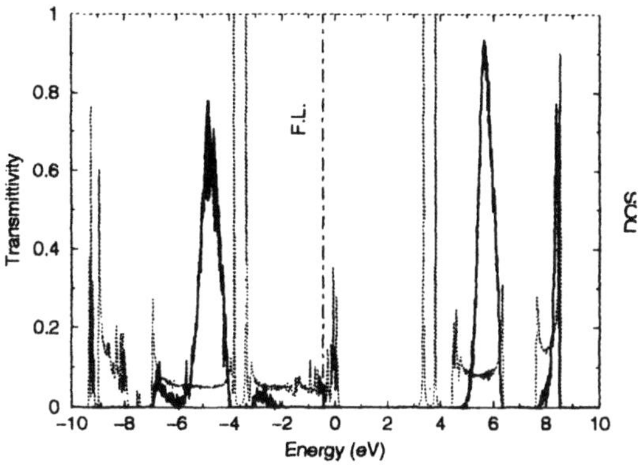

Fig. 3.32. Transmittivity (*solid line*) and DOS (*dotted line*, arbitrary units) of a single strand of fully protonated disordered emeraldine bipolaronic lattice

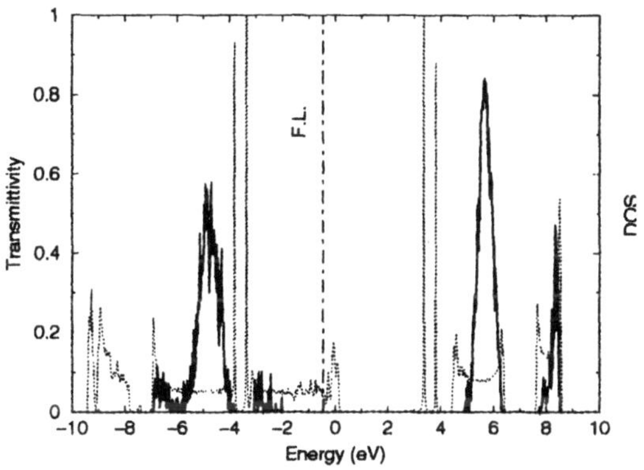

Fig. 3.33. Transmittivity (*solid line*) and DOS (*dotted line*, arbitrary units) of two interacting strands of fully protonated disordered emeraldine bipolaronic lattice. The interchain interaction value is −0.4 eV. (From [6])

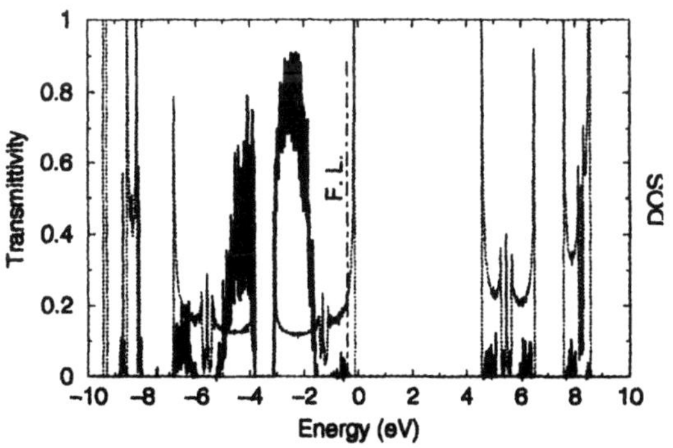

Fig. 3.34. Transmittivity (*solid line*) and DOS (*dotted line*, arbitrary units) of the polaronic lattice with 90% crystallinity. (From [6])

is about 200–300 Å (see [40]). In both situations the Fermi level penetrates into regions where only localized states are present, and it is far from regions of high transmittivity.

In contrast, in the PL of emeraldine salt, the Fermi level penetrates into a region of the spectrum characterized by high transmittivity both in the ordered fully protonated sample and in the presence of weak disorder, as shown in Fig. 3.34. At a higher degree of disorder the Fermi level merges into a multitude of localized states, thus making the polymer non-conducting.

3.9 Conclusions

In this chapter it has been shown that appropriately implemented renormalization and recursion methods are particularly convenient to study the electronic states and transport of conjugated polymers.

Once mapped into effective strictly one-dimensional lattices, the electronic band structure of the polymer can be evaluated by means of simple effective tight-binding Hamiltonians defined in terms of the site and hopping energies of the original system. Basic elements of scattering theory can then be applied to determine the transmittivity of single or coupled strands of polymeric chains, also in the presence of disorder.

Several examples of realistic polymers such as polyacenes, polythiophenes and polyanilines have been provided in order to illustrate the main features and feasibility of the real-space iterative methods.

Acknowledgments. This work was supported in part by MURST and INFM.

References

1. R. Haydock, V. Heine, M.J. Kelly, J. Phys. C: Solid State Phys. **5**, 2845 (1972); **8**, 2591 (1975)
2. G. Grosso, G. Pastori Parravicini, *Solid State Physics* (Academic Press, London 2000)
3. G. Grosso, G. Pastori Parravicini, Adv. Chem. Phys. **62**, 81; **62**, 133 (1985)
4. P. Giannozzi, G. Grosso, S. Moroni, G. Pastori Parravicini, Appl. Num. Math. **4**, 273 (1988)
5. P. Vignolo, R. Farchioni, G. Grosso, Phys. Stat. Solidi (b) **223**, 853 (2001)
6. R. Farchioni, P. Vignolo, G. Grosso, Phys. Rev. B **60**, 15705 (1999)
7. D. Baeriswyl, D.K. Campbell, S. Mazumdar, *Conjugated conducting polymers*, ed. by H. Kiess (Springer, Berlin 1992)
8. R. Farchioni, G. Grosso, G. Pastori Parravicini, Phys. Rev. B **53**, 4294 (1996)
9. G. Grosso, G. Pastori Parravicini, A. Testa, Phys. Rev. B **32**, 627 (1985)
10. E.N. Economou, *Green's Functions in Quantum Physics* (Springer, Berlin, 1983)

11. G. Grosso, G. Pastori Parravicini, A. Testa, Phys. Rev. B **32**, 627 (1985)
12. A. Cordelli, G. Grosso, G. Pastori Parravicini, Phys. Rev. B **33**, 2154 (1988)
13. P.D. Kirkman, J.B. Pendry, J. Phys. C: Solid State Phys. **17** 4327 (1984)
14. P. Vignolo, R. Farchioni, G. Grosso, Phys. Rev. B **59**, 16065 (1999)
15. R. Farchioni, G. Grosso, P. Vignolo, Phys. Rev. B **62**, 12565 (2000)
16. P. Vignolo, A. Minguzzi, M.P. Tosi, Phys. Rev. Lett. **85**, 2850 (2000)
17. C.R. Wu, J.O. Nilsson, O. Inganäs, W.R. Salaneck, J.E. Österholm, J.L. Brédas, Synth. Met. **21**, 197 (1987)
18. S. Hotta, S.D.D.V. Rughooputh, A.J. Heeger, F. Wudl, Macromolecules **20**, 212 (1987)
19. D.S. Boudreaux, R.R. Chance, J.F. Wolf, L.W. Shacklette, J. Chem. Phys. **85**, 4584 (1986)
20. J.L. Brédas, C. Quattrocchi, J. Libert, A.G. MacDiarmid, J.M. Ginder, A.J. Epstein, Phys. Rev. B **44**, 6002 (1991)
21. J. Libert, J.L. Brédas, A.J. Epstein, Phys. Rev. B **51**, 5711 (1995)
22. J. Libert, J. Cornil, D.A. dos Santos, J.L. Brédas, Phys. Rev. B **56**, 8638 (1997)
23. Y. Cao, P. Smith, A.J. Heeger, Synth. Met. **32**, 263 (1989)
24. R.P. MacCall, J.M. Ginder, J.M. Leng, H.J. Ye, S.K. Manohar, J.G. Masters, G.E. Asturias, A.G. MacDiarmid, A.J. Epstein, Phys. Rev. B **41**, 5202 (1990)
25. B. Sjögren, S. Stafström, J. Chem. Phys. **88**, 3840 (1988)
26. C.B. Duke, A. Paton, E.M. Conwell, W.R. Salaneck, I. Lundström, J. Chem. Phys. **86**, 3414 (1987)
27. E.M. Conwell, C.B. Duke, A. Paton, S. Jeyadev, J. Chem. Phys. **88**, 3331 (1988)
28. S. Stafström, J.L. Brédas, A.J. Epstein, H.S. Woo, D.B. Tanner, W.S. Huang, A.G. MacDiarmid, Phys. Rev. Lett. **59**, 1464 (1987)
29. Z.H. Wang, C. Li, E.M. Scherr, A.G. MacDiarmid, A.J. Epstein, Phys. Rev. Lett. **66**, 1745 (1991)
30. Y. Cao, Synth. Met. **35**, 319 (1990)
31. A.G. MacDiarmid, J.C. Chiang, A.F. Richter, A. J. Epstein, Synth. Met. **18**, 285 (1987)
32. Z.L. Mišković, R.A. English, S.G. Davison, F.O. Goodman, Phys. Rev. B **54**, 255 (1996)
33. H. Mizes, E. Conwell, Phys. Rev. B **44**, 3963 (1991)
34. C. Berthod, F. Gagel, K. Maschke, Phys. Rev. B **50**, 18299 (1994)
35. D.H. Dunlap, H.-L. Wu, P. Phillips, Phys. Rev. Lett. **65**, 88 (1990)
36. H.-L. Wu, P. Phillips, Phys. Rev. Lett. **66**, 1366 (1991); P. Phillips, H. -L. Wu, Science **252**, 1805 (1991)
37. F.M. Izrailev, T. Kottos, G.P. Tsironis, Phys. Rev. B **52**, 3274 (1995)
38. R. Farchioni, G. Grosso, Phys. Rev. B **56**, 1170 (1997)
39. W. Zheng, M. Angelopoulos, A.J. Epstein, A.G. MacDiarmid, Macromolecules **30**, 2953 (1997)
40. F. Zuo, M. Angelopoulos, A.G. MacDiarmid, A.J. Epstein, Phys. Rev. B **36**, 3475 (1987); **39**, 3570 (1989)

4 Photophysics of Conducting Polymers

E.M. Conwell

4.1 Introduction

Conducting polymers may be described as quasi-one-dimensional semiconductors. The gaps are typically such that in undoped material the absorption edge is in the visible or near infrared. In Table 4.1 we list most of the polymers to be discussed, abbreviations for their names and their repeat units or monomers. With the exception of polydiacetylene, which is highly crystalline, a second important characteristic of the conducting polymers is disorder. Although the chains may be quite long, typically the conjugation is interrupted by defects, which may be curves or bends, foreign atoms, crosslinks, etc. An average conjugation length in poly(p-phenylene vinylene), PPV, for example, deduced from many different types of evidence, is 6 to 8 monomers. Even so, the levels are grouped into bands and in many contexts it is useful to think of the band structure in the long conjugation length limit. In other contexts it may be more useful to think of the polymer as an assembly of linear molecules of various lengths (oligomers). As will be seen, there are important differences between the optical properties of trans-polyacetylene, which has a degenerate ground state [1], and non-degenerate ground state (NDGS) polymers, which category includes almost all known conjugated polymers other than trans-polyacetylene. The optical properties of the NDGS polymers have many similarities. Because the photophysical properties of PPV have received the most attention, we will most frequently cite its properties as an example.

Illumination of a polymer sample with light past the absorption edge can result in excitation of a variety of entities or "particles". Because it is not always clear whether a given entity is a primary or secondary excitation, we list and describe all the possibilities in Sect. 4.2. In Sect. 4.3 we describe the optical absorption observed under low excitation. The case of PPV will be discussed in detail, because of both its practical importance and the extensive measurements and theoretical analysis that have been carried out for this case. Section 4.4 is devoted to the properties of the lowest lying singlet exciton. The very large Sect. 4.5, discusses the properties of excitations involving interchain interactions: polaron pairs, excimers and aggregates. Section 4.6 deals with single-chain excitations other than singlet excitons. With all this as background we go on to discuss photogeneration and photoconductivity in Sect. 4.7. Sect. 4.8 is devoted to conducting polymer lasers. Because the literature on these subjects is already voluminous, no attempt has been made to cover the optical properties of doped polymers, polymers in solution, and oligomers, i.e., molecules with only a few repetition units or monomers.

4.2 Excitations Created by Light

Upon absorption of a photon past the absorption edge, an electron is raised to the conduction band and a hole is left behind in the valence band. Many possible fates for the resulting electron and hole are detailed in this section.

4.2.1 Polarons

Formation of a polaron requires that the electron and hole be separated sufficiently so that they do not interact to form an exciton. This may happen by the electron and hole, within femtoseconds of creation, separating onto different chains. Alternatively, the electron or hole may be quickly captured, within femtoseconds, by an impurity or defect that provides a lower energy level. A much studied example, to be discussed later, is provided by the addition of C_{60}, a strong electron acceptor, to the polymer. When there is an isolated electron or hole on a chain, as has been shown by calculations of the dynamics for a carrier added to a polyacetylene chain with an even number of sites, the chain proceeds to relax or deform within the next $\sim$100 fs so as to lower the energy of the carrier [2]. The charge and the resulting deformation constitute a polaron, to be denoted P^+ or P^- according to its sign. Polarons are formed similarly on all other conducting polymers. As will be discussed later, polarons are not generated, however, in very high electric fields.

That polarons are generated by light has been well established for PPV (see for example Murata et al. [3]). Murata et al. studied light-generated spins by electron spin resonance and found that the ESR spectra are well accounted for using the polaron spin density for PPV calculated theoretically by use of the Pople–Parr–Pariser model [4]. From these calculations they determined that the extent of the polaron in PPV is 3 or 4 monomers, or 2 to 2.5 nm. The extent of the polaron in polyacetylene is similar [5].

According to the SSH Hamiltonian [1] or other tight-binding Hamiltonians, polarons provide two energy levels located symmetrically about the center of the gap, as shown schematically in Fig. 4.1a. The location of the polaron levels relative to the conduction band edge or valence band edge depends on the chain length. In the limit of conjugation length much larger than the extent of the deformation, for polyacetylene or PPV the levels are calculated to be 0.15 eV [6] or 0.2 eV [7] from the nearest band edge; for shorter conjugation lengths they are further away. (For calculations of the distance between the polaron level and the band edge as a function of conjugation length in PPV see [7].)

Table 4.1. The polymers discussed, their abbreviations and their repeat units or monomers

Structure		Chemical Name	Acronym
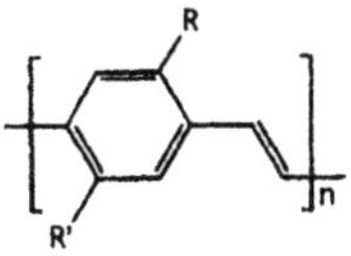	R = 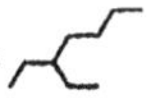 R' = 	poly(2-butyl-5(2'-ethylhexyl)-1,4-phenylene vinylene)	BuEH-PPV
	R = R' = $-O-CH_3$	poly(2,5-dimethoxy-1,4 phenylene vinylene)	DM-PPV (dimethoxy-PPV)
	R = $-O-CH_3$ R' = H	poly(2-methoxy-1,4 phenylene vinylene)	M-PPV (methoxy-PPV)
	R =  R' = $-O-CH_3$	poly[2-methoxy-5-(2'-ethyl-hexyloxy)-1,4-phenylene vinylene]	MEH-PPV
	R = R' = H	poly(p-phenylene vinylene)	PPV
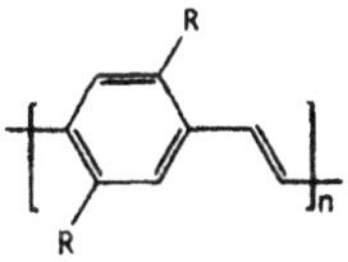	R = 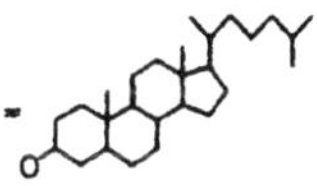	poly(2,5-bis cholestanoxy-1,4 phenylene vinylene)	BCHA-PPV
	R = 	poly{di[(S)-2-methylbutoxy]-1,4 phenylene vinylene}	DMB-PPV
	R = 	poly(2,5-dioctoxy-1,4 phenylene vinylene)	DOO-PPV
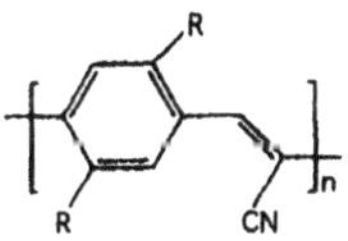	R = 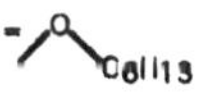	poly(2,5-hexyloxy-1,4-phenylene cyanovinylene)	CN-PPV
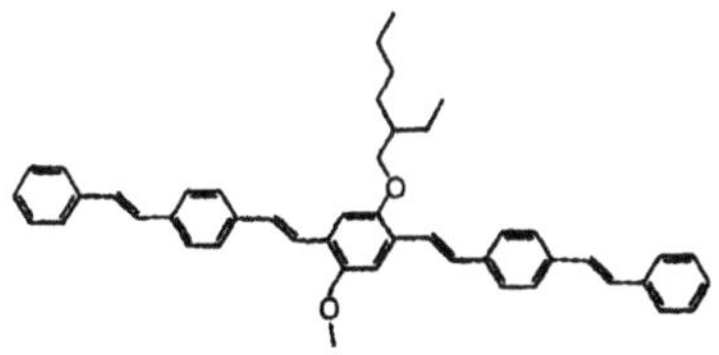		2-methoxy-5-(2'-ethylhexyloxy)-distryryl benzene	MEH-DSB
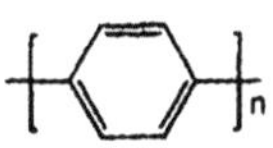		poly(p-phenylene)	PPP
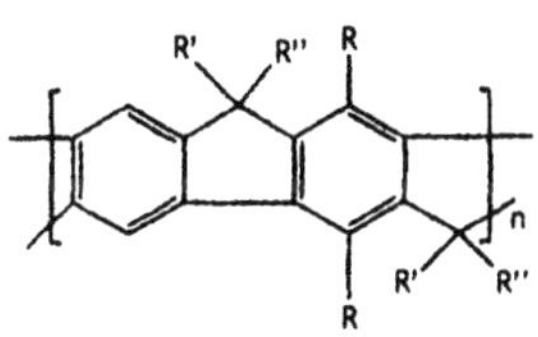	R = n-C_6H_{13} R' = 1,4-C_6H_4-n-$C_{10}H_{21}$ R'' = H	Ladder-type poly (p-phenylene)	L-PPP
	R = n-C_6H_{13} R' = 1,4-C_6H_4-n-$C_{10}H_{21}$ R'' = CH_3	Methyl substituted ladder-type poly (p-phenylene)	MeL-PPP

Table 4.1 (continued)

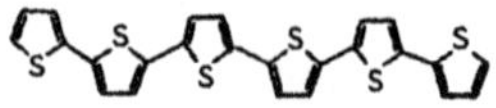	α-sexithiophene	α-6T
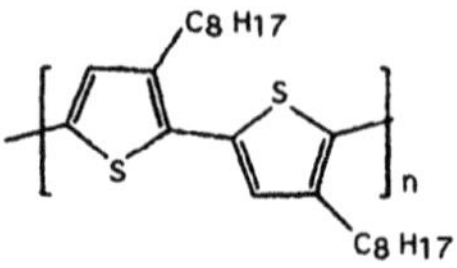	poly(3-octyl)-thiophene	P3OT
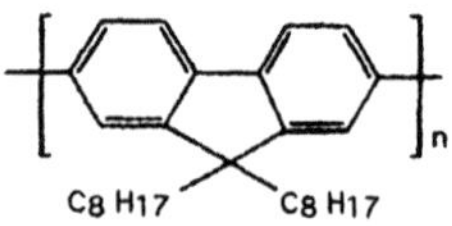	poly(dioctyl-(fluorene)	PFO
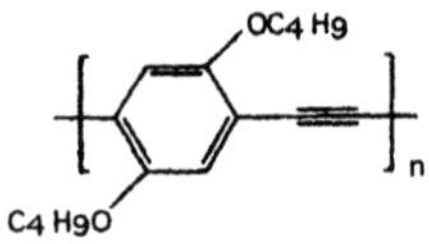	poly(2,5-dibutoxy p-phenylene ethynylene)	DBO-PPE
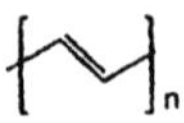	trans-polyacetylene	t-(CH)x
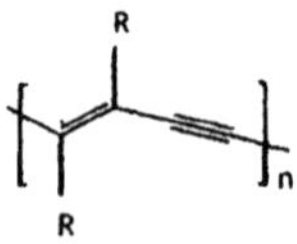	polydiacetylene	PDA
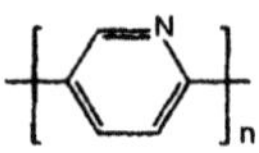	poly(p-pyridine)	PPy
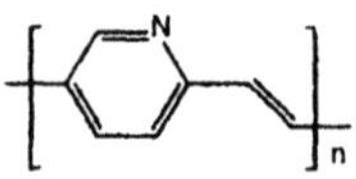	poly(p-pyridine vinylene)	PPyV
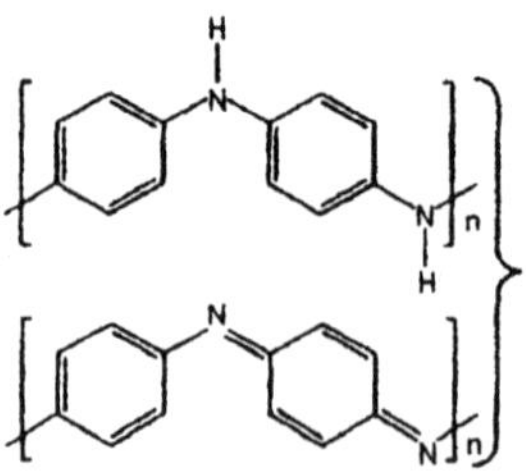 The repeat units of which the various forms of polyaniline are composed	polyaniline	PANI

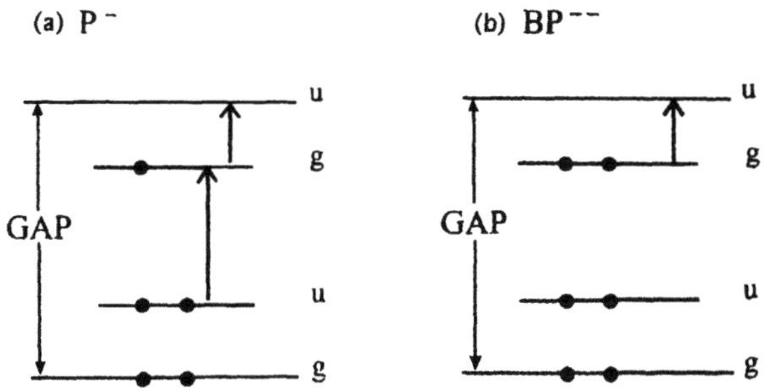

Fig. 4.1. Energy levels of the polaron (**a**) and bipolaron (**b**) and their allowed transitions. The parities of the levels are indicated by u for ungerade (odd) and g for gerade (even)

Selection rules that determine the possible transitions among the levels shown must take into account the fact that the conducting polymers and related oligomers have inversion symmetry. (Inversion symmetry seems to hold even when the sidegroups would cause it to be violated.) This symmetry results in electronic states having either even or odd parity, denoted by A_g and B_u, respectively. In an oligomer, states of even and odd parity alternate with increasing energy. The requirement of parity change in an optical dipole transition results in only two of the three possible transitions being allowed, as is shown in Fig. 4.1a. (This was apparently not generally realized prior to the paper of Poplawski et al. [8].) It has been verified that the dipole selection rule is also obeyed reasonably well in polymers [9].

4.2.2 Singlet Excitons

A likely fate for an electron–hole pair generated by light is that the electron and hole form an exciton, i.e., they remain on the same chain bound to each other by their electrostatic attraction and to a lesser extent by the (relatively small) chain deformation or relaxation they cause. These excitons are singlets because the electron and hole are created with opposite spin. Note that in general more than one type of singlet exciton may be created because of the existence of multiple conduction and valence bands. Thus, a photon whose energy is of the order of the difference between, say, the top of the highest valence band and the bottom of the lowest conduction band will create an exciton with different effective mass and binding energy than a photon whose energy is of the order of the energy difference between the top of the valence band and the next higher conduction band. The former excitons have the lowest energy and we will be almost exclusively concerned with them. By comparison of PPV oligomers of different lengths with the emission from the polymer, it has been deduced that the lowest energy exciton is quite extended in PPV, of the order of 6 monomers or ~4 nm [10]. Excitons in PDA have been found to be similarly extended [11].

The Exciton Binding Energy

The binding energy ε_b of the lowest singlet exciton in PPV has been a matter of great controversy, with one school of thought maintaining $\varepsilon_b \leq k_B T$, another that it is a few tenths of an eV and a third school that it is about 1 eV. Part of the difficulty arises from the differences between the polymers and ordered three-dimensional semiconductors. For the latter the definition of ε_b is straightforward. Simply $\varepsilon_b = E_g - E_a$, where E_g is the energy gap and E_a the absorption edge. However, for the polymers, below a long-chain limit of the order of a few times the average conjugation length, E_g and E_a are functions of the conjugation length; they increase as the conjugation length decreases. This difficulty can be overcome by specifying ε_b for the long-chain limit.

A more basic difficulty with defining ε_b for a polymer stems from the relaxation and consequent energy change that occur when an electron is added to or taken from a chain. In experiments where electrons or holes are inserted into a polymer from an electrochemical cell or an electrode, it is expected that insertion occurs into a P^- or P^+ level which, as discussed above, has lower energy than the corresponding band edge [12–14]. This can be seen to be the result of quantum lattice fluctuations, which prepare the appropriate configurations in the polymer into which electrons in the Fermi sea of the contact can tunnel. The bandgap measured by the difference between insertion voltages for electrons and holes is then the energy difference between the polaron levels P^- and P^+ rather than E_g. This energy difference is called the "single particle energy gap" by Campbell et al. [12]. With electron–hole symmetry, which holds reasonably well for PPV [15], the single particle energy gap is smaller than E_g by twice the distance between the polaron level and the closer band edge, ~0.4 eV for long chains.

In a light emitting diode (LED), electrons inserted at the cathode go into P^- states, holes inserted at the anode into P^+ states. An electric field sweeps them toward each other. When a P^+ and a P^- meet they may form an exciton. It is clear that an exciton formed in this way has the same properties as a photogenerated exciton; the emission from the LED is the same as that seen in photoluminescence. Correspondingly, when an exciton in a conducting polymer is dissociated, a pair of polarons is formed. The exciton binding energy may be defined as the energy required to dissociate the exciton into a pair of separated polarons.

In a number of publications [16–18] it has been argued that for PPV $\varepsilon_b \leq k_B T$, meaning essentially that the exciton is not bound and that photon absorption leads directly to a free electron and hole. The strongest argument given for this view is that the difference in insertion voltages for electrons and holes in an electrochemical cell based on PPV is 2.4 eV [19], which is also the absorption edge of PPV. According to the discussion of the preceding paragraphs, however, this observation is precisely evidence for an exciton binding energy of ~0.4 eV. From this discussion it is apparent that energy-level calculations, such those of Mazumdar et al. (see, for instance, [20]), that do not take relaxation into account may be correct but cannot give a correct binding energy because they do not allow for the relaxation energy of the two polarons.

It has been pointed out that a value of 0.4 eV for ε_b is more appropriately obtained from the data of Campbell et al. [12] than the value of 0.2 eV they cite [14]. To obtain ε_b it is necessary to subtract from the single particle energy gap the creation energy of a relaxed exciton. Campbell et al. obtained 2.45 eV for the single particle energy gap of MEH-PPV and took as the creation energy the first peak in the absorption, 2.25 eV. More reasonably, as discussed above, the creation energy should be taken as the long-chain limiting value, which corresponds to the absorption edge, 2.1 eV for MEH-PPV. This leads to a binding energy of 0.35 eV. We consider that the evidence for such an exciton binding energy for PPV and derivatives, rather than one of $\sim k_B T$ or 1 eV, is very strong. Additional evidence will be presented below. According to the discussion above, with the absorption edge at 2.4 eV, an exciton binding energy of 0.4 eV, and the correction due to the 0.2 eV between each of the polaron levels and the band edge, the long-chain energy gap E_g of PPV is 3.2 eV. This might be called the unrelaxed energy gap.

It is significant that an exciton binding energy close to 0.4 eV, specifically 0.5 eV, has also been found for several polydiacetylenes [11] and methyl-substituted ladder PPP, MeLPPP [21].

Symmetry of the Excited States and Fluorescence

To discuss excitons in polyacetylene we must consider an additional complication, i.e., that the transition that would yield the exciton with the lowest energy may be forbidden due to symmetry. Because the ground state is a singlet with A_g parity, if the lowest singlet excited state were A_g rather than B_u, excitons would form only in the transition to the B_u level. This is of practical importance because strong fluorescence can only occur if the lowest energy singlet is dipole-allowed [22]. The relative location of the $1B_u$ and the lowest A_g ($2A_g$) energy levels is determined by a sensitive interplay between electron–electron interactions and the size of bond alternation in the π-electron transfer integral along the chain [23]. From the fact that strong fluorescence is observed for PPV, PT, PDA and various of their derivatives, we judge that their lowest excited state is B_u and the description of the exciton given above holds for these cases. In $(CH)_x$, on the other hand, only weak fluorescence is observed. This has been attributed to the fact that strong electron–electron correlation results in the lowest A_g excited state lying below the lowest B_u [24]. Some B_u excitons presumably relax to the excited A_g state, and the selection rule prohibiting the transition from there to the ground state may be violated to a small extent because symmetry is perturbed by the lattice vibrations.

4.2.3 Biexcitons

The biexciton, which may be characterized as an excitonic molecule, has been extensively studied in inorganic semiconductors. It has been predicted that the biexciton would be a stable excitation in π-conjugated polymers, resulting from the excitation of a second exciton close to the first on a given conjugation segment [25]. As will be discussed further below, experimental studies on a PPV oligomer have detected biexcitons in dilute solution and in films, but in the latter case they were unstable against decay on a sub-ps time scale to form interchain species [26].

4.2.4 Triplet Excitons

Although the excitons created directly by exposure to above-bandgap light are singlets, a number of processes, to be discussed below, can convert them into triplets. The triplets are in general more strongly bound than singlets and have longer lifetimes. Existence of triplets after photoexcitation has been established for PT, PPV and some PDAs. They will be discussed further in Sect. 4.6.2.

4.2.5 Bipolarons

A bipolaron consists of two like charges with opposite spins bound together within the same conjugation length. The bipolaron occupies two levels in the gap, as seen in Fig. 4.1b. It has been accepted that bipolarons exist in doped NDGS polymers, the evidence being that, although doping can increase the conductivity by orders of magnitude, it gives rise to only a small change, if any, in the magnetic susceptibility. Another possibility, pointed out by Lane et al. [27], is spin pairing between like-charged polarons on different chains. Spin pairing has been observed in sexithiophene [28]. In a doped film the repulsion of the two like charges making up the bipolaron would clearly be overcome by the nearby oppositely charged ions that donated the charges to the chain. Calculations for the bipolaron have shown, however, that its stability is questionable in the absence of oppositely charged ions [29].

Many researchers, having found in the absorption of the photogenerated entities in PPV two peaks that behaved similarly in terms of intensity, temperature and frequency dependence, have suggested that they are due to bipolarons. However, the selection rules discussed earlier in connection with the polaron dictate that there be only one strong transition for the bipolaron, shown in Fig. 4.1b. As noted above, it has been verified for a number of oligomers and polymers that the dipole selection rules hold [9]. Recently, Lane et al. [9], finding that they could also distinguish between polarons and bipolarons by spin signatures, identified a peak found in photoinduced absorption of a number of polymers as due to bipolarons. This would indicate that the bipolaron is stable in some PPV derivatives at least. For this conclusion to hold, however, it is necessary to prove that the bipolaron is not stabilized in these cases by an impurity ion or defect of some kind.

4.2.6 Excimers

An excimer is a complex between an excited state of a molecule and a molecule of the same species that is not excited, i.e., in its ground state. (Cases in which the electronic wavefunction is delocalized over two or more chains in the ground state as well as in the excited state are called aggregates and will be discussed further below.) The excited state is conveniently created by light. Although only one of the molecules is initially excited, it is apparent that the wavefunction of the complex must be symmetric or antisymmetric in the coordinates because the two molecules are identical. The stability of the excimer arises from the overlap of π-orbitals of the two molecules. It follows that the molecules must be close enough for sufficient interaction but not so close that the ground states form a new stable molecule. The ideal separation is in the range 0.3 to 0.4 nm. In that respect conducting polymers are good candidates for excimers because at least one interchain distance is usually of that order. For good interaction of the π-orbitals the chains should also be parallel over a distance of at least a few monomers, this distance being within a single conjugation length, so that the chains stay in appropriate registry.

An important characteristic of the excimer is that its emission has a lower frequency than that from one of the molecules excited in isolation, because of the lower energy in the excimer state and because in the final state, where both molecules are in their ground state, they repel each other. This characteristic is also found in an aggregate. As will be discussed further below, excimer emission characteristically has a long lifetime because the symmetry of the electronic wavefunction in the excimer state tends to be such that the radiative transition to the ground state is forbidden.

Although it is not obvious that the disordered polymers we deal with should be able to satisfy the conditions to form an excimer, there is emission from CN-PPV and MEH-PPV thin films, as will be detailed below, that has the properties described above for excimer emission. It is reasonable to conclude that excimers are formed in these two cases at least.

4.2.7 Aggregates

As noted above, aggregates differ from excimers in that the electronic wavefunction is delocalized over two or more chains in the ground state as well as in the excited state. Thus, an aggregate should provide additional absorption, which should be detectable by comparison with absorption in the dissolved polymer, as well as the additional emission described in the last section for an excimer.

4.2.8 Polaron Pairs

In principle, portions of chains that do not have the required length or registry may give rise to an interchain excitation that is not as symmetric as the excimer. An example might be the situation of an electron and hole on non-parallel nearest-neighbor chains. Because each of the charges must give rise to the characteristic chain deformation, this excitation is called a polaron pair. It is bound by the Coulomb attraction of the pair, which is considerable when the chains are some tenths of a nanometer apart. This excitation may also be called a charge-transfer exciton or an indirect exciton. We will take up the evidence that it is photogenerated in PPV and in the other conducting polymers.

4.2.9 Solitons

It can be inferred from its representation in Table 4.1 that trans-polyacetylene would have the same properties if the order of the single and double bonds were reversed, i.e., if the double bonds slanted up in going from left to right rather than down, as shown in the Table. This degeneracy is special to trans-polyacetylene; it does not occur for any of the other polymers shown in the Table, and indeed for very few of the conjugated polymers. The degeneracy gives rise to the possibility of a special kind of defect or excitation, the soliton S and the antisoliton $\bar{S}$. As shown schematically in Fig. 4.2, the S or $\bar{S}$ is a portion of chain or a domain wall separating two regions of different bond alternation (i.e., in the B region the pattern is as shown in Table 4.1, while in the A region single and double bonds are interchanged). Actually, this picture is oversimplified in that S and $\bar{S}$ are not one or two sites long but of the order of 10 times that. Apart from the difference in the bonds portrayed in Fig. 4.2, S and $\bar{S}$ have the same properties, so we will describe them in terms of S. There is a localized electron state associated with the soliton which, for a symmetric Hamiltonian such as the SSH Hamiltonian [1], is found to lie at midgap. The soliton can be neutral with spin 1/2 (one electron in the state at midgap) or charged positively or negatively with spin zero (corresponding to zero or two electrons, respectively, in the midgap level). There are various kinds of experimental evidence for the existence of solitons in t-$(CH)_x$. Electron spin resonance measurements on undoped t-$(CH)_x$ made by the Shirakawa process indicate the presence of about 1 spin per 1000 to 3000 atoms, the spin being characterized by the g-value of a π electron [30]. Various types of magnetic resonance experiments indicate that these spins move readily along the chains, as would be expected for solitons. Measurements of their diffusion constant [31, 32] are in good agreement with the theoretical calculation [33, 34]. For further discussion see the review of Conwell [35].

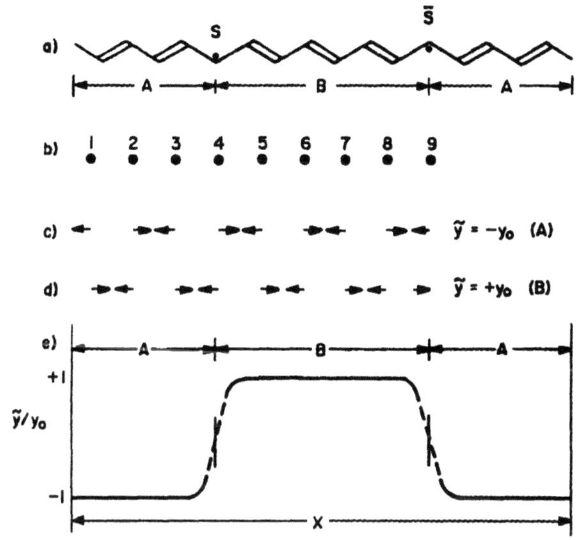

Fig. 4.2. (**a**) Schematic of the soliton S and antisoliton $\bar{S}$ on a trans-polyacetylene chain. (**b**) Uniform, i.e., undimerized, chain, each point representing a CH group. (**c**) The displacements of the uniform chain required to form arrangement A. (**d**) The displacements of the uniform chain required to form arrangement B. (e) Representation of the chain conformation in (**a**) in terms of the order parameter, $\tilde{y}$. (After [35])

4.3 Optical Absorption–Theory and Experiment

Beyond the rapid rise with increasing frequency at the absorption edge, the absorption has a characteristic variation with frequency for each polymer. The absorption is to some extent sample-dependent, as will be noted, and for some properties shows small differences between the polymer and its derivatives with the same backbone. A great deal of attention, both theoretical and experimental, has been paid to PPV and its derivatives due to their possible applications as light emitters. Their optical properties have also been the subject of many controversies. In this section the discussion will be focussed on the optical absorption of PPV and its derivatives.

4.3.1 Experimental Data

A set of data comparing the absorption for a gel-processed, highly oriented sample of MEH-PPV, a blend, with that of a spin-cast film is shown in Fig. 4.3. The solid and dotted lines represent the absorption of the blend for light polarized parallel and perpendicular to the draw axis, respectively. Comparison of the data for the blend with those of the unoriented, spin-cast sample, represented by the dashed line, shows that, despite considerable differences, the four-peak structure for unpolarized light is maintained. The peaks for the blend are, however, shifted to somewhat lower frequencies, in line with the longer conjugation lengths expected for the blend. The absorption spectra of PPV itself show a similar set of four peaks, although, at least in some samples, the peak A_2 at 3.7 eV is not as pronounced. The absorption spectrum was compared to the results of a number of theoretical calculations, described in the next section.

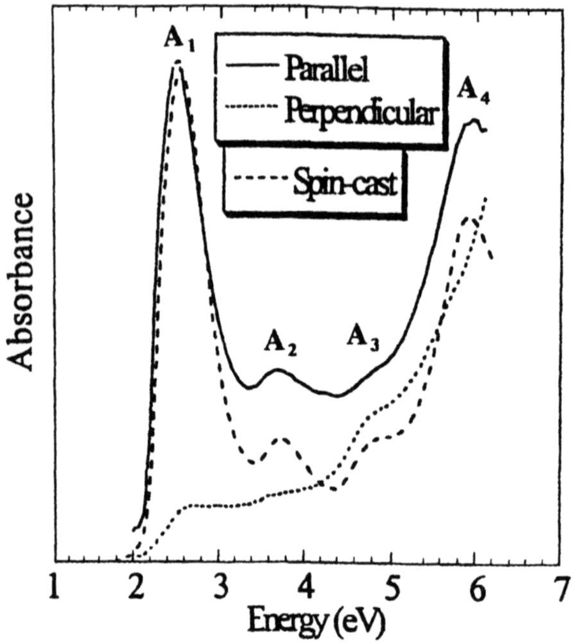

Fig. 4.3. Polarized absorption spectra for an oriented MEH-PPV/polyethylene blend for light polarized parallel to and perpendicular to the draw axis. Also shown is the absorption spectrum of a disordered, spin-cast film of MEH-PPV. (Adapted from [36])

4.3.2 Theory of Optical Absorption in PPV

There have been many calculations of the band structure of PPV. The only three-dimensional (3-D) calculation, based on the X-ray structure determined from polycrystalline samples, used local density functional (LDF) theory [15]. Given one π electron on each carbon atom and 8 carbons in a monomer, there are 8 π bands, constituting 4 valence bands and 4 conduction bands. In the usual 1-D calculation there is exact symmetry about the middle of the gap, but in the LDF calculation the symmetry is less exact. Otherwise, the differences between the 3-D and the various 1-D calculations along the chain direction are not large. The highest valence band, labelled D_1, is delocalized along the chain with a width of 2.5 eV according to the 3-D calculation. In the next lower band, L_1, the wavefunctions are essentially localized on each monomer and the bandwidth is quite small. Below this there are two delocalized bands, D_2 and D_3. The valence and conduction bands resulting from the 1-D band structure calculation of Brazovskii et al. [37] are shown in Fig. 4.4. The results are fairly similar to those obtained for the chain direction by the more accurate LDF calculation, with the major difference being that the LDF calculation gives the maximum (minimum) energy of D_1 (D_1^*) and the minimum (maximum) energy of D_2 (D_2^*) at the edge of the Brillouin zone rather than at the center.

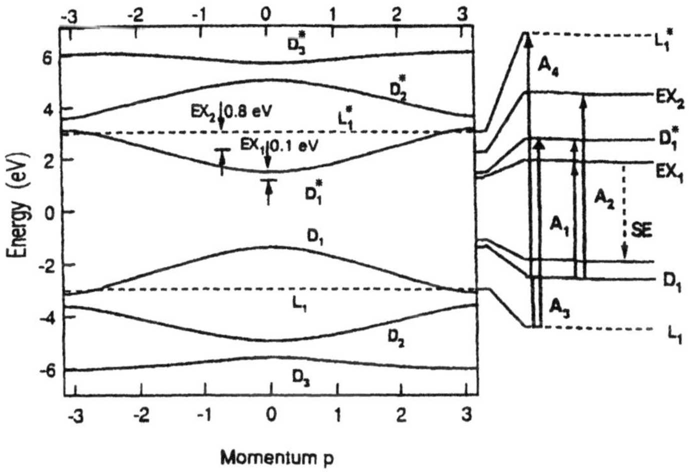

Fig. 4.4. Electronic states, excitons, and optical transitions calculated for an infinite PPV chain. As shown at the right, the peak A_1 corresponds to the transition $D_1 \rightarrow D_1^*$, A_3 to $L_1 \rightarrow D_1^*$ or $D_1 \rightarrow L_1^*$ and A_4 to $L_1 \rightarrow L_1^*$. The exciton EX_1 is associated with the transition $D_1 \rightarrow D_1^*$, while the exciton EX_2 is associated with $D_1 \rightarrow L_1^*$ or $L_1 \rightarrow D_1^*$. (Adapted from [37])

There have been several major attempts to calculate the absorption spectrum of PPV, these requiring that values of various parameters be estimated or chosen to obtain an ultimate fit to the experimental data. The Rice–Gartstein (RG) calculation was based on making the connection with the known spectrum, eigenfunctions and eigenvalues of benzene [38]. The calculation was designed to describe a group of conducting polymers that include phenyl rings, notably PPP and PANI, as well as PPV. The model describes a chain of monomers, each with 2 degenerate hole states and 2 degenerate electron states based on the states for benzene. The transfer or hopping integral was taken to describe intermonomer hopping for the one hole state and the one electron state that have non-vanishing amplitude at the sites connecting the monomers. Long-range Coulomb terms were included to allow the formation of excitons. With guidance from the values for benzene, the parameters were largely chosen to fit the experimental absorption spectrum for PPV. The resulting calculated spectrum for a long chain of PPV had only three peaks, the peak at 3.7 eV being missing. The presence of this peak in the experimental data was attributed to short chains (finite box effect) or to the breaking of charge–conjugation symmetry resulting from derivatizing the polymer [39]. The binding energy of the lowest exciton was estimated in [39] to lie between 0.2 and 0.4 eV.

A semiconductor band model was also used in the calculations of Brazovskii et al. [37]. This model was, however, specifically tailored to PPV, with different transfer integrals within the ring, within the dimer and between ring and dimer. The Hamiltonian contained no Coulomb terms but Coulomb corrections due to electron–hole interactions were introduced semi-empirically to shift the centers of the bands for agreement with optical data. The resulting band structure is shown in Fig. 4.4. Exciton binding energies were estimated in separate calculations using effective mass theory. The resulting binding energy of the exciton associated with the D_1–D_1^* transition, EX_1 in Fig. 4.4, was found to be 0.1–0.2 eV [37]. An earlier calculation using effective mass theory and based on the LDF band structure of PPV gave, however, 0.4 eV [40]. Brazovskii et al. [37] also estimated with effective mass theory the binding energy of the exciton, denoted EX_2 in Fig. 4.4, associated with the D_1 to L_1^* or L_1 to D_1^* transition. With either the electron or the hole in a localized state (L_1^* or L_1, respectively), this exciton is expected to be much more localized and more tightly bound. Guided by the experimental optical absorption, Brazovskii et al. found an effective length for this exciton of ~0.8 nm and a binding energy of 0.8 eV. As indicated in Fig. 4.4, they associated this exciton with the transition A_2.

A series of papers by Mazumdar and colleagues, and by Abe and colleagues, contending that the optical absorption of PPV and derivatives could not be explained within one-electron theories, were based on a Coulomb-correlated model, specifically the Pariser–Parr–Pople model for π-conjugated systems. As indicated in Sect. 4.2.2 it is essential to include electron–electron correlation to obtain the correct order of the excited states in polyacetylene, and, in fact, in linear polyenes in general. In the calculations including electron–electron correlations, various parameters, such as the on-site repulsion U and the dielectric constant in the expression for the Coulomb potential, were determined by the ultimate fit of the theory to experimental optical data rather than by using the conventional semi-empirical parameterization of the Pariser–Parr–Pople model from atomic data. The conclusion from these calculations was an exciton binding energy of 0.9 eV ([41] and references therein) or 0.8 eV ([42] and references therein). The values of U for the best fit were 8 eV for the former calculations and 4.8 eV for the latter, which the authors classified as in the intermediate range. Chandross and Mazumdar found a good fit to the peaks A_1, A_3 and A_4, but concluded that

A_2, the 3.7 eV peak, would only be seen in samples with short chains. Shimoi and Abe found four peaks, but they did not fit the experimental data well.

Quantum chemical calculations of the absorption were carried out for PPV oligomers up to 5 rings. Electronic correlation effects were accounted for with configuration interaction schemes. To simulate absorption spectra, the transition dipole moments and transition energies calculated by the INDO/SCI scheme were used to calculate the polarizability. To take account of vibronic coupling, Franck–Condon factors were included. For PPV these calculations resulted in a peak at 3.6 eV that grew weaker with increasing length. This was not the case for substituted oligomers, but the peak found for that case had perpendicular rather than parallel polarization. (For further details of these calculations and results see [43].) This group also calculated the exciton binding energy. Concluding that there is very little lattice contribution to the binding, they calculated the electronic contribution using density matrix renormalization group calculations within the extended Hubbard–Peierls model. For values of the Hubbard U between 1.5 and 5.0, in units of the transfer integral t, they found ε_b values ranging from 0.12 to 0.37 eV. Quantum chemistry calculations for oligomers which predicted five peaks were carried out in [44].

Of the theoretical treatments discussed, the only one that allows a peak at 3.7 eV in PPV samples with long conjugation lengths is that of Brazovskii et al. Of course, in not attempting an overall calculation of optical absorption versus frequency they did not make some of the simplifications characterizing the other calculations described. Also, they incorporated empirical information to correct their band structure calculation. Their focus on experimental information has allowed them, however, to accomplish what appears to be a convincing explanation of the origin of the peaks in the optical absorption of PPV. Their explanation is shown diagrammatically in Fig. 4.4. The peak A_2 is attributed to the transition D_1 to L_1^* and L_1 to D_1^*. There remains, however, a discrepancy between the results of Brazovskii et al. and the data of Miller et al., which is that the measurements give the polarization of the peak at 3.7 eV as parallel to the chains, whereas Brazovskii et al. predict polarization perpendicular to the chains.

We return to the question of the binding energy of the lowest energy exciton in PPV. As has been seen, and will be further documented, there are results that suggest a low binding energy, 0.1–0.2 eV; an intermediate value, ~0.4 eV; and a large value, ~1 eV. None of the polymer calculations described in this section takes into account the relaxation effects discussed in Sect. 4.2.2. According to that discussion, if there were no relaxation effects the exciton binding energy would be larger than 0.4 eV by twice the energy difference between the polaron level and the nearest band edge, thus ~0.8 eV. On this basis, the calculations of Chandross and Mazumdar, and those of Shimoi and Abe, must be judged to give correct results, making a case for the importance of including electron–electron correlations, but they did not calculate the exciton binding energy as the term is most reasonably defined, because of the omission of relaxation effects.

4.4 Properties of the Lowest Lying Singlet Exciton

4.4.1 Vibronic Structure

As is evident from the above discussion, photons with energy above the absorption edge frequently create excitons. In Sects. 4.2.2 and 4.3 we discussed their binding energy, concluding that it is ~0.4 eV. The excitons may decay radiatively or non-radiatively. A typical lifetime at room temperature for singlet excitons in PPV, for example, is ~300 ps. As will be seen, the decay may not be exponential, indicating that more than one process is involved. Analysis of the geometry of the ground state, labelled S_0, and the lowest excited state, S_1, the singlet exciton in PPV, shows that they have different equilibrium structures [43]. As a consequence, it is expected that electronic excitations of the chains will be coupled to vibrations of the lattice, leading to a vibronic structure in emission and absorption. This structure is shown in Fig. 4.5. It is much more visible at low temperatures and particularly in emission because, as will be discussed in the next section, the excitons tend to migrate to the lowest energy sites before emitting. The phonons involved are principally those involving stretching vibrations of the ring, with energies ~0.2 eV. Calculations of the line shape for the linear absorption taking into account the vibronic coupling have led to good agreement between theory and experiment [43].

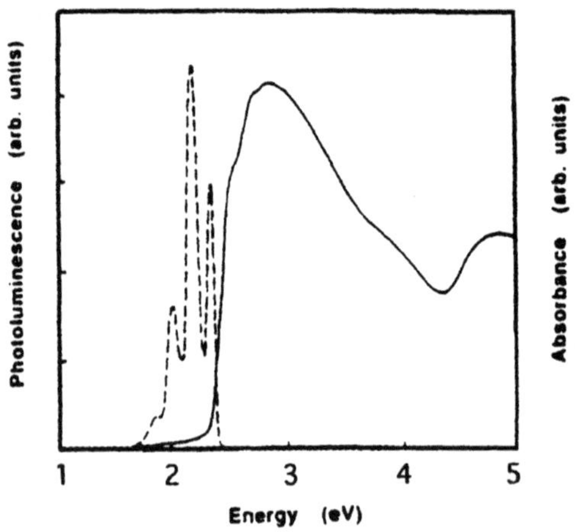

Fig. 4.5. Room temperature absorption (*solid line*) and 80 K emission spectra (*dotted line*) from a PPV thin film. (After [45])

4.4.2 Localization Energy

It is significant that the emission, or photoluminescence, as shown in Fig. 4.5, is over a narrower range of frequencies than the absorption associated with this exciton. This fact, plus the observation that there is a red shift of the emission observable over a few ps after excitation, has led to the conclusion that after creation the excitons migrate to sites where their energy is lower, specifically to chain segments with longer conjugation lengths, before decaying [46]. The latter process has been studied in a number of so-called site-selective fluorescence experiments and the accompanying theory. In these experiments the emission spectrum is measured as a function of excitation frequency. A threshold, known as the localization energy, occurs at a frequency determined by the relative rates of exciton migration and decay. For the exciting photon energy above the localization energy, the emission spectrum is independent of photon energy because excitons quickly migrate to the lower energy sites before radiating. As they move to lower energy sites, the probability of finding still lower energy sites to hop to decreases, until the hopping rate becomes comparable to the exciton decay rate. This defines the localization energy; below this excitation energy the excitons do not migrate, and the emission energy is expected to decrease as the excitation energy decreases, emission occurring from the absorbing site [47–49].

4.4.3 Decay due to Photooxidation

It has been found that the luminescence of PPV and derivatives decreases rapidly on exposure to air. This has been traced to photooxidation, which introduces carbonyls that cause the excitons to decay [50, 51]. The decay occurs by the electron from the exciton being captured by the electronegative carbonyl oxygen, thereby dissociating the exciton. It has also been found that carbonyl defects may be introduced in the thermal conversion of PPV from the precursor polymer, particularly if high conversion temperatures are used [52].

4.4.4 Dissociation in High Electric Fields

Another important characteristic of excitons is that they can be dissociated, resulting in quenching of the photoluminescence, by a high electric field. This has been demonstrated by experiments on poly(phenyl-p-phenylene vinylene), PPPV, blended with polycarbonate, PC, and on MEH-PPV. In a field of 2×10^6 V/cm, 40% of the luminescence in MEH-PPV was quenched and the quenching increased rapidly with higher fields [53]. Similar results were found for the PPPV/PC blends [54, 55]. The quenching evolves on a ps time scale. For the case of excitation close to the absorption edge, the quenching has been modeled by Monte Carlo simulation as the result of field-induced exciton dissociation into geminate electron–hole pairs while the exciton is undergoing a random walk among polymer segments. The fit of the results to experimental data yielded an exciton binding energy of 0.4 eV [56] or 0.3±0.1 eV [57].

Significantly, the degree of quenching dissociation was found to increase with the exciting photon energy. This has been attributed to a mechanism in which the photon energy in excess of the singlet exciton energy is converted into local thermal energy that assists in the dissociation [58]. This work is discussed further in Sect. 4.7.

4.5 Excitations Involving Interchain Interactions

Transient photoinduced absorption, PA, has been very widely studied in an attempt to determine which excitations, in addition to excitons, are created by light. In the experiments a pump laser creates a population of excited-state species in a sample. A probe pulse, mechanically delayed, measures the resulting change in transmission. The knowledge of the wavelengths absorbed, plus information on the time variation of the excitation population obtained by varying the delay time of the probe, are available to identify the excitation. Another technique that has been very useful in determining the nature of the excitations created by light is optically detected magnetic resonance, ODMR.

4.5.1 Photoinduced Absorption in PPV and its Derivatives

A great deal of PA data has been obtained for both solutions and thin films, with widely diverging deductions therefrom. The most thorough study has been done by the Bell Labs group. Their results for stretch-oriented methoxy-PPV films are shown in Fig. 4.6. The bands peaking at 1.5 and 0.5 eV were found to have identical decay dynamics, as shown in the inset to Fig. 4.6, and are thus presumably derived from the same species. Identification of the species is made difficult by the fact that several species, notably singlet excitons, triplet excitons and bipolarons, have been found to show PA in the neighborhood of 1.5 eV. Nevertheless, it has been concluded that the data rule out these three possibilities. The photogenerated species is not triplet excitons because the mid-infrared band seen here is not present in the triplet spectrum, as determined by millisecond PA detected magnetic resonance [59]. Bipolarons are ruled out because their generation should depend on intensity, which was found not to be the case in the Bell Labs study, and also because the PA shown in Fig. 4.6 is already seen in the subpicosecond time range and it is unlikely that bipolarons could be generated so quickly. Polarons are also ruled out by the intensity independence of the decay dynamics because free polaron recombination should be non-geminate. Also, the small size of PPV photoconductivity at low fields indicates low polaron generation, whereas the photogenerated species giving rise to the PA is created with a quantum yield estimated to be 80% to 90% [61]. It has also been suggested that singlet

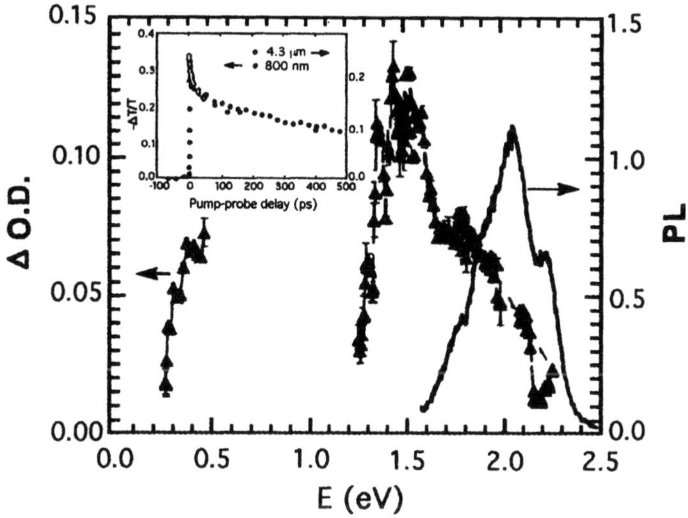

Fig. 4.6. Transient PA spectra (200 ps after 532 nm photoexcitation) in methoxy-PPV (*solid triangles*) with PL (*solid line*) shown for reference. Inset shows the decay dynamics for the mid-infrared band (4.3 μm, *open circles*) and the near-infrared band (800 nm, *solid circles*). (Adapted from [60])

excitons are the species responsible for the PA. As noted above, PA of singlet excitons shows a peak at 1.5 eV in dilute solutions of DOO-PPV, but it differs from the data of Fig. 4.6 in that there is a second peak at 0.8 eV, with no visible absorption below 0.7 eV [62]. Also photoluminescence, PL, in the same sample for which the data of Fig. 4.6 were taken decayed to an undetectable level in 300 ps, while the PA, despite a rapid initial decay, is still large and decreasing very slowly at 500 ps.

4.5.2 Debate About Polaron Pairs

Having ruled out other possibilities, Hsu et al. suggested that the species responsible for the PA was "polaron pairs". The existence of polaron pairs had been proposed earlier to account for a magnetic field dependence found for the photoconductivity of PPV [63]. Polaron pairs had also been proposed as an intermediate state in the photogeneration of charge carriers from excitons [64]. In subsequent investigations on PPV itself Yan et al., identifying the excitons by the fact that they showed stimulated emission, proved that the large PA found for that case also could not be due to excitons because there was a competition between the species responsible for the PA and the excitons; the stimulated emission and the PA were found to have different temporal behavior and action spectra [61]. A more definitive demonstration that the PA is due to interchain pairs came from comparison of the photoexcitation dynamics of a soluble derivative of PPV, MEH-PPV, in films and in dilute solution. Although there was PA peaking at $\sim$ 1.5 eV in both cases (as noted above excitons also absorb at 1.5 eV), PL in the solution decayed exponentially, corresponding to the usual exciton lifetime of $\sim$300 ps, while the PL in the film has a long tail extending to tens of nanoseconds. Another difference between film and solution was in the amount of stimulated emission, which was vanishingly small in the films but large in the solutions [65]. Correspondingly, the quantum efficiency for creating PL was much smaller in the films ($\sim$10%) than in solution ($\sim$30%). Both of these phenomena, the Bell Labs group concluded, indicate that a large proportion of the incident photons produce interchain pairs rather than excitons in the films.

Quite a different conclusion was reached by the Cambridge group from measurements on their PPV samples. Their reasoning was based on the assumption that the PL efficiency Q, defined as the number of photons emitted per absorbed photon, is the product of η, the fraction of absorbed photons that form singlet excitons, and the efficiency for radiative decay of the singlets. The latter quantity was taken as the actual singlet lifetime τ divided by the singlet radiative lifetime τ_r. To apply this relation for the determination of η, Greenham et al. measured Q by using an integrating sphere to collect all the emitted light and obtained τ from the observed exponential decay of the PL for the same sample [66]. The results were $Q = 0.27$ and $\tau = 320$ ps, respectively. With τ_r taken as 1.2 ns from data of Yan et al. on a PPV sample with a low density of oxygen defects, these numbers yielded $\eta \approx 1$. Thus, the conclusion of Greenham et al. was that essentially every photon, rather than only $\sim$20% of them, produces a singlet exciton. This obviously cast considerable doubt on the existence of polaron pairs in PPV and its derivatives. It may be noted that in later work the Cambridge group concluded, more conservatively, that "singlet excitons make up at least 50% of the initially created photoexcitations in pristine PPV" [67].

4.5.3 Discovery of Excimers in PPV Derivatives

An important contribution to the controversy over the existence of interchain pairs in PPV was the finding of Samuel et al. of clear evidence for such pairs in CN-PPV [68]. Comparing PL from solutions and thin films of CN-PPV, they found the peak of the emission shifted down by $\sim$1/2 eV in the films. Also, the emission was broader and lacking the structure seen in the emission from solution. More striking was the lifetime of the emission in the films, which was much longer than that in solution, 5.6 ns compared to 0.9 ns in solution. The properties just described are characteristic of excimers or perhaps aggregates, as discussed in Sects. 4.2.6 and 4.2.7. The finding that interchain interactions are important in at least one PPV derivative raises many questions. Given the evidence cited above for interchain interactions in PPV and MEH-PPV why do they not show the lower frequency emission characteristic of excimers? Why do some samples of PPV apparently show long-lived excitations while others do not? The answer to the second question is by no means clear, but it has become well documented that the degree of interchain interaction is quite sample dependent, as will be discussed in Sect. 4.5.11.

4.5.4 Structures of MEH-PPV and CN-PPV

A partial answer to the first question above comes from comparing the probable structures of MEH-PPV and CN-PPV. Although polymers such as CN-PPV and MEH-PPV are expected to be amorphous, determination of the lowest energy structures in these two cases should give at least the relative values of the ground state spacing between chains and a good idea of the most typical structures. Because of their planar backbones and long alkoxy sidegroups, their low-energy configurations in a film have the planes of the backbones parallel to each other. The results of such a determination are shown in Fig. 4.7. For CN-PPV the lowest energy structure has an interchain distance of 0.33 nm, with interchain registry such that the cyano group in one chain overlaps the edge of a ring in the nearest chain. For MEH-PPV there is a distance of 0.405 nm between chains, with registry such that the exocyclic double bond on one chain sits over a ring of the nearest neighbor [69]. The shorter distance between CN-PPV chains is expected because of the high electron affinity of the cyano group. With the matrix element for emission decreasing strongly with decreasing overlap of the π wavefunctions of the atoms in an interchain pair, the probability for emission from an MEH-PPV pair was found to be smaller than that from CN-PPV by a factor of $\sim$20 [69].

The factor of 20 was calculated with the assumption that the spacing of chains in the ground state is maintained in the excimer. It is likely that the interchain spacing decreases somewhat. The decrease is expected to be small because known excimers have a spacing between partners ranging from 0.3

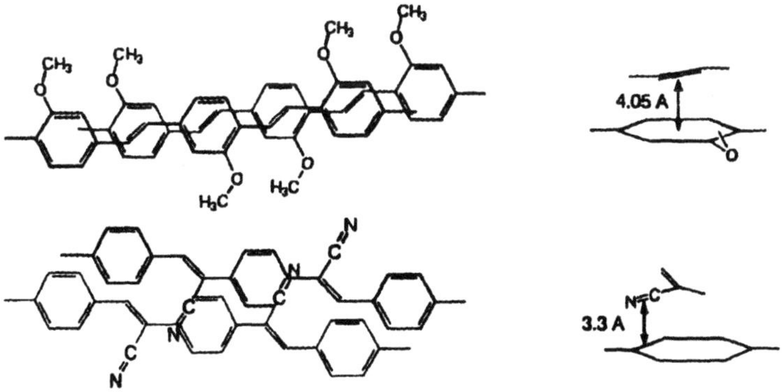

Fig. 4.7. Schematic of chain packing for the Monte Carlo global minimum in MEH-PPV (*top*) and CN-PPV (*bottom*) as viewed perpendicular to the plane of the polymer backbone. Most of the side chains are not shown for clarity. (After [69])

to 0.35 nm (see, for example, [70]). The conclusion of Meng that the distance between chains required for a stable excimer is 0.23 nm for either of the PPV derivatives concerned [71] must be incorrect, because this distance is less than the van der Waals radius of carbon, 0.3 nm. This error could result from an underestimate of the strength of the interchain coupling, $t_{\perp}$, which is difficult to calculate from first principles. In any case, on the evidence of a smaller shift in the emission peak for MEH-PPV excimers than for CN-PPV excimers, as compared with the peak in dilute solution, the spacing remains larger and the emission probability considerably smaller for MEH-PPV.

4.5.5 Effect of High-Intensity Irradiation

The question of the origin of interchain pairs was also addressed by studies on a five-ring PPV oligomer 2-methoxy-5-(2'-ethylhexyloxy)-distyryl benzene, MEH-DSB. Both solutions and films of this oligomer exhibit a PA band with dynamics matching those of the stimulated emission, demonstrating that these features arise from intrachain excitons. The films exhibit an additional PA band with shorter wavelength and different dynamics that has been attributed to non-emissive interchain excitons. Correlations in the dynamics of the two PA features, and the fact that the intensity of the PA band at shorter wavelength depends quadratically on the intensity of the longer wavelength PA, indicate that the formation of the interchain species is mediated by intrachain two-exciton states [72]. Although this is true for these oligomers, it is not true for the polymers, where the corresponding behavior is not found. Nevertheless, it suggests that photogeneration of interchain species requires high pump intensities, which indeed were used in the experiments of Hsu et al. and Yan et al. but not in the experiments of Greenham et al.

4.5.6 Excimers in MEH-PPV

In a series of later experiments Jakubiak et al. showed that a high proportion of photons create excimers in MEH-PPV even in the low excitation density regime [73]. The experiments were done on films spun from THF, tetrahydrofuran, a solvent that was found to give a high quantum yield of excitons on the evidence that films spun from THF have low lasing threshold [74,75]. The emission of these excimers had not been seen sooner because it is quite small at room temperature and masked by the larger exciton emission during the lifetime of the exciton. It is more easily seen at 77 K, where non-radiative transitions are considerably decreased, microseconds after the excitation pulse. At 77 K the excimer emission has ~425 ns lifetime, very much longer than the ~1 ns lifetime of the excitons. The spectrum observed microseconds after excitation at 77 K was shown to resemble the emission spectrum obtained from the temporal tail (4–12 ns) of the 300 K emission, confirming that there is also some excimer emission at room temperature. From their data Jakubiak et al. concluded that at least 45% of incident photons create excimers in MEH-PPV rather than excitons [76].

Another point made by the experiments on MEH-PPV is that the relation frequently cited (see, for example, Greenham et al. [66]), between the number of photons emitted per absorbed photon, or PL efficiency Q, and the emission lifetime τ, i.e., $Q = \eta\tau/\tau_r$ where, η is the fraction of absorbed photons that produce singlet excitons and τ_r is the radiative lifetime, may not be correct [77]. As seen in the above discussion, Q is wavelength dependent, excimers with different luminescence efficiency than excitons being created at long wavelengths, whereas τ is not. The problem with the relation cited between Q and τ is that it assumes only singlet excitons are created. Conceivably, this is true in some samples of PPV and its derivatives, but it is clearly not true in many others.

4.5.7 The Case of PPV

Although PA in PPV also shows a much longer decay time than PL, there has been no evidence for excimer emission from PPV at atmospheric pressure. This suggests that interchain excitations in PPV take the form of polaron pairs rather than excimers, as might be expected from its structure. Unlike MEH-PPV and the other derivatives with long side chains, which are amorphous, PPV tends to crystallize in a herringbone structure. Neighboring chains that are parallel to each other are quite close, with a perpendicular distance of only ~0.3 nm, but π overlap is not increased commensurately because the chains are slipped relative to each other. The closest neighbors, the two chains in the unit cell, are approximately at right angles to each other. Thus, excimer formation is not favored. The symmetrized wavefunction for a pair of polarons, or a charge-transfer exciton, has very little net dipole moment, allowing little radiation.

Calculations have been carried out for a polaron pair made up of the two chains in a unit cell of PPV [78] using a tight-binding Hamiltonian similar in spirit to the SSH Hamiltonian but with parameters chosen to fit quantities obtained from experiment or quantum chemical calculations for PPV. The gap was chosen to be 2.8 eV, obtained from the absorption edge at 2.4 eV and the exciton binding energy of 0.4 eV. The interchain coupling was calculated using the X-ray-determined, structure of PPV and the interaction of the two chains in a unit cell obtained from the LDF band structure calculation. With these parameters, the polaron level on a long chain is calculated to be 0.2 eV away from the band edge, in agreement with quantum chemical calculations [6]. The main effect of the Coulomb interaction between the electron and the hole is to raise the levels of the chain with the electron relative to those of the chain with the hole. The calculations predicted a peak at 0.4 eV due to the transition between either polaron and the nearest band edge, and, in their original version, another peak at ~1.4 eV due to the interchain transition between the positive and negative polarons when they are directly opposite each other. Allowing for polaron motion, each moving in the potential well due to the other, results in a spread of the absorption frequency comparable to what is observed experimentally. It was later realized that the interchain transition would be quite weak because the matrix element would be very small for a polaron pair, not having the exciton wavefunction mixed in. This led to the conclusion that the higher energy peak must also be due to an intrachain transition [79]. To calculate the energy of this transition accurately would require the inclusion of Coulomb effects. However, it is clear that the resulting absorption would be quite similar to that of an isolated polaron, although not identical because the relative positions of the levels may be shifted by about 0.1 to 0.2 eV due to the different shifts of the band and polaron levels. This is in general agreement with the experimental results of Lane et al. to be discussed in Sect. 4.5.8 [80]. Detailed comparison with their results is not possible because the experiments were not carried out on PPV. Also, the calculated results for polaron pairs in PPV are in fair agreement with the experimental results shown in Fig. 4.6.

The calculations also gave the result that the energy required to create such a polaron pair is ~0.3 eV less than the energy required to create an exciton. Since annihilation of the pair requires back transfer of the electron or hole and subsequent recombination, the large energy difference between the pair and an exciton can explain the long decay time of the PA. Back transfer to form excitons may also explain the long tail found in the PL [77].

Application to PPV of high pressure was found to change the character of the emission. At 50 kbar, the emission, although considerably diminished, was clearly structureless and red-shifted, thus apparently having changed from exciton emission to excimer emission [81]. This situation is similar to that found in anthracene and naphthalene crystals. These also have the herringbone structure and do not show excimer emission at normal pressure, but do show it at high pressure [82]. This has been attributed to dislocations in the crystals which result in the formation of some pairs of chains with a configuration almost suitable for excimer emission. Presumably, the spacing of the pair is not small enough at atmospheric pressure but becomes so at high pressure. Alternatively, the high pressure could create defects appropriate for excimer formation.

4.5.8 Polaron Pairs in Other Polymers

Polaron-pair photoexcitations have been studied in a variety of π-conjugated polymers by PA and ODMR. The combination of the two techniques made it possible to distinguish between the absorption of polaron pairs and that of isolated polarons. Using these techniques, Lane et al. determined that both polaron pairs and polarons are characterized by two absorption bands [80]. They found that the high-energy PA band of the pairs is blue-shifted with respect to that of isolated polarons and the low-energy PA band of the pairs is relatively weak. Their results for DOO-PPV doped with C_{60} to create polaron pairs are shown in Fig. 4.8. They conclude that the results agree with the theory of Mizes and Conwell if the transition that transfers an electron between the partially occupied levels of the two polarons is weak or forbidden; that the latter is the case was noted in Sect. 4.5.6.

Another case in which there is good evidence for the existence of polaron pairs is trans-polyacetylene [83]. Like PPV, it tends to crystallize in a herringbone structure. Also, it is not soluble, so it is not possible to compare solution and film properties. There are two cogent arguments, based on the properties of the photoconductivity of t-$(CH)_x$, that illumination with above-bandgap light produces polaron pairs. As will be discussed in detail below in the section on photoconductivity and photogeneration, illumination of oriented samples with the electric vector of the light either parallel or perpendicular to the chains produces solitons with a subpicosecond lifetime [84]. The rapid decay of these solitons led to the conclusion that only ~1% of the photons create photocarriers (see, for example, [85]). Additional evidence for the correctness of this conclusion is that the mobility deduced from this number of carriers and the measured photoconductivity is a few cm^2/Vs, in agreement with the ps mobility deduced from the decay of light-induced dichroism [86], with the calculated mobilities of drifting polarons and solitons in t-$(CH)_x$ [34,87], and with the mobility deduced from the ps photoconductivity [88].

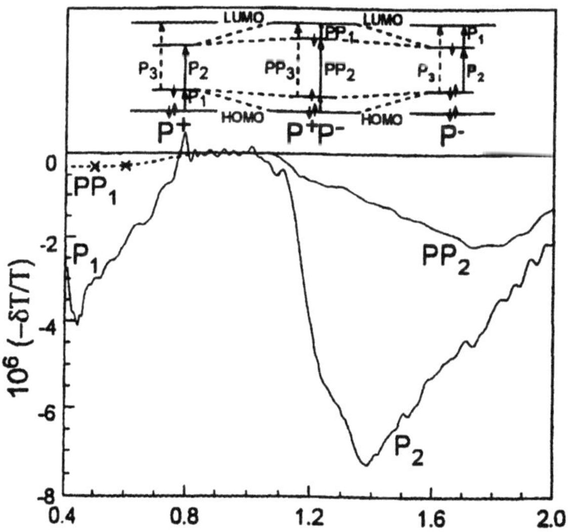

Fig. 4.8. PA spectra detected by magnetic resonance for isolated polarons (P_1 and P_2) and polaron pairs (PP_1 and PP_2) in DOO-PPV doped with C_{60}. Also shown are schematic diagrams and optical transitions of isolated polarons and polaron pairs. (Adapted from [9])

For the electric vector $\boldsymbol{E}$ perpendicular to the chains, where it is possible to create an electron and a hole on separate chains, in addition to the short-lived solitons, an excitation with a long lifetime is created. This excitation, showing little decay in 300 ps, is created by ~40% of the photons [84]. If these excitations were individual polarons, as suggested by Rothberg et al., the yield of carriers contributing to PC of a sample of t-$(CH)_x$ with arbitrary orientation should be much larger than 1% per incident photon. Also, PC for $\boldsymbol{E}$ perpendicular to the chains should be very much greater than PC for $\boldsymbol{E}$ parallel to the chains, which produced only short-lived solitons. However, PC for the $\perp\boldsymbol{E}$ vector was only 1.7 times as large as that for the $\|\boldsymbol{E}$ vector [89]. As pointed out earlier, these discrepancies are resolved if the polarons created by the $\perp\boldsymbol{E}$ vector are mainly bound in pairs by their Coulomb attraction and thus unable to contribute to PC [83,90].

Measurements of PA in t-$(CH)_x$ are available only in the range 0.3–0.5 eV [84]. The PA is continuous in this range of energies but shows two local maxima, at 0.35 and 0.45 eV, the latter being somewhat higher. Calculation of this absorption for a P^+P^- pair on adjacent chains in t-$(CH)_x$, using the SSH Hamiltonian plus the Coulomb potential and an interchain hopping term, gives the result that it ranges from 0.38 to 0.36 eV for chains from 200 sites down to ~100 and then increases with decreasing chain length to 0.6 eV at 30 sites [83]. It is noteworthy that absorption and resonant Raman spectra in t-$(CH)_x$ have been fitted well using a bimodal Gaussian distribution of conjugation lengths, centered at chains of 30 CHs and 200 CHs. According to the calculations, then, the local maximum at 0.38 eV can be assigned to conjugation lengths from 200 down to ~100 sites. The maximum at 0.45 eV could be obtained with a Gaussian centered at ~45 sites.

Another polymer for which there is evidence of polaron pairs is polythiophene, PT. A soluble derivative of PT is poly(3-octylthiophene), abbreviated P3OT. For P3OT a long-lived PA was found in thin films but not in dilute solutions [91]. The long-lived PA had a maximum at 1.2 eV and a smaller maximum at 1.9 eV. In this study, data were not taken at lower frequencies. PA on PT showed three maxima in the long-lived excitations, at 0.5, 1.35 and 1.8 eV [92]. The 1.8 eV band was identified as due to polarons. The two lower-frequency bands were found to decay together and were therefore assigned to bipolarons. In light of later knowledge, the bipolaron should show only one absorption band, and it therefore seems reasonable to assign these two absorptions to polaron pairs. The lower of the two PA peaks more or less agrees with the lower peak found in PPV and t-$(CH)_x$, as expected. The upper PA peak would be expected to be about the same as seen in PPV and PPV derivatives because the gaps are similar. It is in good agreement with that for PPV, as seen in Fig. 4.6, but in less good agreement with results for other PPV derivatives, such as those shown in Fig. 4.7, which have the second peak at ~1.8 eV. These differences are not understood.

Femtosecond pump-probe experiments, similar to those of the Bell Labs group on PPV and its derivatives, have been carried out on ladder-type poly(paraphenylene), L-PPP. In agreement with the results of the Bell Labs group, stimulated emission, which was attributed to singlet exciton recombination, and PA, which was taken to be due to polaron pairs (although it might have been to excimers), were found [93]. Thus, polaron pairs may exist in the same sample as aggregates.

Extensive ODMR investigations of J. Shinar and colleagues have provided insight into the interactions and influences on each other of the photogenerated excitations in many conducting polymers [94]. Excitation of all polymer films, solutions and blends at wavelengths of 458 nm and greater were found to yield a narrow PL-enhancing polaron resonance. This was attributed to enhanced polaron-pair recombination, removing a source of singlet exciton quenching. At shorter exciting wavelengths, in poly(3-alkylthiophenes) and all but the most ordered and pristine PPVs, a PL-quenching polaron resonance is found. This is attributed to enhanced fusion of free polarons to bipolarons, which also act to quench singlet excitons.

4.5.9 Some Properties of Excimers

The Long Lifetime

Both excimers and polaron pairs involve an electron on one chain and a hole on a neighboring chain. The excimer is emissive because its wavefunction also includes a contribution from an exciton on each chain, giving it a net dipole moment. As noted in Sect. 4.2.1, due to inversion symmetry the electronic states of these polymers must have either even or odd parity. A symmetrized wavefunction may be made up of a symmetric or antisymmetric combination of exciton wavefunctions on each chain plus an admixture of a symmetric or antisymmetric combination of the wavefunctions of the two possible charge-transfer excitons, i.e., the hole on chain 1, electron on chain 2 and the hole on chain 2, electron on chain 1. It is found that the sum of the antisymmetric combinations leads to an even parity wavefunction, while the sum of symmetric combinations leads to an odd parity wavefunction. When the amount of admixture of charge-transfer wavefunctions is varied to find the minimum energy of the excimer, it is found that the even parity wavefunction has the lower energy [71,95]. This would result in the transition from the excimer to the ground state of the polymer, which also has even parity, being forbidden. Emission is made possible by the thermal vibrations, which mix in states of odd parity. Nevertheless, this accounts for the very long lifetime of the excimer state, as seen particularly in MEH-PPV [73].

Emission-Frequency Dependence on the Distance Between Chains

Because the chains are in a thin film, they may not be able to move on optical excitation to obtain the most favorable separation for the excimer. As a result the emission frequency must vary from that of the exciton (chains far apart) to a minimum frequency characteristic of the excimer with optimal spacing. To study the variation of the emission frequency with the distance between chains, it is necessary to have a value for $t_\perp$. Wu and Conwell took $t_\perp$ as the value required to give the observed shift in the peak emission of the excimer relative to that of the exciton, with the interchain distance of the excimer taken as 0.34 nm, the value calculated for the ground-state spacing between CN-PPV chains. The result is a smooth variation of the emission frequency from 2.3 eV (the exciton frequency) at a distance of 0.5 nm to ~1.75 eV at 0.33 nm [96]. Note that the variation of emission energy with separation of the chains shown for MEH-PPV in this publication is incorrect because it was based on incorrect experimental data in the literature for the shift. From the results of Jakubiak et al. cited above the shift in the peak emission in going from exciton to excimer is larger in MEH-PPV than was found earlier.

Dissociation of Excimers

Dissociation of excimers may occur due to high electric field or thermal energy. Observations of the PL decay of MEH-PPV in high electric fields indicate that the excimer, identified by its long lifetime, is less susceptible to quenching than the exciton [73]. Evidence for thermally activated excimer dissociation to form single-chain excitons, some of which subsequently decay radiatively, has been obtained for CN-PPV [97] and MEH-PPV [73]. It was earlier postulated that the long tail in PL observed for PPV is due to dissociation of polaron pairs to reform single-chain excitons [77]. These dissociation phenomena are all in agreement with the theoretical finding of a binding energy for a polaron pair, and thus presumably for an excimer, larger than that for an exciton by Mizes and Conwell [78].

4.5.10 Aggregates

The observation of strongly downshifted, featureless emission from a film, as compared with that from a dilute solution of the same polymer, may not signify that the excitation in the film is an excimer. The emission may also result from an aggregate, in which two or more chains interact sufficiently so that the electronic wavefunctions are delocalized over them in the ground state as well as in excited states. As a consequence the aggregate emission may be directly excited optically, while that of the excimer cannot be. Aggregates in conducting polymer films were first recognized in ladder-PPP (L-PPP) [98]. A low-energy (yellow) emission band was observed in the PL spectrum of a solid film, while dilute L-PPP shows only blue PL. The absorption spectrum in the film exhibits a pronounced low-energy tail, extending to the yellow, that is not seen in the solution. Excitation in the tail produces the yellow emission. This emission is therefore assigned to aggregate states formed by portions of different polymer chains that are closely packed in the L-PPP film. There is a great variation in the ratio of yellow to blue emission from film to film. It is interesting that the weight of the aggregate band relative to that due to isolated excitons is found to be much larger in emission than in absorption. This demonstrates highly efficient transfer of excitons into aggregate states.

Phenomena similar to those just described for L-PPP have been seen also in the pyridine-based polymers poly(p-pyridyl vinylene), poly(p-pyridine), and poly(p-pyridyl vinylene p-phenylene vinylene) (PPyVPV) [99,100]. The PL of solution samples is characterized by high quantum efficiency (>70% in PPyVPV) with a radiative lifetime of 1 ns, while film samples display strongly red-shifted, featureless emission with low quantum yield (<20%) and much longer decay times. The low quantum yield of the film is attributed to the long radiative lifetime, which allows more time for non-radiative recombination. The long decay time suggests that the radiative transition is forbidden, as discussed above for the excimer. Near-field scanning optical microscopy (NSOM) measurements on PPyV films indicate that emission is localized in partially aligned regions of the film of domain size ~200 nm [101]. Although the domain size is too large to correspond to a single aggregate, it is reasonable that aggregates would be more likely to form in more ordered regions of the film. Interestingly, the effects described above are not seen in powders of the pyridine-based polymers, which can be expected to be more disordered.

Other polymers for which aggregates have been reported are a polyfluorene [102] and dialkoxy-p-phenylenevinylene and dialkoxy-thiophene in solution [103]. In the polyfluorene each pair of phenylene rings is bridged by a carbon atom that may also carry additional substituents. The aggregate work was reported in PFO, a dioctyl-substituted polyfluorene. With appropriate treatment, as will be discussed in the next section, PFO shows the properties of additional low-frequency absorption, as compared with solution, in the form of a small distinct peak; red-shifted emission which can be directly excited; and emission that indicates rapid excitation transfer from directly excited on-chain excitons into the aggregates. As might be expected from the similarity in structure, with appropriate treatment PFO has similar properties to well-ordered L-PPPs, such as very-well-resolved vibronic structure in emission and a small Stokes shift. With other treatments, PFO shows no low-energy aggregate peak, but a very long lifetime, presumably due to excimers.

4.5.11 Factors Affecting the Formation of Polaron Pairs, Excimers and Aggregates

It is apparent from the above discussion that the formation of polaron pairs, which are non-radiative, results in degradation of the quantum efficiency of luminescence. Even though they are radiative, excimers and aggregates, particularly the former, also degrade the quantum efficiency because the longer radiative lifetime allows more non-radiative transitions. It is apparent that formation of these interchain entities depends on the physical structure of the film. Undoubtedly, improvements have been made. Denton et al. [67] note that refinements in PPV synthesis and processing have made possible the observation of stimulated emission in that material, whereas only PA was detected in earlier experiments at Cambridge [104]. Two factors that clearly can be used to influence structure are heat treatment and, in the case of soluble polymers, choice of solvent.

Experiments on PFO have been particularly informative in these regards because its physical structure can be modified without changes in chemical structure [102]. The modifications can be achieved by varying the solvent and the thermal treatment. It is generally understood that in poor solvents polymer chains are forced into close proximity by unfavorable interactions with the solvent, encouraging the growth of excimers and aggregates. For PFO, thermal treatment consisting of cooling to liquid nitrogen and then allowing the sample to warm up slowly to room temperature was shown to give rise to the low-energy absorption peak characteristic of the aggregate. The heat treated sample showed clearly resolved vibronic peaks in emission, indicating that the heat treatment produced a well-ordered state with increased conjugation length.

A number of methods have been used to reduce excimer and aggregate formation. Existence of an excimer requires essentially a pair of conjugated chain segments of length a few monomers or more on adjacent chains that are: (1) parallel to each other, (2) in appropriate interchain registry (which may be different from that of the ground state), and (3) close enough (≤ 0.4 nm) to have sufficient overlap of the π wavefunctions on the two segments. It appears that conditions (1) and (2) need not be satisfied for a polaron pair. The conditions must also be satisfied for an aggregate, although in general more than two chains are involved, but the interchain distance must be smaller to achieve interaction in the ground states. Many researchers have demonstrated improved PL, and presumably a decrease in excimer or aggregate formation, by designing samples to violate these conditions. Use of a good solvent, which tends to keep the chains apart, is one clear method. Dilution, which can also be achieved by making a blend in an inert polymer, has been shown to increase PL efficiency of many polymers. Making a blend in a suitable polymer, such as ultrahigh molecular weight polyethylene, has been shown, in fact, to yield a system with considerably reduced disorder [105]. Short conjugation lengths, which lead to poor interchain registry, have been shown to increase PL. The short conjugation length may be achieved, for example, in PPV by incomplete conversion of the precursor polymer [106] or by insertion of sections of another polymer into the backbone. For example, insertion into the L-PPP backbone of short segments of PPP, which introduce a twisting angle, was found to eliminate the yellow emission due to aggregate formation in that material [107,108]. Another demonstration that poor interchain registry works against pair creation was the observation of increased PL in PPV samples engineered to have a portion of the usual trans-linkages replaced by randomly occurring cis-linkages. A novel method used for reducing aggregates in PPyVPV polymers is to attach continuous chains, termed "straps", to the 2,5 ortho positions of the phenylene rings [109].

4.6 Single Chain Excitations Other than Singlet Excitons

One of the most important single chain excitations, the polaron, has already been discussed in Sects. 4.2.1 and 4.2.8. Further discussion of polarons will be postponed to the section on photoconductivity. Photogeneration of solitons and polarons in polyacetylene will also be discussed in the section on photoconductivity.

4.6.1 Bipolarons

Many groups, finding in PA two peaks that behaved similarly with respect to decay rate and/or temperature dependence, attributed them to bipolarons [110–113]. Because, as noted in Sect. 4.2.5, bipolarons should show only one absorption peak, it is now clear that the peaks must be due to polarons or polaron pairs, probably the latter.

A particularly extensive study of infrared absorption and steady-state PA in PPV and a number of its alkoxy derivatives, including MEH-PPV, carried out by Voss et al. [114], also found two correlated absorption peaks and concluded that they represented bipolarons. In this instance sufficient data were given to make a strong case that they were observing polaron pairs. Their studies included the IRAV, which are infrared modes made active by the breaking of the translational symmetry of a chain due to self-localized charges such as solitons, polarons and bipolarons [115]. One such mode occurs for each Raman-active optical phonon band. The frequencies of the IRAV depend on the pinning potential that resists the translation of the charges. In a system with no pinning, the lowest frequency mode would be at zero frequency. In a doped sample the pinning is due to the donor or acceptor ions. In trans-polyacetylene, for example, the pinned mode in a moderately doped sample is centered at about 900 cm^{-1}. For heavy doping it shifts down to 500 cm^{-1} [116], reflecting the effective lowering of the potential barrier confining the carrier motion when the potential wells due to adjacent ions overlap [117]. The pinning frequency for polyacetylene samples where the charges are photogenerated was also found to be 500 cm^{-1} [118]. Since in the case of photogeneration there should be no ions to restrict the motion, the closeness of the two values was surprising. A similar situation was found for PPV and its derivatives by Voss et al.; the frequencies of the IRAV for the photogenerated defects are quite close to those of defects resulting from doping. For the mode coming from the trans-vinylene C-C stretch, for example, the two frequencies are 1275 and 1281 cm^{-1}, respectively.

It was suggested that the photoexcitations Voss et al. were studying were actually polaron pairs [79]. The arguments given would hold for excimers also. For dimethoxy-PPV, for example, their PA peaks were found to correspond almost exactly to the peaks found in ps PA for methoxy-PPV by Hsu et al. [60] and attributed by them to polaron pairs. The peaks for the other derivatives studied by Voss et al. occur at similar frequencies. Additional evidence for the identification of the excitations as polaron pairs rather than bipolarons comes from the properties of the IRAV detailed above. The fact that the IRAV arising from the photoexcitations have almost the same frequency as the IRAV arising from excitations introduced by doping can be explained by the pinning provided by the oppositely charged polaron in the pair; this pinning should be comparable to that provided by a doping ion. Additional evidence that polaron pairs are generated rather than bipolarons comes from the observation of Voss et al. that there is a red shift between photoinduced and doping-induced modes that increases with the length of the side chain, going, for example, from 6 cm^{-1} in PPV to 136 cm^{-1} for the mode associated with the trans-vinylene C-C stretch in DOO-PPV. Voss et al. attributed the increase in red shift to the decrease in Coulomb binding between bipolarons on adjacent chains as the interchain distance grows due to the longer sidegroups. For the case of photogeneration, however, the likelihood of there being a bipolaron on an adjacent chain is quite small. A reasonable explanation is that the decreased pinning is due to the decreased Coulomb binding between the two polarons of the pair as the interchain distance grows.

The large pinning found for photoinduced charges in trans-polyacetylene can also be attributed to this mechanism. Evidence has been given that there is strong photogeneration of polaron pairs in trans-polyacetylene [83]. This will be discussed in Sect. 4.7.4.

4.6.2 Triplet Excitons

A peak generally seen in the PA of PPV at ~1.5 eV is attributed to a transition within the triplet manifold [110]. This identification was reinforced by PADMR measurements on PPV [62]. The peak due to the transition in the triplet manifold was also found at 1.5 eV for poly(3octylthiophene) [119]. The original excitations being singlet excitons, it was concluded that the triplet excitons arose due to intersystem crossing, ISC. The time for ISC was determined to be 1.2 ns for poly(3octylthiophene) from the time delay for the rise of triplet–triplet absorption [119]. This time was unaffected by defects or impurities, consistent with the coupling of singlet and triplet manifolds being due to the relatively large spin-orbit coupling of sulfur. The time for ISC in PPV films was estimated to be 4 ns [62], the ISC, in the absence of sulfur, perhaps enhanced by the presence of paramagnetic O_2 molecules [120].

Measurements on MeL-PPP, because of its relatively high interchain order and small defect density, yielded additional information. The strong triplet–

triplet absorption seen in PA in this material is at 1.3 eV [121]. Using a technique that involves measurement of the photomodulation signal far from the steady state, Wohlgenannt et al. have been able to determine the photogeneration quantum efficiency for the various photoexcitations in MeL-PPP. For triplets they find a step function increase at the optical gap, followed by a flat region and then a rise starting at ~3.2 eV that ends in a plateau at 3.7 eV. They suggest that this indicates that triplet generation occurs by two processes. The first process is associated with thermalized singlet excitons and therefore has a step-function energy dependence at the optical gap as does the PL. They identify this process as ISC. The process with an onset at 3.2 eV they attribute to hot excitons, specifically to singlet exciton fission into two triplets. From the onset of this process at 3.2 eV, they deduce that the triplet energy is 1.6 eV. ODMR studies show that the triplet exciton is quite small, ~0.3 nm, thus essentially confined to a phenylene ring [122]. The strong localization is consistent with the deep energy level.

Combination of two triplet excitons to produce a singlet has also been observed. Evidence comes from the delayed fluorescence of PPV and poly(p-phenylene ethynylene), PPE, observed at 20 K [123]. With the excitation varied at frequency ω, the delayed photoluminescence was found to have a frequency component at 2ω, and this was attributed to triplet–triplet annihilation. The triplets may have originated from interchain polaron pairs, which, although they are presumably created as spin singlets, may end up as triplet polaron pairs due to hyperfine interaction of the electron spin with magnetic moments of protons [124]. At the low temperatures at which the experiments were done, a singlet pair could not make the transition to a singlet exciton because the energy of the pair is lower. However, triplet pairs could make the transition to triplet excitons because, as seen above, the triplet exciton energy is much lower. The delayed fluorescence from this source was 3% of the total fluorescence in PPV. The experiments also gave values for the triplet lifetime due to monomolecular decay as 70 μs in PPV.

Triplet excitons have also been studied in the pyridine-based polymers discussed in Sect. 4.5.10. The pyridine-based polymers differ from polymers such as PPV and PPP in that they possess (n, π^*) states resulting from the promotion of a lone-pair electron from the nitrogen heteroatom to the π^* backbone. Quantum chemical calculations indicate that the lowest (n, π^*) states should be located more than 1 eV above the lowest singlet (π, π^*) transitions if the polymer is planar [100]. This is significant because systems in which an (n, π^*) state lies below the lowest (π, π^*) singlet generally display only weak luminescence because ISC between the (n, π^*) and (π, π^*) states is allowed to first order in the spin-orbit interaction [125]. Nevertheless, the presence of (n, π^*) states leads to enhanced ISC to triplet excitons, particularly for the powder form of the pyridine-based polymers, e.g., PPyV [126]. In the film form, as discussed earlier, aggregates play a key role. PA studies show direct conversion of singlet to triplet excitons on the ps time scale. For

PpyV powders the triplet–triplet absorption is seen at 1.8 eV, corresponding to exciton states with lifetimes ~540 μs [126].

4.6.3 Biexcitons

The biexciton is a bound state of two excitons. The existence of stable biexcitons in π-conjugated polymers was predicted on the basis of a Coulomb correlated model [127]. The Coulomb interaction parameters were characterized by the authors as realistic (on-site repulsion $U = 7.5$ eV). It was suggested that the high-energy ps PA found in films is due to the transition from the exciton to the biexciton. As discussed in Sect. 4.5.5, fs pump–probe experiments were carried out by Klimov et al. on the five-ring oligomer MEH-DSB to detect biexcitons. In solution at low intensity they saw a single PA peak at 1.4 eV and stimulated emission (SE), both of which can be explained to be due to singlet excitons. With increasing pump intensity they found an additional fast component in SE with amplitude proportional to the square of original, slower component. They infer that the fast SE component originates from doubly excited molecules, i.e., intrachain two exciton states or biexcitons. The enhanced radiative decay can be understood in terms of the superradiant emission of two coherent oscillators. In films they found the same 1.4 eV PA as was seen in solution, and attributed to singlets, and an additional PA at 1.8 eV. The latter had no SE associated with it and Klimov et al. attributed it to polaron pairs. They concluded that in the films the biexciton is unstable and decays on a ps time scale into polaron pairs [26].

4.7 Photoconductivity

As discussed in Sect. 4.2.2, it has been argued by some workers in the field that absorption in PPV of a photon with energy past the absorption edge creates a free electron and hole, i.e., the exciton binding energy $\varepsilon_b \le k_B T$. A good deal of the earlier section of this review, however, has been devoted to showing that the binding energy in PPV and other NDGS polymers is distinctly greater than $k_B T$, ~0.4 eV. Thus, the possible carriers of photocurrent in NDGS polymers are polarons and bipolarons rather than free electrons and holes. In degenerate ground state polymers charged solitons may also contribute to the photocurrent.

It has been proposed that the fast transient photoconductivity, PC, in conducting polymers could be explained as a displacement current generated by the electric-field-induced polarization of bound excitons rather then due to the motion of current carriers [63]. This proposal has been refuted by a number of arguments, chief among them being that the required exciton polarizabilities are far too large in comparison with the measured value in single crystal polydiacetylene, which should fairly well represent the value in PPV, etc. [128]. A second argument against the PC being due to a displacement current is that it cannot explain the large increase in the magnitude and the decay time of transient PC in MEH-PPV and other polymers on addition of as little as 1% C_{60}. C_{60} acts as electron acceptor to the photoexcited state of MEH-PPV, leading to ultrafast electron transfer and the creation of free hole polarons [129–131].

There has been a great deal of discussion on the process of photogeneration and the relative amounts of different types of carriers generated. This will be taken up Sect. 4.7.1 for PPV. In Sect. 4.7.2 we will discuss transient and steady-state PC in PPV. That will be followed by sections on L-PPP and t-$(CH)_x$.

4.7.1 Photogeneration in PPV and Its Derivatives

It has been directly demonstrated experimentally that the carriers involved in PC are derived primarily from exciton dissociation. Specifically, reduction of the exciton population (by SE), which was demonstrated by a decrease in PL, was shown to reduce PC [132]. The fact that PC is seen at the optical absorption edge has been one of the arguments for the point of view that free carriers are directly excited at the absorption edge. If the binding energy of the exciton were greater than $k_B T$, it was argued, an activation energy should be required for PC, which experimentally is not the case for PPV. (PC onset in the polydiacetylenes is above the absorption edge by ~0.5 eV, the exciton binding energy, but the situation there is quite different [133]).

The variation of PC with photon frequency for MEH-PPV is shown in Fig. 4.9. The PC rises abruptly at the absorption edge, then varies little with frequency up to ~0.7 eV past the threshold before the steady-state PC starts increasing rapidly. Significantly, at and near the absorption edge, where the absorption is quite high, the yield of photocarriers per incident photon is quite small, typically 10^{-2} or 10^{-3}, even as little as $<10^{-5}$ for some polymers, for fields below 10^4 V/cm. This feature is quite different from what is seen in 3-D photoconductors, where the yield in this range is typically of the order of unity. It has been suggested by a number of workers that the weak PC between the onset and the rapid rise is due to exciton dissociation at surfaces and interfaces [20, 133]. That implies that the weak, photon-energy-independent PC between the absorption edge and the rapid rise is not characteristic of the bulk crystal and may be called *extrinsic*, while the large, steady-state PC at higher photon energies is termed *intrinsic*. The photon energy at which intrinsic PC begins to dominate the extrinsic PC depends, of course, on how large the extrinsic conductivity is for the particular sample, in turn determined by the details of the contact interfaces, the concentration of acceptor-type impurities and defects, etc.

The rapid rise in PC is seen to begin at ~2.7 eV in the data shown in Fig. 4.9. Thus, the excess energy beyond the exciton creation energy, or the absorption edge, to create an intrinsic carrier is ~0.7 or 0.8 eV. The rapid rise in PC begins at ~3 eV for another MEH-PPV sample (Chandross et al. [20]).

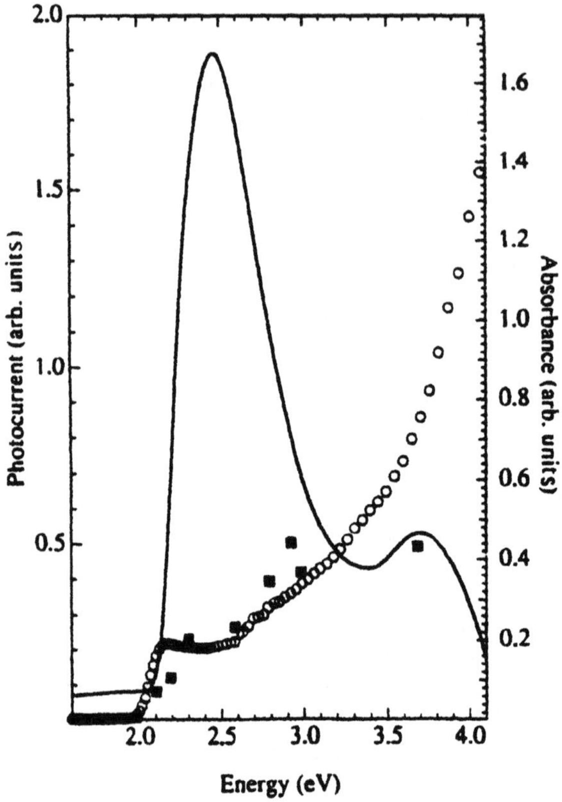

Fig. 4.9. Steady-state (○) and transient (■) photoconductivity excitation spectra and absorption spectrum (*solid line*) of a cast film of MEH-PPV at room temperature. (After [16])

Apparently, the sample of Chandross et al. has greater extrinsic conductivity. It is significant, however, that the beginning of the rapid rise for both samples is at essentially the photon energy corresponding to the distance between the valence-band edge and the conduction-band edge for MEH-PPV, as will now be shown. The exciton binding energy and the distance the polaron level is below the band edge are expected to be about the same for PPV and MEH-PPV. By the arguments of Sect. 4.2.2 the band edge of MEH-PPV should then be below that of PPV by the amount that its absorption edge is lower, ~0.4 eV, thus ~2.8 eV above the valence-band edge. This is the energy required to create a free electron and hole in MEH-PPV, albeit with zero kinetic energy. With increasing photon energy, the free electron and hole have more kinetic energy and an increasingly better chance of getting away from each other, thus avoiding exciton formation and becoming photocurrent carriers. Thus, the beginning of the rapid rise in PC occurs for photon energy in the neighborhood of what we earlier called the unrelaxed gap.

That the weak PC observed close to the threshold is actually extrinsic, specifically due to exciton dissociation at an electrode, has been carefully documented for the particular case of a PPV derivative with ether linkages in the backbone, to be denoted PPV-ether. This derivative can be prepared with much greater purity than PPV made by a standard precursor method. To demonstrate the extrinsic nature, a sample was prepared in an LED configuration with the PPV-ether sandwiched between indium tin oxide, ITO, and aluminum electrodes. The forward photocurrent was measured upon irradiating through the positively biased ITO electrode with a photon frequency at the peak of the absorption. When an 8 to 10 nm thick SiO layer was inserted between the ITO and the PPV-ether to prevent photoinjection of positive photocarriers, it was found that the forward injection current was reduced by typically three orders of magnitude [134]. It is clear that the major portion of the photocurrent in the absence of the SiO layer was extrinsic, due to dissociation of excitons at the metal–polymer interface. Current in the reverse direction was the same with and without the SiO layer because the other contact was non-injecting. All currents were linear in the light intensity.

Eliminating extrinsic conductivity by the SiO layer in the sandwich, Barth et al. went on to study the field and temperature dependence of the intrinsic PC. For many materials the variation of intrinsic PC with electric field intensity, temperature and photon frequency has been shown to be in good agreement with Onsager theory of geminate pair recombination [135]. The theory starts from the premise that a photon of sufficient energy has generated a Coulombically bound electron and hole (geminate pair) which thermalize at a separation distance denoted r_0. This initial separation distance increases with photon energy. Whether they dissociate into a pair of free carriers or recombine to form an exciton is determined by r_0 and by their subsequent diffusion in the electric field, which is a superposition of the Coulomb well due to their attraction and the applied field. The theory predicts that electron–hole dissociation is an activated process, with the activation energy decreasing with increasing field. The field dependence predicted is quite steep over a wide range of fields. The experiments of Barth and Bässler showed that over a wide range of fields and temperatures the yield of free carriers in steady-state PC for PPV-ether and another PPV derivative, PPV-amine, in the intrinsic range was in fair agreement with Onsager theory [136]. Better agreement could be obtained by modifying the Onsager theory to allow for the energetic randomness of the hopping sites in the polymers, as discussed by Barth and Bässler. (For a general discussion of hopping in a disordered medium see, for example, [137].)

There is another factor operative in the variation of photogeneration yield with photon frequency. As shown in Fig. 4.3 the absorption of MEH-PPV has four peaks at photon frequencies in the range 2 to 6.5 eV. Measurements of PC of MEH-PPV as a function of photon frequency showed a peak associated with the beginning of each absorption peak, the feature being quite small for peaks A_1, A_2 and A_4 but larger for A_3 [138]. Each of the absorption peaks represents a band and, as discussed in Sect. 4.2.2, has associated with it an exciton. Wavefunctions of the excitons associated with the four peaks, and of the exciton associated with a peak at still higher frequency, A_5, have been calculated [138]. It was found that the excitons associated with A_1, A_2, and A_4, where the electron and hole both have delocalized wavefunctions or both have localized wavefunctions (see Sect. 4.3.2), have the two carriers spatially overlapped, whereas in A_3 and A_5, corresponding to transitions between one localized state and one delocalized state, the electron and hole are further separated. This can account for the greater photoconductive yield observed at the photon frequency corresponding to the beginning of A_3, thus to the exciton associated with that band [138]. This effect superimposes some structure on the extrinsic and intrinsic regions.

4.7.2 Transient Versus Steady-State Photoconductivity in PPV and Its Derivatives

To address the question of what happens at the earliest times after photoexcitation, extensive studies of transient PC were carried out on PPV [139]. Very similar results were obtained from measurements of PC in MEH-PPV, except that the peak transient PC was an order of magnitude smaller [140]. One may speculate that this is the result of more carriers being geminately bound, in excimers in MEH-PPV. The transient PC was initiated by an excitation pulse of width 20 to 30 ps and its decay measured as a function of time. The excitation was at 2.92 eV, close to the peak of the absorption. The transient peak PC for an incident photon flux of $\sim 10^{15}/\mathrm{cm}^2$ was 0.25 S/cm at a field of 2×10^4 V/cm. This is comparable to the peak transient PC found in polyacetylene. It was found that there is an initial fast response with decay time of ~100 ps, followed by a slower component with decay time of ~600 ps. Lee et al. note that they were not able to resolve the initial time response of the photocurrent, concluding that the initial decay time might be much less than 100 ps. The magnitude of the fast component was independent of temperature and linearly dependent on light intensity. It was also linear in electric field up to fields of a few times 10^4 V/cm [141]. With an estimate of the carrier concentration, a lower limit for the mobility in the linear field range was obtained as 0.6 $\mathrm{cm}^2/\mathrm{Vs}$, in good agreement with the measured polaron drift mobility in PPV [142]. At very short times drift along the chains is expected to be dominant. At longer times, after polarons have drifted to the end of their initial conjugation lengths, hopping becomes the rate-limiting step. The mobility in the hopping regime is orders of magnitude smaller. The increase in mobility at high fields may be attributed to the field assisting hopping polarons to get over barriers due to energy differences between hopping sites.

The lack of temperature dependence and the linear proportionality to the applied field of the fast component are clearly in disagreement with the Onsager theory, and were taken as evidence that free carriers were generated rather than excitons, thus that the exciton binding energy $< k_B T$. In light of the discussion of the previous section, however, a more reasonable conclusion is that the fast component represents extrinsic PC. In agreement with this, as seen in Fig. 4.9, the transient PC does not show the large rise at photon energies past 3 eV that is shown by the steady-state PC. The slow component, which is temperature dependent, corresponding to an activation energy of 100 meV, could represent intrinsic PC. Its activation energy was similar to the activation energy observed for steady-state PC, 140 meV. The implication is that the process of carrier generation by the Onsager mechanism takes a time of the order of nanoseconds rather than picoseconds for the conditions in PPV at room temperature.

Finding that the slow component in the ps transient PC varied as the square root of the light intensity, Lee et al. assigned the slow component to bipolarons [139]. In previous studies of the steady-state PC of PPV, this group had found very slow (stretched exponential) relaxation [17]. Because bipolarons are expected to have much lower mobility, they interpreted their results as indicating that the photocurrent is carried by mobile polarons near the surface but bipolarons, formed by fusion of like-sign polarons, within the bulk. At the time of this interpretation it was reinforced by the belief that the two peaks seen in PA of PPV were due to bipolarons. As discussed in Sect. 4.6.1, this is not correct. It was suggested in that section that the two PA peaks are due to polaron pairs (see Sect. 4.5.7). The photocarriers are thus polarons because the pairs are not expected to be mobile. The extrinsic PC might include a contribution from hole polarons coming from excitons dissociated by carbonyls, a common defect in PPV (see Sect. 4.4.3). The persistent PC might be due to the slow recombination of holes with the electrons trapped on the carbonyls, or other deep traps.

As has been mentioned many times in this review, there has been a series of papers devoted to the proposition that free carriers are directly generated by photons past the absorption edge, the exciton binding energy being less than $k_B T$. Evidence that convincingly refutes the arguments for this proposition has been detailed. One piece of evidence not yet refuted is the recent finding that in thin films of MEH-PPV, ~100 nm in thickness, the steady-state PC is only weakly dependent on temperature [143]. A possible explanation is that PC of this sample is still in the extrinsic range for the particular photon frequency, electric field intensity and temperature employed in the measurements. It must also be remembered that there are significant deviations from the predictions of Onsager theory for these polymers due to their disorder. For the thicker samples in the study of Moses et al. an activated PC was found, in agreement with the Onsager model.

4.7.3 Photogeneration and Photoconductivity in L-PPP

Photoconduction was studied in thin films of methyl-substituted L-PPP, MeL-PPP, sandwiched between ITO and Al electrodes. This material is advantageous to work with because the methyl groups prevent formation of aggregates. PC was found to be largely similar to that in PPV and derivatives.

There was an extrinsic contribution consisting of hole injection arising from dissociation of excitons at the ITO and Al electrodes. Another contribution, found at long wavelengths, i.e., close to the absorption edge, and showing little wavelength dependence, was attributed to defect-sensitized dissociation of thermalized excitons. For short wavelengths, $\leq$320 nm, there was a rapid rise of PC with increasing photon energy, which was therefore identified as intrinsic PC. The onset of the intrinsic region was found to be at an excess energy of 1.1$\pm$0.1 eV beyond the exciton energy or the absorption edge [144]. As noted by the authors this excess energy is larger than the value of 0.7 or 0.8 eV found for PPV derivatives.

The quantum efficiency for polaron photogeneration in MeL-PPP was determined directly by measurements on the PA peaks identified as due to polarons [121]. The results of this group differed somewhat from those of Barth et al. [144]. They found the region just above the absorption edge in which the quantum efficiency varies little with photon energy to be quite short, $\sim$0.5 eV rather than 1.1 eV. They were in agreement with Barth et al. [144] in finding the quantum yield in the intrinsic region, where it is rising rapidly, to be quite weakly dependent on temperature. This is, of course, in disagreement with the results obtained for PPV and at variance with the Onsager mechanism for photogeneration.

Two different mechanisms have been proposed to account for the temperature-independent intrinsic photogeneration of MeL-PPP. In one mechanism the photon energy in excess of the exciton binding energy is considered to be distributed among the vibrations of the chain segment housing the exciton, thus raising the local temperature above ambient for the short time before the excess energy decays. The combination of this heating and a high enough electric field are supposed to overcome the Coulomb binding of the electron and hole and allow them to form an on-chain geminate pair of free polarons. The polarons are then free to jump to other conjugation lengths and carry photocurrent [58].

The other mechanism proposed to explain the temperature independence of the intrinsic photogeneration in MeL-PPP invokes interchain tunneling of the electron on the hot exciton [121]. This mechanism, of course, creates a polaron pair rather than free polarons. The authors suggest that the pairs will dissociate without an activation energy being required. Their reasoning is based on the Coulomb binding of the pair being only 0.2 eV due to an interchain distance of 2 nm. The latter is an unrealistically large interchain distance, however. A more likely value is $\sim$0.6 nm, for which the Coulomb binding of the pair would be 0.7 eV, making it unlikely that they could contribute to PC at the usual fields and temperatures.

4.7.4 Photogeneration of Solitons and Polarons and Photoconduction in Trans-Polyacetylene

A theoretical study by Su and Schrieffer predicted that, when absorption of a photon created an electron and a hole on the same t-$(CH)_x$ chain, rather than forming an exciton they would evolve into a soliton–antisoliton pair within $\sim 10^{-13}$ s [2]. This finding stimulated many experimental studies. A study of the PA that resulted from illumination with above-bandgap photons found peaks at 1.35 eV and ~0.5 eV [145]. The excitation associated with the 1.35 eV peak was found to be neutral, while that associated with the lower energy peak, later found by more accurate measurements to be at 0.43 eV, was charged [146].

Several kinds of evidence established that the charged excitations were solitons, in agreement with the prediction of Su and Schrieffer. One piece of evidence was that the production of these excitations also gave rise to absorption lines further in the infrared. These absorptions were identified as IRAV (see Sect. 4.6.1), having frequencies close to the IRAV found in doped trans-polyacetylene where the donated electrons and holes go into soliton states [118, 147]. In the paper by Vardeny et al. only two IRAV frequencies, at 1370 and 1260 cm^{-1}, were found, and it was concluded that the photogenerated solitons are not pinned. As noted in Sect. 4.6.1, Blanchet et al. also found the pinned mode for these solitons, at a frequency of 500 cm^{-1}. The origin of the pinning is not clear. One possibility is that it is due to geminate soliton–antisoliton pairs. Additional evidence for the production of charged solitons was obtained by Flood et al. [148], who showed, using electron spin resonance, that the photoinduced excitations do not have spin. As for the neutral photoexcitation found at 1.35 eV, it was widely speculated that it was a neutral soliton. This was finally established by Wei et al., who, using the ADMR technique, showed that this excitation has spin 1/2 [59].

Picosecond PA measurements yielded the information that the charged solitons generated as described by Su and Schrieffer have a very short lifetime, of the order of a ps [149]. It was suggested by Kivelson and Wu that, when a pair of oppositely charged solitons is created, before they have a chance to separate, their interaction will convert them to a pair of neutral solitons [150]. This accounts for the short lifetime of the Su–Schrieffer solitons and the observation of the PA (at 1.35 eV) due to neutral solitons. The fact that no rise time for the 1.35 eV absorption was seen in experiments with fs resolution [151] was rationalized to be the result of both the extremely rapid generation and rapid decay of the neutral solitons [149].

Experiments with fs resolution were carried out on oriented samples of t-$(CH)_x$ when such samples became available. It had been speculated that light polarized perpendicular to the chains might create an electron and hole on different chains, in which case each would evolve into a polaron rather than a soliton. Being on different chains, the polarons could be expected to have a longer lifetime. The experiment showed that light polarized perpendicular to the chains resulted in a PA with the same subpicosecond lifetime and absorption spectrum, both peaking at ~0.43 eV, as light polarized parallel to the chains [84]. Thus, light polarized perpendicular to the chains also creates primarily intrachain excitations with substantial quantum yield. (This had been predicted by D. Baeriswyl and K. Maki [152].) For the case of light polarized perpendicular to the chains, however, as was pointed out in Sect. 4.5.8, the PA did not decay to zero but rather to ~40% and persisted at almost the same level for 300 ps, as long as it was observed. The PA spectrum associated with the long-lived excitation did not change from 5 to 100 ps after the light was turned off. Relative to the PA of the charged soliton it was shifted to the infrared with broad local maxima at 0.35 and 0.45 eV. Rothberg et al. assigned this long-lived excitation to polarons. Although some polarons are undoubtedly generated, as discussed in Sect. 4.5.8, most of these excitations must be polaron pairs. Maxima in the PA at 0.35 and 0.45 eV can be accounted for as arising from a bimodal distribution of chain lengths [83]. Such a bimodal distribution was invoked earlier to account for the observed infrared absorption of t-$(CH)_x$ [153].

4.7.5 Photoconductivity in Trans-Polyacetylene

As was described above for PPV, the PC of t-$(CH)_x$ consists of a fast and a slow component. Particularly in light of the results described in Sect. 4.7.4 on photogeneration of solitons, it is desirable to study the fast component with high temporal resolution. After measurements by a number of groups with time resolution of ~100 ps [85,89,154–157], measurements were carried out with a time resolution of ~5 ps using a cross-correlation technique [88]. The results of those measurements will be discussed in the next section and the results of the studies of the slow PC in the following section.

Fast Photoconductivity in Trans-Polyacetylene

The samples on which Reichenbach et al. performed their measurements were oriented. With their higher temporal resolution, they determined that the decay time for the fast photocurrent due to excitation by light with electric vector perpendicular to the chains was 4.5 ps. Since, as discussed in Sect. 4.7.4, this mode of excitation produced polarons with lifetime much greater than 4.5 ps and solitons with lifetime less than 1 ps, it is reasonable to assume that, at least at short times, PC resulting from this type of excitation is by polarons only. It follows that the polarons are highly mobile for a time of ~4.5 ps and much less mobile after that. This can be understood if 4.5 ps represents the average time for the polaron to drift the distance of a conjugation length along the chain on which it has been created. When it reaches the end of the conjugation length, to continue carrying current it must hop to another conjugation length. After that its mobility is determined by the hopping rate and is orders of magnitude smaller than the drift mobility. From the knowledge of the average conjugation length, $\ell \cong 30$ CH units or ~3.6 nm, the electric field intensity $E = 2\times10^4$ V/cm and the drift time t_d of 4.5 ps, we obtain the polaron drift mobility μ ($= \ell/Et_d$) as 4 cm^2/Vs. This is in good agreement with the value of polaron mobility calculated for polarons scattered by acoustic modes, which should be the situation at room temperature [87]. With the electric vector of the light parallel to the chains, only solitons are generated. Reichenbach et al. found a decay time for this case similar to that obtained for polarons [88], leading to a similar mobility. This mobility value is in reasonable agreement with the values of soliton mobility cited in Sect. 4.5.8.

Another characteristic of the ps PC of t-$(CH)_x$, found in all the references cited, was very weak dependence on temperature, from 300 K down to 10 K. This was taken to show that the carriers are "hot" [85], because when excited by photons with energy well above the absorption edge they are created with high kinetic energy. However, this excess energy can be dissipated by phonon emission. A calculation of the rate of energy loss to acoustic phonons of hot solitons and polarons shows that the thermalization times are fractions of a ps, too short for any hot carrier effects to be seen in the PC experiments described [158]. Although we do not have much information about the temperature dependence of polaron mobility, it is reasonable to expect it to be similar to that of solitons. It can be documented that solitons have weak temperature dependence over most of the range concerned, from 300 K down to ~75 K, because the soliton mobility can be calculated from the soliton diffusion rate measured by magnetic resonance experiments [31] using the Einstein relation [159]. The weak temperature dependence results from the opposite effects on temperature dependence of phonon scattering and defect scattering, the latter dominating the low temperature behavior because few phonons are excited at low temperatures [34]. As for the remaining low temperature discrepancy, i.e., for temperatures below ~75 K, it was pointed out by Reichenbach et al. [88] that for the high photon fluxes used, especially for the parallel pump case, the samples could be heated by hundreds of degrees.

With the electric field perpendicular to the chains, the ps PC is much smaller and behaves quite differently. From room temperature down to 160 K the PC shows an activated behavior with activation energy 0.048 eV. Below this, down to the lowest temperature for which data are given, 120 K, the PC is temperature independent [160]. A theoretical calculation of the mobility perpendicular to the chains, treating it as phonon-assisted hopping, could account for the temperature dependence in the activated region [161–163]. The lack of temperature dependence at lower temperatures was attributed to the assisting phonons having only zero point energy.

Slow Photoconductivity in Trans-Polyacetylene

The slow component is orders of magnitude smaller than the fast component and does not decay exponentially. It includes carriers that persist for a duration of the order of seconds. It is activated with an activation energy of 0.2 eV [155, 164]. The existence of an activation energy and the fact that the exciting phonon energy was 2.3 eV, almost 1 eV above the absorption edge, in the latter case suggests that, like PPV, the PC in this range is determined by the Onsager mechanism. Consistent with this, Bleier et al. found some electric field dependence of the photocurrent.

Although the fast component of the PC is due to polarons, there is a question as to whether the carriers involved in the slow photoconductivity are solitons or polarons. In time-resolved PA experiments Rothberg et al. found a time-delayed (by 40 to 50 ps) contribution to the 0.43 eV peak attributed to charged solitons [149, 165]. This implies a conversion of polarons to charged solitons. There are two possible mechanisms for this conversion: (1) a polaron meets another polaron of like charge and the pair is transformed into a pair of charged solitons; and (2) a polaron meets a neutral soliton and transfers its charge to the soliton. Both seem to occur, and which of them predominates appears to be sample-dependent [166]. Rothberg et al. suggested that the second mechanism was operative in their samples. Strong support for the first mechanism has been given by some experimental studies [166, 167]. In any case, it has been shown that there is good consistency between the steady-state photocurrent and the PA at 0.43 eV [168], establishing that solitons constitute at least a sizeable fraction of the slow and steady-state photocurrents. This explains the observation of Flood et al. [148], given earlier, that the light-induced electron spin resonance, which reflects the spin population at relatively long times after the exciting pulse, does not show spins.

4.8 Laser Action

The use of SE in conducting polymers was introduced, as discussed in Sect. 4.5.2, to identify excitons in PPV derivatives for the purpose of studying their behavior [169]. SE was first used to create a conducting polymer laser, consisting of MEH-PPV in an appropriate solvent, in 1992 [170]. Several years later the first thin-film conducting polymer laser was made by optically pumping a thin film of PPV in a microcavity [171]. This was followed shortly by the demonstration of gain narrowing on optical pumping in thin films of PPV derivatives and PPP derivatives [172]. A polymer showing particular promise is poly(2-butyl, 5-(2'-ethyl-hexyl)-1,4-phenylene vinylene), BuEH-PPV, with a PL efficiency of 75% in solution and 62% in neat films. The PL spectrum of BuEH-PPV as a function of the pump energy per pulse is shown in Fig. 4.10. As the pump energy is increased, a narrow peak arises out of the broad emission spectrum. The collapse of the line width at very low pump energies, less than 1 μJ/pulse, implies a very large cross section for SE. Similar line narrowing at comparable pump energies has been found in many other conducting polymers (Hide et al. [172]) The small blue shift of the emission peak at the highest pumping energy seen in Fig. 4.10 suggests that the SE

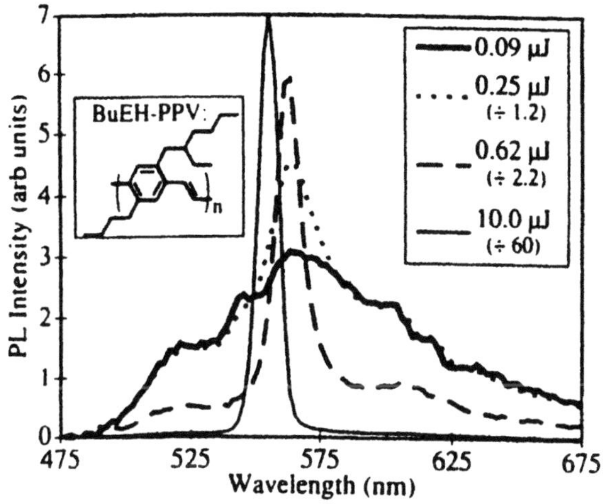

Fig. 4.10. Photoluminescence spectra of 210 nm thick film of BuEH-PPV spun on glass at various pump energies both above and below the lasing threshold. Inset: chemical structure of BuEH-PPV. (After [74])

occurs more rapidly than the spectral diffusion due to exciton hopping to longer chains that takes place within picoseconds after photoexcitation (see Sect. 4.4.2).

As might be expected from the discussion in Sect. 4.5.11 of interchain interactions, the line-narrowing is affected by the solvent from which the film is cast. This effect was strong in MEH-PPV, although much less so in BuEH-PPV (see, for example, [74]). Under strong pumping, MEH-PPV films cast from tetrahydrofuran, THF, showed a much narrower linewidth (17 nm) than those cast from aromatic solvents chlorobenzene, CB, and p-xylene (~50 nm). Also, the threshold for line narrowing was smaller by a factor 3 for THF than for the aromatic solvents. Diaz-Garcia et al. [74] suggest that the differences are due to interchain interactions, which, as seen earlier, decrease the probability of radiative transitions. With the chains being more rigid in the aromatic solvents, there is likely to be more π–π interaction between chains, this being further enhanced by the slow evaporation of these high boiling temperature solvents which gives the chains time to order before their configuration is frozen. In THF, which is fairly polar, the chains are likely to interact less, because they are kinked (which is evidenced by the blue shift of absorption in this solvent) and because the solvent evaporates rapidly.

It is of interest to compare the energy threshold for gain narrowing with that of a dye used in a dye laser. In Fig. 4.11 the evolution of the line width with pump-pulse energy of the BuEH-PPV sample of Fig. 4.11 is compared with that of the laser dye DCM suspended in a polystyrene film with optical density comparable to that of the polymer sample. Although the DCM concentrations used were more than two orders of magnitude higher than in typical dye lasers, the threshold for gain narrowing is more than 1000 times

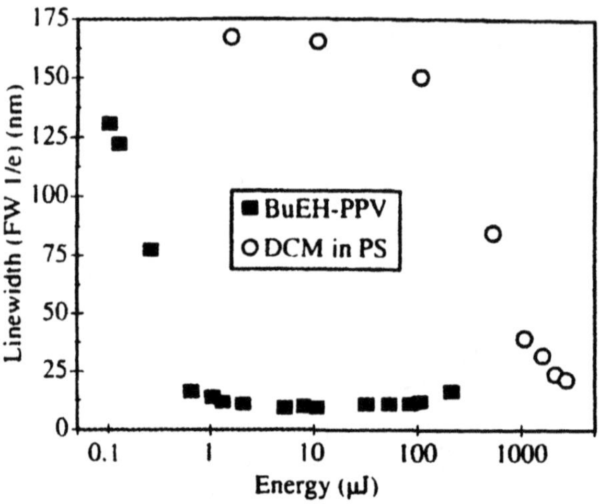

Fig. 4.11. Evolution of the line width versus pump-pulse energy for the BuEH-PPV sample of Fig. 4.10 (■) and a 2.6% by weight film of DCM in polystyrene (○) with a comparable optical density. (After [74])

as great as that of the BuEH-PPV. The advantage of the polymer arises from the greater density of chromophores in the solid film.

The mechanism of spectral narrowing has been a subject of debate. For the demonstration of spectral narrowing the polymer was usually in the form of a thin film on a glass substrate in air. Because the index of refraction of the various polymers used was higher than that of the glass, as well as that of air, waveguiding in the polymer is possible if the film is thick enough to support a guided mode. Preparing films of various thicknesses, Hide et al. showed that a well-defined minimum thickness was required for spectral narrowing to occur [172]. This thickness was found to coincide with the minimum thickness required for a guided mode to exist, given the indices of refraction of the polymer, substrate and air. Thus, some sort of rudimentary waveguiding, which confines the emitted photons, was apparently necessary for spectral narrowing to occur. This suggests that the narrowing is due to amplification of spontaneous emission, ASE, as the light travels down the guide. The narrowing occurs because the amplification is maximum at the wavelength where the net gain is the highest.

Other reports have attributed the line narrowing to superradiance, SR [173,174]. In this effect, spontaneous emission occurs on a time scale shorter than the normal fluorescence lifetime as a result of cooperation between the emitting dipoles. There have also been proposals that both effects occur, with SR being amplified by SE as it travels down a waveguide [175], and that the spectral narrowing is due to interacting excitons, specifically biexcitons [176] or condensed excitons [177]. Further investigations of spectral narrowing established that the mechanism is indeed ASE [178,179]. Denton et al. point out that the ultrafast relaxation of the excited states in the polymers precludes the coherence required for SR.

A detailed picture of the gain mechanism in the polymers was provided by the work of Schweitzer et al. [180]. The polymer they studied was MeL-PPP. Exciting in the low-energy tail of the transition, they saw the onset of some strong resonant modes at energies that differed from the excitation photon energy by amounts equal to the energies corresponding to molecular vibrations. This demonstrated that vibrational levels are involved in the amplification process. They view the system as a four-level laser. Optical pumping excites an electron from the ground state, $S_{0,0}$, into a vibrational level in the next higher electronic state, S_1. Ultrafast cooling (by phonon emission) brings the electron to the lowest vibrational level of this electronic state, $S_{1,0}$. Thus, there is a population inversion between $S_{1,0}$ and the vibrational levels of the lower electronic state, S_0. Emission, which may be spontaneous or stimulated, then takes the electron into one of the vibrational states of S_0, after which there is a second rapid thermalization process to take the system back into the ground state. The ultrafast thermalization processes among the vibrational levels allow the polymer to act as a gain medium.

Great interest has been expressed in obtaining an electrically pumped conducting polymer laser, which is potentially more useful than an optically pumped laser. It is anticipated that for an electrically pumped laser much higher currents and electric fields would be required than are currently used in LEDs. For the photon densities in the luminescent films at the threshold for spectral narrowing, under the assumption of a 5% internal quantum efficiency for electroluminescence, the transient current density necessary to reach the threshold could be in excess of thousands of amperes/cm^2 [172]. A transient current density of 1000 A/cm^2 has been obtained in a PPV LED with ITO and Ca electrodes [181]. To achieve this required 45 V in pulses of width 100–200 ns across a PPV emitting region 140 nm wide, thus an average field of $\sim 3\times 10^6$ V/cm. Tessler et al. estimated the carrier density as a function of current density by measuring the change in reflection at 1.6 eV as a function of current density and using an optical cross-section per injected charge to convert the reflection change into charge density. They concluded that the mobility at the high current density might be as high as 0.1 cm^2/Vs. Unfortunately, the procedure for obtaining the charge density is questionable. As has been discussed earlier, absorption and reflection at 1.6 eV in PPV are due to singlet excitons, triplet excitons and polaron pairs. The optical cross-section was obtained under the assumption that the entity responsible for the absorption was the bipolaron [182].

Currents close to 1000 A/cm^2 were also obtained by Chayet et al. [183]. In this case it required 400 V across 50 nm of PPV, for an average field of 8×10^7 V/cm. Injection was undoubtedly less efficient than for the LED used by Tessler et al. [181], one reason being the use of an Al cathode, which has a much larger work function than Ca. At fields above 2×10^7 V/cm a new blue-UV emission was found in addition to the characteristic green emission of the exciton. The new emission grew rapidly as the field increased. Also, at an average field close to 6×10^7 V/cm, a large increase in current was found which was attributed to impact ionization (Chayet et al. [183]). Although the source of the new emission is not yet clear, it is apparent that its excitation requires highly energetic electrons (and/or holes).

Large increases of current carrier energy in a high electric field are familiar in groups IV and III-V crystalline semiconductors. The origin of these effects is the high carrier mobility μ, because the rate at which an electric field feeds energy into a carrier with charge e is $e\mu E^2$. Whereas in low fields μ for these crystalline semiconductors is in the range 10^3 to 10^5 cm^2/Vs, for the polarons that carry the current in LEDs, μ is $\sim10^{-5}$ cm^2/Vs, a hopping mobility. It has been shown, however, that in a high-enough electric field the polaron is dissociated because the field causes the electron to move faster than the chain deformation characteristic of the polaron can move. For polyacetylene in fields $\geq10^6$ V/cm, an already-formed polaron will be dissociated [184]. It is expected that a similar field would dissociate a polaron in PPV. Once the electron is free in the conduction band, its mobility is much larger and it could be heated strongly by the fields that were applied by Chayet et al. With reference to the band structure of PPV, it has been shown that the threshold for impact ionization for an electron moving along the chain direction is 3.7 eV [185]. Attaining this energy requires Zener tunneling from the lowest conduction band D_1^* to the next delocalized band D_2^* (see Fig. 4.4). At the high fields applied by Chayet et al., it is likely that an electron can achieve the energy required for impact ionization before it loses its energy in phonon emission [186].

References

1. W.P. Su, J.R. Schrieffer, A.J. Heeger, Phys. Rev. B **22**, 2099 (1980)
2. W.P. Su, J.R. Schrieffer, Proc. Natl. Acad. Sci. **77**, 5626 (1980)
3. K. Murata, Y. Shimoi, S. Abe, S. Kuroda, T. Noguchi, T. Ohnishi, Chem. Phys. **227**, 191 (1998)
4. Y. Shimoi, S. Abe, S. Kuroda, K. Murata, Solid State Commun. **95**, 137 (1995)
5. K. Fesser, A.R. Bishop and D.K. Campbell, Phys. Rev. B **27**, 4804 (1983)
6. J. Cornil, D. Beljonne, J.L. Bredas, J. Chem. Phys. **103**, 842 (1995)
7. H.A. Mizes, E.M. Conwell, Synth. Met. **68**, 145 (1995)
8. J. Poplawski, E. Ehrenfreund, J. Cornil, J.L. Brédas, R. Pugh, M. Ibrahim, A.J. Frank, Mol. Cryst. Liq. Cryst. **256**, 407 (1994)
9. P.A. Lane, X. Wei, Z.V. Vardeny, Synth. Met. **84**, 521 (1997)
10. S.C. Graham, D.D.C. Bradley, R.H. Friend, Synth. Met. **41–43**, 1277 (1991)
11. G. Weiser, Phys. Rev. B **45**, 14076 (1992)
12. I.H. Campbell, T.W. Hagler, D.L. Smith, and J.P. Ferraris, Phys. Rev. Lett. **76**, 1900 (1996)
13. J.-L. Brédas, J. Cornil, A.J. Heeger, Adv. Mater. **8**, 447 (1996)
14. E.M. Conwell, Synth. Met. **83**, 101 (1996)
15. P. Gomes da Costa, R.G. Dandrea, E.M. Conwell, Phys. Rev. B **47**, 1800 (1993)

16. K. Pakbaz, C.H. Lee, A.J. Heeger, T.W. Hagler, D. McBranch, Synth. Met. **64**, 295 (1994)
17. C.H. Lee, G. Yu, A.J. Heeger, Phys. Rev. B **47**, 15543 (1993)
18. T.W. Hagler, K. Pakbaz, A.J. Heeger, Phys. Rev. B **49**, 10968 (1994)
19. Q. Pei, G. Yu, C. Zhang, Y. Yang, A.J. Heeger, Science **269**, 1086 (1995)
20. M. Chandross, S. Mazumdar, S. Jeglinski, X. Wei, Z.V. Vardeny, E.W. Kwock, T.M. Miller, Phys. Rev. B **50**, 14702 (1994)
21. G. Meinhardt, A. Horvath, G. Weiser, G. Leising, Synth. Met. **84**, 669 (1997)
22. Z.G. Soos, S. Etemad, D.S. Galvão, S. Ramasesha, Chem. Phys. Lett. **194**, 341 (1992)
23. Z.G. Soos, S. Ramasesha, D.S. Galvão, Phys. Rev. Lett. **71**, 1609 (1993)
24. T. Watanabe, T. Kubo, K. Ishikawa, H. Takezoe, A. Fukuda, Jpn. J. Appl. Phys. **33**, L48 (1994)
25. S. Mazumdar, F. Guo, Synth. Met. **78**, 187 (1996)
26. V.I. Klimov, D.W. McBranch, N. Barashkov, J. Ferraris, Phys. Rev. B **58**, 7654 (1998)
27. P.A. Lane, X. Wei, Z.V. Vardeny, Phys. Rev. Lett. **77**, 1544 (1996)
28. M.G. Hill et al., J. Am. Chem. Soc. **114**, 2728 (1992)
29. Y. Shimoi, S. Abe, Phys. Rev. B **50**, 14781 (1994)
30. B.R. Weinberger, J. Kaufer, A.J. Heeger, A. Pron, A.G. MacDiarmid, J. Chem. Phys. **72**, (1980)
31. M. Nechtschein, F. Devreux, F. Genoud, M. Gugliemi, K. Holczer, Phys. Rev. B **27**, 61 (1983)
32. K. Mizoguchi, K. Kume, H. Shirakawa, *Abstracts of International Conference on Science and Technology of Synthetic Metals*, Kyoto (1986) pp. 55
33. S. Jeyadev, E.M. Conwell, Phys. Rev. Lett. **58**, 258 (1987)
34. S. Jeyadev, E.M. Conwell, Phys. Rev. B **36**, 3284 (1987)
35. E.M. Conwell, IEEE Trans. On Electrical Insulation, EI-22, 591 (1987)
36. E.K. Miller, D. Yoshida, C.Y. Yang, A.J. Heeger, Phys. Rev. B **59**, 4661 (1999)
37. S. Brazovskii, N. Kirova, A.R. Bishop, V. Klimov, D. McBranch, N.N. Barashkov, J.P. Ferraris, Opt. Mater. **9**, 472 (1998)
38. M.J. Rice, Yu.N. Gartstein, Phys. Rev. Lett. **73**, 2504 (1994)
39. Yu.N. Gartstein, M.J. Rice, E.M. Conwell, Phys. Rev. B **52**, 1683 (1995)
40. P. Gomes da Costa, E.M. Conwell, Phys. Rev. B **48**, 1993 (1993)
41. M. Chandross, A. Mazumdar, Phys. Rev. B **55**, 1497 (1997)
42. Y. Shimoi, S. Abe, Synth. Met. **78**, 219 (1996)
43. D. Beljonne, J. Cornil, D.A. dos Santos, Z. Shuai, J.L. Brédas, In: *Conjugated Polymers: Molecular Exciton vs. Semiconductor Band Model*, ed. by N.S. Sariciftci (World Scientific, Singapore 1997) Chapter 19
44. S. Mukamel, S. Tretiak, T. Wagersreiter, V. Chernyak, Science **277**, 781 (1997)
45. D.D.C. Bradley, A.R. Brown, P.L. Burn, J.H. Burroughes, R.H. Friend, A.B. Holmes, K.D. Mackay, R.N. Marks., Synth. Met. **41–43**, 3135 (1991)
46. R. Kersting, U. Lemmer, R.F. Mahrt, K. Leo, H. Kurz, H. Bässler, E.O. Gobel, Phys. Rev. Lett. **70**, 3820 (1993)
47. U. Rauscher, H. Bässler, D. D. C. Bradley, M. Hennecke, Phys. Rev. B **42**, 9830 (1990)
48. H. Bässler, M. Grailberger, R.F. Mahrt, J.M. Oberski, G. Weiser, Synth. Met. **49–50**, 341 (1992)

49. S. Heun, R.F. Mahrt, A. Greiner, U. Lemmer, H. Bässler, D.A. Halliday, D.D.C. Bradley, P.L. Burn, A.B. Holmes, J. Phys. Condens. Matter **5**, 247 (1993)
50. M. Yan, L.J. Rothberg, F. Papadimitrakopoulos, M.E. Galvin, T.M. Miller, Phys. Rev. Lett. **73**, 744 (1994)
51. F. Papadimitrakopoulos, M. Yan, L.J. Rothberg, H.E. Katz, M.E. Galvin, Mol. Cryst. Liq. Cryst. **256**, 663 (1994)
52. F. Papadimitrakopoulos, K. Konstadinidis, T.M. Miller, R. Opila, E.A. Chandross, M.E. Galvin, Chem. Mater. **6**, 1563 (1994)
53. M.I. Khan, G.C. Bazan, Z.D. Popovic, Chem. Phys. Lett. **298**, 309 (1998)
54. R. Kersting, U. Lemmer, M. Deussen, H.J. Bakker, R.F. Mahrt, H. Kurz, V.I. Arkhipov, H. Bässler, E.O. Gobel, Phys. Rev. Lett. **73**, 1440 (1994)
55. M. Deussen, M. Scheidler, H. Bässler, Synth. Met. **73**, 123 (1995)
56. V.I. Arkhipov, H. Bässler, M. Deussen, E.O. Gobel, R. Kersting, H. Kurz, U. Lemmer, R.F. Mahrt, Phys. Rev. B **52**, 4932 (1995)
57. M. Scheidler, U. Lemmer, R. Kersting. S. Karg, W. Riess, B. Cleve, R.F. Mahrt, H. Kurz, H. Bässler, E.O. Gobel, P. Thomas, Phys. Rev. B **54**, 5536 (1996)
58. V.I. Arkhipov, E.V. Emelianova, H. Bässler, Phys. Rev. Lett. **82**, 1321 (1999)
59. X. Wei, B.C. Hess, Z.V. Vardeny, F. Wudl, Phys. Rev. Lett. **68**, 666 (1992)
60. J.W.P. Hsu, M. Yan, T.M. Jedju, L.J. Rothberg, B.R. Hsieh, Phys. Rev. B **49**, 712 (1994)
61. M. Yan, L.J. Rothberg, F. Papadimitrakopoulos, M.E. Galvin, T.D. Miller, Phys. Rev. Lett. **72**, 1104 (1994)
62. S.V. Frolov, M. Liess, P.A. Lane, W. Gellermann, Z.V. Vardeny, M. Ozaki, K. Yoshino, Phys. Rev. Lett. **78**, 4285 (1997)
63. E.L. Frankevich, A.A. Lymarev, I. Sokolik, F.E. Karasz, S. Blumstengel, R. Baughman, H.H. Horhold, Phys. Rev. **46**, 9320 (1992)
64. M. Gailberger, H. Bässler, Phys. Rev. B **44**, 8643 (1991)
65. M. Yan, L.J. Rothberg, E.W. Kwock, T.M. Miller, Phys. Rev. Lett. **75**, 1992 (1995)
66. N.C. Greenham, I.D.W. Samuel, G.R. Hayes, R.T. Phillips, Y.A.R.R. Kessener, S.C. Moratti, A.B. Holmes, R.H. Friend, Chem. Phys. Lett. **241**, 89 (1995)
67. G.J. Denton, N. Tessler, N.T. Harrison, R.H. Friend, Phys. Rev. Lett. **78**, 733 (1997)
68. I.D.W. Samuel, A.G. Rumbles, C.J. Collison, Phys. Rev. B **54**, R11573 (1995)
69. E.M. Conwell, J. Perlstein, S. Shaik, Phys. Rev. B **54**, R2308 (1996)
70. J. Michl, V. Bonacic-Koutecky, *Electronic Aspects of Organic Photochemistry* (Wiley, New York 1990)
71. H.-F. Meng, Phys. Rev. B **58**, 3888 (1998)
72. V.I. Klimov, D.W. McBranch, N.N. Barashkov, J.P. Ferraris, Chem. Phys. Lett. **227**, 109 (1997)
73. R. Jakubiak, C.J. Collison, W.C. Wan, L.J. Rothberg, B.R. Hsieh, J. Phys. Chem. A**103**, 2394 (1999)
74. M.A. Diaz-Garcia, F. Hide, B.J. Schwartz, M. Andersson, Q. Pei, A.J. Heeger, Synth. Met. **84**, 455 (1997)
75. W. Holzer, A. Penzkofer, S.H. Gong, D.D.C. Bradley, X. Long, A. Bleyer, Chem. Phys. **224**, 315 (1997)

76. R. Jakubiak, L.J. Rothberg, W. Wan, B. Hsieh, Synth. Met. **101**, 230 (1999)
77. L.J. Rothberg, M. Yan, F. Papadimitrakopoulos, M.E. Galvin, E.W. Kwock. T.M. Miller, Synth. Met. **80**, 41 (1996)
78. H.A. Mizes, E.M. Conwell, Phys. Rev. B **50**, 11243 (1994)
79. E.M. Conwell, H.A. Mizes, Synth. Met. **78**, 201 (1996)
80. P.A. Lane, X. Wei, Z.V. Vardeny, Phys. Rev. B **56**, 4626 (1997)
81. S. Webster, D.N. Batchelder, Polymer **37**, 4961 (1996)
82. J.M. Thomas, E.L. Evans, J.O. Williams, Proc. R. Soc. Lond. Ser. A **331**, 417 (1972)
83. E.M. Conwell, H.A. Mizes, Phys. Rev. B **51**, 6953 (1995)
84. L. Rothberg, T.M. Jedju, S. Etemad, G.L. Baker, Phys. Rev. Lett. **65**, 100 (1990)
85. M. Sinclair, D. Moses, A.J. Heeger, Solid State Commun. **59**, 343 (1986)
86. Z. Vardeny, J. Strait, D. Moses, T.-C. Chung, A.J. Heeger, Phys. Rev. Lett. **49**, 1657 (1982)
87. S. Jeyadev, E.M. Conwell, Phys. Rev. B **35**, 6253 (1987)
88. J. Reichenbach, M. Kaiser, S. Roth, Phys. Rev. B **48**, 14104 (1993)
89. S.D. Phillips, A.J. Heeger, Phys. Rev. B **38**, 6211 (1988)
90. E.M. Conwell, H.A. Mizes, Synth. Met. **69**, 613 (1995)
91. B. Kraabel, D. McBranch, N.S. Sariciftci, D. Moses, and A.J. Heeger, Mol. Cryst. Liq. Cryst. **256**, 733 (1994)
92. G.S. Kanner, X. Wei, B.C. Hess, L.R. Chen, Z.V. Vardeny, Phys. Rev. Lett. **69**, 538 (1992)
93. T. Pauck, R. Hennig, M. Perner, U. Lemmer, R.F. Mahrt, U. Scherf, K. Mullen, H. Bässler, E.O. Gobel, Chem. Phys. Lett. **244**, 171 (1995)
94. J. Shinar, In: *Handbook of Organic Conductive Molecules and Polymers*: Vol. 3, ed. by S. Nalwa (Wiley, New York 1997)
95. E.M. Conwell, Phys. Rev. B **57**, 14200 (1998)
96. M.W. Wu, E.M. Conwell, Phys. Rev. B **56** R10060 (1997)
97. G. Rumbles, I.D.W. Samuel, C.J. Collison, P.F. Miller, S.C. Moratti and A.B. Holmes, Synth. Met. **101**, 158 (1999)
98. U. Lemmer, S. Heun, R.F. Mahrt, U. Scherf, M. Hopmeier, U. Siegner, E.O. Gobel, K. Mullen, H. Bässler, Chem. Phys. Lett. **240**, 373 (1995)
99. J.W. Blatchford, S.W. Jessen, L.-B. Lin, T.L. Gustaffson, D.-K. Fu, H.-L. Wang, T.M. Swager, A.G. MacDiarmid, A.J. Epstein, Phys. Rev. B **54**, 9180 (1996)
100. A.J. Epstein, J.W. Blatchford, Y.Z. Wang, S.W. Jessen, D.D. Gebler, L.B. Lin, T.L. Gustafson, H.-L. Wang, Y.W. Park, T.M. Swager, A.G. MacDiarmid, Synth. Met. **78**, 253 (1996)
101. J.W. Blatchford, T.L. Gustafson, A.J. Epstein, S.A. Vanden Bout, J. Kerimo, D. Higgins, P.F. Barbara, D.K. Fu, T.M. Swager, A.G. MacDiarmid, Phys. Rev. B **54**, R3683 (1996)
102. D.D.C. Bradley, M. Grell, X. Long, H. Mellor, A. Grice, SPIE **3145**, 254 (1997)
103. B.M.W. Langeveld-Voss, E. Peeters, R.A. Janssen, E.W. Meijer, Synth. Met. **84**, 611 (1997)
104. I.D.W. Samuel, F. Raksi, D.D.C. Bradley, R.H. Friend, P.L. Burn, A.B. Holmes, H. Murataa, T. Tsutsui, S. Saito, Synth. Met. **55**, 15 (1993)
105. T.W. Hagler, K. Pakbaz, K.F. Voss, A.J. Heeger, Phys. Rev. B **44**, 8652 (1991)

106. I.D.W. Samuel, B. Crystall, G. Rumbles, P.L. Burn, A.B. Holmes, R.H. Friend, Synth. Met. **54**, 281 (1993)
107. J. Huber et al., Acta Polym. **45**, 244 (1994)
108. J. Gruner, P. Hamer, R.H. Friend, H. Huber, U. Scherf, A.B. Holmes, Adv. Mater. **6**, 748 (1994)
109. M. Fahlman, D.D. Gebler, N. Piskun, T.M. Swager, A.J. Epstein, J. Chem. Phys. **109**, 2031 (1998)
110. N.F. Colaneri, D.D.C. Bradley, R.H. Friend, P.L. Burn, A.B. Holmes, C.W. Spangler, Phys. Rev. B **42**, 11670 (1990)
111. H.S. Woo, S.C. Graham, D.A. Halliday, D.D.C. Bradley, R.H. Friend, P.L. Burn, Phys. Rev. B **46**, 7379 (1992)
112. A.R. Brown, K. Pichler, N.C. Greenham, D.D.C. Bradley, R.H. Friend, A.B. Holmes, Chem. Phys. Lett. **210**, 61 (1993)
113. J.M. Leng, S. Jeglinski, X. Wei, R.E. Benner, Z.V. Vardeny, Phys. Rev. Lett. **72**, 156 (1994)
114. K.F. Voss, C.M. Foster, L. Smilowitz, D. Mihailovic, S. Askari, G. Srdanov, Z. Ni, S. Shi, A.J. Heeger, F. Wudl, Phys. Rev. B **43**, 5109 (1991)
115. B. Horovitz, Solid State Commun. **41**, 729 (1982)
116. Y.H. Kim, A.J. Heeger, Phys. Rev. B **40**, 8393 (1989)
117. E.M. Conwell, H.A. Mizes, Synth. Met. **65**, 203 (1994)
118. G.B. Blanchet, C.R. Fincher, T.C. Chung, A.J. Heeger, Phys. Rev. Lett. **50**, 1938 (1983)
119. B. Kraabel, D. Moses, A.J. Heeger, J. Chem. Phys. **103**, 5102 (1995)
120. E. Frankevich, A. Zakhidov, K. Yoshino, Y. Maruyama, K. Yakushi, Phys. Rev. B **53**, 4498 (1996)
121. M. Wohlgenannt, W. Graupner, G. Leising, Z.V. Vardeny, Phys. Rev. Lett. **82**, 3344 (1999)
122. L.S. Swanson, P.A. Lane, J. Shinar, F. Wudl, Phys. Rev. B **44**, 10617 (1991)
123. J. Partee, E.L. Frankevich, B. Uhlhorn, J. Shinar, Y. Ding, T.J. Barton, Phys. Rev. Lett. **82**, 3673 (1999)
124. E.L. Frankevich, I.A. Sokolik, A.A. Lymarev, Mol. Cryst. Liq. Cryst. **175**, 41 (1989)
125. M. El-Sayed, J. Chem. Phys. **38**, 2834 (1963)
126. S.W. Jessen, J.W. Blatchford, L.-B. Lin, T.L. Gustafson, J. Partee. J. Shinar, D.-K. Fu, M.J. Marsella, T.M. Swager, A.G. MacDiarmid, A.J. Epstein, Synth. Met. **84**, 501 (1997)
127. F. Guo, M. Chandross, S. Mazumdar, Phys. Rev. Lett. **74**, 2086 (1995)
128. D. Moses, D. Comoretto, C.H. Lee, A.J. Heeger, Synth. Met. **84**, 559 (1997)
129. N.S. Sariciftci, L. Smilowitz, A.J. Heeger, F. Wudl, Science **258**, 1474 (1992)
130. L. Smilowitz, N.S. Sariciftci, R. Wu, C. Gettinger, A.J. Heeger, F. Wudl, Phys. Rev. B **47**, 13835 (1993)
131. B. Kraabel, C.H. Lee, D. McBranch, D. Moses, N.S. Sariciftci, A.J. Heeger, Chem. Phys. Lett. **213**, 389 (1993)
132. L.J. Rothberg, M. Yan, A.W.P. Fung, Y.M. Jedju, E.W. Kwock, M.E. Galvin, Synth. Met. **84**, 537 (1997)
133. K. Pichler, D.A. Halliday, D.D.C. Bradley, P.L. Burn, R.H. Friend, A.B. Holmes, J. Phys.: Condens. Matter **5**, 7155 (1993)
134. S. Barth, H. Bässler, H. Rost, H.H. Horhold, Phys. Rev. B **56**, 3844 (1997)
135. L. Onsager, Phys. Rev. **54**, 554 (1938)

136. S. Barth and H. Bässler, Phys. Rev. Lett. **79**, 4445 (1997)
137. H. Bässler, Phys. Stat. Sol. B **175**, 15 (1993)
138. A. Kohler, D.A. dos Santos, D. Beljonne, Z. Shuai, J.-L. Brédas, A.B. Holmes, A. Kraus, K. Mullen, R.H. Friend, Nature **392**, 903 (1998)
139. C.H. Lee, G. Yu, D. Moses, A.J. Heeger, Phys. Rev. B **49**, 2396 (1994)
140. C.H. Lee, G. Yu, N.S. Sariciftci, A.J. Heeger, C. Zhang, Synth. Met. **75**, 127 (1995)
141. D. Moses, H. Okumoto, D. Comoretto, C.H. Lee, A.J. Heeger, T. Ohnishi, T. Noguchi, Synth. Met. **84**, 539 (1997)
142. R.J.O.M. Hoofman, M.P. de Haas, L.D.A. Siebbeles, J.M. Warman, Nature **392**, 54 (1998)
143. D. Moses, J. Wang, G. Yu, A.J. Heeger, Phys. Rev. Lett. **80**, 2685 (1998)
144. S. Barth, H. Bässler, U. Scherf, K. Mullen, Chem. Phys. Lett. **288**, 147 (1998)
145. J. Orenstein, G.L. Baker, Phys. Rev. Lett. **49**, 1043 (1982)
146. J. Orenstein, G.L. Baker, A. Vardeny, J. Phys. (Paris) **44**, C3 (1983)
147. Z. Vardeny, J. Orenstein, G.L. Baker, Phys. Rev. Lett. **50**, 2032 (1983)
148. J.D. Flood, E. Ehrenfreund, A.J. Heeger, A.G. MacDiarmid, Solid State Commun. **44**, 1055 (1982)
149. L. Rothberg, T.M. Jedju, S. Etemad, G.L. Baker, Phys. Rev. Lett. **57**, 3229 (1986)
150. S. Kivelson, W-K. Wu, Phys. Rev. B **34**, 5423 (1986)
151. C.V. Shank, R. Yen, R.L. Fork, J. Orenstein and G.L. Baker, Phys. Rev. Lett. **49**, 1660 (1982)
152. D. Baeriswyl, K. Maki, Phys. Rev. B **38**, 8135 (1988)
153. G.P. Brivio, E. Mulazzi, Phys. Rev. B **30**, 876 (1984)
154. M. Sinclair, D. Moses, R.H. Friend, A.J. Heeger, Phys. Rev. B **36**, 4296 (1987)
155. H. Bleier, S. Roth, Y.Q. Shen, D. Schaafer-Siebert, G. Leising, Phys. Rev. B **38**, 6031 (1988)
156. D.D.C. Bradley, Y.Q. Shen, H. Bleier, S. Roth, J. Phys. C **21**, L515 (1988)
157. A.D. Walser, R. Dorsinville, R. Tubino, R.R. Alfano, Phys. Rev. B **43**, 7194 (1001)
158. S. Jeyadev, E.M. Conwell, Phys. Rev. B **35**, 5917 (1987)
159. E.M. Conwell, S. Jeyadev, Mol. Cryst. Liq. Cryst. **160**, 443 (1988)
160. A. Walser, A. Seas, R. Dorsinville, R.R. Alfano, R. Tubino, Solid State Commun. **67**, 333 (1988)
161. H.-Y. Choi, E.M. Conwell, Synth Met. **41–43**, 3667 (1991)
162. H.-Y. Choi, E.M. Conwell, Mol. Cryst. Liq. Cryst. **104**, 23 (1991)
163. E.M. Conwell, H.-Y. Choi, S. Jeyadev, J. Phys. Chem. **96**, 2827 (1992)
164. S. Etemad, T.Mitani, M. Ozaki, T.C. Chung, A.J. Heeger, A.G. MacDiarmid, Solid State Commun. **40**, 75 (1981)
165. L. Rothberg , T.M. Jedju, S. Etemad, G.L. Baker, Phys. Rev. B **36**, 7529 (1987)
166. F. Moraes, Y.W. Park, A.J. Heeger, Synth. Met. **13**, 113 (1986)
167. N.F. Colaneri, R.H. Friend, H.E. Schaffer, A.J. Heeger, Phys. Rev. B **38**, 3960 (1988)
168. P.D. Townsend and R.H. Friend, Synth. Met. **17**, 631 (1987)
169. M. Yan, L.J. Rothberg, F. Papadimitrakopoulos, M.E. Galvin, T.M. Miller, Phys. Rev. Lett. **72**, 1104 (1994)
170. D. Moses, Appl. Phys. Lett. **60**, 3215 (1992)

171. N. Tessler, G.J. Denton, R.H. Friend, Nature **382**, 695 (1996)
172. F. Hide, M.A. Diaz-Garcia, B.J. Schwartz, M.R. Anderson, Q. Pei, A.J. Heeger, Science **273**, 1833 (1996)
173. S.V. Frolov, M. Ozaki, W. Gellermann, Z.V. Vardeny, K. Yoshino, Jpn. J. Appl. Phys. **35**, L1371 (1996)
174. S.V. Frolov, W. Gellermann, M. Ozaki, K. Yoshino, Z.V. Vardeny, Phys. Rev. Lett. **78**, 729 (1997)
175. X. Long, A. Malinowski, D.D.C. Bradley, M. Inbasekaran, E.P. Woo, Chem. Phys. Lett. **272**, 6 (1997)
176. G. Kranzelbinder, M. Nisoli, S. Stagira, S. de Silvestri, G. Lanzani, K. Mullen, U. Scherf, W. Graupner, G. Leising, Appl. Phys. Lett. **71**, 2725 (1997)
177. G. Gelinck, J. Warman, M. Remners, D. Neher, Chem. Phys. Lett. **265**, 320 (1997)
178. M.D. McGehee, R. Gupta, S. Veenstra, E. Kirk Miller, M.A. Diaz-Garcia, A.J. Heeger, Phys. Rev. B **58**, 7035 (1998)
179. G.J. Denton, N. Tessler, M.A. Stevens, R.H. Friend, Adv. Mater. **9**, 547 (1997)
180. B. Schweitzer, G. Wegmann, H. Giessen, D. Hertel, H. Bässler, R.F. Mahrt, U. Scherf, K. Mullen, Appl. Phys. Lett. **72**, 2933 (1998)
181. N. Tessler, N.T. Harrison, R.H. Friend, Adv. Mater. **10**, 64 (1998)
182. M.G. Harrison, K.E. Ziemelis, R.H. Friend, P.L. Burn, A.B. Holmes, Synth. Met. **55**, 218 (1993)
183. H. Chayet, R. Pogreb, D. Davidov, Phys. Rev. B **56**, R12702 (1997)
184. S.V. Rakhmanova, E.M. Conwell, Appl. Phys. Lett. **75**, 1518 (1999)
185. E.M. Conwell, Phys. Rev. B **57**, R12670 (1998)
186. E.M. Conwell, S.V. Rakhmanova, Israel Journal of Chemistry, to be published

5 Photoelectron Spectroscopy Studies on the Electronic and Chemical Structures of Conjugated Polymers

M. Fahlman and W.R. Salaneck

5.1 Introduction

The existence of polymers achieved general acceptance in the early years of this century. Widely known as "plastics", polymers became popular as structural material after World War II, and can be found in almost all products in present-day society, be it in clothes, furniture, home appliances, cars, airplanes, electronics, etc. The idea of utilizing the electrically conducting properties of polymers was proposed as early as in the 1960s [1], but conjugated polymers were mostly ignored until 1977, when it was discovered that trans-polyacetylene could be doped and thereby transformed into a good electrical conductor [2]. Interest in the field of conjugated polymers has grown dramatically ever since. Early on, most of the research effort was concentrated on the conducting properties of the doped conjugated polymers [3], and conjugated polymers were often referred to as "conducting polymers". Using oriented samples of doped trans-polyacetylene, conductivities almost comparable to that of a single crystal of copper was reached [4,5]. The hope was to replace existing conducting materials with the polymers, and although not realized in full, conducting polymers are now used as both transparent conductors [6] and conducting textiles [7]. The mechanism for electron conduction in these materials is still under intense study, and reviews on the topic can be found elsewhere [8].

The interest in conjugated polymers also has gradually expanded to include the properties of their semiconducting pristine state. Conjugated polymers and oligomers [9] now are used in the development of variety of organic-based applications: light-emitting devices [10], anti-corrosion coatings [11], photo- voltaic devices [12,13], rechargeable batteries [14], transistors and integrated circuits [15–18], and lasers [19–22].

Perhaps the most promising application is the polymer light-emitting device (LED), given the current high interest in developing ultra thin computer monitors and television sets, i.e., flat-panel displays. The use of polymers as light-emitting material was proposed as early as 1969 [23], and electroluminescence by various organic materials was demonstrated in the following decades [24,25]. In 1990, results on LED with poly(p-vinylene), PPV, as the emitting layer were published [26], and this marked the start of a rapid expansion in this field. Products based on polymer LEDs are scheduled to appear in 2001, while polymer-based integrated circuits are further away from true commercialization [27]. Knowledge of the electronic and chemical structure of the light-emitting polymers as well as the polymer/metal interfaces occurring at the electrode contacts has played a key role throughout the development of this technology. In this chapter, we will give an overview of some polymers and metal/polymer interfaces that are commonly used in LEDs.

Another interesting application for conjugated polymers and oligomers that also illustrates the importance of the metal/polymer interface is anti-corrosion coatings. Chromate- and zinc-containing paints are commonly used today to prevent corrosion of metals, aluminum and steel in particular [28]. However, environmental considerations might reduce or even rule out the use of chromate-based paints in the future, which has created interest in the development of new anti-corrosion coatings. Polyaniline [29–32] and polypyrrole [33] coatings are being considered as possible candidates and other electroactive polymers also may prove to be effective. These coatings can be electro-deposited [29], applied from solutions [31,32] or applied as paint [11]. With this application in mind, we will discuss the interactions between polyaniline and some metals: iron and aluminum. For a general reference on anti-corrosion coatings using conducting polymers, see [34].

5.2 Conjugated Polymers

The emergence of the field of conjugated polymers came about in a rather inauspicious way. In the early 1970s, a graduate student at the Tokyo Institute of Technology accidentally discovered a new way to synthesize trans-polyacetylene, producing free-standing silvery films instead of the usual dark powder. The films were still found to be insulating [35]. However, when exposed to dopants, increase in conductivity of up to nine orders of magnitude was achieved [2], and a new era in the field of conjugated polymers was born.

Linear polymers are macromolecules consisting of a great number of repeating units, which are coupled to each other forming a chain. Such linear chains can also crosslink and form more complex, three-dimensional structures, which is the norm in most commercial polymers, since this improves the mechanical properties of the materials. The repeating units can of course include any group of atoms, but for conjugated polymers the repeat units primarily consist of carbons.

Standard, insulating organic polymers have saturated (sp^3 hybridized) carbons along the backbone of the chain. In organic conjugated polymers, the backbones consist of sp^2 hybridized carbons. This electron configuration results in three σ-bonding electrons, the 2s, $2p_x$ and $2p_y$ electrons, and a remaining $2p_z$ electron. Two adjacent $2p_z$ orbitals can overlap, forming a π-bond (see Fig. 5.1).

In a conjugated polymer, the $2p_z$ electron orbitals overlap along the backbone, which gives rise to the delocalization of the π-electron system. Of

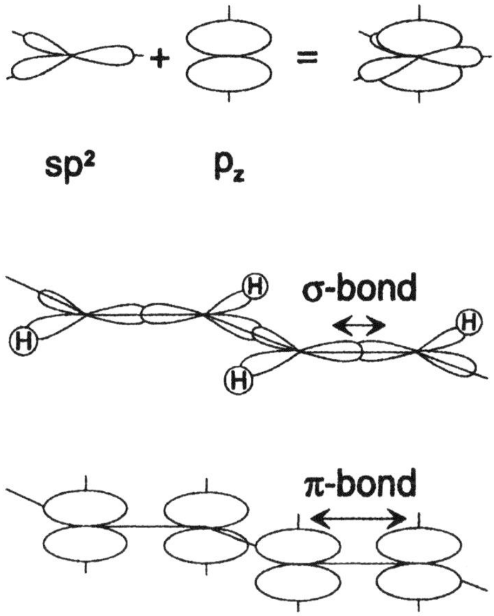

Fig. 5.1. A simple representation of σ-bonds and π-bonds

course, one also can have the case where the $2p_z$ electron orbitals do not overlap along the backbone (being orthogonal to each other), and in that case, the resulting π-bands are flat, i.e., show no dispersion. A more proper name for these materials would hence be π-conjugated polymers.

Many different types of conjugated polymers exist that have been synthesized, reflecting the vast number of different types of applications for which they are intended. In this chapter we review results on PPV and polyaniline, which are depicted in Fig. 5.2. In conjugated polymers, charge-transfer doping and optical excitations induce large (local) modifications of both the electronic and the geometric structure of the polymer. In the field of conjugated polymers, doping means charge transfer, be it by oxidation (p-type doping) or reduction (n-type doping), as well as the insertion of a counter ion in the vicinity of the polymer. This counter ion helps screen the charge on the polymer chain and ensures charge neutrality for the system as a whole.

The vast majority of conjugated polymers, however, have a non-degenerate ground state. For this class of polymers, optical excitation or doping-induced charge injection also causes a reversal of the bond alternation, though of a different type than that in the case of degenerate ground-state polymers. These bond-alternation reversals are created in pairs, i.e., a transition back to the original bond-alternation conformation will occur along the chain. Since the ground state is non-degenerate, the area between the reversed ones will have higher energy than the ground-state conformation. This will lead to an attraction between the defects, minimizing the region of higher energy. The resulting pair of interacting bond-alternation transitions is therefore viewed as one unit. Those that are singly charged are called polarons, P^-(negative

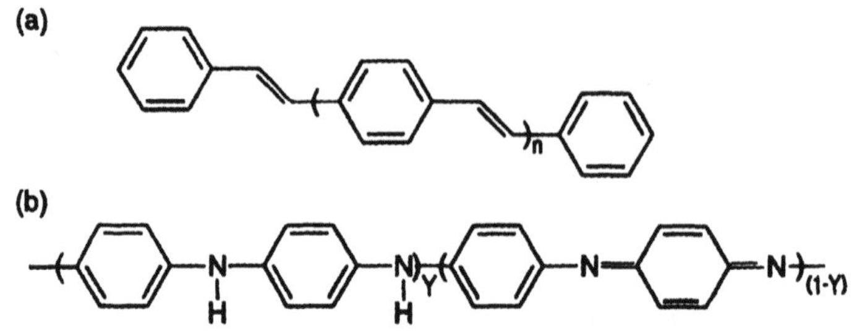

Fig. 5.2. The chemical structure of (**a**) poly(p-phenylene vinylene) and (**b**) polyaniline

charge) and P^+ (positive charge) [36]. When two polarons interact, they form a doubly charged state. These states are called bipolarons, BP^{++} and BP^{--} [37]. The nature of the charge carriers, polarons or bipolarons, depends on the particular system and density of charge [38–42]. Even for those cases where the bipolarons are energetically more favorable than polarons, the first charge induced on a non-degenerate polymer chain must be in the form of a polaron. In the electronic structure, these defects correspond to two new localized states in the band gap, because polarons have density of states at the Fermi level, whereas bipolarons do not. When enough charge is introduced upon a polymer chain, the localized polaron or bipolaron states will begin to overlap and form bands. An extensive overview of charge carriers in conjugated polymers can be found in [43].

5.3 Photoelectron Spectroscopy

Photoelectron spectroscopy is based on the photoelectric effect. When a molecule (or an atom) is irradiated by light there is a finite chance that the species will become excited. If the photon energy is high enough, the molecule may become ionized, emitting an electron, i.e., photoemission. Such a process is illustrated by the equation:

$$M_0 + h\nu \rightarrow M_+^* + e^- \ . \tag{5.1}$$

In the equation, M_0 represents the isolated neutral molecule, $h\nu$ the photon causing the ionization, M_+^* the positive molecular ion in the excited state and e^- the photoemitted electron. Energy must be conserved, resulting in the equation:

$$E_0 + h\nu = E_+^* + E_k \ , \tag{5.2}$$

where E_0 is the total energy of the neutral molecule, $h\nu$ the energy of the photon, E_+^* the total energy of the ionized (excited) molecule, and E_k is the kinetic energy of the emitted electron. Hence, the binding energy of a particular electron, $E_{\rm B}^{\rm V}$, as referred to the vacuum level, can be written as follows (Koopmans' theorem [44]):

$$E_{\rm B}^{\rm V} = E_{+}^{*} - E_0 = h\nu - E_k \ . \tag{5.3}$$

This is the basic equation used for the interpretation of photoelectron spectra. Since the energy of the photons used is known and the kinetic energy of the emitted electrons can be measured, it is possible to obtain the electron binding energies. Note that the measured binding energy of an electron corresponds to the energy difference between the initial ground state of the neutral molecule and the final, possibly excited state of the ionized molecule.

For the basic interpretation of photoelectron spectra, one normally relies upon the one-electron model of the neutral molecule [45, 46]. The basic assumption of this model is that an electron moves in a time-averaged potential of the other electrons instead of an instantaneous potential. This approximation results in the one-electron Hartree–Fock equation. The electrons also are considered to feel the same potential, regardless of whether they are in the ground state or in an excited state. Hence, neither relaxation nor polarization effects are taken into account in this model. In photoelectron spectroscopy, electrons are transferred from discrete initial states into continuum final states, which gives an expression for the photoinduced interband transition cross section:

$$\sigma(\omega) \approx 4\pi^2 \alpha\, \omega_{ki} |\langle k \,|\, \boldsymbol{r} \cdot \boldsymbol{\xi} \,|\, i \rangle|^2 \rho(\varepsilon_k) \ , \tag{5.4}$$

where α is the fine structure constant, i the initial state, k the final state, $\hbar\omega_{ki}$ the energy difference between the k and i states, $\hbar\omega$ the photon energy, $\boldsymbol{r}$ the position vector of the electron and $\boldsymbol{\xi}$ the polarization vector of the electric field (assuming transverse plane waves) and $\rho(\varepsilon_k)$ is the density of states in the continuum that has energy $\varepsilon_k = \hbar\omega + \varepsilon_i$ [47].

The one-electron picture will thus result in a photoelectron spectrum with a number of sharp peaks corresponding to the occupied levels. Note that the one-electron model gives a one-to-one correspondence between the peaks in the photoelectron spectrum of a molecule (polymer) and the energy levels (bands) in the ground state [44]. The one-electron model, though useful, does not adequately describe many of the features that appear in a photoelectron spectrum, so a brief discussion involving phenomena where the one-electron picture no longer holds is given below.

For molecular solids, there are two types of relaxation effects that occur during the photoemission process. The *intermolecular* relaxation is caused by the electronic polarization of the surrounding molecules. These molecules then help to screen the hole created by the photoelectron emission event. This will cause a shift in the measured binding energy by approximately 1–2 eV towards lower binding energy, as compared to an isolated molecule [48]. The *intramolecular* relaxation takes place in all molecules and atoms, regardless of whether they are isolated or not. Although the process is "quantum mechanical", in a classical argument, a photoemitted electron leaves the molecule within 10^{-15} s. Upon the creation of a hole, the remaining electrons relax in about 10^{-16} s in order to screen it. The relaxation time for the nucleus, however, is on the order of 10^{-13} s. This means that the electronic structure will have time to relax before the photoemitted electron leaves, whereas the nucleus remains frozen during the process [49]. This is, in essence, the Born–Oppenheimer approximation, which states that the electronic part and the nuclear part of a system can be treated separately. The fact that the electronic structure has time to relax gives rise to other interesting features in the photoelectron spectra. As mentioned previously, the measured binding energy of an electron corresponds to the energy difference between the initial ground state of the neutral molecule and the final (excited) state of the ionized molecule. In fact, all possible final states corresponding to a specific initial state will, in principle, show up in the photoelectron spectrum, leading to a breakdown of the one-electron picture [46]. When an absorbed photon causes an electron to be emitted, hence creating a hole, the surrounding electron cloud will contract in order to screen the positive charge thus created, i.e., intramolecular relaxation. This contraction, however, may give rise to "shock-waves" in the electronic cloud leading to excitation across the forbidden band gap, *shake-up*, and/or ionization, *shake-off*, of the molecule.

5.3.1 X-Ray Photoelectron Spectroscopy

The modern-day use of photoelectron spectroscopy began with the development of electron spectroscopy for chemical analysis, ESCA [50]. This method also is known as X-ray photoelectron spectroscopy (XPS), since it uses X-ray photon sources. The most commonly used X-ray lines are the $MgK\alpha_{1,2}$ (1253.6 eV) and the $AlK\alpha_{1,2}$ (1486.6 eV) lines, with full width of half maxima (FWHM) of about ~0.8 eV and ~1.2 eV respectively [50]. Due to the large line width of the $AlK\alpha_{1,2}$ source, it is often equipped with a monochromator in order to improve the energy resolution. The high photon energy of these sources gives access to the deep-lying core levels as well as the valence region, unlike the case for ultraviolet light. Since the electronic structure, the core levels and valence region, is unique for all atomic and molecular species, X-ray photoelectron spectroscopy can thus be used as a "fingerprint" method in analyzing the sample's chemical composition in the surface region. A quantitative analysis of the composition of the sample is possible, but limited in its exactness due to inelastic scattering of electrons in the sample and cross-section effects. Information about the chemical environment can also be obtained, since the electronic structure in both the core level and valence-band regions is affected by bonding and other chemical reactions such as charge transfer. The shake-up features of a core-level spectrum yield information about the unoccupied levels as well. This often proves highly useful when studying the interactions between metals and conjugated polymers [32, 51], since the shake-up region can also be used as a fingerprint of certain molecules and molecular constituents.

5.3.2 Ultraviolet Photoelectron Spectroscopy

As mentioned above, photoelectron spectroscopy on the valence region is possible using XPS. However, the cross section for valence electrons is an order of magnitude lower than that for core electrons, which dramatically increases the time needed for a spectrum. Also, the resolution of the experiment is more important when studying the valence region, since the states/bands tend to be much closer spaced.

Ultraviolet photon sources are commonly used when studying the valence electronic structure . The photoionization cross sections for C 2p and C 2s atomic orbitals, which make up the valence electronic structure in organic molecules, are 10^3 to 10^5 times larger for photon energies used in UPS than in XPS [52]. The most common ultraviolet photon source is the He resonance lamp, providing photons of 21.2 eV (HeI) and 40.8 eV (HeII) in energy. The line width (FWHM) of the HeI and HeII lines is ~30 meV, much narrower than that of the X-ray photon sources.

Photoelectron spectroscopy of solid samples is carried out with the binding energy given relative to the Fermi level. However, when interpreting photoelectron spectra, it is often necessary to take into account the changes in work function of the sample that occur upon doping or other modifications to the sample. This is important not only when studying valence-band features, but also for core-level spectroscopy as well, since the chemical shifts are only meaningful when related to the vacuum level.

By measuring the position of the zero kinetic energy cut-off of the secondary electron distribution, E_{co}, the work function of a sample may be obtained, enabling the adjustment of the binding energies so that the vacuum level is taken as the reference. The work function of the sample, ϕ_s, can be derived from the following equation

$$\phi_s = h\nu - (E_{co} - E_F) . \quad (5.5)$$

Clearly, each method (XPS and UPS), has its own advantages and disadvantages. Ideally, one would like to have a photon source whose photon energy could be varied in order to take advantage of cross-section effects, give high-enough energy to reach the lowest lying core-levels, while at the same time retain the excellent line width of the He photon sources.

Synchrotron radiation light provides all of that and more, and it is briefly discussed below.

5.3.3 Synchrotron Radiation

Synchrotron radiation light is produced by having electrons (or positrons) circulate in an ultra-high vacuum storage ring at relativistic speeds, i.e., close to that of light. When charged particles are thus accelerated, they emit electromagnetic radiation with wavelengths ranging between mm (IR) and Å (hard X-rays).

Synchrotron radiation light has many desirable properties besides it continuous and large spectral range. It is very intense, i.e., high photon flux, brightness and brilliance. In fact, the brilliance of synchrotron radiation can be larger than that of conventional X-ray sources by a factor of 10^{11}. By using soft X-ray monochromators, line widths comparable to those of the He lines, and better, can be obtained. The synchrotron light also is polarized and has a pulsed time structure, which allows one to perform time resolved experiments and coincidence studies. The light is plane polarized in the horizontal plane, but out of the horizontal plane it is elliptically polarized tending to circularly polarized for large angles. The polarized nature of the light allows for angle-resolved photoelectron spectroscopy measurements of atomic and molecular systems [53]. Also, the linear polarized light enables mapping of the one-dimensional electron-band structure of ordered conjugated polymer samples [54]. General information about synchrotron radiation and its applications can be found in [55, 56].

5.4 Pristine Polymers

5.4.1 Poly(p-phenylene vinylene)s (PPVs)

As noted previously, what sets the conjugated polymers apart from the more conventional polymers is their π-electronic structure (delocalized) along the polymer backbone and the novel forms of charge carriers that are created upon doping. Knowledge about the frontier π-electronic structure, i.e., the bands surrounding the Fermi level, is particularly important for application such as LEDs and transistors, where electrons and holes are injected into the polymers. PPV and its derivatives are commonly used as light-emitting material in polymer LEDs, so we will use this polymer to discuss the frontier π-electronic structure and the influence that substituents have on it.

The chemical structure of PPV and some substituted PPVs are depicted in Fig. 5.3. The repeat unit of PPV consists of a benzene ring and a vinylene group. In Fig. 5.4a the degenerate e_{1g} state of benzene is shown. One of the two degenerate halves of the wavefunction of the e_{1g} state does not have nodes at the para-positions of the benzene ring and therefore overlaps with the highest π-state of the vinylene group. This interaction leads to the dispersed highest lying π-band of PPV (see Fig. 5.4b). The half of the wavefunction of the degenerate e_{1g} state that does have nodes at the para-positions of the benzene ring and, therefore, does not overlap with the highest π-state of the vinylene group, makes up the second highest lying π-band of PPV. Since there is no overlap of the wavefunctions along the backbone, this band is flat (no dispersion) as depicted in Fig. 5.4c.

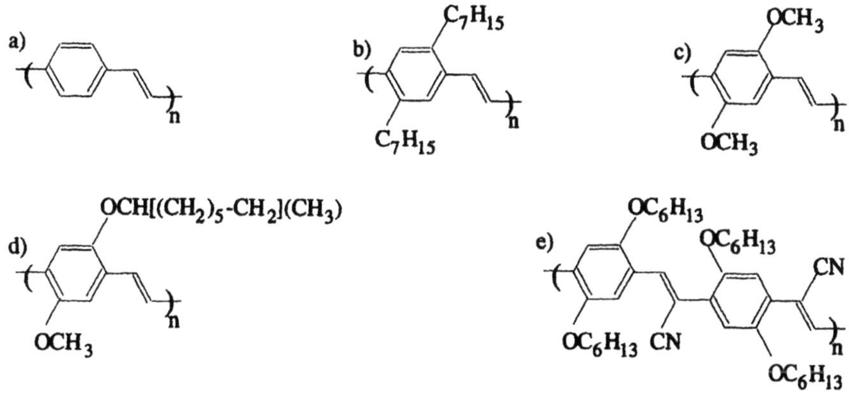

Fig. 5.3. The chemical structure of (**a**) poly(p-phenylene vinylene), (**b**) poly(2,5-diheptyl-1,4-phenylene vinylene), (**c**) poly(2,5-dimethoxy-1,4-phenylene vinylene), (**d**) poly(2-methoxy-5-(2'ethylhexoxy)-1,4-phenylene vinylene) and (**e**) poly(2,5,2',5'-tetrahexyloxy-8,7'-dicyano-diparaphenylene vinylene) (PPV). (From [57] and [61])

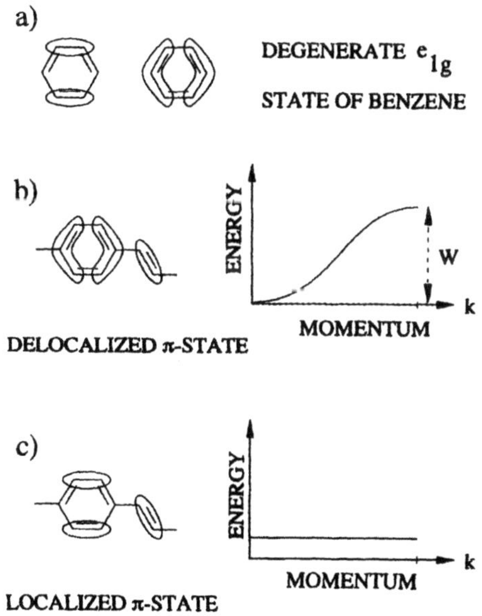

Fig. 5.4. (**a**) The degenerate benzene e_{1g} state, (**b**) the origin of the dispersed valence π-band and (**c**) the origin of the flat (valence-1) π-band

In Fig. 5.5a [57], an experimental HeI spectrum of PPV is compared with theoretically derived density of valence states (DOVS). Since the focus is on the frontier π-electronic structure, we only show the two highest lying bands.

Peak 1 is derived from the upper part of the highest lying π-band. As mentioned previously, this band originates from states delocalized along the

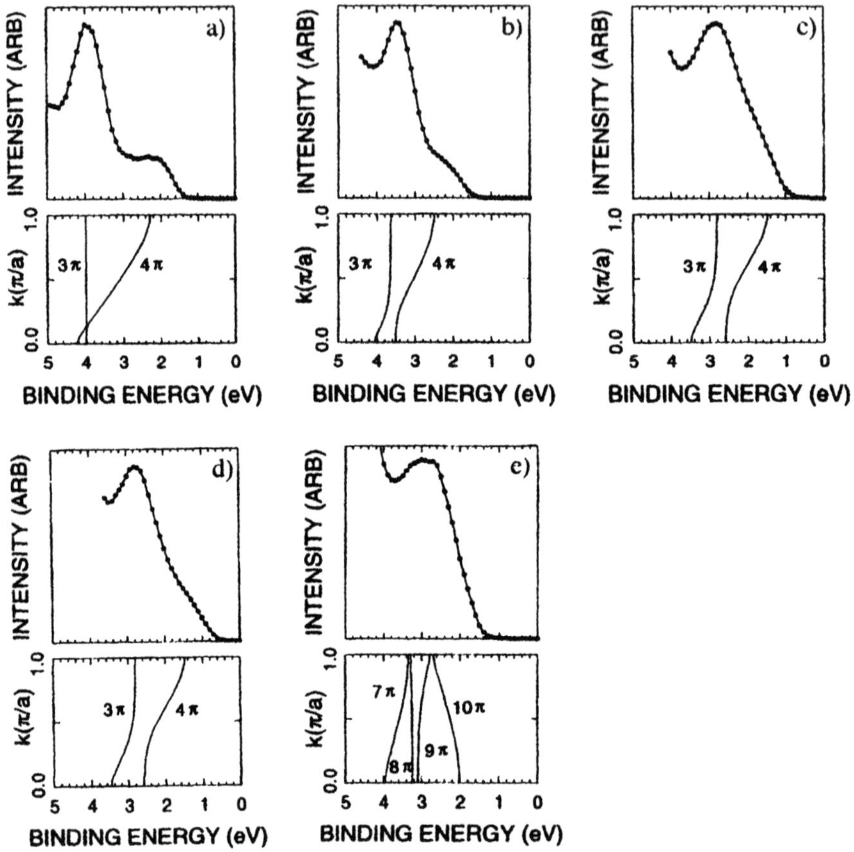

Fig. 5.5. Comparison of the top part of the HeI UPS spectrum and the highest occupied (frontier) π-bands for (**a**) poly(p-phenylene vinylene), (**b**) poly(2,5-diheptyl-1,4-phenylene vinylene), (**c**) poly(2,5-dimethoxy-1,4-phenylene vinylene), (**d**) poly(2-methoxy-5-(2'ethylhexoxy)-1,4-phenylene vinylene) and (**e**) poly(2,5,2',5'-tetrahexyloxy-8,7'-dicyano-diparaphenylene vinylene). (From [57] and [61])

polymer backbone. Peak 2 is derived from the lower part of the highest π-band and the second highest π-band, the latter being flat since it is created by states localized on the ortho carbons of the phenylene rings. The difference between the locations of the two peaks reflects the difference between the top part of band 4π and the flat band 3π, i.e., about 1.8 eV, with the total bandwidth being 2.0 eV.

When substituents are added to the backbone, the π-electronic structure and hence the π-bands can be affected in both a direct and indirect way. The direct or primary effect occurs when the interaction between wavefunctions of the substituent and the backbone either stabilizes or destabilizes the π-

bands, increasing or decreasing the binding energy respectively. The indirect or secondary effect occurs if the substituent induces a change in the torsion angle between the benzene rings and the vinylene groups. An increase of the torsion angle will lead to a decreased overlap between the benzene and vinylene π-orbitals. This decreases the conjugation along the backbone and decreases the dispersion of the band, making it flatter. On the other hand, a decrease of the torsion angle leads to an increased overlap of the π-orbitals, and hence a larger dispersion of the highest lying π-band, as shown in Fig. 5.6.

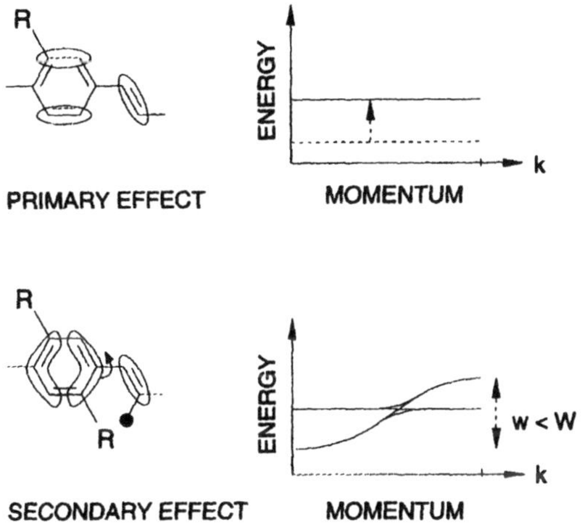

Fig. 5.6. The primary (substituent) and secondary (steric) effects on the frontier π-electronic structure in poly(p-phenylene vinylene)

Thus, by clever use of substituents on the backbone, one can alter both the shape and energy of the frontier π-bands of a particular conjugated polymer, and thereby optimize its performance as the active material in a potential device. We illustrate these effects with UPS spectra of the PPV derivatives depicted in Fig. 5.3b–e.

In diheptyl-substituted poly(p-phenylene vinylene) (DHPPV), the steric hindrance between the first CH_2 unit of the alkyl chains and the hydrogen of the vinylene moieties creates a large torsion angle between the phenylene rings and vinylene moieties along the polymer backbone. This angle is roughly 34° according to semi-empirical calculations [57], as compared to the near coplanar conformation (7°) of PPV. Angles as large as 34° would decrease the p_z overlap along the backbone and create a nearly flat valence band. However, solid-state packing effects in DHPPV films are expected to drive the geometry into a more planar conformation. The effect of the alkyl substituents on the frontier electronic structure is clearly visible in Fig. 5.5b [57], where the HeI spectrum of DHPPV is compared to calculated DOVS. Peak 2 in the UPS spectrum is shifted towards lower binding energies, 3.5 eV from 4.0 eV in PPV. The calculations show that the positions of band 3π and the flat part of band 4π are shifted towards lower binding energies by about

0.5 eV as compared to unsubstituted PPV, in excellent agreement with the experimental results. According to the Valence Effective Hamiltonian (VEH) wavefunctions, this shift is caused by anti-bonding interactions between the first carbon of the aliphatic chain and the ring orthocarbon to which it is attached. The total bandwidth is less than that of unsubstituted PPV, signifying an increase in torsion angle as compared to PPV. However, since the valence band shows dispersion, solid-state packing has driven the conformation into significantly smaller torsion angles than those calculated for the gas/liquid phase (34°).

Similar effects occur for dimethoxy-substituted poly(p-phenylene vinylene) (DMeOPPV). This is evident from Fig. 5.5c [57], where the upper two calculated energy bands of DMeOPPV are compared with the corresponding UPS spectrum. The oxygen atoms strongly destabilize the highest two occupied bands, the flat parts of the two bands being shifted approximately 1.2 to 1.4 eV towards lower binding energies as compared with the results for unsubstituted PPV, as is shown in the figure. This destabilization is explained by the anti-bonding character of the interaction between the phenylene ring and the methoxy wavefunctions for the two bands [58]. The avoided crossing between the two bands increases due to the presence of the alkoxy groups, causing an even larger separation of the two bands [58]. The flat parts of bands 4π and 3π lead to a broader peak in the calculated DOVS compared to that for DHPPV and PPV, which is in excellent agreement with the experimental spectra. Because of this broadening, the peak closest to the Fermi energy (that was observed in PPV and DHPPV) is unresolved in the experimental spectrum of DMeOPPV. The total bandwidth remains 2 eV, as in PPV and DHPPV, which is evidence for a coplanar conformation in the solid state. Indeed, instead of having a steric interaction between the substituent groups and the hydrogen of the vinylene moieties, semiempirical calculations show that weak hydrogen bonds are formed between the methoxy oxygen and the vinylene hydrogen [59]. This drives the geometry of the polymer chains into coplanar conformations, in agreement with the experimental results.

The length of the substituent chains does not seem to strongly affect the stabilization or destabilization of the frontier π-bands, nor does the chain length significantly alter the torsion angles. The alkoxy-substituted PPV shown in Fig. 5.3d, has a longer substituent chain than for DMeOPPV. However, Fig. 5.5d [57] clearly demonstrates that the frontier electronic structure is virtually identical to that of DMeOPPV. The calculated π-bands of the two polymers are found to be essentially identical in shape, position and the total bandwidth. The two UPS spectra are also nearly identical, as predicted by theory. A difference in length of the substituent groups does affect the π-bands at higher binding energies, but instead can only affect the characteristics of the upper π-bands if they induce different ring torsion angles along the backbone. However, if one chooses substituents with an alkoxy oxygen at the ortho position(s) of the phenylene rings, the weak hydrogen bonding mentioned previously is expected to keep the conformation roughly coplanar. Thus, aggregation and hence aggregation-induced exciton quenching in polymer solids can be reduced by using long, bulky alkoxy substituents without significantly changing the valence and conduction bands of the polymer [60].

Alkyl and alkoxy substituents serve to destabilize the frontier π-electron structure, decreasing the binding energy of the bands surrounding the Fermi level. Although such substituents are often needed to make conjugated polymers soluble, decreased electron affinities hamper electron injection into these materials, thus creating problems in such applications as LEDs. In order to enhance the electron injection, substituents that stabilize the frontier π-electronic structure and increase the electron affinity are needed. In Fig. 5.3e the geometry of a cyano- and alkoxy-substituted PPV is shown (CN-PPV). The alkoxy groups make the polymer soluble and the cyano group acts as an electron withdrawer, stabilizing the π-electronic structure. Note that the repeat unit of this polymer consists of two phenylene-vinylene segments instead of one as in the previous polymers. In Fig. 5.5e [61] the UPS spectra of the highest lying features and their associated π-bands are depicted. Compared with unsubstituted PPV, the flat branches of these bands are shifted to lower binding energies by about 0.8–1.0 eV. This is due to two competing effects. First, there is the anti-bonding interaction between the wavefunctions of the hexyloxy groups and the phenylene rings, which destabilizes the flat parts of the bands and moves them to lower binding energies by roughly 1.2–1.4 eV. However, the positions relative to the Fermi energy of the flat branches of these π-bands and the peak derived from them are also shifted roughly 0.4 eV towards higher binding energies, as compared to dialkoxy-substituted PPV, due to the presence of the cyano groups. This latter shift can be understood in a simple way, by taking into account the acceptor nature of the cyano groups, which provokes an overall stabilization of the π-bands, and, therefore, a shift to higher binding energies [61].

Comprehensive photoelectron studies of PPV and its derivatives discussed in this chapter can be found elsewhere in the literature [57,61,62]. The theoretical DOVS were obtained from VEH calculations [63,64] on AM1-derived geometries [65]. The method of combining AM1 and VEH calculations to produce DOVS for conjugated polymers is well established and known to yield excellent results [66,67].

5.4.2 Polyanilines

Polyaniline is unusual in that it can exist in three different stable forms, all of them insulating. The three different states are depicted in Fig. 5.7. In emeraldine base (a), roughly 50% of the nitrogen is protonated, in leucoemeraldine base (b) all nitrogen are protonated and in pernigraniline base (c) no protonation of the nitrogen is present. There is also a highly conducting form of polyaniline, emeraldine salt, which can be achieved by protonation of emeraldine base or p-doping of leucoemeraldine base [68].

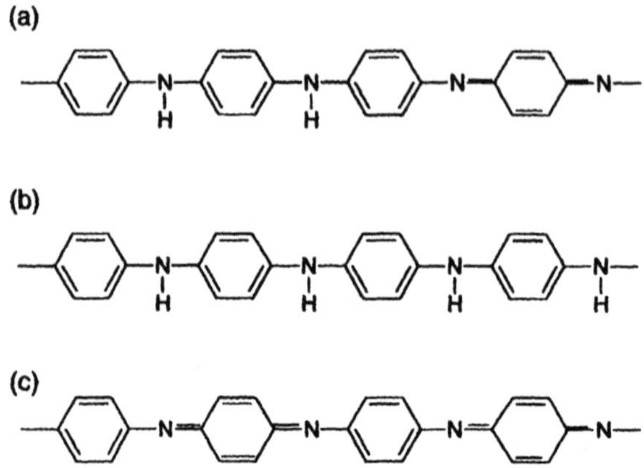

Fig. 5.7. The chemical structure of (**a**) emeraldine base polyaniline, (**b**) leucoemeraldine base polyaniline and (**c**) pernigraniline base polyaniline

The electronic structures of the three base forms are significantly different, both in terms of the core levels and the valence levels. The amine nitrogen of the reduced form (leucoemeraldine base) has an N 1s core-level binding energy that is ~1 eV higher than that of the imine nitrogen of the fully oxidized form, pernigraniline base. The imine nitrogen has a shake-up feature ~4 eV above the imine N 1s core-level peak. In emeraldine base form, amine and imine nitrogen are present in an approximately 1:1 ratio. Polyaniline has been studied extensively in both its insulating and conducting forms using XPS [69–75]. Here we will focus on the frontier electronic structure as studied by UPS and the combined AM1 semi-empirical and VEH pseudo-potential method.

The valence electronic structure of polyaniline is not only affected by the ratio of amine and imine nitrogen, but there is an indirect effect of ring torsion as well. The ring torsion varies depending on the ratio of quinoid and aromatic ring forms in the material. We illustrate these effects using the emeraldine base and pernigraniline base forms, where UPS spectra are compared with theoretically derived DOVS. For emeraldine base the ring torsion angles were calculated to be 8°, −43°, 16°, −43° respectively, starting with the quinoid ring and going along the chain (see Fig. 5.7a). The ring torsion angles are defined relative to the plane of the nitrogen atoms. It is evident from these results that the steric hindrance between adjacent aromatic rings is significantly less than that for an aromatic ring adjacent to a quinoid ring.

In Fig. 5.8 [76], a UPS spectrum of emeraldine base is compared with the theoretical DOVS. Note that the binding energy is given relative to the vacuum level.
The highest occupied band corresponds to an electronic state that is mainly localized on the amine nitrogens and the aromatic rings they surround. The second highest occupied band corresponds to an electronic state whose wave-

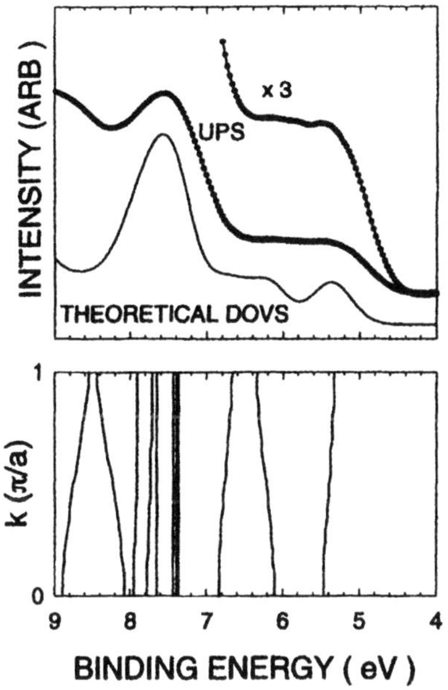

Fig. 5.8. Comparison of the UPS spectrum of emeraldine base polyaniline with the calculated density of valence states and band structure. (From [76])

function is mainly distributed over the amine nitrogens and the phenylene rings. The third highest occupied band is derived from an electronic state whose wavefunction is delocalized over all the nitrogen atoms (imine and amine) as well as the phenylene rings. In the experimental spectrum, these three bands give rise to one broad feature instead of two separate ones, due to the resolution of the spectrometer. The flat bands that create the peak at roughly 8 eV are from states completely localized on the phenylene rings.

Fig. 5.9 depicts the UPS spectrum of pernigraniline base compared with theoretical DOVS [76]. Again, we choose to show only the frontier electronic structure. Here, the ring torsion angles were found to be 2° and −58° out of plane for the alternating quinoid and aromatic rings respectively. The combined bond length and ring torsion dimerization contributes to a formation of a Peierls gap [77]. The frontier electronic structure is similar to that of emeraldine base in respect to the peak at ~8 eV, which is formed by flat occupied bands derived from electronic states localized on the phenylene rings. In fact, the position and shape of this peak will be roughly the same for all three base forms of polyaniline, given the strong localization of the wavefunctions to the phenylene rings, thus making them unaffected by ring torsion.

In addition, since no contributions from nitrogen are present in these bands, neither will the ratio between amine and imine nitrogen significantly affect the position and shape of the bands. The electronic structure does differ significantly for the valence band. The valence band in pernigraniline base is derived from an electronic state mainly localized on the imine nitrogen and the aromatic ring of the repeat unit. This band is shifted towards higher

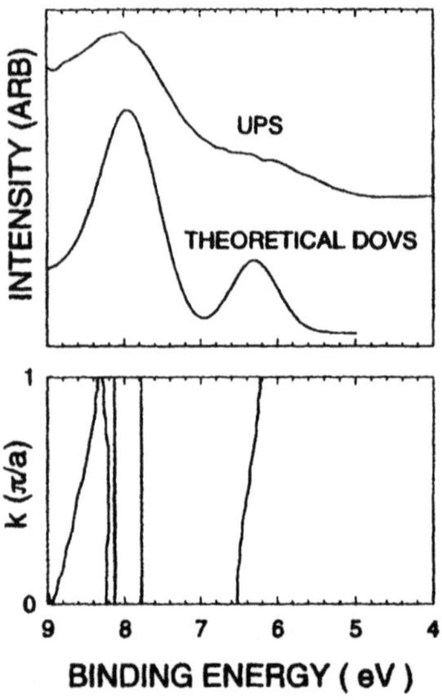

Fig. 5.9. Comparison of the UPS spectrum of pernigraniline base polyaniline with the calculated density of valence states and band structure. (From [76])

binding energies by about 0.8 eV as compared to the valence band of emeraldine base (and thus increases the ionization potential by the same amount). Moreover, unlike emeraldine base, in which there are three bands in the region between 5 to 7 eV, in pernigraniline base only the valence band occurs in this region. That affects the shape of the low binding energy shoulder, making it less broad compared to the corresponding feature in emeraldine base, as is evident in Figs. 5.8 and 5.9. The excellent agreement between experiment and theory yet again demonstrates the power of analyzing UPS spectra with AM1/VEH-derived DOVS.

5.5 Metal/Polymer Interfaces

Understanding the nature of metal/polymer interfaces has become increasingly important due the advent of molecular electronic applications where electrical contacts have to be formed by metal deposition onto polymer/oligomer films. The choice of metal is made based upon a wide variety of criteria including charge-injection properties, interface stability (diffusion) and environmental stability. All of those properties can be studied by photoelectron spectroscopy of the metal/polymer interface. Depending on the information sought, the experiments can be carried out with spectra taken between submonolayer deposition steps as the metal contact is built up, or by studying the "buried" interface after a thin metal contact has been formed on top of the polymer film. Of specific interest is the nature of the metal–organic bonds at the interface, i.e., whether they are covalent or ionic. A change in the chemical and electronic structure of the polymer film caused by the interaction with metal atoms can significantly improve or degrade the performance of a device. The possibility of diffusion of the metal atoms can also be probed using XPS.

5.5.1 Metals on Poly(p-phenylene vinylene)s

As stated previously, the various derivatives of PPV are commonly used in the development of polymer LEDs. One of the problems in increasing the efficiency of such devices has been the low rate of electron injection. PPV has a fairly low electron affinity, which creates a large barrier against electron injection from metals such as gold or even aluminum. Using metals with low work functions, alkali metals for instance, would decrease the barrier against electron injection into the polymers and thus in theory improve the device performance. However, merely looking at how the work functions of the metals line up with the conduction bands of the polymers is not enough in this class of materials. Instead, one needs to study the chemistry of the interfaces in detail in order to explain the metal contacts performance or lack thereof.

Aluminum has often been used as the electron injecting contact in polymer LEDs. It has a relatively low work function and is environmentally stable. The interface between aluminum and PPV has been studied using both XPS and quantum chemical modeling. It was found that when the PPV films contained oxygen impurities, especially at the surface layer, the aluminum atoms bonded with the film through Al–O–C links [78, 79]. For "clean" PPV, however, the aluminum atoms preferentially react with vinylene groups, forming covalent bonds with the carbon atoms [80]. This also holds true for dialkoxy-substituted PPV [80]. When reacting with the vinylene carbon atoms, the aluminum introduces sp^3 defects in the polymer backbone and thus breaks the conjugation along the chain. That will, in turn, affect the frontier π-electronic structure and may significantly degrade the charge-injection properties of the system. However, by introducing substituent groups to PPV that are more favorable to reacting with aluminum than the vinylene moieties, the breaking of the conjugation may be avoided. A study on the early stages of interface formation between aluminum and the cyano-substituted polymer shown in Fig. 5.3e, using a combined approach of XPS and quantum chemical calculations, showed that the aluminum atoms preferentially react with the cyano groups [81]. AM1 semi-empirical calculations predicted covalent bonding between aluminum atoms and the cyano carbon and the formation of a cyano-Al complex, in which an aluminum atom bonds to the cyano nitrogen and the cyano carbon. Mulliken population analyses for the case of two aluminum atoms per repeat unit showed that the aluminum atoms donate ~0.35 $|e|$ in charge per atom. The nitrogen atom gains about 0.5 $|e|$ of electronic charge, whereas the cyano carbon and the vinylene carbon attached to the nitrogen gain 0.16 $|e|$ each. This reaction scheme leaves the conjugation along the polymer chain intact, since neither the vinylene moieties nor the phenylene rings are bonding with the aluminum.

The XPS results are in agreement with the theoretical predictions. In Fig. 5.10 [81] the N 1s core-level spectrum of the pristine polymer is compared with a spectrum taken after aluminum deposition equivalent to two aluminum atoms per repeat unit. The spectrum taken after aluminum deposition show a new feature at around 398 eV, shifted 2 eV towards lower binding energies as compared to the N 1s peak of the pristine polymer at 400 eV. This shift is in agreement with charge donation to the nitrogen atoms and a clear indication that the nitrogen of the cyano groups reacts with the Al atoms. Note that main peak at 400 eV still remains, since the XPS method probes depths of up to 100 Å, whereas the Al atoms stay at the surface, showing little diffusion into the sample, when studied by angle-dependent spectroscopy. The C 1s spectrum of the aluminum/polymer interface also show changes as compared to the pristine case, with a weak shoulder appearing at roughly 1.7 eV lower binding energy than the main carbon peak, indicating charge transfer to some carbon atoms. This feature is quite weak, however, since the number of carbons receiving the charge is small compared to the overall number of carbon atoms in the repeat unit (1:21 ratio). The Al 2p peak shows a shift towards higher binding energy as compared to metallic aluminum at the early stages of interface formation, but moves towards a "metallic" binding energy as the coverage increases. This also suggests charge transfer from aluminum to the polymer [81].

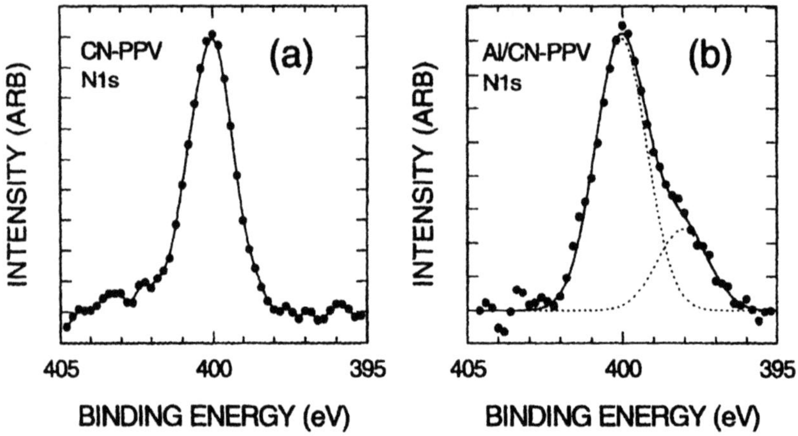

Fig. 5.10. XPS N 1s core level spectra with the Fermi level taken as reference for (**a**) pristine poly(2,5,2',5'-tetrahexyloxy-8,7'-dicyano-diparaphenylene vinylene) and (**b**) interface of Al and poly(2,5,2',5'-tetrahexyloxy-8,7'-dicyano-diparaphenylene vinylene) at approximately two Al atoms per phenylene–vinylene unit. (From [81])

Instead of forming covalent bonds at the polymer interface, (metal) atoms deposited onto the polymer may instead diffuse into the film and cause doping. We illustrate this effect by discussing Na–deposition on PPV [38]. After Na–deposition, the nature of the Na–PPV bond is found to be ionic from the values of the XPS C 1s and Na 1s core-level binding energies. Sodium was also found to rapidly diffuse into the polymer film immediately upon deposition. Hence, sodium does not form a stable interface with PPV, but instead n-dopes the polymer. The doping level is defined as the Na/repeat-unit ratio, given in %, determined from the intensities of the C 1s and Na 1s XPS peaks. Upon doping, there is an immediate decrease in the work function of PPV and at a 100% doping level, the work function is decreased by more than 2 eV.

At intermediate doping levels, two new states appear in the previously empty electron energy gap (the "band gap"), and their intensities in the UPS spectra increase uniformly with doping, as shown in Fig. 5.11. At maximum doping, i.e., near 100%, the two peaks are separated by 2.0 eV, with the lower binding energy peak positioned at 3.2 eV from the vacuum level. There are two ways PPV, a non-degenerate ground-state polymer, can accept the added charges: by forming two polaron bands or two bipolaron bands in the previously forbidden band gap. Theoretical results are used to differentiate between the two cases. According to results from VEH calculations, the

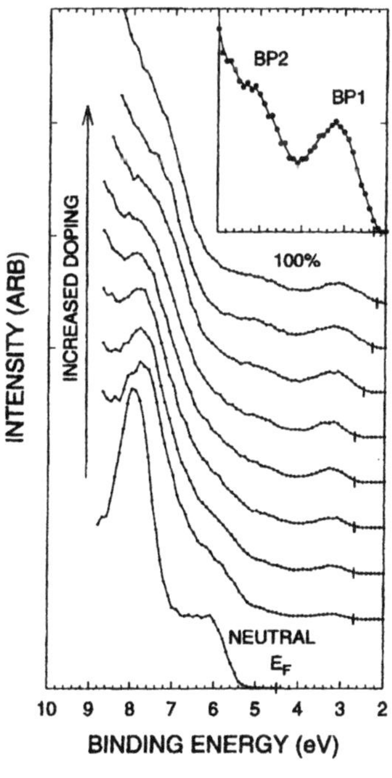

Fig. 5.11. HeI UPS spectra of pristine and increasingly Na-doped poly(p-phenylene vinylene) with the vacuum level taken as reference. (From [38])

two bipolaron bands generated at 100% doping will be separated from each other, as well as from the conduction and valence bands. These bands will lead to two bipolaron peaks in the UPS spectrum and no finite density of states at the Fermi level. In contrast, the polaron bands at 100% doping are predicted to fuse with the conduction and valence bands, which would lead to a large broadening of the polaron peaks in the UPS spectrum and a significant density of states present at the Fermi level. The latter scenario is clearly not the case, as is evident from Fig. 5.11 [38]. The apparent small density of states at the Fermi level in the UPS spectrum at 100% doping is a function of the resolution and broadening in the spectra, and is not to be considered "real". Hence, a bipolaron lattice is formed in this case at high doping levels. As mentioned in Sect. 5.1, the balance between bipolarons and polarons at high doping levels is very delicate and change of dopant or modification of the polymer may alter the nature of the doped states. Even if there isn't a change from bipolaron to polaron bands, substitution of the polymer backbone can have a significant effect on the shape and position of the bipolaron bands, as illustrated by the cyano- and dihexyloxy-substituted PPV depicted in Fig. 5.3e. As in PPV, there is an immediate shift of the work function upon sodium deposition, and two well-defined peaks are visible in the previously forbidden band gap at intermediate to high (100%) doping levels.

There are, however, some differences between the results of Na-doped cyano-substituted PPV and PPV. The experimental peak-to-peak splitting of the bipolaron peaks in CN-PPV is about 1.05 eV, as compared to 2.0 eV for Na-doped PPV at a 100% doping level, as seen in Fig. 5.12 [61]. This is caused by a stronger confinement of the bipolaron charge carriers in the cyano-substituted polymer. At a doping level of 100%, i.e., one unit charge per phenylene–vinylene unit, the geometry of the phenylene ring adjacent

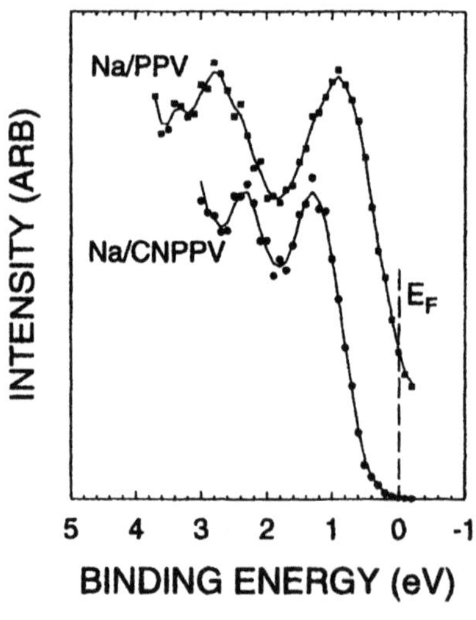

Fig. 5.12. HeI UPS spectra of the bipolaron states of poly(p-phenylene vinylene) and poly(2,5,2',5'-tetrahexyloxy-8,7'-dicyano-diparaphenylene vinylene) at a doping level of one Na atom per phenylene–vinylene unit. (From [61])

to the cyano groups is only slightly affected (a maximum difference in bond lengths within the ring of about 0.03 Å). However, the bond lengths in the vinylene groups and the other phenylene ring strongly evolve, with the rings assuming a strongly quinoid structure (a maximum difference in bond lengths within the ring of about 0.10 Å). This indicates that the bipolarons are confined on cyano–vinylene–ring–vinylene–cyano segments along the polymer backbone. This is in sharp contrast to the case for Na-doped PPV, where at a doping level of 100%, all phenylene rings take on a semi-quinoid structure (a maximum difference in bond lengths within the ring of about 0.07 Å). In Fig. 5.13 the increase in charge induced at 100% doping, as calculated by Mulliken population analysis, is depicted for cyano-substituted PPV (a) and PPV (b). For the cyano-substituted polymer, over 75% of the induced charge is distributed on the segments containing the bipolarons, with the phenylene rings included in those segments accepting nearly twice as much charge as the unperturbed phenylene rings. On the other hand, for the case of fully doped PPV, all phenylene rings accept −0.68 extra charge per ring, whereas the vinylene groups accept −0.32 extra charge each as compared to pristine PPV. Hence, the pinning of charge to the cyano vinylene groups and the phenylene rings to which they are conjugated prevents a uniform distribution of the bipolarons along the polymer backbone, thus causing a confinement of the charge carriers. As a result of that confinement, the bipolaron levels appear deeper in the gap as compared to unsubstituted PPV.

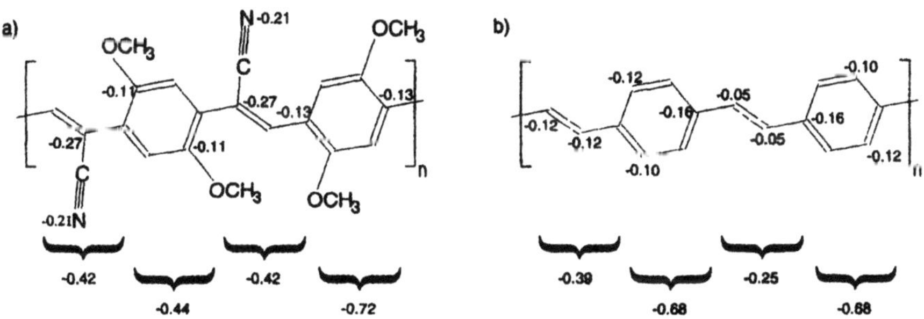

Fig. 5.13. Mulliken population analysis of the charge increase (as compared to the pristine case) at a doping level of one Na atom per phenylene–vinylene unit for (**a**) poly(2,5,2',5'-tetrahexyloxy-8,7'-dicyano-diparaphenylene vinylene) and (**b**) poly(p-phenylene vinylene). (From [61])

5.5.2 Metals on Polyanilines

The interface formation between metals and polyaniline has drawn interest mainly in two applications: anti-corrosion protection and polymer LEDs. Hence, numerous XPS studies have been carried out on various metal/polyaniline interfaces [82–85]. We discuss here some results on XPS studies on the iron/emeraldine base interface. The interface between iron (cold rolled steel) and emeraldine base (EB) was studied by sputter-depositing atoms (>99.2% Fe) from a sputter-cleaned cold rolled steel source onto EB/Au films in UHV, taking XPS core-level spectra after each deposition [32]. The early stages of the resulting Fe/EB interface formation are shown in Fig. 5.14. At low levels of coverage, the Fe $2p_{3/2}$ peak is shifted towards a higher binding energy, 708.2 eV, as compared to 706.7 eV for metallic iron, giving evidence of charge donation from the iron into the EB film. At higher levels of coverage, the iron peak becomes more metallic (707.2 eV), but still shows a slight shift towards higher binding energies as compared to bulk Fe metal. Sputter-deposited iron on Au substrates showed a shift of the Fe $2p_{3/2}$ binding energy to 707.5–707.6 eV at the early stages of interface formation. In Fig. 5.15, the evolution of the N 1s core level is shown at increasing levels of iron coverage on EB. At low levels of coverage, the N 1s core level retains its roughly 1-to-1 ratio of amine (~399.9 eV) and imine (~398.6 eV) nitrogen, as depicted for pristine EB in Fig. 5.15a [32]. The third feature of pristine EB at 402.1 eV is assigned as a shake-up peak originating from the imine nitrogen [86]. The relative intensities of the three peaks for pristine EB are 45%, 48% and 7%, for the imine, amine and shake-up respectively. For low coverage, no significant binding energy shifts in the N 1s core-level peak are detected, although the intensity of the shake-up peak at ~402 eV is decreased. This is similar to the initial interface formed for Al on EB, where no significant shift is seen for the imine peak at low coverage (<3 Å) [82]. At high coverage (~10 Å), shown in Fig. 5.15b [32], the relative intensities of the amine and shake-up features are decreased to 31% and 4% respectively, and an increase to 65% in the relative intensity of the imine feature is detected. The imine peak is now slightly shifted to a lower binding energy, 398.2 eV, and there is also a slight broadening of this peak, 2.15 eV (FWHM), as compared to 1.97 eV

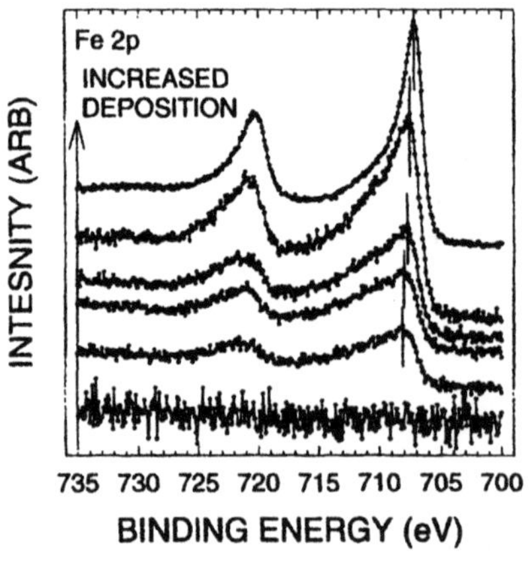

Fig. 5.14. The evolution of the XPS Fe 2p core-level spectrum during increased deposition of iron onto emeraldine base polyaniline. (From [32])

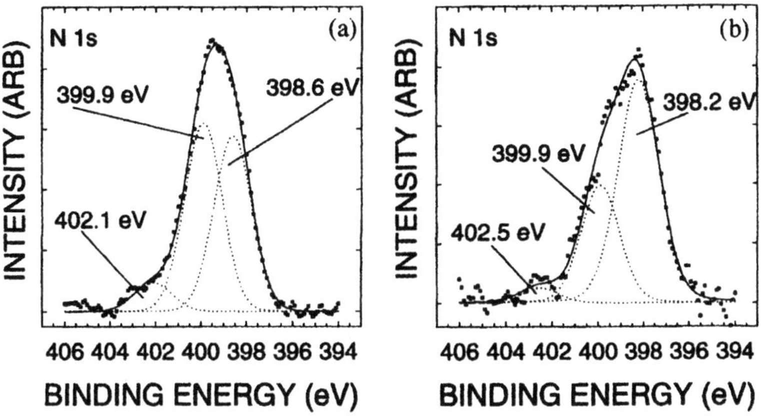

Fig. 5.15. The XPS N 1s core level spectrum for (**a**) pristine emeraldine base polyaniline and (**b**) the interface of iron and emeraldine base polyaniline at high iron coverage (~10 Å). (From [32])

for the amine peak. The initial shift of the Fe 2p core level and the decrease in imine shake-up feature of the N 1s core level at low coverage are proposed to be indicative of iron preferentially reacting with the imine nitrogen of the EB films, possibly forming covalent bonds. The lack of a significant shift in the imine peak position is attributed to the accepted charge being spread out over the imine–quinoid–imine unit, as was the case for Al on EB [82].

The broadening observed for the low binding energy peak (398.2 eV) of the N 1s core level at high iron coverage is suggested to be caused by iron reacting with EB, moving the iron–amine feature of the N 1s spectrum to lower binding energies due to charge transfer. The shift is large enough to place the iron–amine peak close to that of the imine and iron-imine features, thus causing the broadening and slight shift of the low binding energy peak of the N 1s spectrum. This also explains the observed increase of the low binding energy peak intensity and the drop of the amine peak intensity at high levels of coverage. This result is similar to the case of amine nitrogen of EB reacting with aluminum, where a shift towards lower binding energy of 1.7 eV occurs [82].

The evolution of the C 1s core level upon iron deposition shows no shift as expected if the accepted charge is spread out over the imine–quinoid–imine carbon atoms. At high levels of coverage, a new feature appears as a low binding energy shoulder, suggesting that iron has reacted directly with carbon sites. This result also supports the model of initial interface formation where iron preferentially reacts with imine, which is then followed by the less favorable iron–amine and iron–carbon interactions [32].

As for the case of aluminum deposition on PPV, the presence of impurities such as oxygen may significantly affect the nature of the metal/polymer interface. This is illustrated by studies on the aluminum/EB interface, where the presence of oxygen species at the interface causes a formation of N–π–AlO_n complexes, π representing the π-electronic structure of the quinoid rings [83, 84]. The oxygen species may either be present in the vacuum during deposition or be adsorbed in the bulk of the polymer and migrate to the interface [83,84]. This reaction scheme is in contrast to the case of aluminum deposition on EB with a "clean" interface, where aluminum preferentially reacts with the N–quinoid–N segments [82, 87].

5.6 Polymer/Metal Interfaces

The interface formed when an organic molecule or a polymer is deposited onto a metal surface is also of interest in molecular electronics and the study of anti-corrosion coatings. Specifically, in polymer LEDs there are severe problems linked to the interface at the (transparent) hole-injecting electrode, usually indium-tin-oxide (ITO). The degradation of these devices is thought to occur due to chemical interactions at the ITO interface. Those reactions may result in migration of ionic indium species and possibly oxygen into the light-emitting polymer layer, where they act as quenching sites as well as cause p-doping and carbonyl formation [88–92].

5.6.1 Poly(p-phenylene vinylene) on ITO

PPV is insoluble and therefore spin cast onto ITO in soluble precursor forms which are then converted into PPV by heating. However, the conversion process may cause chemical reactions to occur between the polymer and the ITO substrate, which in turn can affect device performance and lifetimes. In one possible scenario, HCl eliminated from the precursor-PPV during the thermal conversion may react with the surface layer of ITO and form $InCl_3$, which in turn may migrate into the polymer and p-dope it, or act as a catalyst in the conversion reaction. A schematic of the conversion process of precursor-PPV is depicted in Fig. 5.16. XPS studies of the interfacial chemistry between PPV and ITO during heat treatment have been carried out to shed light on those processes [93, 94], and we give a short summary here.

Clean ITO surfaces were studied with angle-resolved XPS as well as UPS and were found to consist of an oxygen-rich insulating layer, probably indium hydroxide [94]. Subsequently, the conversion process of precursor-PPV films spin-coated onto ITO was compared to that of films spun onto Si substrates; the Si substrates were chosen due to their flatness and relative inertness. The Cl 2p core level was used to monitor the evolution of Cl^- ions into HCl as the conversion process took place. The Cl 2p core level has a spin-split doublet, Cl $2p_{1/2}$, $2p_{3/2}$, with an energy splitting of 1.6 eV and intensity ratio of 1:2. In Fig. 5.17 [93] the evolution of the Cl 2p core level for precursor-PPV on a Si substrate is depicted. The first scan was taken 10 minutes after the

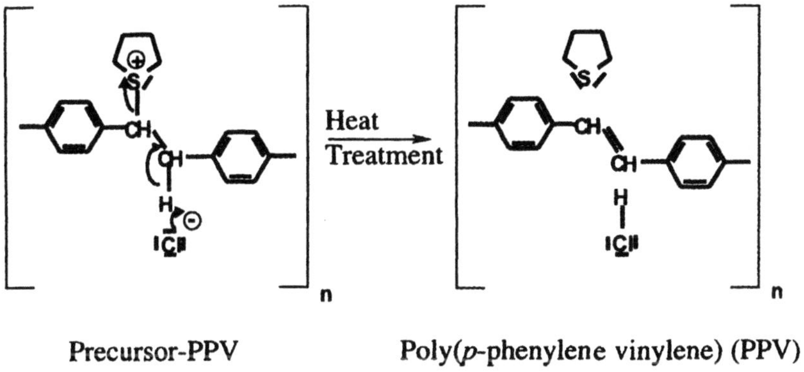

Fig. 5.16. The chemical structures of precursor poly(p-phenylene vinylene) (*left*) and fully conjugated poly(p-phenylene vinylene) (*right*). (From [94])

sample was prepared; the other two scans were taken following the first at 20-min intervals. The core level was deconvoluted into two spin-split doublets; the high binding energy doublet corresponds to HCl, the low binding energy doublet corresponds to the Cl^- ions in the sulfonium chloride salt. The HCl signal increases with time, whereas the Cl^- intensity decreases, signifying a release of the sulfonium groups and a conversion of the precursor into conjugated PPV. A similar set of spectra was taken for precursor-coated ITO at identical time intervals. The Cl 2p core-level spectra for this experiment are depicted in Fig. 5.18 [93]. Here a third doublet appears, with its main

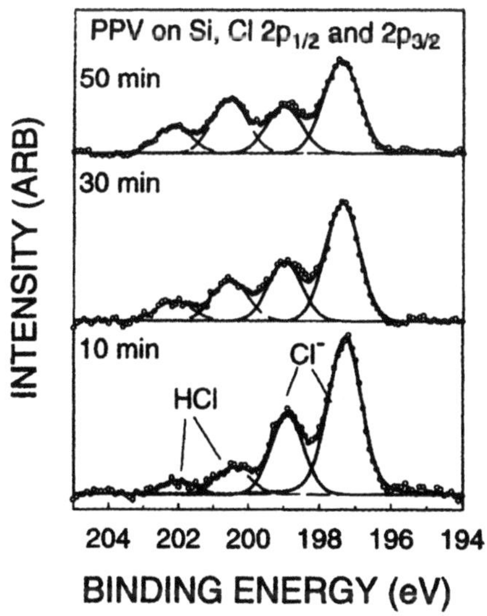

Fig. 5.17. XPS Cl 2p core-level spectra of precursor poly(p-phenylene vinylene) on an Si substrate, taken 10, 30 and 50 min after spin coating. The experimental data was fitted with two spin-split doublets representing Cl in HCl and Cl^-, as depicted. (From [93])

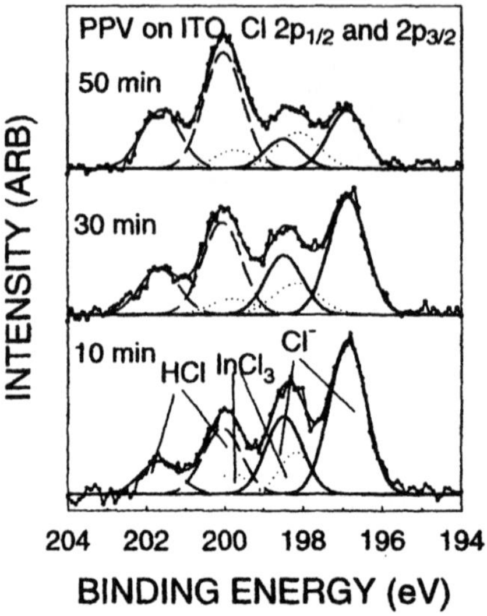

Fig. 5.18. XPS Cl 2p core-level spectra of precursor poly(p-phenylene vinylene) on an ITO substrate taken 10, 30 and 50 min after spin coating. Here the experimental data was fitted with three doublets representing Cl in HCl, $InCl_3$ and Cl^-, as depicted. (From [93])

peak at ~198.2 eV. Reference measurements on $InCl_3$-coated ITO showed that this doublet corresponds to $InCl_3$. Note that $InCl_3$ is formed only 10 min after coating. By comparing Fig. 5.18 with Fig. 5.17, it is evident that the increase of the HCl doublet and decrease of the Cl^- doublet is much more rapid for the case of the ITO substrate than for the Si substrate. The $InCl_3$ seems to acts as a catalyst to the elimination reaction, thus enhancing the rate of conversion [93]. Angle-dependent XPS showed that the $InCl_3$ is situated mainly at the PPV/ITO interface, probably because in vacuum the $InCl_3$ molecules are pumped away from the surface of the polymer faster than they can be replaced from the polymer/metal interface. In fact, PPV films converted on ITO are slightly p-doped due to remaining $InCl_3$ [93].

5.6.2 Polyaniline on ITO

Polyaniline, in both its conducting and insulating forms, has seen use as an interfacial layer between the light-emitting layer and metal electrodes, including ITO [95, 96]. The polyaniline interfacial layer is though to improve hole injection while reducing the degradation of the light-emitting polymer. When studying polymer/metal interfaces, one often makes use of so-called "model molecules" of the polymer, which shares the chemical characteristics of the polymer, and yet are small enough to be condensed in vacuum on the (clean) metal surfaces. By creating the interface in such a way, non-intrinsic impurities can be avoided and the thickness of the organic overlay can be optimized, i.e., it is possible to study just the chemisorbed monolayer of the interface. The latter ensures that all the (molecular) signal in XPS and UPS is from molecules that are interacting with the metal surface. Model molecules of emeraldine base polyaniline were condensed on ozone-cleaned ITO substrates in vacuum and studied by XPS and UPS [97]. The two molecules used, one representing the imine–quinoid–imine parts of the polymer, the other the amine–benzene–amine parts, are depicted in Fig. 5.19 [97].

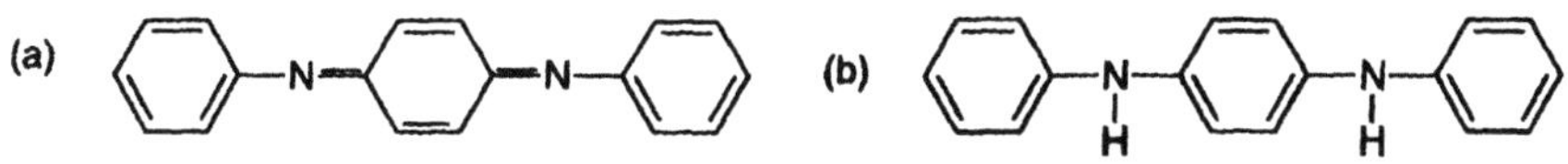

Fig. 5.19. Chemical structure of (**a**) an (imine-type) oxidized phenyl-capped aniline dimer and (**b**) a (amine-type) phenyl-capped aniline dimer. (From [97])

Condensed films of both the amine and imine compounds of different thicknesses were studied and the N 1s core-level spectra of the imine compound are depicted in Fig. 5.20 [97]. In the case of a thick film, where the majority of the molecules in the film are physisorbed and do not interact with the ITO substrate, the N 1s core level consists of a main component at a binding energy of 397.8 eV together with minor peak components at 399.4, 401.6 and 405.1 eV. The latter components are attributed to shake-up transitions and will not be discussed further here. Decreasing the film thickness leads to a broadening of the N 1s peak, with a distinct shoulder appearing at the high binding energy side. When only a chemisorbed monolayer is present on the ITO surface, the N 1s core level consists of the original component (slightly shifted to 397.9 eV) and an additional, high binding energy component at 399.3 eV. The two components are roughly equal in intensity and create a broad, symmetric N 1s peak. This indicates that there is an interaction between one of

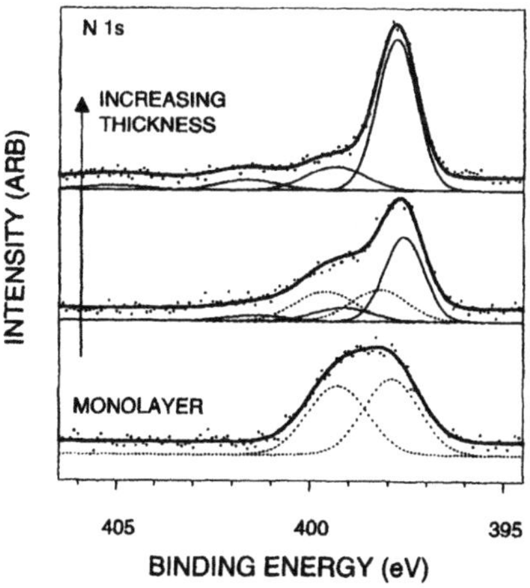

Fig. 5.20. XPS N 1s core-level spectra of oxidized phenyl-capped aniline dimer films of various thicknesses adsorbed on ITO. Fitted peak components: chemisorbed molecules (*dotted lines*) and physisorbed molecules (*solid lines*). (From [97])

the imine nitrogens in the model molecule and the ITO surface; however, the other imine nitrogen apparently does not interact with the surface [97]. In the corresponding series of spectra for the amine model molecule, the N 1s core level remains essentially unchanged (not shown here) [97].

Fig. 5.21 displays a series of UPS spectra of imine model molecule films of various thicknesses (corresponding to the XPS experiments) condensed on ITO [97]. For the thick film, well-resolved peaks appear in the low binding energy range at 6.8, 7.8 and 8.6 eV (denoted as 1, 2 and 3). In the case of the chemisorbed monolayer, however, this structure is completely changed, forming a single, broad peak at 8.2 eV with a shoulder tailing towards low binding energy. Together with the splitting of the N 1s core level observed in XPS, this result clearly indicates a strong interaction of the imine groups with the ITO surface. UPS spectra for a series of increasingly thinner layers of the amine compound on ITO showed a slight broadening and "smearing" of the two features closest to the band gap for the monolayer film as compared to the thick films, indicating that some interaction with the ITO substrates may be occurring (not shown) [97]. However, the modification of the amine spectra was very small compared to tha of the imine compound, as was the case for the N 1s core-level spectra.

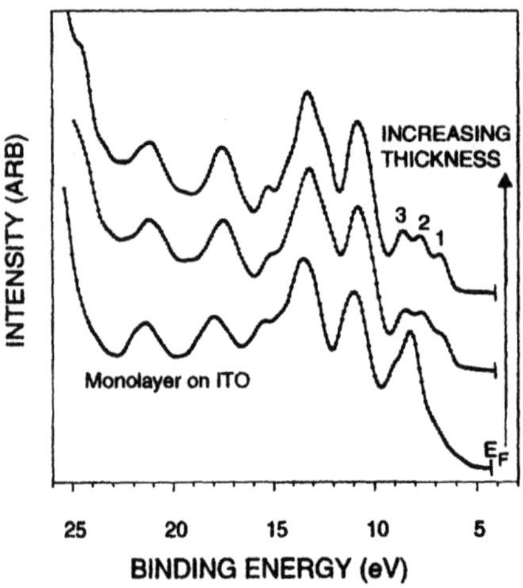

Fig. 5.21. HeII UPS spectra of oxidized phenyl-capped aniline dimer films of various thicknesses adsorbed on ITO. (From [97])

5.7 Summary

A brief outline of the photoelectron spectroscopy technique has been given along with a description of conjugated polymers. A number of examples of the use of photoelectron spectroscopy in studying conjugated polymers and their interaction with metals have been presented. The types of information that one can obtain depending on the photon source have been discussed as have the use of so-called "model" molecules in studying polymer/metal interfaces. The value of combining the experimental results with quantum chemical modeling has been demonstrated. We have also illustrated the ability of photoelectron spectroscopy to solve key questions concerning interfaces in a number of organic-based applications presently being developed.

Acknowledgments. Research on condensed molecular solids and polymers in Linköping is supported in general by grants from the Swedish Natural Science Research Council (NFR), the Swedish Research Council for Engineering Sciences (TFR), two European Commission TMR networks (project number 1354 SELOA and project number 0261 SISITOMAS), and a Brite/EuRam contract (project number 4438 OSCA).

References

1. H. Naarmann, German patent, # 1179715, 1197228, 1179716 (BASF Corp., Germany, 1963)
2. C.K. Chiang, C.R. Fincher, Y.W. Park, A.J. Heeger, H. Shirakawa, E.J. Louis, S.C. Gua, A.G. MacDiarmid, Phys. Rev. Lett. **39**, 1098 (1977)
3. T.A. Skotheim (ed.), *Handbook of Conduting Polymers*, Vol. I and II (Marcel Dekker, New York 1986)
4. H. Naarmann, Synth. Met. **17**, 223 (1987)
5. H. Naarmann, N. Theophilou, Synth. Met. **22**, 1 (1987)
6. S. Roth, W. Graupner, P. McNeillis, Acta Phys. Polonica **87**, 699 (1995)
7. H.H. Kuhn, In: *Intrinsically Conducting Polymers: An emerging technology*, Vol. 246, ed. by M. Aldissi (Kluwer, Dordrecht 1993) pp. 25–34
8. R.S. Kohlman, A.J. Epstein, In: *Handbook of Conducting Polymers-II*, ed. by T. Skotheim, J.R. Reynolds, R.L. Elsenbaumer (Marcel Dekker, New York 1997) pp. 85–121
9. K. Müllen, G. Wegner (eds.) *Electronic Materials: The Oligomer Approach* (Wiley-VCH, Weinheim 1998)
10. R.H. Friend, R.W. Gymer, A.B. Holmes, E.G.J. Staring, R.N. Marks, C. Taliani, D.D.C. Bradley, D.A. dos Santos, J.L. Brédas, M. Lögdlund, W.R. Salaneck, Nature **397**, 121 (1998)
11. B. Wessling, Mater. Corros. **47**, 439 (1996)
12. N.S. Sariciftci, L. Smilowitz, A.J. Heeger, F. Wudl, Science **258**, 1474 (1992)
13. J.J.M. Halls, C.A. Walsh, N.C. Greenham, E.A. Marseglia, R.H. Friend, S.C. Moratti, A.B. Holmes, Nature **376**, 498 (1995)
14. T. Tatsuma, T. Sotumura, T. Sato, D.A. Buttry, N. Oyama, J. Electrochem. Soc **142**, 182 (1995)
15. F. Garnier, R. Hajlaoui, A. Yassar, P. Srivastava, Science **265**, 684 (1994)
16. A.R. Brown, A. Pomp, C.M. Hart, D.M.d. Leeuw, Science **270**, 972 (1995)
17. X. C. Li, H. Sirringhaus, F. Garnier, A.B. Holmes, S.C. Moratti, N. Feeder, W. Clegg, S.J. Teat, R.H. Friend, J. Am. Soc. **120**, 2206 (1998)
18. H. Sirringhaus, N. Tessler, R.H. Friend, Science **280**, 1741 (1998)

19. N. Tessler, G.J. Denton, R.H. Friend, Nature **382**, 695 (1996)
20. S.V. Frolov, M. Ozaki, W. Gellerman, Z.V. Vardeny, K. Yoshino, Jpn. J. Appl. Phys. **35**, L 1371 (1996)
21. F. Hide, M.A. Diaz-Garcia, B. Schwartz, M.R. Andersson, Q. Pei, A.J. Heeger, Science **273**, 1833 (1996)
22. M.A. Diaz-Garcia, F. Hide, B.J. Schwartz, M.R. Andersson, Q.B. Pei, A.J. Heeger, Synth. Met. **84**, 455 (1997)
23. J. Dresner, RCA Rev. **30**, 322 (1969)
24. K. Kaneto, K. Yoshino, K. Kao, Y. Inuishi, Jpn. J. Appl Physics **13**, 1023 (1974)
25. R. H. Partridge, Polymer **24**, 733 (1983)
26. J.H. Burroughes, D.D.C. Bradley, A.R. Brown, R.N. Marks, K. Mackay, R.H. Friend, P.L. Burn, A.B. Holmes, Nature **347**, 539 (1990)
27. G. Cedervall, in Elektroniktidningen 1 (1999), p. 23
28. D.A. Jones, *Principles and prevention of corrosion* (Prentice Hall, Englewood Cliffs 1996)
29. D.W. DeBerry, J. Electrochem. Soc. **132**, 1022 (1985)
30. B. Wessling, Adv. Mater. **6**, 226 (1994)
31. P.J. Kinlen, D.C. Silverman, C.R. Jeffreys, Synth. Met. **85**, 1327 (1997)
32. M. Fahlman, H. Guan, J.A.O. Smallfield, A.J. Epstein, In: *Society of Plastics Engineers Annual Technical Conference*, Vol. 2 (Atlanta 1998) pp. 1238–1242
33. N.V. Krstajic, B.N. Grgur, S.M. Jovanovic, M.V. Vojnovic, Electrochim. Acta **42**, 1685 (1997)
34. W.-K. Lu, S. Basak, R.L. Elsenbaumer, In: *Handbook of Conducting Polymers-II*, ed. by T. Skotheim, J.R. Reynolds, R.L. Elsenbaumer (Marcel Dekker New York, 1997), pp. 881-920
35. T. Ito, H. Shirakawa, S. Ikeda, J. Polymer Sci. Polymer Chem. Ed. **12**, 11 (1974)
36. W.P. Su, J.R. Schrieffer, Proc. Natl. Acad. Sci. USA **77**, 5626 (1980)
37. K. Fesser, A.R. Bishop, D.K. Campbell, Phys. Rev. B **27**, 4804 (1983)
38. M. Fahlman, D. Beljonne, M. Lögdlund, P.L. Burn, A.B. Holmes, R. H. Friend, J.L. Brédas, W.R. Salaneck, Chem. Phys. Lett. **214**, 327 (1993)
39. G. Iucci, K. Xing, M. Lögdlund, M. Fahlman, W.R. Salaneck, Chem. Phys. Lett. **244**, 139 (1995)
40. J. Cornil, D. Beljonne, J.L. Brédas, J. Chem. Phys. **103**, 842 (1995)
41. J.L. Brédas, J. Cornil, A.J. Heeger, Adv. Mat. **8**, 447 (1996)
42. J.L. Brédas, D.A. dos Santos, J. Cornil, D. Beljonne, Z. Shuai, Acc. Chem. Res., in press
43. A.J. Heeger, S. Kivelson, J.R. Schrieffer, W. P. Su, Rev. Mod. Phys. **60**, 781 (1988)
44. T. Koopmans, Physica **1**, 104 (1934)
45. R. Manne, T. Åberg, Chem. Phys. Lett. **7**, 282 (1970)
46. G. Wedin, *Breakdown of the One-electron Pictures in Photoelectron Spectra* (Springer-Verlag, New York 1981)
47. E. Merzbacher, *Quantum Mechanics* (John Wiley & Sons, Singapore 1970)
48. N. Sato, K. Seki, H. Inokuchi, J. Chem. Soc. Faraday Trans. **II 77**, 1621 (1982)
49. H.J. Freund, R.W. Bigelow, Phys. Scripta **17**, 50 (1987)
50. K. Siegbahn, C. Nordling, A. Fahlman, R. Nordberg, K. Hamrin, J. Hedman, G. Johansson, T. Bergmark, S.-E. Karlsson, I. Lindgren, B. Lindgren, *ESCA-Atomic, Molecular and Solid State Structure Studied by Means of Electron Spectroscopy* (Almqvist & Wiksells, Uppsala 1967)

51. P. Dannetun, M. Lögdlund, C. Fredriksson, M. Boman, S. Stafström, W.R. Salaneck, B. E. Kohler, C. Spangler, In: *Polymer-Solid Interfaces*, ed. by J.J. Pireaux, P. Bertrand, J. L. Brédas (IOP Publishing, Bristol 1992) pp. 201–212
52. J.J. Yeh, I. Lindau, Atomic data and Nuclear Data Tables **32**, 1 (1985)
53. E.W. Plummer, W. Eberhardt, In: *Advances in Chemical Physics*, ed. by I. Prigogine, S.A. Rice (John Wiley & Sons, New York 1982) p. 533
54. W.R. Salaneck, M. Fahlman, C. Lapersonne-Meyer, J.L. Fave, M. Schott, M. Lögdlund, J.L. Brédas, Synth. Met. **67**, 309 (1994)
55. W. Eberhardt (ed.), *Applications of Synchrotron Radiation*, Vol. 35, (Springer Verlag, Berlin 1995)
56. H. Winick, S. Doniach (eds.), *Synchrotron Radiation Research*, (Plenum Press, New York 1980)
57. M. Fahlman, O. Lhost, F. Meyers, J.L. Brédas, S.C. Graham, R.H. Friend, P.L. Burn, A.B. Holmes, K. Kaeriyama, Y. Sonoda, M. Lögdlund, S. Stafström, W.R. Salaneck, Macromolecules **28**, 1959 (1995)
58. F. Meyers, A.J. Heeger, J.L. Brédas, J. Chem. Phys. **97**, 2750 (1992)
59. O. Lhost, J.L. Brédas, J. Chem. Phys. **96**, 5279 (1992)
60. M. Fahlman, D.D. Gebler, N. Piskun, T.M. Swager, A.J. Epstein, J. Chem. Phys **109**, 2031 (1998)
61. M. Fahlman, P. Bröms, D.A. dos Santos, S.C. Moratti, N. Johansson, K. Xing, R.H. Friend, A.B. Holmes, J.L. Brédas, W.R. Salaneck, J. Chem. Phys. **102**, 8167 (1995)
62. M. Lögdlund, W.R. Salaneck, F. Meyers, J.L. Brédas, G.A. Arbuckle, R. Friend, A.B. Holmes, G. Froyer, Macromolecules **26**, 3815 (1993)
63. J.L. Brédas, R.R. Chance, R. Silbey, G. Nicolas, P. Durand, J. Chem. Phys. **75**, 255 (1981)
64. J.M. André, J. Dekhalle, J.L. Brédas, *Quantum Chemistry Aided Design of Organic Polymers* (World Scientific, Singapore 1991)
65. M.J.S. Dewar, E.G. Zoebisch, E.F. Healy, J.J.P. Stewart, J. Am. Chem. Soc. **107**, 3902 (1985)
66. M. Lögdlund, P. Dannetun, W.R. Salaneck, In: *Handbook of Conducting Polymers*, ed. by T. Skotheim, J.R. Reynolds, R.L. Elsenbaumer (Marcel Dekker, New York 1997) pp. 667–694
67. W.R. Salaneck, S. Stafström, J.L. Brédas, *Conjugated Polymer Surfaces and Interfaces* (Cambridge University Press, Cambridge 1996)
68. A.G. MacDiarmid, A.J. Epstein, Faraday Discuss. Chem. Soc. **88**, 317 (1989)
69. W.R. Salaneck, I. Lundström, T. Hjertberg, C.B. Duke, E. Conwell, A. Paton, A.G. MacDiarmid, N.L.D. Somasiri, W. S. Huang, A.F. Richter, Synth. Met. **18**, 291 (1987)
70. A.P. Monkman, G.C. Stevens, D. Bloor, J. Phys. D: Appl. Phys. **24**, 738 (1991)
71. B. Sjögren, W.R. Salaneck, S. Stafström, J. Chem. Phys. **97**, 137 (1992)
72. K.G. Neoh, E.T. Kang, K.L. Tan, J. Phys. Chem. **95**, 10151 (1991)
73. K.L. Tan, B.T.G. Tan, E.T. Kang, K.G. Neoh, Phys. Rev. B **39**, 8070 (1989)
74. K.G. Neoh, M.Y. Pun, E.T. Kang, K.L. Tan, Synth. Met. **73**, 209 (1995)
75. E.T. Kang, Z.F. Li, K.G. Neoh, Y.Q. Dong, K.L. Tan, Synth. Met. **92**, 167 (1998)
76. P. Barta, Th. Kugler, W.R. Salaneck, A.P. Monkman, J. Libert, R. Lazzaroni, J.L. Brédas, Synth. Met. **93**, 83 (1998)
77. J.L. Brédas, C. Quattrocchi, J. Libert, A.G. MacDiarmid, J.M. Ginder, A.J. Epstein, Phys. Rev. B **44**, 6002 (1991)

78. K. Konstadinidis, F. Papadimitrakopoulos, M. Galvin, R.L. Opila, J. Appl. Phys. **77**, 5642 (1995)
79. T.P. Nguyen, J.L. Mansot, Thin Solid Films **283**, 135 (1996)
80. M. Lögdlund, J.L. Brédas, J. Chem. Phys. **101**, 4357 (1994)
81. M. Fahlman, W.R. Salaneck, S.C. Moratti, A.B. Holmes, J.L. Brédas, Chem. Eur. J. **3**, 286 (1997)
82. R. Lazzaroni, C. Grégoire, M. Chtaïb, J.J. Pireaux, In: *Polymer-Solid Interfaces* (IOP Publishing Ltd, Bristol 1992) pp. 213
83. S.L. Lim, K.L. Tan, E.T. Kang, Synth. Met. **92**, 213 (1998)
84. S.L. Lim, K.L. Tan, E.T. Kang, J. Vac. Sci. Technol. A **16**, 13 (1998)
85. S.L. Lim, K.L. Tan, E.T. Kang, Langmuir **14**, 5305 (1998)
86. P. Snauwert, R. Lazzaroni, J. Riga, J. Verbist, D. Gonbeau, J. Chem. Phys. **92**, 2187 (1990)
87. A. Calderone, R. Lazzaroni, J.L. Brédas, Phys. Rev. B **49**, 14418 (1994)
88. J.C. Scott, J.H. Kaufman, P.J. Brock, R. DiPietro, J. Salem, J.A. Gotia, J. Appl. Phys. **79**, 2745 (1996)
89. G. Sauer, M. Kilo, M. Hund, A. Wokaun, S. Karg, M. Meier, W. Riess, M. Schwoerer, H. Suzuki, J. Simmerer, H. Meyer, D. Haarer, Fresenius' J. Anal. Chem. **353**, 5 (1995)
90. A.R. Schlatmann, D.W. Floet, A. Hilberer, F. Garten, P.J.M. Smulders, T.M. Klapwijk, G. Hadziioannou, Appl. Phys. Lett. **69**, 1764 (1996)
91. J.C. Carter, I. Grizzi, S.K. Heeks, D.J. Lacey, S.G. Latham, P.G. May, O.R.d.l. Panos, K. Pichler, C.R. Towns, H.F. Wittman, Appl. Phys. Lett. **71**, 34-36 (1997)
92. M.P. de Jong, D.P.L. Simons, M.A. Reijme, L.J. van Ijzendoorn, A.W. Denier van der Gon, M.J.A. de Voigt, H.H. Brongersma, R.W. Gymer, Synth. Met. **110**, 1 (2000)
93. A. Andersson, Th. Kugler, M. Lögdlund, A.B. Holmes, X. Li, W.R. Salaneck, Synth. Met. **106**, 13 (1999)
94. Th. Kugler, A. Andersson, M. Lögdlund, B. Holmes, X. Li, W.R. Salaneck, Synth. Met., **100**, 163 (1999)
95. Y. Yang, A.J. Heeger, Mol. Cryst. Liq. Cryst. **256**, 537 (1994)
96. H.L. Wang, A.G. MacDiarmid, Z.W. Wang, D.D. Gebler, A.J. Epstein, Synth. Met. **67**, 23 (1996)
97. Th. Kugler, M. Lögdlund, W.R. Salaneck, IEEE J. Select. Topics Quantum Elect. **4**, 14 (1998)

Part II

Low Molecular Weight Organic Solids

6 Introduction

N. Karl

6.1 General Remarks

The second part of this book is devoted to low molecular weight organic solids with emphasis on the relations between their material properties and their electrical, optical and optoelectronic properties. Among the wide variety of organic compounds described in the literature only those which display photoconductivity or semiconductivity will be considered. Photoconductivity for visible or near ultraviolet light or (dark) semiconductivity requires energy levels for electrons and holes that can be populated by light of ca. 1 eV to several eV photon energy or through thermal excitation by energies of a few k_BT, provided by the ambient temperature, respectively (note that $k_BT = 25$ meV at room temperature). In addition, a certain degree of intermolecular interaction is necessary to allow local transfer of a charge carrier to a neighboring molecule to occur, leading to incoherent hopping transport, or a delocalized coherent spreading out across several molecules, depending on temperature and interaction strength. Such conditions are met by organic molecules which carry carbon $2p_z$ electronic wavefunctions, most suitably in conjugated π-electron systems, when positioned at not too large intermolecular distances (see Fig. 6.1). Examples are solids of derivatives of ethylene, butadiene, hexatriene and other oligoenes, acetylenes, and aromatic as well as heteroaromatic molecules, deriving from the benzene ring. Some representatives with strong optical absorption in the visible (1.5–3 eV) and high stability are mass products, used technically as pigment dyes.

Historically, photoconduction in organic crystals was discovered with anthracene (see Fig. 6.2) in 1906 by the Italian scientist Pochettino [1]; its dark conductivity was studied (simultaneously with that of silicon!) in 1910 by Konigsberger and Schilling at the University of Freiburg/Brg. in Germany [2], and a prototype of an organic photovoltaic cell was presented 1913 by Vollmer in his habilitation thesis at the University of Leipzig [3]. It is remarkable that in these early investigations measurement of currents as low as 10^{-12}A was already possible, although by tedious reading of the decay of the deflection of electrostatic electrometers.

Anthracene has been used as a model organic photoconductor for quite a while: Crystals were commercially available as the high yield fluorescing material of scintillation detectors for high energy particles.

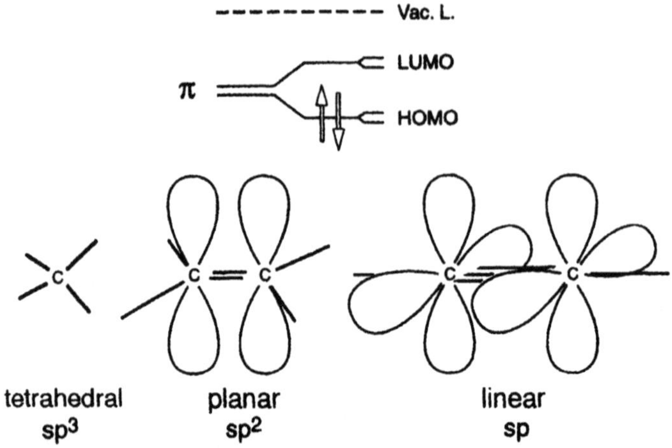

Fig. 6.1. The carbon atom in a tetrahedral sp^3 hybridized single–bond configuration (left) and in C—C sp^2p_z double–bond (center), and sp–p_z–p_y triple–bond pairs (right); (the p_z and p_y wavefunctions are sketched schematically). The interactions in a double bond split the two initially degenerate $2p_z$ energy levels into a bonding and an "anti-bonding" molecular "π" orbital; the two electrons, indicated by the mutual orientation of their spin, occupy the former larger binding energy orbital (HOMO), while the latter remains empty (LUMO). Optical or thermal excitation can promote an electron to an unoccupied level with a certain rate or equilibrium probability, respectively. For *larger* conjugated carbon–carbon double- (or triple-) bond systems, further level splittings occur, which are not indicated here; the lowest unoccupied level (LUMO) exhibits binding energy (relative to the vacuum level) for an (extra) electron. Intermolecular interactions in the solid lead to still further splitting of these molecular levels under formation of narrow bands, as indicated. These are the basis of electrical conduction in organic solids

Soon after the invention of the very efficient purification method of zone-refining by Pfann [4], see also [5], anthracene was obtained in very high purity, allowing the experimental determination of its intrinsic energy levels and of its genuine charge-carrier transport properties [6] (recent data may be found in [7]); at the same time, the effects of impurities by intentional doping were studied (see Chap. 7, Sect. 10). Anthracene in the form of high-purity and perfect single crystals has thus become the prototype of an organic photoconductor, which, at the same time, exhibits one of the lowest intrinsic dark conductivities ($\sigma = 10^{-15}\ \Omega^{-1}\mathrm{cm}^{-1}$, when sufficiently pure and measured under non-injection conditions by a capacitive, contact free method) (see [8]). Fundamental semiconductor properties of anthracene, characteristic of a wider class of condensed aromatic hydrocarbons, such as the different charge-generation mechanisms, charge-carrier transport and recombination, space-charge-limited conduction and trapping effects have been studied in great detail. These properties are described by two monographs on anthracene [9, 10]. (Historical reviews of the entire field may be found in, for example, [11]; the physical fundamentals and a wide variety of experimental results are critically presented in the books by Pope and Swenberg [12,13]; a bibliography of work published before 1984 is available in [14]).

Hand in hand with work on photoconduction, luminescence and absorption properties have been studied; these investigations also greatly profited from the invention of zone-refining, because luminescence spectra and lifetimes are very sensitively influenced by impurities in the materials, as is charge-carrier transport. This is a consequence of the fact that (electrically neutral) excited molecular states – similar to charge carriers – are not strongly localized, but can screen a certain number of unit cells in the crystal lattice by diffusional motion, which considerably increases the probability of energy transfer from these excitons to impurity molecules or to host molecules at distorted lattice sites, with the consequence of localization and spectral changes or shifts of the emission ("sensitized luminescence").

Interest soon also focussed on *theoretical* descriptions of neutral and ionic energy levels in organic molecular solids and on the transport properties for neutral and charged excitations.

From the experimental side, investigations concentrated on other related compounds, such as naphthalene, pyrene, perylene, p-terphenyl, and trans-stilbene (see Fig. 6.2), in the form of high-purity single crystals, and on bigger, lower vapor pressure dye molecules, which promised a higher potential for practical applications – such as, for example, diverse metal-phthalocyanines and perylenetetracarboxylic dianhydride and bisimide derivatives (see Fig. 6.2); however, due to a lack of single crystals, these compounds were mainly used in the form of compressed powders [15] and thin films [16–18]. One important product of the latter scientific activities was the composite photoconductor/polymer transport layer foils used extensively in most modern copiers and laser printers. Present developments are aimed at even more sophisticated applications, such as organic thin-film field-effect transistors [19–27], photovoltaic cells [28–36], organic double injection luminescence displays [37], and laser diodes [38]. The development of thin-film organic light-emitting diodes (OLEDs) and displays (see Chap. 11) has reached the commercial production level quite recently. OLEDs made from low molecular weight materials, often by vapor deposition under high vacuum conditions, are presently in strong competition with polymeric systems, and it can safely be said that, whenever fast, low-loss applications require high charge-carrier mobilities, low molecular weight materials promise higher purity and structural perfection, and hence higher mobilities [39], especially in the form of *epitaxially ordered* thin films.

In what follows, a brief overview of the *specific properties* of the class of van der Waals-bonded low molecular weight π-electron-carrying molecular solids/crystals will be given, including their *purification*, the *growth of single crystals*, and the *nature and characterization of defects.* The neutral and ionic excited states will then briefly be outlined and their dynamical properties

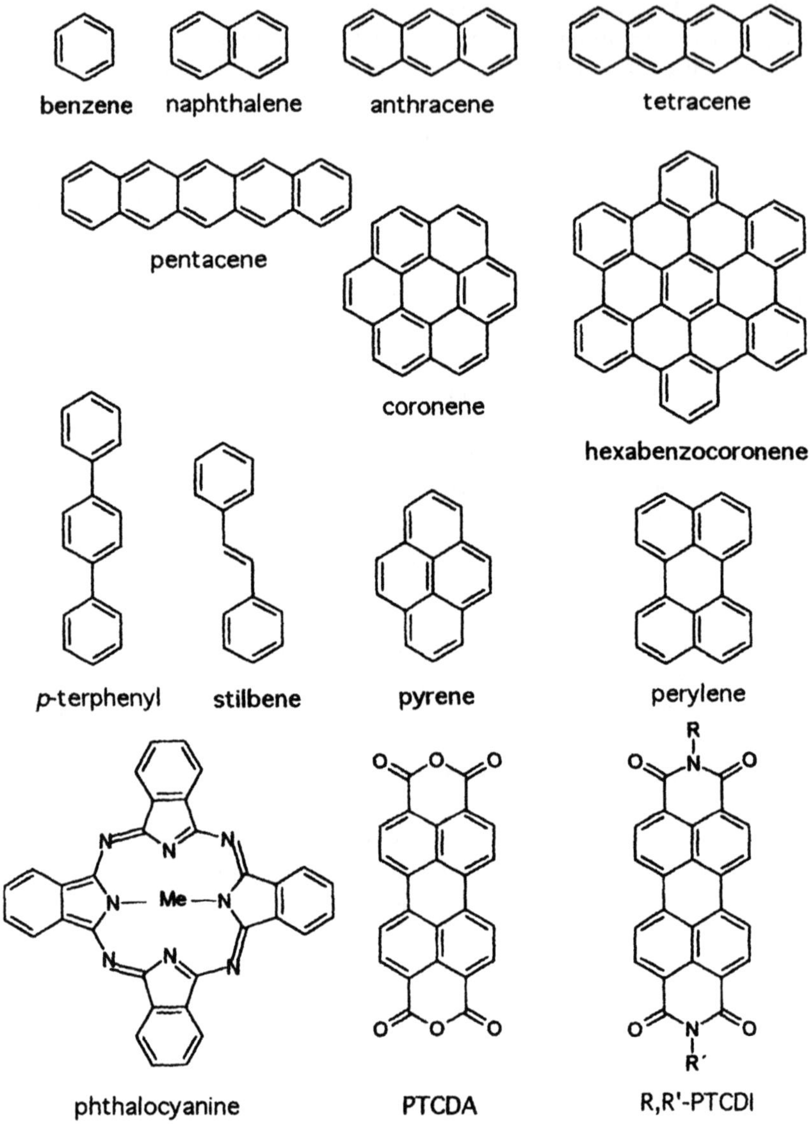

Fig. 6.2. Structural formulae of the molecules mentioned in this chapter (R and R' can be a variety of equal, or different organic substituents, or hydrogen)

summarized. The links between the properties of single crystals and those of thin films, that appear more suited for practical applications, deserve a special mention. At the end of this chapter the basic understanding of this class of materials will be summarized and possible applications outlined.

6.2 Organic Crystals and Materials-Science Aspects

6.2.1 Purification

The need for purification arises from the fact that impurities ("chemical defects") strongly influence the electrical as well as the optical properties of organic solids and strongly deteriorate the wanted properties. This is not very much different from the situation with inorganic semiconductors, where great efforts are necessary to arrive at the required impurity-depletion levels typically in and even below the ppm range (1 ppm $= 10^{-6}$ atomic or molecular fractional content). Subsequent controlled doping allows the establishment of specific properties, such as, for example, *n*- or *p*-type conduction at a well-defined charge-carrier concentration level, connected with a distinct position of the Fermi level and the formation of specific contact properties. Residual impurities may hamper control over these parameters and cause additional charge-carrier scattering (reduction of mobilities) and trapping. Thus, for the model compound anthracene several groups have shown that even ppm traces of the "natural" impurity tetracene (the four ring, next-higher homologue of anthracene [see Fig. 6.2] brought in from the usual starting material, coal tar) form shallow electron as well as hole traps at room temperature, as is shown in Fig. 6.3 (see [6]; see also [40] and the references given therein). This behavior can be understood by considering a simple localized energy level picture; its cause is the difference between the ionization energy and the electron affinity of the free host and guest molecules (see Chap. 7, Sect. 10), where "host" stands for the matrix or crystal and "guest" for the dopant or impurity. In the same way, any impurity with a smaller ionization energy and/or larger electron affinity than the host molecules will form trap levels in the energy gap between the transport levels (the valence and conduction "bands"). In addition, a substitutional impurity molecule will rarely fit exactly into the host lattice. Therefore, in most cases local strain is induced which alters the intermolecular distances and, in turn, the interaction energies of neighboring host molecules. These distortions can also lead to trap states. It thus appears as the exception that an impurity molecule will act neither as an electron trap nor as a hole trap.

From these remarks it becomes clear that during crystallization of usually impure starting material either the acceptability of a host material for incorporation of an available impurity – described by the distribution coefficient $k = c_s/c_l$ of the impurity (of concentration c) between the solid (s) and the starting gas/liquid/solution (l) phase – should be low, or the impurity should already have been depleted in the starting material by an efficient purifica-

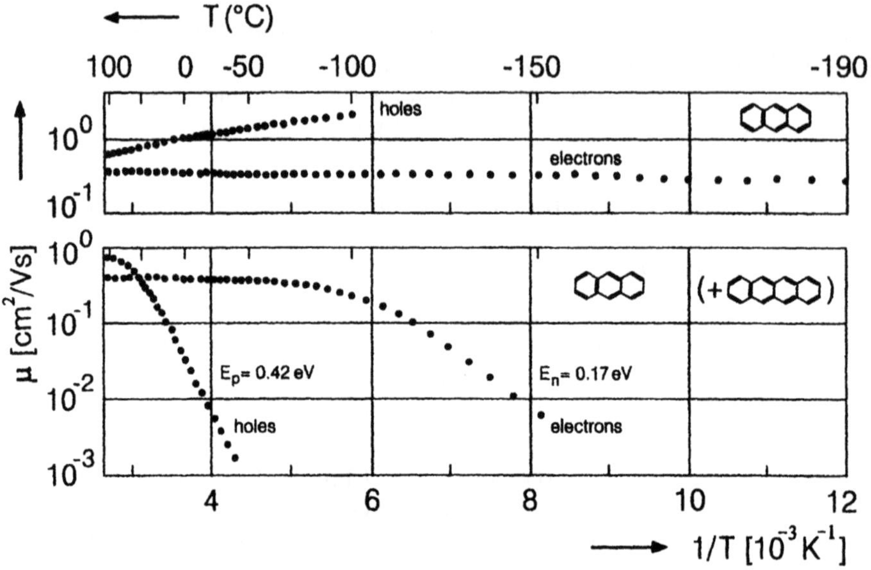

Fig. 6.3. Even a ca. 1 ppm doping by the impurity tetracene drastically reduces the electron and the hole mobilities in an anthracene single crystal (*bottom*). The mobilities of an undoped crystal are also shown (*top*). These are log μ versus inverse temperature Arrhenius plots; the electric field was applied along the crystallographic c^* direction. The lower temperature mobilities are thermally activated and reflect shallow (monoenergetic) tetracene electron and hole traps, as indicated. The same ultrapurified anthracene was used for both crystals. After [6]; see also [41]

tion process (see, e.g., [42]), or avoided by using material from a different preparation route. Except for very unfavorable distribution coefficients (k close to one), which, nevertheless, do occur occasionally (see, e.g., [43, 44]), zone-refining, i.e., repeated recrystallization from the melt by passing molten zones across a rod-shaped "ingot" (see Fig. 6.4) is the most efficient purification method for organic compounds containing chemically similar impurities. If attention is paid to the details (see, e.g., [43, 45]) it is possible to get down to the ppm and even sub-ppm level with elaborate techniques, as has been shown for the example of naphthalene in [42] and [46].

The fundamental condition for the applicability of zone-refining, however, is long-term thermal stability of the compound at a temperature slightly above its melting point (m.p.). This is fulfilled, for example, for the five ring aromatic molecule perylene ($C_{20}H_{12}$, m.p. 274 °C), but not necessarily for larger, higher-melting-point aromatic hydrocarbons. To give an example, coronene ($C_{24}H_{12}$, m.p. 437 °C; see Fig. 6.2), which is light yellow, slowly forms an orange dimer and a red trimer. While highly symmetric mere hydrocarbon-type aromatic molecules are comparatively stable, most chemical substituents of the H atoms lead to increased reactiv-

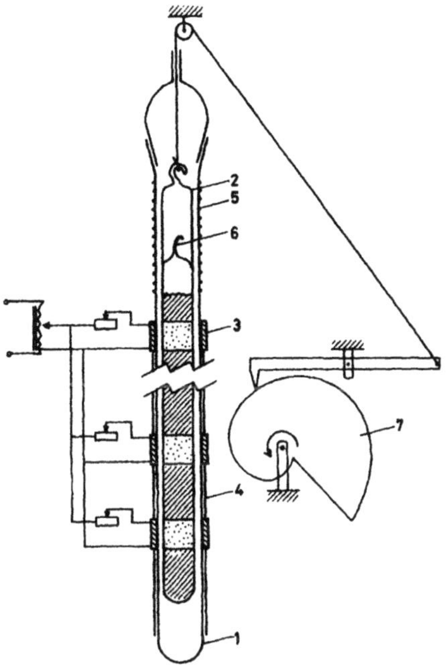

Fig. 6.4. A simple realization of a vertical multistage zone-refiner using the "intermittent" technique. The setup consists of a periodic furnace made from a Pyrex or fused silica glass tube (1), on which ca. 20 electrically powered resistive heating coils (3) and tubular spacers (4) are aligned in a regular sequence; the starting material is sealed into a Pyrex or fused silica zone-refining tube (2) which may be equipped with a vacuum break seal (6) to allow transfer of fractions of the purified material under exclusion of air (see [46]); a rotating Archimedes cam (7) induces an intermittent slow (typically 1 mm/h) up and a fast down motion of the ingot by a distance of one period of the furnace, for every rotation. At every jump of the lever the next heater takes over the previous molten zone (dotted fraction). The furnace (5) prevents material condensation from the vapor phase. (From [43])

ity and hence reduced thermal stability. Concerning zone-refining, the following are possible ways out of the dilemma: (a) Zone-refining of stoichiometric, congruently melting complexes or eutectic mixtures, which obey the schematic solid/liquid phase diagram of a typical 1:1 donor:acceptor (D:A) complex sketched in Fig. 6.5, leaving, however, the problem of separating the wanted compound from the purified binary mixture. (b) Lowering the melting point via minor modification of the molecules by attaching appropriate substituents. Thus, the perylene dye N,N'-bis(1-hexylheptyl)perylene-3,4:9,10-bis(dicarboximide) [47] (see Fig. 6.2, R=R'=-$CH(C_6H_{13})_2$) melts at 161 °C without decomposition, while the parent compound and most other perylene dyes decompose at their usually much higher melting points.

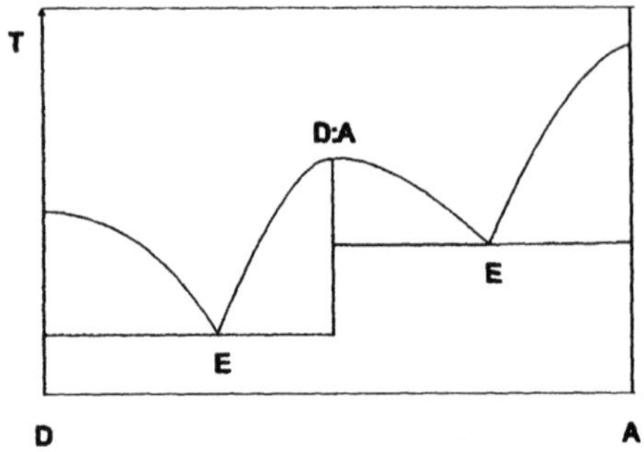

Fig. 6.5. Schematic liquid/solid phase diagram of a typical binary mixture of donor (D) and acceptor (A) molecules which form a stoichiometric 1:1 D:A complex. Zone-refining at the extrema D:A or E may be possible for systems where decomposition occurs at the higher melting point of A

There are *alternative purification methods*, which, however, in most cases are much less efficient than multipass zone-refining; these are the following:

- fractionation by sublimation, applying a temperature-step program or a temperature gradient (see Fig. 6.6)
- fractionation by recrystallization from a suitable solvent under slow evaporation or temperature lowering
- fractionation by chromatographic methods which subject the material under consideration to a sequence of phase equilibria between a solution and an adsorptive surface or thin film (preparative gas chromatography or liquid chromatography).

It is always useful to start ultrapurification from the purest available material, especially with respect to the concentration of the most persistent impurities; this may even require that material from a different source be used or the synthetic route be modified.

Concerning achievement and maintenance of high purity, this section should not be ended without mentioning the frequent interferences caused by oxygen/ambient atmosphere, by light, and by a combination of these (photochemical reactions). Special vacuum break seal techniques (see Fig. 6.4) and the use of orange safety lights may be required.

Altogether, larger molecular weight materials, which deserve preference in practical applications for their larger intermolecular (van der Waals) binding energy, may suffer from the problem that ultrapurification is difficult because, due to thermal decomposition, zone-refining cannot be applied. Therefore, smaller molecules, which melt without decomposition and thus can be obtained in high purity by zone-refining, should not be regarded as historical curiosities, but still be considered as useful model systems for future fundamental investigations.

6.2.2 Growth of Single Crystals

Single crystals of high purity and perfection are ideal objects for studying intrinsic material properties. Organic single crystals can be grown by

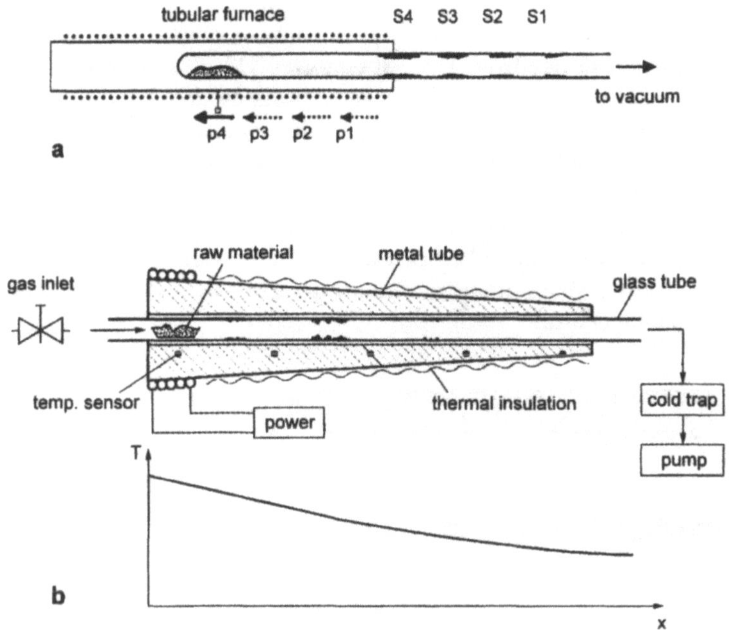

Fig. 6.6. (**a**) Purification by fractional vacuum sublimation in a temperature-step method. The process is started at a temperature T_1 with the furnace mark in position p1 (relative to the fixed sublimation tube), where the (pre)sublimate S1 is deposited. Then, the furnace is shifted to position p2, and the temperature raised to T_2, where the next fraction, S2, is deposited. This process is continued until all positions have been occupied. (**b**) Purification by temperature gradient sublimation. A suitable temperature gradient is established by using a conical tubular metal furnace heated at one end and kept at a constant lower temperature at its other end. Molecules vaporized from the starting material, which is placed at the hottest point, are dragged along with a weak stream of a well-purified inert gas (N_2, Ar), which is adjusted by a needle valve in such a way as to maintain a pressure of ca. 1 mbar upon pumping with a mechanical pump (connected across a cold trap to prevent mutual contamination). The different components of the vaporized material condense at different temperatures and, hence, places, as indicated for two fractions (compare the temperature profile plotted underneath). The procedure may be repeated with the purest fraction as the new starting material. A combination of (**b**) and then (**a**) is often advantageous

a number of different methods. The best way to maintain a high purity level of a given polycrystalline starting material is to base crystal growth on a mere phase transition either from the vapor phase [43, 46, 48] or from the melt [43, 46, 48, 49], in a suitable temperature gradient, forcing directional growth (see Figs. 6.7 and 6.8, respectively). The material is most highly protected in a flame-sealed Pyrex glass or fused silica ampoule, either evacuated or filled with an inert gas atmosphere. To prevent decomposition during sealing, specific refrigeration and cleaning techniques are required. Special attention is necessary with controlled seeding or single seed selection; this is crucial whenever only a small quantity of material is available and should be grown into one single crystal.

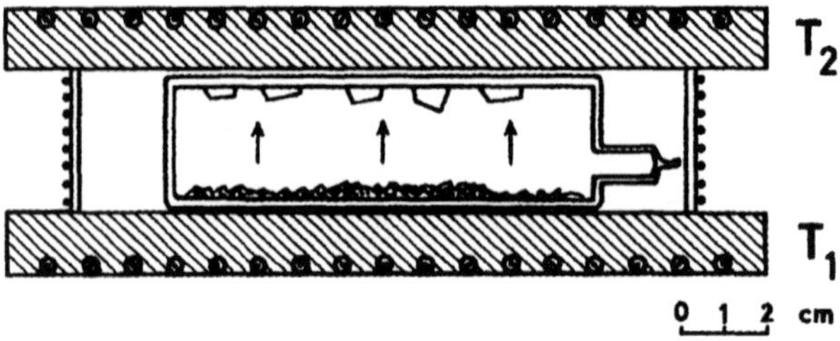

Fig. 6.7. Setup for crystal growth from the vapor phase by the plate sublimation method. In a circular evacuated Pyrex or fused silica glass ampoule, the material to be grown is transported in a small vertical temperature gradient ($T_2 < T_1$). For more details, see [43, 48, 50] and references given therein

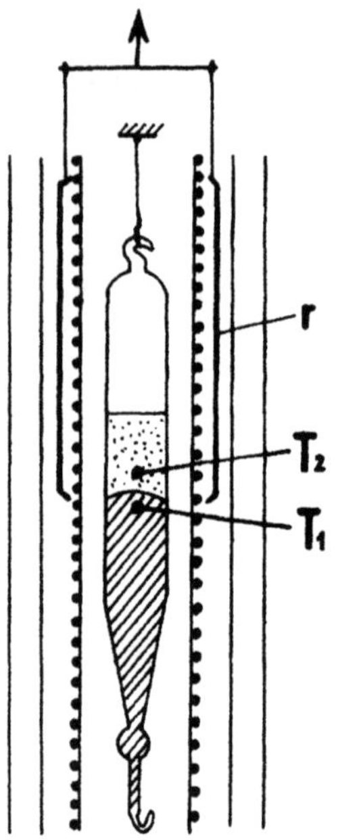

Fig. 6.8. Setup for crystal growth from the melt by the Bridgman–Stockbarger moving-phase-front method. In the "travelling reflector" variant shown here [43], uniformly distributed basic heating is used, leading to a temperature just below the melting point; melting is induced by the radiation reflected back from a mobile semi-transparently gilded glass cylinder, r, which is slowly (0.1–1 mm/h) pulled up to allow a crystal to grow, starting from an intentional or accidental seed with favorable growth direction in the lower tip of the ampoule (From [43]). For alternative methods see [49]

With larger molecules for which undecomposed melting is impossible, sublimation growth is often still feasible. Alternatively, highly ordered [51] or even epitaxial [52, 53] thin films (see Fig. 6.9) can be grown by molecular beam deposition under high or ultrahigh vacuum (see also [17, 18, 54–64]). Whenever the sublimation temperature also exceeds the limit of thermal stability, auxiliary phases which permit growth at lower temperatures are required, such as a solvent or a flux medium (a solid which allows formation of a solution above its melting point).

Once a bulk or epitaxial single crystal has been obtained, a number of questions arise regarding sufficiently cautious handling, determination of the crystallographic phase actually obtained and of the orientation of the crystallographic coordinate system, controlled oriented cutting of samples, and last, but not least, characterization of the physical perfection and chemical purity; all of these topics require the appropriate equipment and experience. Some of these topics will be addressed in the following sections.

6.2.3 Handling, Cutting, Phase Assignment, and Orienting of Single Crystals

Organic molecular crystals usually have only low hardness and often exhibit a propensity for gliding. Hints on cautious handling, cutting and polishing techniques may be found in [43]. Assignment of the crystallographic phase is most easily and reliably achieved by X-ray powder diffraction in Bragg–Brentano geometry. Optical methods may be applicable if all possible phases are known.

Most organic crystal structures are highly anisotropic. Preparation and characterization of samples in specific crystallographic orientations is therefore mandatory for studying fundamental anisotropic properties, such as, e.g., conductivity or charge-carrier mobilities (see Chap. 8 and [65]). Crystal orientation can start from macroscopic features, such as natural faces, cleavage plane(s), optical dichroism, or the orientation of the optical refractive index ellipsoid (indicatrix), observed conoscopically with polarized light. A precise unambiguous orientation, however, requires diffraction methods, such as the determination of the (monochromatic) Bragg–Brentano specular reflection angle Θ at a set of lattice planes parallel to the given surface plane whose Miller indices are wanted, or taking a Laue diffraction diagram in polychromatic radiation at perpendicular incidence to the face in question. As organic crystals are usually of low symmetry, Laue orientation requires the comparison with calculated Laue patterns in most cases [43, 66, 67]. Because irradiation with X-rays regularly introduces *chemical defects*, the precise sample orientation should be determined after the electrical or optical measurements.

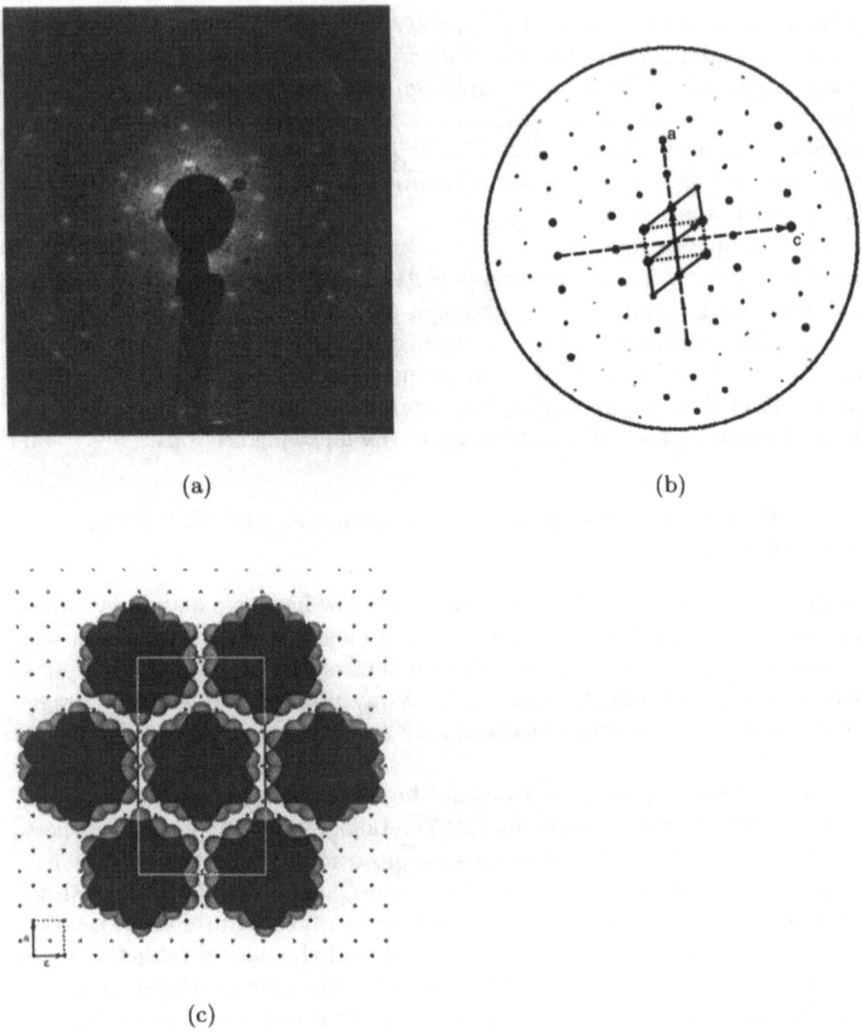

Fig. 6.9. Epitaxial growth of a monolayer of hexabenzocoronene (see Fig. 6.2) on the (010) surface of a freshly cleaved germanium-sulfide (GeS) semiconductor crystal [53]. (**a**) Reciprocal-space low-energy electron-diffraction (LEED) pattern; (**b**) diffraction pattern (which coincides with the measured one) calculated from the real-space structural model shown in (**c**)

6.2.4 Assessment of Sample Quality: Defects and Their Characterization

Imperfections of a sample are conveniently classified as chemical or physical defects. A chemical defect is caused by a foreign molecule (an impurity or a dopant), whereas a physical defect occurs in the pure crystal as a distortion of the regular three-dimensionally translation-symmetric array of the composing molecules. Physical defects can also be induced by chemical defects distorting their neighborhood.

The first aim in materials science will always be to reduce the concentration of the accidental chemical defects, using one of the purification methods mentioned above. A severe problem, however, is to analytically detect traces of residual impurities with a sensitivity down to the 1 ppm level or even below, especially when these are structurally closely related to the host molecule (and therefore left back during prepurification). High-resolution (capillary) gas chromatography, HRGC, and analytical high-pressure liquid chromatography have the power to separate and sensitively detect such impurities. HRGC has the separation power to even resolve isotopic isomers, e.g. anthracene, $C_{14}H_{10}$, and perdeuterated anthracene, $C_{14}D_{10}$ [68]. A large dynamic range which extends down to a few ppm, however, requires a high loading and thus compromises concerning the separation efficiency. While these methods are sensitive enough to detect small amounts/concentrations, they are, however, rather unspecific and barely suitable to identify the chemical nature of an unknown impurity.

Mass spectrometry can give detailed molecular information, especially when combined with separating analytical methods such as HRGC or HPLC, but its sensitivity does not usually extend down to 1 ppm for mixtures of related molecules. The best methods to check purity levels are based on those physical effects and quantities which are strongly influenced by traces of impurities. Sensitive effects are especially those which are connected with transport and trapping of mobile excited states (singlet S_1 and triplet T_1 excitons, charge carriers): sensitized fluorescence (observed best at liquid He temperature), lifetime and spectra of delayed fluorescence caused by the mutual annihilation of two triplet excitons (see, e.g., [43, 44]), and shallow and deep trapping effects of charge carriers at impurities (see Fig. 6.3 and [42]), including thermal and optical trap release spectroscopies (see Chap. 7, Sect. 10 and [40, 43]).

Physical defects can be classified into "zero-dimensional" or "point" defects (vacancies, i.e., lacking molecules, and interstitial molecules), "one-dimensional" or "line" defects (dislocation lines), and "three-dimensional" defects, such as disordered clusters, grains under (small) statistic misalignment, inclusions of a polymorph, twinning, etc. Most of these defects give rise to local strain fields. Experimentally, little is known about point defects in organic crystals, but there is a thermodynamically governed minimal concentration caused by the entropy term in free energy; higher concentrations can be frozen in. Crystals can be free of dislocations, but this is very rare. Dislocations can easily be made visible by X-ray topographic methods, due to their strain field. This topic has been treated extensively in the literature (see [69–71] and references therein). In exceptional cases organic crystals with large (1 cm^2) dislocation-free areas have been obtained (e.g., 2,3-dimethylnaphthalene, grown from the melt by the Bridgman method [44, 69]).

Three-dimensional defects can also be seen by X-ray topography if not too small in size. X-ray (or neutron) rocking width, describing the angular spread of reciprocal lattice vectors, is a very distinctive measure of crystal quality, which should be determined and communicated regularly. It is worth mentioning that even for vapor-deposited thin films rather small rocking widths can be obtained under suitable deposition conditions (e.g., 0.02° for diindenoperylene and 0.2° for perylene-tetracarboxylic-dianhydride, PTCDA, see Fig. 6.2) for the reciprocal lattice vector perpendicular to the thin film [72].

6.3 Excited States of Organic Crystals

Three principal types of electronically excited states can be populated in a pure organic crystal (see Fig. 6.10):

1. *Neutral excited states*, either with no dipole moment or with an intramolecular dipole moment created in the excitation process or with alteration of an already present (ground state) dipole moment. (Notice that these stationary dipole moments should not be mixed up with the [intermediate] transition dipole moment.) These electronic excitations can be in a singlet or in a triplet spin configuration.
2. *Ionic pair states*, where a full elementary charge is transferred to a neighboring molecule, giving rise to an excited-state *intermolecular* dipole moment (intermediate situations with partial "charge transfer" also occur). The electron and the hole are bound by Coulomb interaction and can move only in a correlated way. These states are usually called charge-transfer (CT) states, sometimes "geminate pairs". Depending on the relative spin orientation of the electron and the hole state, singlet and triplet CT states can be distinquished.
3. *Ionic single charge states*, where a radical (i.e., spin = 1/2) cation is formed by removing an electron to infinity or a radical anion is formed by loading an extra electron onto a molecule.

While the creation of a neutral excited state is a "monomolecular" reaction, that of an ionic pair state is a "bimolecular" one, and it is frequently not possible to treat the positive and the negative charge, which have been separated out of a neutral ground state, separately (except for large separation distances). The correct description is in a two-particle energy diagram, rather: "energy-of-an-electron-in-the-presence-of-a-hole" (at a distance $\Delta\mathbf{r}$), "$E_{\mathrm{eiph}}(\Delta\mathbf{r})$" (see [41], Fig. 2). Since the dielectric constants of organic solids

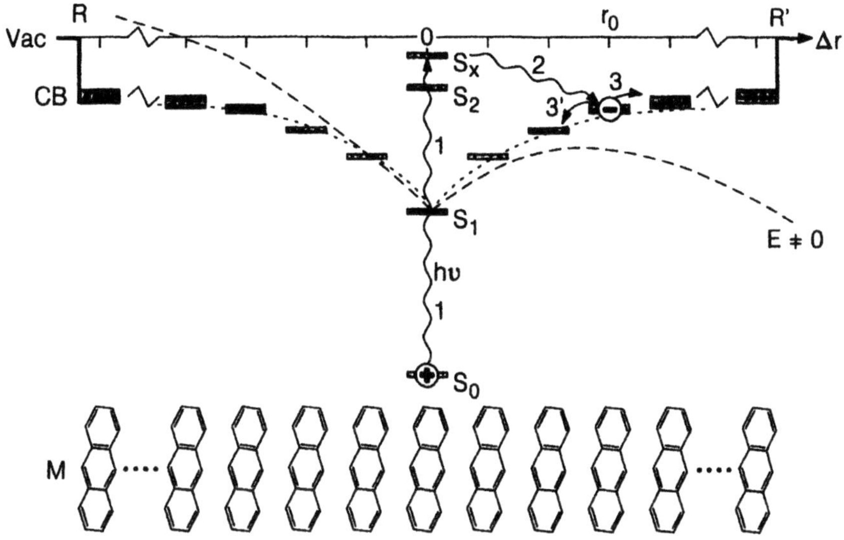

Fig. 6.10. A schematic of the different neutral and ionic excited states in a photoconductive organic molecular crystal (anthracene in the example), explained for light-induced charge separation. The binding energy of an electron at position Δr (in units of the intermolecular separation in the crystal) near a hole at (*fixed*) position $r = 0$ (abscissa) is plotted on a downward electron-in-the-presence-of-a-hole binding energy scale for different electron–hole separations Δr. The vacuum level, V_{ac}, has been taken as the zero-energy reference. For $\Delta r = 0$ (path 1) neutral electronically excited molecular states S_1, S_2, and S_x can be populated; in a periodic crystal these are mobile Frenkel exciton states (the higher states $x > 1$ have usually short lifetimes, however). With a certain probability, the higher states can autoionize (path 2), instead of relaxing to S_1, and thus indirectly populate a "geminate pair" Wannier-type state at, e.g., a separation r_0. There may also be a certain probability of direct optical charge-transfer population by absorbed photons of an appropriate energy. When Δr is sufficiently large to allow the Coulombic attraction to be negleted, the excitation can be described by two independent single charge states: For the electron to arrive there (in the conduction band, CB, position R', path 3) thermal energy and/or bending of the potential curve by an applied electric field (indicated by the dashed line) are necessary to compete with attraction under very efficient kinetic energy loss and final geminate recombination (path 3'). Alternatively, single sign charges can be injected from an electrode or from vacuum (not plotted in the Figure)

are usually small, there is little screening of the Coulombic interaction–charge carriers "feel" each other across considerable distances; this interaction is supported by a fairly small coherent delocalization of charge carriers, which is frequently the case for and typical of these materials. There has been considerable confusion in the literature concerning the necessity of two-particle descriptions. If, however, an extra electron (or hole) is added (e.g., by injection from "outside", path 3) into a molecular solid (or more than one, but in sufficient dilution), the usual one-particle picture applies.

To understand the electronic energy levels in the organic solid, it is useful to start from the electronic states of the *isolated free molecule* in vacuum (which we assume to carry N/2 conjugated double bonds): The intramolecular interactions between the atoms lead to a splitting of the initially degenerate N carbon $2p_z$ electronic levels and to a delocalization of these "π" electrons over the molecule (see Fig. 6.1). The resulting bonding orbitals take up these electrons, while the anti-bonding orbitals remain empty. On a downward positive electron binding energy scale there is a "highest" occupied molecular orbital (HOMO) and a "lowest" unoccupied molecular orbital (LUMO), with an energy gap in between of typically around one or a few eV.

In the most simple approximation of an energy-level description, internal electron–electron interactions are dropped, with the consequence that the LUMO describes *at the same time* the lowest energetic position of an electron excited out of one of the occupied lower levels of a neutral molecule *and* the lowest possible energetic position of an extra electron brought in from "outside". The crude approximation made here and frequently used (albeit not always in a well-reflected way) is to take the same energy levels for two completely different situations: for the interaction of N nuclei with $n + N$ electrons (where n is the number of σ electrons) in the neutral state and for the interaction with $n + N + 1$ electrons in the radical anion state (and even for describing the $n + N - 1$ radical cation hole state).

The *molecular* (gas phase) π-electron energy levels are important because they are the basis for an approximate description of those energy levels of the crystal in which transport of excitons and charge-carriers occurs. In the crystal the electronic systems of neighboring molecules are usually only in weak interactions, a fact reflected by the weak van der Waals cohesive energy (for example, this is ca. 2 eV for the total of 14 carbon and 10 hydrogen atoms of the anthracene molecule inside a crystal as compared to 9.3 eV[1] for a single Si atom inside a silicon crystal). Therefore the concept of the persistence of the molecular states in the organic solid, with only second-order changes of the energy levels by the intermolecular interactions (*the so-called oriented gas model*), is a reasonable approximation – very different from the situation that holds for the main valence-bonded silicon atoms in the silicon crystal.

The changes upon condensation from the gas into the solid state comprise a moderate splitting of the molecular energy levels by intermolecular

[1] Twice the sublimation energy into free atoms.

interactions into narrow "bands"[2] and a general shift due to the fact that a neutral excited molecule or a molecular ion is embedded in an interacting polarizable environment in the crystal. It turns out that the specific properties of the surrounding molecules are less important, but what counts is their average polarizability, which is more determined by the gross volume density of (mainly) the π-electrons than by details of the molecular structure. Hence, the "site shift" of *neutral* excited states in the crystal with respect to their energetic position in the free molecule is expected to be a rather universal, unspecific quantity for entire classes of molecular solids. Similar arguments hold for the "polarization shift", $-P_p$ for the hole and $+P_n$ for the electron state, described by the shift of the HOMO level and the LUMO level of the free ion, when brought from vacuum into a neutral crystal (see Fig. 6.11). Indeed, experimental polarization shifts of a great number of aromatic molecules are rather similar, with P_p around 1.7 eV and P_n slightly less [73]; a large amount of data are listed in [74, 75], but see also [65].

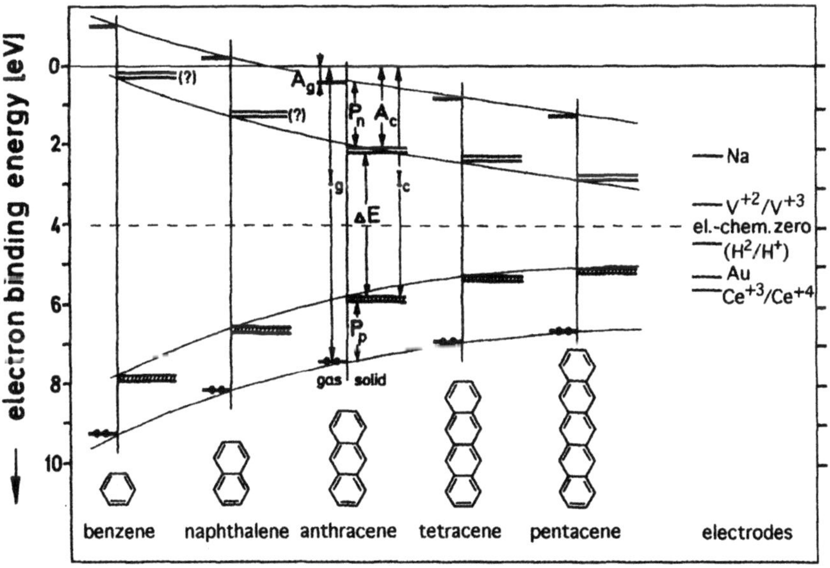

Fig. 6.11. Energy levels of (*single*) ionic states of the acenes from benzene to pentacene for the free molecule in the gas (g) state (*left*), and for the crystal (c) (*right*), defined by the electron-binding energy relative to the vacuum level. In addition, the Fermi levels of selected metals and the redox potential of electrolytic redox systems are indicated, together with the zero point of the electrochemical scale. Further explanations are given in the text. (From [41], updated)

[2] The term "band" here means an energetic interval only, and not necessarily a band model description in terms of coherent Bloch wave transport.

The HOMO level of a free molecule as well as that of a crystal can be measured as the ionization threshold energy I_g or I_c, respectively (g for gas, c for crystal), by UV or X-ray photoelectron spectroscopy (UPS and XPS, respectively), or by photoelectron yield spectroscopy. It is, however, important to be aware that not necessarily the full amount of interatomic and intermolecular (i.e., vibronic and phononic) relaxation energy is contained in the measured photoemission threshold energy. The (empty) LUMO level of the free molecule is similarly characterized by its binding energy A_g for an electron (e.g., with respect to the vacuum level), usually for an extra electron (see above), and – at least in principle – is experimentally accessible by photoelectron spectroscopy as well. Reaching a sufficient population density is, however, not a trivial matter. Inverse photoelectron spectroscopy (IPES) and investigation of the temperature dependence of the chemical equilibrium of the electron-transfer reaction of the gas molecules in contact with, for example, alkali atoms are alternative methods to determine A_g. For A_g data and references see [65]. The solid-state electron affinity A_c is more difficult to measure, but IPES has been applied succesfully in a few cases.

In the solid the gap between the highest occupied level and the lowest unoccupied level is $\Delta E = I_c - A_c$. The width of this region with zero density of states ("band gap") decreases as the number of interacting π electrons increases (see Fig. 6.11); for a comparison the Fermi levels of two extreme metals, Na and Au, are indicated, which can be used as electron- and hole-injecting electrodes, respectively. The redox potentials of the extreme electrochemical redox pairs V^{+2}/V^{+3} and Ce^{+3}/Ce^{+4} are also indicated.

The Fermi level of an organic semiconductor is often not found in the middle between the HOMO and the LUMO level of the solid, as would be expected. This is a consequence of accidental doping. Therefore, contact barrier heights can often not be predicted, but need to be determined experimentally, e.g., by UPS; interfacial and surface dipole layers need additional attention, see, e.g., [76]. Ohmic contacts require the Fermi level be shifted into a band, which not only implies that the problem of high-concentration doping be solved, but also that an analytical control be available.

More details on the description of energy levels in organic molecules and in molecular solids are presented in Chaps. 7, 9 and 11 for one-component materials and in Chap. 10 for hetero-interfaces and for donor:acceptor complexes. For an introduction, the books by Pope and Swenberg [12,13] deserve a special mention.

6.4 Dynamical and Transport Properties of Excited States in Crystals

Thermally and optically induced excitations with immediate or subsequent partial or full CT to neighboring molecules of the same or of different chemical nature ("homo-CT" or "hetero-CT") are the intrinsic source of thermally activated dark conductivity and of optically induced photoconductivity of organic solids, although complete separation of the initial "geminate" electron–hole pair may need additional energy in subsequent processes (Fig. 6.10). Thus, usually a thermal and/or electric field activation is necessary for the final separation of the geminate pair [77, 78]. Even metallic-like conduction, with a partly filled band even at $T \sim 0$ K, can occur in two-component organic crystals, when partial or full CT occurs in every crystallographic unit cell in a counterbalancing of the required energy by (among other causes) the gain of Madelung energy of the ionic lattice formed ("radical ion salts"). Even transitions to superconductivity occur for a number of radical ion salts, with the highest transition temperatures ($T_c \sim 30$ K) observed for alkalimetal$^+$ C_{60} radical anion salts (see [79]). These topics are not among the subjects of this book. More information can be found in the references [80–87].

Another source of mobile charge carriers is injection of extra electrons or holes from (or emission of electrons to) the outside (electrodes, vacuum), giving rise to (extrinsically induced) space-charge conduction in situations with a large dielectric relaxation time. (Note that properly injected electrons can move in an *insulating* crystal in a high-lying, otherwise empty conduction band, and injected holes in an otherwise filled-up valence band, but these charges are sources of an electrical field and thus distort an externally applied field). Charge-compensated, extra charge carrier, bipolar transport is possible, but usually associated with strong Langevin-type recombination of the barely screened electron and hole in the overlap region, leading to a population of neutral exciton levels. Irrespective of whether formed by electron–hole recombination or by optical population, the singlet states usually decay within a short lifetime of typically a few nanoseconds, often with high radiative yield; the triplet states can live between milliseconds and seconds, and predominance of radiative decay over non-radiative transitions is the exception for triplets.

Electronic and singlet or triplet excitonic transport is primarily (i.e., at temperature $T = 0$ K) a consequence of the (weak) interactions between initially degenerate levels on neighboring molecules, giving rise to tunneling transition probabilities, to dipolar transfer or to exchange, respectively. However, relaxation of the local environment may lead to self-trapping, and transport steps may require either waiting times, until thermal fluctuations have created a resonant situation on the next molecule, or a thermal excitation of the self-trapped state. These are the basic concepts of hopping-type transport models. In contrast, band-type transport requires sufficiently short-time and weak local interactions, associated with a high diffusivity or mobility based on a certain degree of delocalization, in order not to allow phononic (lattice), vibronic (intramolecular) and electronic polarization interactions to lead to localization in reducing the extension of the exciton- or charge-carrier wave packets to "one lattice constant" before the charge carrier has moved away (i.e., to a breakdown of the coherent wave picture). Band transport is indeed a typical situation for crystalline high-purity aromatic hydrocarbons at low temperature [42,88,89], but does not usually apply at higher temperatures. There are still fundamental unsolved problems, however, with a correct theoretical description. Macroscopic transport is based on either diffusional motion in a concentration gradient (excitons, charge carriers) or drift motion in an electrical potential gradient (charge carriers) and described by the principal components and orientation of the diffusion constant and mobility tensor, D and μ, respectively, defined by the equations for the current density

$$\boldsymbol{j}_D = -e\boldsymbol{D}\boldsymbol{\nabla} n \qquad \boldsymbol{j}_\mu = -en\boldsymbol{\mu}\boldsymbol{\nabla} U \tag{6.1}$$

where n is the concentration of holes (or electrons), U the potential ($-\boldsymbol{\nabla} U$ the local electric field) and e the electronic charge unit (e and $\boldsymbol{\mu}$ are to be taken positive for holes, negative for electrons).

As a consequence of the comparatively weak intermolecular interactions, excited states in perfect organic crystals are only moderately split to form more or less narrow energy bands and delocalized (as a consequence of the translational invariance of the crystal lattice) only to some extent at sufficiently low temperatures. The degree of localization is determined by those processes which destroy coherence lifetime in a given quantum state: scattering by phonons and interactions with static distortions of the translational symmetry, on the one hand, and by the degree of "electron–phonon interaction", on the other, describing the local relaxation of the excited state by inducing phononic, vibronic, and electronic polarization. Band models describing motion in Bloch-type wave packets which extend over a finite number of translational periods of the lattice may thus only be appropriate to describe excitons and electrons/holes at low temperatures. Whenever the scattering-rate-induced energetic uncertainty becomes larger than the bandwidth, the appropriate description is given by hopping-type models.

It is a signature of coherent band transport to be "thermally deactivated", i.e., to be fast at low temperature and to increasingly suffer from collision with thermally excited vibrations as the temperature is raised (see Chap. 8). In contrast, for hopping transport it is typical that an exponential thermal activation is generally involved, which dominates the overall temperature dependence. The details may be much more complicated, depending on the type of hopping model that applies.

Disorder finally leads to a variation width of molecular energy levels and to a distribution of interaction distances and hence interaction energies and transition probabilities ("diagonal" and "off-diagonal" disorder in the density matrix representation). For sufficient disorder, transport becomes "dispersive"; there is no universal reasonable average drift velocity in an external field; and charge-carrier mobility is no longer a suitable material property, because, for a given charge-carrier packet, mobility decreases gradually with time, in that deeper and deeper states are visited, and thus also depends on the distance traveled. A distribution of shallow impurity levels may behave similarly, whereas a single impurity level may lead to a well-defined average mobility (see Fig. 6.3 and the references given).

For practical applications of organic electronic materials where dynamical quantities and loss factors count, the exciton diffusion tensor components (and lifetimes, and hence diffusion lengths) and the charge-carrier mobility tensor components are among the most important material properties. Therefore, sample purity and structural perfection deserve utmost attention. This will be another aspect of the review on charge-carrier mobility given in Chap. 8.

6.5 Organic Non-polymeric Thin Films

6.5.1 Current Developments

Whereas pure and high-perfection single crystals are ideal model systems for scientific studies of fundamental properties and relations, they are too sensitive, bulky, inflexible and expensive for most practical applications. A great deal of activity has therefore recently been concentrated on the preparation of organic thin films and on the evaluation of their electronic and optoelectronic properties. In Part II of this book only low molecular weight materials will be considered (for polymers see Part I). Organic thin films can be made by printing, casting and spin-coating methods, by adsorption, e.g., from solution, and by vacuum vapor deposition, including molecular beam deposition and epitaxy under high or ultra-high vacuum conditions. The latter is definitely the most sophisticated method; it not only promises the utmost crystallographic order and perfection, as well as chemical purity, but also provides the potential for tailoring complex layered architectures [17,18,52–64,90–92]. It is thus hoped that well-ordered organic thin films may attain electronic transport properties which approach those of the best single crystals. Recent experiments clearly support this expectation [51,72]. It is also expected that hetero-multilayers allow detection of entirely new phenomena, which may lead to unexpected applications. The review by Agranovich et al. on excitons in multilayers (Chap. 10) and the review by Bulović et al. culminating in thin-film organic light-emitting devices (Chap. 11) are vivid expositions of the current trends in this field. Organic field-effect transistors are presently under intensive exploitation [19–27,72]; thin-film photovoltaic cells have been demonstrated [28–36]; and organic diodes have been shown to possess gigahertz switching capabilities [93].

Very exciting results have quite recently been obtained with high-perfection thin vapor-grown *single crystals* of materials such as anthracene, tetracene, pentacene, perylene, and C_{60}-fullerene in field effect transistor electrode geometry: high-field effect mobilities at low temperatures, approaching 10^5 cm^2/Vs at 1.5 K [94, 95], h/e^2 integer and fractional quantum Hall effect resistance plateaus [95], switching by the gate field from an insulating to a superconducting state [96], and lasing via population of excited states by recombination of injected electrons and holes [38].

It is not unreasonable to expect that these crystals might be replaceable by high quality epitaxial organic thin films in the future.

6.5.2 Possible Future Trends

The prospect of including self-ordering properties of more complex molecules to obtain even more complex structures and functions is an especially challenging aspect of organic materials. It is also hoped that the dimensions of molecular electronic logical units can finally be reduced down to the molecular level. The recent invention of scanning microscopy methods which allow individual molecules to be addressed with even submolecular resolution has delivered unexpected tools to proceed into these dimensions.

References

1. A. Pochettino, Atti Acad. Lincei Rend. **XV** (1), 355 (1906) and **XV** (2), 171 (1906)
2. J. Koenigsberger, K Schilling, Ann. Physik **32**,179 (1910)
3. M. Volmer, Ann. Physik **40**, 775 (1913)
4. W.G. Pfann, Trans. AIME **194**, J. Metals **4**, 747 (1952)
5. W.G. Pfann, *Zone Melting* (J. Wiley, New York 1958)
6. K.H. Probst, N. Karl, Phys. Stat. Sol. A **27**, 499 (1975); see also erratum Phys. Stat. Sol. A **31**, 793 (1975)
7. N. Karl, Mol. Cryst. Liq. Cryst. **357**, 1 (2001)
8. N. Karl, H. Rohrbacher, D. Siebert, Phys. Stat. Sol. (a) **4**, 105 (1971) (the communicated off diagonal elements of the ε tensor should be multiplied by a factor of 2 for the conventional form of presentation)
9. N. Karl, Habilitation Thesis, University of Stuttgart (1975)
10. R.G. Kepler, In: *Treatise on Solid State Chemistry*, Vol. 3, ed. by N.B. Hannay (Plenum Press, New York 1976) pp. 615
11. H. Inokuchi, M. Sano, Y. Maruyama, N. Sato (eds.) *Proceedings of the Oji Internat. Seminar on Organic Semiconductors – 40 Years*, Mol. Cryst. Liq. Cryst. **171**, pp. 1–356 (1989)
12. M. Pope, Ch.E. Swenberg, *Electronic Processes in Organic Crystals* (Clarendon Press/Oxford University Press, Oxford/New York 1982)
13. M. Pope, Ch.E. Swenberg, *Electronic Processes in Organic Crystals and Polymers*, (Oxford Sci. Publ., New York 1999)
14. A.J. Hodina, Synth. Met. **10**, B55; B99; B155 (1984)
15. H. Meier, *Organic Semiconductors, Dark and Photoconductivity of Organic Solids* (Verlag Chemie, Weinheim 1974)
16. C. Hamann (ed.) *Organische Festkörper und organische dünne Schichten* (Akad. Verlagsges. Geest & Portig, Leipzig 1978)
17. T. Kobayashi, In: *Crystals, Growth, Properties and Applications*, Vol. 13, ed. by N. Karl (Springer, Heidelberg 1991), pp. 1–63
18. S.R. Forrest, Chem. Rev. **97**, 1793 (1997)

19. J.G. Laquindanum, H.E. Katz, A.J. Lovinger, A. Dodabalapur, Chem. Mater. **8**, 2542 (1996)
20. D.J. Gundlach, Y.-Y. Lin, T.N. Jackson, S.F. Nelson, D.G. Schlom, IEEE Electron Device Lett. **18**, 87 (1997)
21. Y.-Y. Lin, D.J. Gundlach, S.F. Nelson, T.N. Jackson, IEEE Electron Device Lett. **18**, 607 (1997)
22. C.D. Dimitrakopoulos, B.K. Furman, F. Graham, S. Hedge, S. Purushothaman, Synth. Met. **92**, 47 (1998)
23. G. Horowitz, Adv. Mater. **10**, 365 (1998)
24. S.F. Nelson, Y.-Y. Lin, D.J. Gundlach, T.N. Jackson, Appl. Phys. Lett. **72**, 1854 (1998)
25. L. Torsi, A. Dodabalapur, L.J. Rothenberg, A.W.P. Fung, H.E. Katz, Phys. Rev. B **57**, 2271 (1998)
26. J.H. Schön, Ch. Kloc, R.A. Laudise, B. Batlogg, Appl. Phys. Lett. **73**, 3574 (1998)
27. Ch.D. Dimitrakopoulos, I. Kymissis, S. Purushothaman, D.A. Neumayer, P.R. Duncombe, R.B. Laibowitz, Adv. Mater. **11**, 1372 (1999)
28. D.L. Morel, A.K. Ghosh, T. Feng, E.L. Stogryn, P.E. Purwin, R.F. Shaw, C. Fishman, Appl. Phys. Lett. **32**, 495 (1978)
29. A.K. Ghosh, T. Feng, J. Appl. Phys. **49**, 5982 (1978)
30. G.A. Chamberlain, Solar Cells **8**, 47 (1983)
31. C.W. Tang, Appl. Phys. Lett. **48**, 183 (1986)
32. D. Wöhrle, D. Meissner, Adv. Mater. **3**, 129 (1991)
33. G. Yu, J. Gao, J.C. Hummelen, F. Wudl, A.J. Heeger, Science **270**, 1789 (1995)
34. U. Bach, D. Lupo, P. Comte, J.E. Moser, F. Weissörtel, J. Salbeck, H. Spreitzer, M. Grätzel, Nature **395**, 583 (1998)
35. M. Granström, K. Petritsch, A.C. Arias, A. Lux, M.R. Andersson, R.H. Friend, Nature **395**, 257 (1998)
36. J. Gao, G. Yu, A.J. Heeger, Adv. Mater. **10**, 692 (1998)
37. J. Kalinowski, J. Phys. D: Appl. Phys. **32**, R179 (1999)
38. J.H. Schön, Ch. Kloc, A. Dodabalapur, B. Batlogg, Science **289**, 599 (2000)
39. N. Karl, K.-H. Kraft, J. Marktanner, M. Münch, F. Schatz, R. Stehle, H.-M. Uhde, J. Vac. Sci. Technol. A**17**, 2318 (1999)
40. N. Karl, In: *Defect Control in Semiconductors*, Vol. II, ed. by K. Sumino (Elsevier, North Holland, Amsterdam 1990), pp. 1725–1746
41. N. Karl, In: *Festkörperprobleme/Adv. Solid State Physics* XIV, ed. by H.J. Queisser (Pergamon/Vieweg, Headington/Braunschweig 1974) pp. 261–290
42. W. Warta, R. Stehle, N. Karl, Appl. Phys. A**36**, 163 (1985)
43. N. Karl, In: *Crystals, Growth, Properties, and Applications*, Vol. 4, ed. by H.C. Freyhardt (Springer-Verlag, Berlin 1980) pp. 1–100
44. N. Karl, Mater. Sci. **10**, 365 (Technical Univ. of Wroclaw, Poland, 1984)
45. N. Karl, J. Cryst. Growth **51**, 509 (1981)
46. N. Karl, Mol. Cryst. Liq. Cryst. **171**, 157 (1989)
47. S. Demming, W. Langhals, Chem. Ber. **121**, 225 (1988)
48. N. Karl, Inst. Phys. Conf. Ser. **103**, 107 (1989)
49. G.J. Sloan, A.R. McGhie, *Techniques of Melt Crystallization*, Vol. XIX of *Techniques of Chemistry*, ed. by W.H. Sounders (J. Wiley, New York 1988)
50. N. Karl, J. Cryst. Growth **99**, 1009 (1990)
51. N. Karl, J. Marktanner, Mol. Cryst. Liq. Cryst. **315**, 163 (1998)

52. U. Zimmermann, N. Karl, Surf. Sci. **268**, 296 (1992)
53. N. Karl, Ch. Günther, Cryst. Res. Technol. **34**, 243 (1999)
54. T. Fritz, M. Hara, W. Knoll, H. Sasabe, Mol. Cryst. Liq. Cryst. **235**, 269 (1994)
55. A Hoshino, S. Isoda, H. Kurata, T. Kobayashi, J. Appl. Phys. **76**, 4113 (1994)
56. A Hoshino, S. Isoda, H. Kurata, T. Kobayashi, J. Cryst. Growth **146**, 636 (1995)
57. A. Koma, Progr. Cryst. Growth Charact. **30**, 129 (1995)
58. T.J. Schuerlein, A. Schmidt, P.A. Lee, K.W. Nebesny, N.R. Armstrong, Jpn. J. Appl. Phys. **34**, 3837 (1995)
59. A. Schmidt, T.J. Schuerlein, G.E. Collins, N.R. Armstrong, J. Phys. Chem. **99**, 11770 (1995)
60. W. Gebauer, M. Bässler, R. Fink, M. Sokolowski, E. Umbach, Chem. Phys. Lett. **266**, 177 (1997)
61. C. Seidel, J. Poppensieker, H. Fuchs, Surf. Sci. **408**, 223 (1998)
62. E. Umbach, K. Glöckler, M. Sokolowski, Surf. Sci. **402–404**, 20 (1998)
63. C. Seidel, H. Kopf, H. Fuchs, Phys. Rev. B **60**, 14341 (1999)
64. T. Schmitz-Hübsch, F. Sellam, R. Staub, M. Törker, T. Fritz, Ch. Kübel, M. Müllen, K. Leo, Surf. Sci. **445**, 358 (2000)
65. N.Karl, In: *Landolt-Börnstein* New Series, Vol 17i, ed. by O. Madelung, M. Schulz, H. Weiss (Springer Verlag, Heidelberg 1985) pp. 106–218
66. N. Karl, H. Port, W. Schrof, Mol. Cryst. Liq. Cryst. **78**, 55 (1981)
67. D. Massa, N. Karl, Mol. Cryst. Liq. Cryst. **175**, 93 (1989)
68. N. Karl, unpublished results
69. H. Klapper, In: *Crystals, Growth, Properties and Applications* Vol. 13, ed. by N. Karl (Springer Verlag, Berlin 1991) pp. 109–162
70. H. Klapper, In: *X-ray and Neutron Dynamical Diffraction: Theory and Applications*, ed. by A. Authier et al. (Plenum Press, New York 1996) pp. 137–146 and 167–176
71. H. Klapper, Mater. Sci. Forum **276–277**, 291 (1998)
72. M. Münch, N. Karl, to be published; M. Münch, PhD thesis dissertation, University of Stuttgart (2001)
73. I. Eisenstein, R.W. Munn, Chem. Phys. **77**, 47 (1983)
74. N. Sato, K. Seki, H. Inokuchi, J. Chem. Soc. Faraday Trans. **2**, 77, 1621 (1981)
75. N. Sato, Synth. Met. **64**, 133 (1994)
76. H. Ishii, K. Sugiyama, E. Ito, K. Seki, Adv. Mater. **11**, 605 (1999)
77. R.H. Batt, C.L. Braun, J.F. Hornig, J. Chem. Phys. **49**, 1967 (1968)
78. R.H. Batt, C.L. Braun, J.F. Hornig, Appl. Optics, Suppl. 3: Electrophotography, 20 (1969)
79. M. Baenitz, In: *Festkörperprobleme/Adv. Solid State Phys.* 38, ed. by B. Kramer, (Vieweg, Braunschweig 1999) pp. 471–483
80. D. Jérome, H.J. Schulz, Adv. Phys. **31**, 299 (1982)
81. S. Kagoshima, H. Nagasawa, T. Sambongi, *One-dimensional Conductors* (Springer, Berlin, Heidelberg 1988)
82. J.-P. Farges (ed.) *Organic Conductors* (Marcel Dekker, New York 1994)
83. D. Jérome (ed.), *Common Trends in Synthetic Metals and High-Tc Superconductors* J. Physique **6**, 12 (1996) (I.F. Schegolev's Memorial Volume)
84. M. Lang, Superconduct. Review **2**, 1 (1996); also as Habilitationsschrift, Technische Universität Darmstadt (1996)
85. H.S. Nalwa (ed.) *Handbook of Organic Conductive Molecules and Polymers* (John Wiley, New York, 1996)

86. J. Wosnitza, *Fermi Surfaces of Low-Dimensional Organic Metals and Superconductors* (Springer, Berlin 1996)
87. T. Ishiguro, K.Yamaji, G. Saito, *Organic Superconductors*, 2nd edn. (Springer, Berlin 1998)
88. N. Karl, J. Marktanner, R. Stehle, W. Warta, Synth. Met. **41–43**, 2473 (1991)
89. W. Warta, N. Karl, Phys. Rev. B **32**, 1172 (1985)
90. C. Ludwig, B. Gompf, W. Glatz, J. Petersen, W. Eisenmenger, M. Möbus, U. Zimmermann, N. Karl, Z. Phys. B **86**, 397 (1992)
91. C. Ludwig, B. Gompf, J. Petersen, R. Strohmaier, W. Eisenmenger, Z. Phys. B **93**, 365 (1994)
92. R. Strohmaier, C. Ludwig, J. Petersen, B. Gompf, W. Eisenmenger, Surf. Sci. **351**, 292 (1996)
93. P. Urbach, D. Ammermann, W. Kowalsky, *Int. Conf. On Solid State Devices and Materials, Yokohama, Japan, Conference Proceedings* (1996) p. 586
94. J.H. Schön, *"High Mobilities in Organic Semiconductors: Basic Science and Technology"*, E-MRS Spring Meeting, Strassbourg, June 2000, Conference Proceedings Paper I-VI.2
95. J.H. Schön, Ch. Kloc, B. Batlogg, Science **288**, 2338 (2000)
96. J.H. Schön, Ch. Kloc, B. Batlogg, Nature **406**, 702 (2000)

7 Electronic States in Organic Molecular Crystals

V. Čápek and I. Muzikante

7.1 Introduction

Organic molecular crystals (OMC) (see, e.g., [1]) are specific in several respects. The basic feature distinguishing these solids from their standard inorganic counterparts is the fact that the main composition units are organic molecules which, in the condensed matter state, preserve their physical identity well. This is due to the following facts:

- Atoms in the individual molecules are bound by forces which have mostly a covalent character. This is in contrast with the binding forces among the molecules, which are usually of the van der Waals type. The remarkable difference between the strengths of the covalent and dispersion forces implies that, in many respects, the molecules keep behaving as though they were in a gas phase even though they are well oriented and have a well-defined position in space (model of an oriented molecular gas).
- As far as the repulsive forces between individual molecules do not allow overlap of electronic clouds in the molecular ground states, the ground state of such molecular crystals can usually be successfully built up of ground states of the individual molecules. This allows *inter alia* ascribing, in the crudest approximation, of electrons to their molecules, i.e., full neglecting the exchange of electrons between different molecules.

However, one should realize that this picture is often too idealized. Namely:

- Higher excited states of the individual molecules (if they are still not ionized in the gas phase) may already overlap. This may cause plenty of effects like photoconductivity and implies that the above characteristics of the organic molecular crystals may well only apply to, for example, the electronically unexcited states.
- There may be plenty of mechanisms intermixing the ground and excited states of the molecules in the crystalline phase. Note that the van der Waals dispersion forces are based on correlated fluctuations of (electric) dipole momenta of individual molecules. In order to describe these fluctuations, one must include the off-diagonal elements of the dipole momentum operator, i.e., the possibility of a virtual excitation of the molecules to their excited states which could be appreciably less localized.

Another, potentially even more important mechanism involving higher excited states of isolated molecules might be connected with a possible presence of, for example, a static dipole momentum of the molecules in the gas phase (polar molecules). Upon formation of the molecular crystal, the electronic states of the molecules could become appreciably influenced by the static polarization of the surrounding molecules. The resulting induced (additional) polarization means, with respect to the gas-phase molecular states, inclusion of gas-phase excited states. Even if we refrain from considering the undeformed gas-phase electronic molecular states, it is easy to see that the distorted electronic states in fact define a "different" molecule existing now in the crystalline phase.

- On the top of all that, one should realize that hydrogen bonds among, for example, elongated molecules, molecular tails, etc., in the crystalline phase, or other even weaker forces acting on relatively soft parts of the molecules may appreciably distort even their skeletons in the crystal. Thus, in the solid phase, we may have still more or less individual molecules but they may appreciably differ from their gas-phase counterparts.

So as to avoid all these complications, we shall limit our attention here to just the case of crystals composed of relatively small molecules (dimensions comparable with lattice parameters) which are neutral and non-polar in their ground states, corresponding well to those in the crystalline phase under investigation, and whose electronic (ground and first few excited) states on neighboring molecules practically do not overlap. Polyacene crystals are a typical (but not the only) type of OMC of this family.

7.2 Model

For simplicity, let us ignore external fields. Let, in the non-relativistic approximation,

$$H_{\mathrm{E}}(r) = \sum_i -\frac{\hbar^2}{2m}\Delta_{\boldsymbol{r}_i} + \frac{1}{2}\sum_{i\neq j}\frac{e^2}{|\boldsymbol{r}_i - \boldsymbol{r}_j|}\,, \tag{7.1}$$

$$H_{\mathrm{N}}(R) = \sum_i -\frac{\hbar^2}{2M_i}\Delta_{\boldsymbol{R}_i} + \frac{1}{2}\sum_{i\neq j}\frac{Z_i Z_j e^2}{|\boldsymbol{R}_i - \boldsymbol{R}_j|}\,, \tag{7.2}$$

and

$$H_{\mathrm{EN}}(r,R) = -\sum_{i,j}\frac{Z_j e^2}{|\boldsymbol{r}_i - \boldsymbol{R}_j|} \tag{7.3}$$

be respectively the Hamiltonians of the electrons, nuclei and the electron–nucleus interaction. Here r and R represent all the electron and nuclear coordinates $\boldsymbol{r}_i$ and $\boldsymbol{R}_j$. Then $H_{\mathrm{E}}(r)+H_{\mathrm{EN}}(R)$ may be taken as an electron Hamiltonian for a fixed configuration of nuclei. For our purposes, let us rewrite it as

$$H_{\mathrm{E}}(r) + H_{\mathrm{EN}}(R) = h(r,R) + v(r), \quad v(r) = \frac{1}{2}\sum_{i\neq j} \frac{e^2}{|\boldsymbol{r}_i - \boldsymbol{r}_j|}\,. \tag{7.4}$$

This means that $h(r,R)$ is the Hamiltonian of the electrons not-interacting mutually but moving in the Coulomb field of all nuclei. Further, let $|p\pi\rangle$ ($|q\kappa\rangle$ etc.) designate, for example, the πth Hartree–Fock electron spin-orbital of the isolated pth molecule. It is here that we tacitly (and at least for a while) assume that the neighboring spin-orbitals practically do not overlap, i.e., that we may ascribe electrons to their molecules. Here, we must also decide whether we work in the so-called crude (rigid, diabatic) or moving (adiabatic) bases. As the former approach is technically simpler, we choose it here. We should add that it may be handicapped physically (more complicated description of, for example, electronically excited states of vibrationally oscillating molecules) but is, on the other hand, technically unambiguous [2]. This means that $|p\pi\rangle$ are defined for a prescribed configuration of the nuclear skeleton of the pth molecule. In case these spin-orbitals are not sufficiently non-overlapping with their neighboring molecules, they should, for our purposes below, be first pre-orthogonalized. Next, for products of two such spin-orbitals $\langle p\pi, q\kappa| = (|p\pi, q\kappa\rangle)^*$ (the asterisk designating the complex conjugate). Then, by the rules of the second quantization,

$$\begin{aligned} H_{\mathrm{E}}(r)+H_{\mathrm{EN}}(r,R) &= \sum_{p\iota} h_{p\iota,p\iota} + \frac{1}{2}\sum_{p\iota_1,q\iota_2} (v_{p\iota_1,q\iota_2,p\iota_1,q\iota_2} - v_{p\iota_1,q\iota_2,q\iota_2,p\iota_1}) \\ &+ \sum_{p\pi,q\kappa}\left\{h_{p\pi,q\kappa} + \sum_{l\iota}(v_{p\pi,l\iota,q\kappa,l\iota} - v_{p\pi,l\iota,l\iota,q\kappa})\right\}\mathcal{N}\left[a^{\dagger}_{p\pi}a_{q\kappa}\right] \\ &+ \frac{1}{2}\sum_{p\pi,q\kappa,r\rho,s\sigma} v_{p\pi,q\kappa,r\rho,s\sigma}\mathcal{N}\left[a^{\dagger}_{p\pi}a^{\dagger}_{q\kappa}a_{s\sigma}a_{r\rho}\right]\,. \end{aligned} \tag{7.5}$$

In connection with this form of $H_{\mathrm{E}}(r)+H_{\mathrm{EN}}(r,R)$, the following points should be stressed:

- In (7.5) as well as below, the summation over the ι or ι_i indices is for the occupied spin-orbitals in the ground state of the molecules in question. This is introduced artificially in connection with introduction of the normal product symbol $\mathcal{N}$, designating normal ordering with respect to these ground-state molecular configurations. In other words, for the purposes of ordering, the operators in the argument of $\mathcal{N}\ldots$, the electron creation operators in empty spin-orbitals as well as the electron annihilation operators (hole creation operators) in the occupied spin-orbitals are considered the excitation creation operators. Similarly, the electron annihilation operators in the empty spin-orbitals as well as the electron creation operators in the occupied spin-orbitals (hole annihilation operators) are considered the excitation annihilation operators. The ordering provided by $\mathcal{N}\ldots$ means that (with respect to the signs given by anticommutational relations among electron and hole creation and annihilation operators) all the excitation creation operators are placed on the left of the excitation annihilation operators (mutual order of, e.g., the latter being arbitrary).

- The first two terms on the right-hand side of (7.5) designated henceforth as $E_{\mathrm{SCF}}(R)$ represent, up to the nucleus–nucleus interaction, just a very crude estimate of the mean binding energy of the OMC, even if the positions of the nuclei are optimized. The point is that the van der Waals forces are still not included.
- In contrast to the matrix elements $h_{a\alpha,b\beta} = \langle a\alpha|(-\hbar^2/(2m)\Delta_{\boldsymbol{r}} - \sum_j Z_j e^2/|\boldsymbol{r}-\boldsymbol{R}_j|)|b\beta\rangle$, the elements $v_{a\alpha,b\beta,c\gamma,d\delta} = \langle a\alpha, b\beta|(e^2/|\boldsymbol{r}-\boldsymbol{r}'|)|c\gamma, d\delta\rangle$ do not depend on R (i.e. $\boldsymbol{R}_1, \boldsymbol{R}_2$ etc.). This is because the spin-orbitals have been chosen here to be R-independent.

Thus, we can complement (7.5) by adding $H_{\mathrm{N}}(R)$ to get the total Hamiltonian. Designating, by the superscript index $\cdots^{(i)}$, $i = 0, 1, \ldots$, the ith order expansion contribution in powers of the lattice point displacements $R - R^{(0)}$ from their equilibrium positions, we get by the expansion

$$\begin{aligned} H &= H_{\mathrm{E}}(r) + H_{\mathrm{N}}(R) + H_{\mathrm{EN}}(r,R) \\ &= E_{\mathrm{SCF}}^{(0)} + \frac{1}{2}\sum_{i\neq j}\frac{Z_i Z_j e^2}{|\boldsymbol{R}_i^{(0)} - \boldsymbol{R}_j^{(0)}|} + H_{\mathrm{ph}} + H_{\mathrm{e}} + H_{\mathrm{e-ph}}\,, \end{aligned} \tag{7.6}$$

where the first two terms in the second equality determine the zero-order binding energy of OMC,

$$\begin{aligned} H_{\mathrm{ph}} \equiv H_{\mathrm{vibr}} &= \sum_i -\frac{\hbar^2}{2M_i}\Delta_{\boldsymbol{R}_i} + \frac{1}{2}\sum_{i\neq j}\left[\frac{Z_i Z_j e^2}{|\boldsymbol{R}_i - \boldsymbol{R}_j|} - \frac{Z_i Z_j e^2}{|\boldsymbol{R}_i^{(0)} - \boldsymbol{R}_j^{(0)}|}\right] \\ &\quad + E_{\mathrm{SCF}}(R) - E_{\mathrm{SCF}}^{(0)} \approx \sum_r \hbar\omega_r (b_r^\dagger b_r + \frac{1}{2})\,, \end{aligned} \tag{7.7}$$

$$\begin{aligned} H_{\mathrm{e}} &= \sum_{p\pi,q\kappa} h_{p\pi,q\kappa}^{(0)} \mathcal{N}\left[a_{p\pi}^\dagger a_{q\kappa}\right] + \sum_{p\pi,q\kappa,l\iota} (v_{p\pi,l\iota,q\kappa,l\iota} - v_{p\pi,l\iota,l\iota,q\kappa})\mathcal{N}\left[a_{p\pi}^\dagger a_{q\kappa}\right] \\ &\quad + \frac{1}{2}\sum_{p\pi,q\kappa,r\rho,s\sigma} v_{p\pi,q\kappa,r\rho,s\sigma}\mathcal{N}\left[a_{p\pi}^\dagger a_{q\kappa}^\dagger a_{s\sigma} a_{r\rho}\right]\,, \end{aligned} \tag{7.8}$$

and

$$H_{\mathrm{e-ph}} = \sum_{i=1}^{+\infty} \sum_{p\pi,q\kappa} h_{p\pi,q\kappa}^{(i)} \mathcal{N}\left[a_{p\pi}^{\dagger} a_{q\kappa}\right] . \tag{7.9}$$

In (7.7), $\sum_r \ldots$ indicates summation over all the OMC vibrational (lattice, intramolecular, intermolecular optical, librational) modes. ω_r then is the harmonic frequency of the rth mode while $b_r^{\dagger}$ and b_r designate the standard phonon creation and annihilation operators in the mode r, with standard Bose–Einstein commutational relations. The last approximate equality in (7.7) clearly applies just in the harmonic approximation. As for the summation $\sum_{i=1}^{+\infty} \ldots$ (summation over all but the zero-order Taylor expansion terms) in (7.9), it physically designates partitioning of the coupling to the single-phonon ($i = 1$), two-phonon ($i = 2$), etc., terms as the displacement vectors $\boldsymbol{R}_j - \boldsymbol{R}_j^{(0)}$ are linear in $b_r^{\dagger}$ and b_r.

Although (7.6) with (7.7–7.9) may already serve as a good basis for discussion of electronic states in OMC, further manipulation is still physically required. The point is that (7.5) does not properly distinguish the electron states above occupied valence states from those above empty valence ones. The former states designate those of an electron in a molecule which thus becomes negatively charged (i.e., the conduction electron). The latter states are standardly taken as the simplest form of the electron–hole pair, where the (excited) electron remains on the same molecule. We are thus led to the definition of the Frenkel exciton states. Hence, a redefinition of all the creation and annihilation operators is required by the physics of the problem.

In order to make the formal technical manipulations as simple as possible, we shall limit our attention to just one valence ($\pi, \kappa, \ldots = 0$) and one conduction spin-orbital ($\pi, \kappa, \ldots = 1$) at each molecule. (Generalizations to an arbitrary number of the spin-orbitals is straightforward.) First, we define the products

$$\bar{a}_p^{\dagger} = a_{p1}^{\dagger} a_{p0} \quad \text{and} \quad \bar{a}_p = a_{p0}^{\dagger} a_{p1} \tag{7.10}$$

as the Frenkel exciton creation $\bar{a}_p^{\dagger}$ and annihilation $\bar{a}_p$ operators. Clearly, from the electron Fermi anti-commutational relations ($\{C, D\} \equiv CD + DC$)

$$\left\{a_{p\pi}, a_{q\kappa}\right\} = \left\{a_{p\pi}^{\dagger}, a_{q\kappa}^{\dagger}\right\} = 0 \qquad \text{and} \qquad \left\{a_{p\pi}, a_{q\kappa}^{\dagger}\right\} = \delta_{p\pi,q\kappa} \tag{7.11}$$

we easily get the exciton Pauli-like (anti- and) commutational relations

$$\begin{gathered}
\left[\bar{a}_p, \bar{a}_q\right] = \left[\bar{a}_p, \bar{a}_q^{\dagger}\right] = \left[\bar{a}_p^{\dagger}, \bar{a}_q^{\dagger}\right] = 0,\ p \neq q, \quad \left\{\bar{a}_p, \bar{a}_p\right\} = \left\{\bar{a}_p^{\dagger}, \bar{a}_p^{\dagger}\right\} = 0\,, \\
\left\{\bar{a}_p, \bar{a}_p^{\dagger}\right\} = \begin{cases} 1 & \text{for neutral molecular states,} \\ 0 & \text{for ionized molecular states.} \end{cases}
\end{gathered} \tag{7.12}$$

With only unity on the right-hand side of the last equality, such relations are often called Pauli relations. Thus, the Frenkel excitons are neither fermions nor bosons; working with just neutral molecular states, one could call them paulions instead.

As for other (than in (7.10)) combinations of the electron creation and annihilation operators, one can always rewrite them using identities

$$a_{p1}^\dagger = a_{p1}^\dagger(a_{p0}a_{p0}^\dagger + a_{p0}^\dagger a_{p0}), \qquad a_{p1} = a_{p1}(a_{p0}a_{p0}^\dagger + a_{p0}^\dagger a_{p0}),$$
$$a_{p0}^\dagger = a_{p0}^\dagger(a_{p1}a_{p1}^\dagger + a_{p1}^\dagger a_{p1}), \qquad a_{p0} = a_{p0}(a_{p1}a_{p1}^\dagger + a_{p1}^\dagger a_{p1}). \tag{7.13}$$

Then we can introduce the new electron creation and annihilation operators

$$A_p^\dagger = a_{p1}^\dagger a_{p0}^\dagger a_{p0} \quad \text{and} \quad A_p = a_{p1}a_{p0}^\dagger a_{p0}. \tag{7.14}$$

(Clearly, these operators create and annihilate one electron in the conduction band at site p above an occupied valence spin-orbital at the same molecule). Similarly, we introduce the new hole creation and annihilation operators

$$B_p^\dagger = a_{p0}a_{p1}a_{p1}^\dagger \quad \text{and} \quad B_p = a_{p0}^\dagger a_{p1}a_{p1}^\dagger, \tag{7.15}$$

which create and annihilate a hole below an empty spin-orbital at the same molecule. Worth noticing now is that in (7.13), the remaining terms can be now rewritten as

$$\begin{aligned}
a_{p1}^\dagger a_{p0}a_{p0}^\dagger &= a_{p1}^\dagger a_{p0}a_{p0}^\dagger(a_{p1}a_{p1}^\dagger + a_{p1}^\dagger a_{p1}) = \bar{a}_p^\dagger B_p, \\
a_{p1}a_{p0}a_{p0}^\dagger &= a_{p0}(a_{p1}a_{p1}^\dagger + a_{p1}^\dagger a_{p1})a_{p0}^\dagger a_{p1} = B_p^\dagger \bar{a}_p, \\
a_{p0}^\dagger a_{p1}^\dagger a_{p1} &= -a_{p1}^\dagger(a_{p0}a_{p0}^\dagger + a_{p0}^\dagger a_{p0})a_{p0}^\dagger a_{p1} = -A_p^\dagger \bar{a}_p, \\
a_{p0}a_{p1}^\dagger a_{p1} &= -a_{p1}^\dagger a_{p0}a_{p1}(a_{p0}a_{p0}^\dagger + a_{p0}^\dagger a_{p0}) = -\bar{a}_p^\dagger A_p.
\end{aligned} \tag{7.16}$$

Commutational and anti-commutational relations (in addition to those in the first row of (7.12)) of these operators read

$$\{A_p, A_{p'}\} = \left\{A_p^\dagger, A_{p'}^\dagger\right\} = 0,$$

$$\left\{A_p, A_{p'}^\dagger\right\} = \delta_{p,p'}a_{p0}^\dagger a_{p0} = \delta_{p,p'}(A_p^\dagger \bar{a}_p \bar{a}_p^\dagger A_p + B_p B_p^\dagger),$$

$$\{B_p, B_{p'}\} = \left\{B_p^\dagger, B_{p'}^\dagger\right\} = 0,$$

$$\left\{B_p, B_{p'}^\dagger\right\} = \delta_{p,p'} a_{p0} a_{p0}^\dagger = \delta_{p,p'} (\bar{a}_p^\dagger A_p A_p^\dagger \bar{a}_p + B_p^\dagger B_p) \ ,$$

$$\left[\bar{a}_p, \bar{a}_{p'}^\dagger\right] = \delta_{p,p'} (a_{p0}^\dagger a_{p0} - a_{p1}^\dagger a_{p1})$$

$$= \delta_{p,p'} (A_p^\dagger (\bar{a}_p \bar{a}_p^\dagger - 1) A_p + B_p B_p^\dagger - \bar{a}_p^\dagger B_p B_p^\dagger \bar{a}_p) \ ,$$

$$\{A_p, B_{p'}\} = \left\{A_p^\dagger, B_{p'}^\dagger\right\} = \left\{A_p, B_{p'}^\dagger\right\} = 0 \ ,$$

$$[A_p, \bar{a}_{p'}] = \left[A_p^\dagger, \bar{a}_{p'}^\dagger\right] = 0 \ , \quad \left[A_p, \bar{a}_{p'}^\dagger\right] = -\delta_{p,p'} \bar{a}_p^\dagger A_p \ ,$$

$$[B_p, \bar{a}_{p'}] = \left[B_p^\dagger, \bar{a}_{p'}^\dagger\right] = 0, \quad \left[B_p, \bar{a}_{p'}^\dagger\right] = -\delta_{p,p'} \bar{a}_p^\dagger B_p \ . \tag{7.17}$$

(This implies *inter alia* that $A_p \bar{a}_p^\dagger = B_p \bar{a}_p^\dagger = 0$.) The relations in (7.17) show that the algebra of the new operators is closed. Anyway, the right-hand sides of these relations are standard just for $p \neq p'$. Thus, these new operators can be used as creation and annihilation operators of independent excitations just in connection with different sites. Even with this limitation, however, the physics of the problem becomes more transparent with the use of the above new operators (7.10), (7.14) and (7.15) below.

One can now use the transformation to the new operators of the type (7.16) to transform the Hamiltonian of the electron system (7.8) and its coupling to the bath (7.9). Clearly, plenty of terms result which can be classified according to the new excitations (conduction band electrons, holes or Frenkel excitons) whose operators enter. Thus,

$$H_{\text{e}} = H_{\text{exc}} + H_{\text{el}} + H_{\text{h}} + H_{\text{exc-el}} + H_{\text{exc-h}} + H_{\text{el-h}} + H_{\text{exc-el-h}} \tag{7.18}$$

with, for example, the pure Frenkel exciton Hamiltonian

$$H_{\text{exc}} \approx \sum_p \left[(h^{(0)}_{p1,p1} - h^{(0)}_{p0,p0}) \bar{a}_p^\dagger \bar{a}_p + h^{(0)}_{p1,p0} \bar{a}_p^\dagger + h^{(0)}_{p0,p1} \bar{a}_p \right]$$

$$+ \frac{1}{2} \sum_{p \neq q} \left[v_{p0,q0,p1,q1} \bar{a}_p \bar{a}_q + v_{p1,q1,p0,q0} \bar{a}_p^\dagger \bar{a}_q^\dagger + v_{p0,q1,p1,q0} \bar{a}_q^\dagger \bar{a}_p + v_{p1,q0,p0,q1} \bar{a}_p^\dagger \bar{a}_q \right]$$

$$+ \frac{1}{2} \sum_{p \neq q} \left[v_{p0,q0,p0,q0} + v_{p1,q1,p1,q1} - v_{p0,q1,p0,q1} - v_{p1,q0,p1,q0} \right] \bar{a}_p^\dagger \bar{a}_p \bar{a}_q^\dagger \bar{a}_q$$

$$+ \frac{1}{2} \sum_{p \neq q} \left\{ \left[v_{p0,q1,p1,q1} - v_{p0,q0,p1,q0} \right] \bar{a}_q^\dagger \bar{a}_q \bar{a}_p + \left[v_{p1,q1,p0,q1} - v_{p1,q0,p0,q0} \right] \bar{a}_q^\dagger \bar{a}_q \bar{a}_p^\dagger \right.$$

$$\left. + \left[v_{p1,q0,p1,q1} - v_{p0,q0,p0,q1} \right] \bar{a}_p^\dagger \bar{a}_p \bar{a}_q + \left[v_{p1,q1,p1,q0} - v_{p0,q1,p0,q0} \right] \bar{a}_p^\dagger \bar{a}_p \bar{a}_q^\dagger \right\}$$

$$+\sum_{p}\sum_{l(\neq p)}\sum_{\iota}[v_{p1,l\iota,p1,l\iota}-v_{p0,l\iota,p0,l\iota}]\,\bar{a}_p^\dagger\bar{a}_p+\sum_{p}\sum_{\iota(\neq 0)}[v_{p1,p\iota,p1,p\iota}$$

$$-v_{p1,p\iota,p\iota,p1}-v_{p0,p\iota,p0,p\iota}+v_{p0,p\iota,p\iota,p0}]\,\bar{a}_p^\dagger\bar{a}_p+\sum_{p}\sum_{\iota}\left\{\left[\sum_{l}v_{p1,l\iota,p0,l\iota}\right.\right.$$

$$\left.\left.-v_{p1,p\iota,p\iota,p0}\right]\bar{a}_p^\dagger+\left[\sum_{l}v_{p0,l\iota,p1,l\iota}-v_{p0,p\iota,p\iota,p1}\right]\bar{a}_p\right\}. \tag{7.19}$$

(Here, double-overlap integrals like $v_{p0,q0,q1,p1}, p \neq q$, have been omitted assuming, as typical of OMC, that spin-orbitals on neighboring molecules practically do not overlap.) Similar results are found for other terms in (7.18). In (7.19), we have also included all the relevant terms which result from the electron–electron coupling $v(r)$ in (7.8). The reason is that some of them will be important below. In contrast, for other terms in (7.18), we shall only indicate the presence of (less important) contributions which come from $v(r)$ in (7.8). So, the Hamiltonian of the conduction band electrons reads

$$H_{\mathrm{el}}=\sum_{p,q}h^{(0)}_{p1,q1}A_p^\dagger A_q+\sum v_{\dots,\dots}, \tag{7.20}$$

and the hole Hamiltonian is

$$H_{\mathrm{h}}=-\sum_{p,q}h^{(0)}_{p0,q0}B_q^\dagger B_p+\sum v_{\dots,\dots}. \tag{7.21}$$

Further terms already describe a coupling between (or among) the excitations. So

$$H_{\mathrm{exc-el}}=\sum_{p\neq q}\left[h^{(0)}_{p0,q0}\bar{a}_q^\dagger A_p^\dagger A_q\bar{a}_p-h^{(0)}_{p1,q0}\bar{a}_q^\dagger A_p^\dagger A_q-h^{(0)}_{p0,q1}A_p^\dagger A_q\bar{a}_p\right]+\sum v_{\dots,\dots}, \tag{7.22}$$

$$H_{\mathrm{exc-h}}=\sum_{p\neq q}\left[-h^{(0)}_{p1,q1}\bar{a}_p^\dagger B_q^\dagger B_p\bar{a}_q-h^{(0)}_{p1,q0}\bar{a}_p^\dagger B_q^\dagger B_p-h^{(0)}_{p0,q1}B_q^\dagger B_p\bar{a}_q\right]+\sum v_{\dots,\dots}, \tag{7.23}$$

$$H_{\mathrm{el-h}}=\sum_{p\neq q}\left[h^{(0)}_{p1,q0}A_p^\dagger B_q^\dagger+h^{(0)}_{p0,q1}B_pA_q\right]+\sum v^{(0)}_{\dots}\dots, \tag{7.24}$$

$$H_{\mathrm{exc-el-h}}=\sum_{p\neq q}\left[-h^{(0)}_{p0,q0}\bar{a}_q^\dagger B_pA_q-h^{(0)}_{p0,q0}A_p^\dagger B_q^\dagger\bar{a}_p+h^{(0)}_{p1,q1}\bar{a}_p^\dagger B_pA_q\right.$$
$$\left.+h^{(0)}_{p1,q1}A_p^\dagger B_q^\dagger\bar{a}_q-h^{(0)}_{p1,q0}\bar{a}_p^\dagger\bar{a}_q^\dagger B_pA_q-h^{(0)}_{p0,q1}A_p^\dagger B_q^\dagger\bar{a}_p\bar{a}_q\right]+\sum v_{\dots,\dots}. \tag{7.25}$$

One should realize in detail the physical meaning of the individual terms above. For instance, the first term on the right-hand side of (7.22) means, in the old language, an electron transfer from the spin-orbital $q0$ to the spin orbital $p0$ provided that both $p1$- and $q1$-spin-orbitals are occupied. Similar transformation should also be performed for the individual terms in the electron coupling to vibrations (phonons) (7.9). We shall, however, avoid this point now as it is unimportant for our purposes here and return to it later in connection with the notion of polaron. Let us only stress that our spin-orbitals are defined for a static (rigid) configuration of the atoms. Hence, the matrix elements of $v(r)$ are independent of the atom displacements, i.e., in the electron coupling to phonons (7.9), there are only terms resulting from the single-particle part of the electron Hamiltonian $h(r)$. As a consequence, within our formulation, starting from the crude (rigid, diabatic) basis, there is no chance to obtain, for example, a site off-diagonal exciton–phonon coupling (the term $\mathcal{N}\left[a^\dagger_{p\pi}a_{q\kappa}\right]$ in (7.9) does not yield, for $p \neq q$, terms $\sim \bar{a}^\dagger_p\bar{a}_q$ upon using the above transformation to the new operators (7.10), (7.14) and (7.15)). It does not mean that there are no phonon-assisted transitions of real Frenkel-exciton-like excitations owing to the coupling to the phonon system (7.9) in nature. It only means that

- within our picture, the excitons, electrons, holes etc., are, in general somewhat different excitations than those within formalism of the adiabatic (moving) basis, where the spin-orbitals depend on the atomic configuration [2], and
- in both the adiabatic and the rigid bases (used here), the real excitations are not merely excitons (conduction-band electrons or holes) but rather more complex objects.

7.3 Eigen- or Quasiparticle States

Looking at the structure of the Hamiltonian, we can see that it is a very complicated form of a general many-body problem. Although we have introduced our conduction-band electrons and holes and the Frenkel excitons as excitations above the ground state, it is technically impossible to try to find real eigenstates of (7.6) analytically. As an illustration, even the "vacuum" state $|\Psi_0\rangle$ without any (conduction-band) electrons, holes, excitons or phonons, defined by

$$A_p|\Psi_0\rangle = B_p|\Psi_0\rangle = \bar{a}_p|\Psi_0\rangle = b_r|\Psi_0\rangle = 0 \qquad (7.26)$$

for any site p and any r, is no eigenstate of the total Hamiltonian (7.6). The main reason being the terms in the Hamiltonian which do not preserve the number of elementary excitations and create them even in the "vacuum" state $|\Psi_0\rangle$. We shall see their importance and some physical implications below. Here, let us just add a few words about implications of the above complexity we shall discuss in more detail further on.

Our aim is to describe the electronic states in OMC. Owing to the many-body character of the problem, we shall never be able to find real eigenstates of the above Hamiltonian of OMC. Fortunately, posing any question connected to a physical observable (quantum-statistical mean values of operators) which could be quantified by experiment, we could, in principle, work from any basis in the corresponding mathematical Hilbert space. The reason for this being that quantum-statistical mean values reduce to a trace operation,

$$\langle \ldots \rangle(t) = \mathrm{Tr}(\rho(t) \ldots) \; , \tag{7.27}$$

where $\rho(t)$ is the density matrix of the system under investigation. As the trace $\mathrm{Tr}(\ldots)$ is independent of the basis used, one could use, in principle, any. That is why, for example, theories of the energy (charge etc.) transport coexist that work with polarons as well as bare particles, i.e., seemingly fully contradictory pictures [2]. On the other hand, for a physical understanding of the processes in OMC, and for a justification of the simplifying models or approximations, the question of proper (electron as well as vibrational) states is fully legitimate and unavoidable. Here, as usual in the many-body problems, we shall speak about states which behave for a finite (but sufficiently long) interval of time as if they were real eigenstates. These so-called quasiparticle states, may, by their structure, appreciably differ from (to us, for technical reasons, unaccessible) real eigenstates. On the other hand, they yield a very important physical insight. For our purposes here, we shall start with rather special model situations.

7.4 Excitons

One usually understands excitons as quasiparticles resulting from absorption of light quanta in condensed matter. Instead, we shall start here by showing what their role is in the formation problem of OMC when no light quanta enter.

As implicitly indicated above, the exciton-non-conserving terms, such as those on the right-hand side of (7.19), do not necessarily mean finite lifetime effects for excitons. In fact, as we are going to show now, they simply imply that the number of the Frenkel excitons (as defined above) may fluctuate, causing, for instance, the van der Waals coupling among the molecules forming OMC that keeps the OMC together.

For simplicity, we shall first try to approximate the ground state of OMC by assuming no conduction-band electrons, no holes or phonons, and at most a slightly fluctuating mean number of excitons. In connection with that, we omit all but the exciton part of the Hamiltonian. In addition, we shall for simplicity keep just the most important term, i.e., the second term on the right-hand side of (7.19). Within the four matrix elements in this term, we can then perform the standard multipole expansion of the electrostatics. Keeping, for our qualitative purposes, just the first non-zero term, we get

$$H_{\text{exc}} \approx \frac{1}{2} \sum_{p \neq q} \left[\frac{\hat{\boldsymbol{d}}_p \hat{\boldsymbol{d}}_q}{|\boldsymbol{R}_{pq}|^3} - 3 \frac{(\hat{\boldsymbol{d}}_p \boldsymbol{R}_{pq}) \cdot (\hat{\boldsymbol{d}}_q \boldsymbol{R}_{pq})}{|\boldsymbol{R}_{pq}|^5} \right] , \tag{7.28}$$

where

$$\hat{\boldsymbol{d}}_p = \langle p0|\boldsymbol{d}|p1\rangle \bar{a}_p + \langle p1|\boldsymbol{d}|p0\rangle \bar{a}_p^\dagger \tag{7.29}$$

and $\boldsymbol{R}_{pq} = \boldsymbol{R}_p - \boldsymbol{R}_q$ is the relative position vector of, e.g., the centres of gravity of the molecules p and q. In (7.29), $\boldsymbol{d} = e\boldsymbol{r}$, where $\boldsymbol{r}$ is the electronic coordinate over which the integration is performed in the matrix elements in question. Thus, $\hat{\boldsymbol{d}}_p$ designates the (electronic or total) dipole momentum operator of the molecule p. Notice that $\langle p\pi|\boldsymbol{d}_p|p\pi\rangle = 0$, as we have assumed our molecules to be non-polar. Thus, for example, also

$$\langle \Psi_0|\hat{\boldsymbol{d}}_p|\Psi_0\rangle = 0 \tag{7.30}$$

(see (7.26)). Mean values of $\hat{\boldsymbol{d}}_p$ in states with uncertain (fluctuating) values of the Frenkel excitons may, however, become non-zero. This is the way which can be used, as we shall show below, to describe polarized states of the molecules in OMC.

Clearly, the exciton-non-conserving form of (7.28) represents the dipole-dipole coupling among the molecules. Assuming the zero-order wavefunction of the ground state as a trial function of the variational principle in form of $|\Psi_0\rangle$ from (7.26), no binding occurs. This results from (7.30), which implies that also the mean value of (7.28) in this state is zero. Only in the next (second) order of the perturbation theory (or allowing fluctuations of the number of excitons in the trial wavefunction) does the dipole–dipole coupling yield the standard (and always attractive) van der Waals coupling of the molecules in OMC [2].

Let us now turn our attention to so-called single-exciton states which describe the situation after excitation of a single electron from the (formerly occupied) spin-orbital $p0$ to the (so far unoccupied) orbital $p1$. (The spin orientations of $p0$ and $p1$ may easily be the same after absorption of a photon. The resulting exciton is then called a singlet, in contrast to the triplet excitons, where the spin orientation changes upon excitation.) Limiting our attention to, so-called monoexcited states $|\Psi_p\rangle = \bar{a}_p^\dagger|\Psi_0\rangle$ (state of OMC with a single exciton at site p), we obtain from (7.19)

$$\langle \Psi_p|H_{\text{e}}|\Psi_q\rangle = \delta_{pq} E_p^{\text{exc}} + (1 - \delta_{pq}) M_{pq} ,$$

$$E_p^{\rm exc} \approx h^{(0)}_{p1,p1} - h^{(0)}_{p0,p0} + \sum_{\iota(\neq 0)} \left[v_{p1,p\iota,p1,p\iota} - v_{p1,p\iota,p\iota,p1} - v_{p0,p\iota,p0,p\iota} \right.$$
$$\left. + v_{p0,p\iota,p\iota,p0} \right] + \sum_{l(\neq p)} \sum_{\iota} \left[v_{p1,l\iota,p1,l\iota} - v_{p0,l\iota,p0,l\iota} \right],$$
$$M_{pq} = v_{p1,q0,p0,q1} \, . \tag{7.31}$$

It is worth noticing here that the exciton-non-conservation terms in (7.19) do not contribute. In order to describe their influence, the set of the mono-excited states $|\Psi_p\rangle$ would have to be extended to also include states with a different (including zero) number of the excited molecules. In connection with the dominant character of the single-molecular excitation energies as compared to other combinations of the matrix elements, the effect of the exciton-non-conserving terms (i.e. the exciton number fluctuation on the single-excited states) would here be negligible. Let us add that the exciton-non-conserving terms from (7.19) could also be removed by a canonical transformation.

Worth noting is the structure of $E_p^{\rm exc}$ in (7.31). The difference $h^{(0)}_{p1,p1} - h^{(0)}_{p0,p0}$ describes the excitation energy of the molecule p in the gas phase, corrected, however, by inclusion of the Coulomb interaction of the electron of the molecule p (raised from the spin-orbital $p0$ to $p1$) with nuclei of other molecules. This influence is incorporated into $h(r)$ above and can be of course different on the $p1$ and $p0$ spin-orbitals. The term

$$\sum_{\iota(\neq 0)} \left[v_{p1,p\iota,p1,p\iota} - v_{p1,p\iota,p\iota,p1} - v_{p0,p\iota,p0,p\iota} + v_{p0,p\iota,p\iota,p0} \right]$$

represents the difference of the (direct as well as exchange) interactions of the raised electron with other electrons on the same molecule in the spin-orbitals $p1$ and $p0$. The difference between the interactions of the electron in these two spin-orbitals with electrons of other molecules is represented by the last term in $E_p^{\rm exc}$. All these terms, although having clear physical meanings, are insufficient in determining the trial wavefunctions and energies due to the mixing term $M_{pq} = v_{p1,q0,p0,q1}$. This term describes the excitation transfer between two molecules which is not accompanied by the corresponding charge transfer. Multipole expansion in the matrix element $v_{p1,q0,p0,q1}$ again reveals a whole series of contributions starting from the one which is due to the dipole–dipole coupling between the molecules p and q. This element is the basis of the standard Förster theory of the excitation transfer [1–3]. As we are only interested in finding eigenstates of (7.19) (notice that we fully ignore, for example, the role of phonons), the matrix (7.31) should be diagonalized. In a general molecular condensate, this is a task solvable only numerically. In periodic OMC, one can use the periodic boundary conditions to solve the problem. As we only have one molecule per elementary cell in standard solid-state physics [4, 5], the index p designates both the molecule and the cell. One can then introduce the wavevectors $\boldsymbol{k}$ from the first Brillouin zone. The eigenstates then read

$$|\Psi_{\boldsymbol{k}}\rangle = \frac{1}{\sqrt{N}} \sum_p \mathrm{e}^{i\boldsymbol{k}\boldsymbol{R}_p^{(0)}} |\Psi_p\rangle \ , \tag{7.32}$$

and the corresponding exciton excitation energies are

$$\epsilon_{\boldsymbol{k}}^{\text{exc}} = \sum_q M_{pq} \exp[-i\boldsymbol{k}(\boldsymbol{R}_p^{(0)} - \boldsymbol{R}_q^{(0)})] \ . \tag{7.33}$$

Hence, the possibility that the exciton will be transferred in the lattice leads to the dependence of the exciton energy on $\boldsymbol{k}$, i.e. to the formation of the exciton band. Direct optical absorption without, for example, phonon participation is only allowed to the singlet excitonic state with $\boldsymbol{k} = 0$ (standing exciton). The optical absorption band due to molecular excitations to just one excited spin-orbital is therefore much narrower than that of the exciton energies. Inclusion of phonons would, however, appreciably broaden the optical absorption band [1]. In the case where we have M molecules per elementary cell, invoking symmetry reduces the problem to a secular equation of the order of M only. Hence, M exciton bands and (in general, up to) M optical absorption bands result as a consequence of the photon absorption and a singlet exciton creation in OMC (Davydov doublet for $M = 2$). We shall return to the role of excitons in connection with the so-called Toyozawa polarons.

7.5 Electron and Hole States in OMC

In this section, we will refer only to conduction-band electrons, as the same applies for holes. Again, let us ignore phonons and start first from the single-electron state $|\Phi_p\rangle = A_p^\dagger|\Psi_0\rangle$ (a conduction-band electron sitting at molecule p, converting it thus to a negative molecular ion). Taking just the first (dominant) term in (7.20) and omitting (for our qualitative purposes) the remaining contributions (indicated as $\sum v_{\dots\dots}$ in (7.20)) either renormalizing the first term or contributing to more-electron processes, we get

$$\langle\Phi_p|H_{\text{el}}|\Phi_q\rangle = h_{p1,q1}^{(0)} \ . \tag{7.34}$$

In periodic OMC, this matrix can be diagonalized utilizing periodic boundary conditions. For instance, if we have just one molecule per elementary cell, the resulting wavefunction is characterized by a electron wavevector $\boldsymbol{k}$ from the first Brillouin zone and reads

$$|\Phi_{\boldsymbol{k}}\rangle = \frac{1}{\sqrt{N}} \sum_p \mathrm{e}^{i\boldsymbol{k}\boldsymbol{R}_p^{(0)}} |\Phi_p\rangle \ , \tag{7.35}$$

while the band energy reads

$$\epsilon_{\boldsymbol{k}}^{\rm el} = \sum_q h_{p1,q1}^{(0)} \exp[-i\boldsymbol{k}(\boldsymbol{R}_p^{(0)} - \boldsymbol{R}_q^{(0)})] \; . \tag{7.36}$$

(In the case with M molecules in one elementary cell, we simply get M bands of such plane-wave states after diagonalization of the corresponding $M \times M$ secular problem.) As with the exciton state (7.32), (7.35) describes the state where the electron resides on one molecule only, but this may be an arbitrary one in the OMC in question.

This approach provides the basis of the so-called band theory of OMC which now is only of historical significance [2–6]. The point is that the bandwidth given by the $\boldsymbol{k}$ dispersion of $\epsilon_{\boldsymbol{k}}^{\rm el}$ is much less than corrections (shifts, deformation and level broadening) due to the interaction of the conduction-band electron (or, correspondingly, hole) with other quasiparticles (excitons, phonons). In other words, before the electron can jump to a neighboring molecule (as a consequence of integrals $h_{p1,q1}^{(0)}$, $p \neq q$ in (7.20) or their values renormalized by, for example, the Coulomb terms), it will certainly undergo a scattering by excitons, phonons, etc., omitted so far. Thus, all the dispersion of $\epsilon_{\boldsymbol{k}}^{\rm el}$ is just academic and has very limited practical meaning.

Let us therefore include the excitons already partly investigated above into the description. Thus, our minimal Hamiltonian now combines the conduction-band electron Hamiltonian (7.20), the exciton Hamiltonian (7.28), and that of the conduction-band electron versus exciton coupling (7.22). As the theory is based on the above observation regarding the unimportant role of the $h_{p1,q1}^{(0)}$, $(p \neq q)$ integrals, we keep just the terms which are of zero-order in the overlap S of neighboring spin-orbitals. After that, a great many of the Coulomb-interaction-induced terms $\sum v_{\dots\dots}$ in (7.22) (and, similarly, (7.23) if the hole state is investigated) disappear. The remaining and, for our problem, relevant terms from (7.20), (7.28), and (7.22) read

$$\begin{aligned} H_{\rm el-pol} \approx{}& \sum_p E_p^{\rm el} A_p^\dagger A_p + \sum_p E_p^{\rm exc} \bar{a}_p^\dagger \bar{a}_p + \frac{1}{2}\sum_{p\neq q} [(v_{p1,q1,p1,q1} \\ &-v_{p0,q1,p0,q1})\bar{a}_p^\dagger \bar{a}_p A_q^\dagger A_q + (v_{p1,q1,p1,q1} - v_{p1,q0,p1,q0}) A_p^\dagger A_p \bar{a}_q^\dagger \bar{a}_q] \\ &+ \frac{1}{2}\sum_{p\neq q} [(v_{p0,q1,p1,q1}\bar{a}_p + v_{p1,q1,p0,q1}\bar{a}_p^\dagger) A_q^\dagger A_q + (v_{p1,q0,p1,q1}\bar{a}_q \\ &+ v_{p1,q1,p1,q0}\bar{a}_q^\dagger) A_p^\dagger A_p] + \frac{1}{2}\sum_{p\neq q}\left[\frac{\hat{\boldsymbol{d}}_p \hat{\boldsymbol{d}}_q}{|\boldsymbol{R}_{pq}|^3} - 3\frac{(\hat{\boldsymbol{d}}_p \boldsymbol{R}_{pq})\cdot(\hat{\boldsymbol{d}}_q \boldsymbol{R}_{pq})}{|\boldsymbol{R}_{pq}|^5}\right] . \end{aligned} \tag{7.37}$$

Here, in an analogy to $E_p^{\rm exc}$ from (7.31), $E_p^{\rm el}$ designates the conduction-band electron energy at site p and includes $h_{p1,p1}^{(0)}$ as well as the corresponding renormalization terms stemming (in the zero-order in S) from the $\sum v_{\dots\dots}\dots$ term in (7.20).

In the next step, we perform yet another simplification. Notice that the problem we tackle now is that of a standing electron surrounded by a cloud of virtual excitons. This exciton cloud describes nothing but a field of neutral molecules (around the molecular ion) which are polarized by the electron charge in the centre. This is sometimes called the electron-polaron problem, in connection with the origin of the polarization around the central molecular ion, i.e. polarization (due to the shift with respect to the molecular skeletons) of the electronic (mainly π) spin-orbitals of the neutral molecules in the field of the molecular ion. This problem was investigated first by Toyozawa [7]. Possible electron energy shifts thus obtained (as compared to (7.36)) may even exceed 1 eV. These surprisingly high values are, however, mainly due to cooperation effects of a relatively large number of neutral molecules. Thus, the mean number of excitons created by the electron charge at any neighboring molecule (i.e. the induced polarization of the molecule) or, in other words, the probability of finding a neutral molecule in the vicinity of the electron excited by its Coulomb field remains relatively small. The third sum on the right-hand side of (7.37) describes a shift of the electron energy induced by the excitation of (thus polarized) neighboring neutral molecules. So, this term should, for each neutral molecule, yield only a small effect. Another interpretation is that it represents shifts of the excitation energies of neutral molecules caused by the Coulomb field of the molecular ion on which the electron resides. Hence, it should only slightly influence the polarizability of the neutral molecules around the molecular ion. For our purposes, we shall disregard this term here.

As for the fourth sum on the right-hand side of (7.37), it represents a very important contribution. Expanding the argument (Coulomb interaction) in the matrix elements $v_{\dots}$ in this term, we find that the first term comes from the charge–dipole interaction. Hence, limiting it to just this term, (7.37) approximately reads

$$H_{\text{el-pol}} \approx \sum_p E_p^{\text{el}} A_p^\dagger A_p + \sum_p E_p^{\text{exc}} \bar{a}_p^\dagger \bar{a}_p + e \sum_{p \neq q} \frac{\hat{\boldsymbol{d}}_q \boldsymbol{R}_{pq}}{|\boldsymbol{R}_{pq}|^3} A_p^\dagger A_p + \frac{1}{2} \sum_{p \neq q} \left[\frac{\hat{\boldsymbol{d}}_p \hat{\boldsymbol{d}}_q}{|\boldsymbol{R}_{pq}|^3} - 3 \frac{(\hat{\boldsymbol{d}}_p \boldsymbol{R}_{pq}) \cdot (\hat{\boldsymbol{d}}_q \boldsymbol{R}_{pq})}{|\boldsymbol{R}_{pq}|^5} \right] . \tag{7.38}$$

Here $e < 0$ is the electron charge. In case of the electron-polaron problem around a hole (positive charge), only the sign of the third term (apart from a different notation) in (7.38) would be changed.

The problem (7.38) has a classical analogy in the problem of finding the energy minimum of a set of classical induced dipoles (replacing neutral molecules) with a prescribed polarizability, interacting mutually via the dipole–dipole coupling and also with a central point charge $\pm e$ representing the molecular ion. Various corrections due to higher multipoles, the non-local character of real molecules, etc., can easily be incorporated. All these classical microelectrostatic problems in fact result from a variational estimate of the energy of our quantum problem (7.38) or its contingent generalizations where quantum fluctuations of the polarization (i.e., of the number of virtual excitons at individual neutral molecules) as well as correlations among induced dipoles replacing the neutral molecules are neglected [2,6,8]. Quantum corrections to this variational estimate including the possible effect of the fluctuating polarization of the central molecular ion have also been performed [9], and they confirmed that the microelectrostatic approach is excellent from the point of view of the quantum theory. Minimum energy $E_{p,\mathrm{min}}$ of the microelectrostatic problem (or the mean energy within the quantum variational problem) is always less than the corresponding value of E_p^{el}, where p is the molecular ion (charge always polarizes its surroundings so as to form a potential well in which it finally resides). Introducing the electron (hole) polarization energy (i.e., the energy of the system around the central electron or hole) as $P_{\mathrm{e,h}} = E_p - E_{p,\mathrm{min}}$, we may obtain, in real OMC, values exceeding even 1 eV [2, 6]. As the polarizations $P_{\mathrm{e,h}}$ can easily be deduced from experiments, many reliable computational methods have been developed to determine them theoretically. For a survey of these see, for example, [2, 6]. Let us note only that unless higher (than dipole) multipoles are included, the theory always yields $P_{\mathrm{e}} = P_{\mathrm{h}}$. Finally, let us be reminded that such electron-polarization effects as discussed here could, in principle, be expected in other solids as well. In OMC, however, they become extraordinarily important for a combination of the following reasons:

- relatively high polarizability of, in particular, π-electron shells of the molecules,
- very small mobility of the charge as a reason for which the electron polarization has time enough to organize around the central charge before the latter leaves for another place in the lattice,
- practically no screening by free carriers, i.e., almost unscreened long-range Coulomb coupling among charges in OMC.

7.6 Charge-Transfer States

By analogy with the above single- (conduction-band-) electron or single-hole states, states of the electron–hole pairs can easily be formed provided that the electron and hole stay far enough apart for both their Coulomb interaction and the overlap of their electron-polarization clouds to become unimportant. The situation drastically changes when these conditions are not fulfilled.

In standard semiconductors, Mott–Wannier (large radius) and intermediate excitons appear as bound states of electron–hole pairs. In OMC, such states also appear. However, the theory is complicated because the electron polarization is shared by both of the particles in the pair and because there is a very small particle mobility. So, almost “frozen” counterparts of the large and intermediate excitons appear in OMC. These states are, however, called charge-transfer states here. With an approximation analogous to the one which gave the electron-polaron Hamiltonian (7.38), i.e., first of all in the zero-order approximation in S (immobile charge), we obtain

$$H_{\mathrm{CT}} \approx \sum_p E_p^{\mathrm{el}} A_p^\dagger A_p + \sum_p E_p^{\mathrm{ho}} B_p^\dagger B_p + \sum_p E_p^{\mathrm{exc}} \bar{a}_p^\dagger \bar{a}_p + e \sum_{p \neq q} \frac{\hat{\boldsymbol{d}}_q \boldsymbol{R}_{pq}}{|\boldsymbol{R}_{pq}|^3} A_p^\dagger A_p$$
$$- e \sum_{p \neq q} \frac{\hat{\boldsymbol{d}}_q \boldsymbol{R}_{pq}}{|\boldsymbol{R}_{pq}|^3} B_p^\dagger B_p + \frac{1}{2} \sum_{p \neq q} \left[\frac{\hat{\boldsymbol{d}}_p \hat{\boldsymbol{d}}_q}{|\boldsymbol{R}_{pq}|^3} - 3 \frac{(\hat{\boldsymbol{d}}_p \boldsymbol{R}_{pq}) \cdot (\hat{\boldsymbol{d}}_q \boldsymbol{R}_{pq})}{|\boldsymbol{R}_{pq}|^5} \right] \tag{7.39}$$

($e < 0$). Here E_p^{ho} is the hole site energy $\approx -h_{p0,p0}^{(0)}$ corrected by inclusion of the Coulomb terms stemming from $\sum v_{\dots} \dots$ in (7.21) for $S = 0$ (compare with the electronic site energy E_p^{el} above). A word of warning, however: Our model used (for the sake of simplicity) here includes just one conduction and one valence spin-orbital. Hence, once we have a conduction-band electron or hole at a molecule (i.e., once both these spin-orbitals are to be occupied or empty, respectively), no exciton can be found at the same site. (This would require, in our simple picture, the $p1$ spin-orbital to be occupied while the $p0$ spin-orbital would have to remain empty.) Hence, the molecular dipole momentum of the molecular ion can neither fluctuate nor become on average non-zero. In other words, we have no possibility to include effects of polarization of the molecular ions themselves. While this problem is most likely unimportant for the electron-polaron problem, around a single molecular ion, it could become important for the above charge-transfer states, where the additional electron and hole are located at neighboring molecules, i.e., the molecular ions themselves could become appreciably polarized. No quantum or classical (microelectrostatic) final study of this effect (polarization of the molecular ions) seems to exist so far. What can be expected on the grounds of the existing microelectrostatic studies [2, 6] mostly ignoring the molecular-ion (static as well as dynamic) polarization is that

- the polarization around a bound charge-transfer (electron–hole pair) state is not a simple sum of the polarization energies of the two isolated components (rather $P_{\mathrm{e-h}} \equiv P_{\mathrm{CT}} \lesssim P_{\mathrm{e}} + P_{\mathrm{h}}$ could be expected), and
- the charge-transfer state polarization energy P_{CT} can be quite important for energy position of the state, i.e., for all effects where these states might play an important role (e.g., charge-carrier recombination).

7.7 Influence of Phonons

So far, we have completely ignored the existence of phonons of any type (intramolecular, lattice, etc.), which, in principle, can influence the electronic states appreciably. For us, the phonons enter around the electronic states as a thermodynamic bath. This bath, however, can play not only its standard passive role (as, for example, a source of thermal energy), but also may be even quite active and important as a responsive medium appreciably, changing its local state in the vicinity of the carrier and, because of that, also properties of the carrier itself (polaron shifts, width of the polaron bands, etc.). As the problem is analogous for electrons, holes, and partly also for excitons, we shall for the simplest model discussion take just one electron interacting with phonons and shall forget for a while about its coupling to other excitations (certainly existing in real OMC). With regard to the coupling to phonons, one should realize the different roles of the individual terms (different i) in (7.9). One usually takes just the term with $i = 1$, proportional to the first powers of the atomic displacements $\boldsymbol{R}_j - \boldsymbol{R}_j^{(0)}$. This term is likely to play the most decisive role in many situations. We shall also limit our discussion to this term only. However, it should be made clear that, for example, symmetry arguments may imply that this term should sometimes be zero (out-of-plane intramolecular vibrations in plane-like molecules). Nonetheless, practically nothing is known about electronic states in OMC when higher-order terms (e.g., quadratic terms with $i = 2$) start to become decisive.

The term with $i = 1$ in (7.9) is also linear in the phonon creation and annihilation operators (as the atomic displacements themselves are linear). Hence, we shall take as the simplest model here

$$
\begin{aligned}
H = &\sum_p E_p^{\mathrm{el}} A_p^\dagger A_p + \sum_{p\neq q} J_{pq} A_p^\dagger A_q \\
&+ \sum_{\boldsymbol{k}} \hbar\omega_{\boldsymbol{k}} b_{\boldsymbol{k}}^\dagger b_{\boldsymbol{k}} + \frac{1}{\sqrt{N}} \sum_{\boldsymbol{k}} \sum_p G_{\boldsymbol{k}}^p \hbar\omega_{\boldsymbol{k}} A_p^\dagger A_p (b_{\boldsymbol{k}} + b_{-\boldsymbol{k}}^\dagger) \, .
\end{aligned} \tag{7.40}
$$

Here, apart from renormalizations owing to the Coulomb terms $v_{\dots}, \dots$, one can see that, for example, $J_{pq} \approx h_{p1,q1}^{(0)}$ $(p \neq q)$ in (7.20). The phonons are taken as usual harmonic and plane-wave-like, with $\boldsymbol{k}$ designating the wavevector. Only one phonon branch is considered for simplicity, and the coupling to phonons is assumed to be site-diagonal with the coupling constants $G_{\boldsymbol{k}}^p = (G_{-\boldsymbol{k}}^p)^*$. (The site off-diagonal coupling is already of the first order in the small parameter S of the overlap between spin-orbitals on neighboring sites.)

Limiting the situation first to just the lowest (zero) order in S, we first set J_{pq}=0, after which (7.40) can be diagonalized. The corresponding single-electron eigenstates are localized small-polaron states[1] characterized by the polaron position (site) p and by set $\{\nu_{\boldsymbol{k}}\}$ of the phonon occupation numbers

---

[1] One should mention that also large and intermediate polarons exist [5]. The term small here means that real values of the hopping (resonance) integrals J_{pq} in OMC remain much less than, for example, the polaron energy shifts – see below or [5] for details.

$\nu_{\boldsymbol{k}} = 0, 1, 2, \ldots$. Explicitly, they read

$$|p, \{\nu_{\boldsymbol{k}}\}\rangle = \prod_{\boldsymbol{k}} \left[\frac{1}{\sqrt{\nu_{\boldsymbol{k}}!}} \left(b^{\dagger}_{\boldsymbol{k}} + \frac{1}{\sqrt{N}} G^{p}_{\boldsymbol{k}} \right)^{\nu_{\boldsymbol{k}}} \right] \exp \left(\frac{1}{\sqrt{N}} \sum_{\boldsymbol{k}} G^{p}_{\boldsymbol{k}} (b_{\boldsymbol{k}} - b^{\dagger}_{-\boldsymbol{k}}) \right) |\Phi_p\rangle \,, \tag{7.41}$$

with the corresponding energies

$$E^{\text{small-polaron}}_{p,\{\nu_{\boldsymbol{k}}\}} = E^{\text{el}}_p - \frac{1}{N} \sum_{\boldsymbol{k}} \hbar\omega_{\boldsymbol{k}} |G^{p}_{\boldsymbol{k}}|^2 + \sum_{\boldsymbol{k}} \hbar\omega_{\boldsymbol{k}} \nu_{\boldsymbol{k}} \,. \tag{7.42}$$

Some similarities may be observed with the above electron-polaron problem resulting from the interaction of the charge carriers with the quantum field of excitons (instead of phonons like here). The term $-\frac{1}{N}\sum_{\boldsymbol{k}} \hbar\omega_{\boldsymbol{k}} |G^{p}_{\boldsymbol{k}}|^2 \equiv -E_{\text{b}} < 0$ in (7.42) is the so-called small-polaron shift and is reminiscent of the electron or hole polarization energy shift to lower energies $-P_{\text{e,h}} < 0$ in the electron polaron problem. The term E^{el}_p in (7.42) has the same meaning (energy of the bare particle in the centre of the polaron) as the corresponding term in, e.g., (7.38). The electron-polaron above is made of a cloud of the virtual excitons (i.e. polarization of electron shells of the neighboring molecules) while the present small polaron (or small phonon-polaron so as to distinguish it from the electron one) is made of the cloud of virtual phonons describing atom shifts (i.e., shifts of the molecules as well as their deformations) in the vicinity of the central charge (conduction-band electron located at site p in (7.41)). There are only two differences as compared with the electron-polaron above:

- There is an additional term, $\sum_{\boldsymbol{k}} \hbar\omega_{\boldsymbol{k}} \nu_{\boldsymbol{k}}$, in (7.42) expressing energy of free phonons in the OMC already deformed in the vicinity of the (phonon) small polaron (notice that $b^{\dagger}_{\boldsymbol{k}} + \frac{1}{\sqrt{N}} G^{p}_{\boldsymbol{k}}$ in (7.41) now plays the role of the phonon creation operator in the lattice deformed by the small polaron at site p). In the electron-polaron problem, we had no such term as we did not add real excitons to the virtual exciton cloud around the molecular ion.
- In the electron-polaron problem, we took just terms of zero order in S into account, i.e., we took the polaron (also its central charge) as immobile. For that, there are good reasons given by a hierarchy of times typical for different processes in OMC [2, 6] (see also below). Moreover, formal inclusion of the possibility that the central charge can move provides technically a very difficult problem. In contrast, with our (phonon) small-polaron problem, the physical situation is (owing to much slower lattice dynamics) less favorable to consider just immobile carriers. Also, technically, the problem of including the carrier ability to move is less difficult. That is why we are now going to include the (so far ignored) term $\sum_{p \neq q} J_{pq} A^{\dagger}_p A_q$ in (7.40).

One should realize that in the polaron problem, we have at least three main parameters for the dimension of energy which define different regimes of the problem:

- The typical value J of $|J_{mn}|$ entering also the band width of the bare particle from which the polaron results,
- the typical phonon energy $\hbar\omega_0$ of the phonon branch in question, and
- the polaron shift (binding energy) E_b appearing physically; however, only if the bare particle (from which the polaron results) is fully immobile (see above).

Thus, we have at least the following six limiting regimes to be at least mentioned here: (a) $J \gg E_b \gg \hbar\omega_0$, (b) $J \gg \hbar\omega_0 \gg E_b$, (c) $\hbar\omega_0 \gg J \gg E_b$, (d) $E_b \gg J \gg \hbar\omega_0$, (e) $E_b \gg \hbar\omega_0 \gg J$, and (f) $\hbar\omega_0 \gg E_b \gg J$. Regimes (a) and (b) can be described as those of a large polaron where approximation of the bare-particle dispersion by a parabolic one with an effective mass of the bare carrier could be meaningful [4, 5]. In the regimes (b) and (c), the expected effect of the coupling to the bath is small. Interesting and little investigated is regime (d), where the coupling to the bath (phonons) is dominant but the carrier tries to escape the polaron well before the lattice starts to respond to its presence. In particular here, effects of dynamic suppression of the (otherwise) strong polaron effects or those of a transition to a self-trapped state [10] could be expected. The regimes (e) and (f) are those of the so-called small polaron expected to become relevant in OMC. Only here (if at all) we can take the so far ignored term $\sum_{p\neq q} J_{pq} A_p^\dagger A_q$ (see the second term on the right-hand side of (7.40)) as a small perturbation. Before doing that, however, let us realize that even the above complicated classification of the possible polaron regimes could be brought into a more detailed form including other relevant parameters of the dimension of energy (or time), such as the product $k_B T$ of the Boltzmann constant and temperature, the phonon-dephasing time (which might appreciably differ from $2\pi/\omega_0$ for other than acoustic phonon modes), the phonon relaxation times, etc. Moreover, intermediate regimes between any two of the above regimes exist.

Keeping by definition the set of phonon occupation numbers $\nu_{\boldsymbol{k}}$ in (7.41) fixed, and trying, with periodic boundary conditions, to diagonalize the matrix of (7.40) between states (7.41) with different p's, we get, in the case of fully periodic OMC, plane wave states. Again, for the simplest situation with just one molecule per elementary cell, they read

$$|\boldsymbol{q}, \{\nu_{\boldsymbol{k}}\}\rangle = \frac{1}{\sqrt{N}} \sum_p \mathrm{e}^{i\boldsymbol{q}\boldsymbol{R}_p^{(0)}} |p, \{\nu_{\boldsymbol{k}}\}\rangle \, . \tag{7.43}$$

For the band energies depending on the wavevector $\boldsymbol{q}$ of the polaron, we obtain (with $E_0 \equiv E_p$ for all p)

$$E_{\boldsymbol{q},\{\nu_{\boldsymbol{k}}\}} = E_0 + J_{\boldsymbol{q},\{\nu_{\boldsymbol{k}}\}} - E_b + \sum_{\nu_{\boldsymbol{k}}} \hbar\omega_{\boldsymbol{k}}\nu_{\boldsymbol{k}} \, , \tag{7.44}$$

$$J_{q,\{\nu_k\}} = \sum_s J_{rs} \cdot e^{-iq(R_r^{(0)} - R_s^{(0)})} \cdot \exp\left(-\frac{1}{N} \sum_k |G_k|^2 \left(1 - \cos[k(R_p^{(0)} - R_q^{(0)})]\right) (1 + 2\nu_k) \right) . \quad (7.45)$$

As the last term on the right-hand side of (7.44) determines the energy of the harmonic phonons propagating in the molecular lattice of OMC already distorted by the polaron, the remaining terms might be taken as the q-dependent energy of the small polaron. Taking the formal average over the state of the phonons (canonical distribution at temperature $T = (k_B\beta)^{-1}$), we obtain for the temperature-dependent small-polaron energy

$$E_q(T) = E_0 - E_b + \sum_s J_{rs} \cdot e^{-iq(R_r^{(0)} - R_s^{(0)})} \cdot \exp\left(-\frac{1}{N} \sum_k |G_k|^2 \left(1 - \cos[k(R_r^{(0)} - R_s^{(0)})]\right) [1 + 2n_B(\hbar\omega_k)] \right) . \quad (7.46)$$

Here $n_B(z) = 1/\left[\exp(\beta z) - 1\right]$ is the standard Bose–Einstein–Planck distribution for phonons. Comparing (7.46) with, for example, (7.36), one can easily see that the down-in-energy small polaron-shift E_b and the Debye–Waller renormalization

$$J_{rs} \rightarrow J_{rs} \exp\left(-\frac{1}{N} \sum_k |G_k|^2 \left[1 - \cos(k(R_r^{(0)} - R_s^{(0)}))\right] [1 + 2n_B(\hbar\omega_k)] \right) \equiv J_{rs} \exp(-2W_{rs}) \quad (7.47)$$

(shrinking) of the small polaron hopping (transfer or resonance) integrals J_{rs}, i.e., shrinking of the small-polaron band width due to overlaps of polaron deformation clouds (lattice deformation) around the polaron, located at two different places appear. When this overlap yielding $\exp(-2W_{rs})$ in (7.47) after averaging over the phonon occupation numbers (i.e., not the overlap of the localized states of the bare carrier located in the middle of the small polaron) becomes negligible, the polaron becomes immobile due to its coupling to the lattice.

Though interesting, these results are far from perfect in several respects. First of all, the small-polaron propagation owing to the term $\sum_{p \neq q} J_{pq} A_p^\dagger A_q$ in (7.40) certainly does not always proceed with only unchanged (constant) values of the phonon occupation numbers as above. The first effect expected is that the polaron changes its momentum upon absorbing or emitting a phonon at the moment of a single transfer $q \rightarrow p$ mediated by the hopping term J_{pq}. Hence, a total (carrier + total phonon) wavevector can be taken as a good quantum number preserving during the time evolution, i.e. as a quantum number characterizing carrier + phonon stationary states. With regard to optimization of the polaron wavefunction, much more intricate variational methods exist, yielding energy corrections, qualitative changes in the wavefunction due to, for example, transition to so-called self-trapped states, etc. (see [11, 12] and papers cited therein). Physically, however, this is only a part of the effects connected with the problem of electron (polaron, exciton) states in OMC.

More complicated problems appear when we realize that we cannot properly describe the carrier (electron, hole, or exciton) states by many-body wavefunctions. The point is that the latter functions can be neither determined well nor identified experimentally in many-body systems and at arbitrary finite (even arbitrarily low) temperature. (In connection with this, let us note that more intricate methods able to describe the polaron phenomena are based on the variational principle which applies only to the total ground state of the carrier-phonon system.) Correspondingly, one should rather work with the notion of the total density matrix of the carrier + the phonon bath

$$\rho_{\text{tot}} = \sum_i |\psi_i\rangle p_i \langle\psi_i| \,, \tag{7.48}$$

where $|\psi_i\rangle$ and p_i are the carrier-phonon eigenstates and the probability of finding them in the quantum ensemble, respectively. This is, however, owing to the practically infinite (for an extended OMC) number of the phonon modes, still impractical. Moreover, in our problem of the electronic states in OMC, any information about phonons is superfluous. That is why one usually gets rid of all the information concerning phonons, introducing the single-carrier density matrix into the site representation

$$\rho_{mn} = \text{Tr}(\rho_{\text{tot}} A_n^\dagger A_m) \,. \tag{7.49}$$

In the case of fully periodic OMC, transformation to the plane-wave representation is straightforward. For instance, with only one molecule per elementary cell, the transformation reads

$$\begin{aligned} \rho_{\boldsymbol{k}_1\boldsymbol{k}_2} &= \frac{1}{N}\sum_{mn} \mathrm{e}^{i\boldsymbol{k}_2\boldsymbol{R}_n^{(0)} - i\boldsymbol{k}_1\boldsymbol{R}_m^{(0)}} \rho_{mn} = \text{Tr}(\rho_{\text{tot}} A_{\boldsymbol{k}_2}^\dagger A_{\boldsymbol{k}_1}) \,, \\ A_{\boldsymbol{k}}^\dagger &= \frac{1}{\sqrt{N}}\sum_m \mathrm{e}^{i\boldsymbol{k}R_m^{(0)}} A_m^\dagger \,, \text{ etc.} \end{aligned} \tag{7.50}$$

Notice here that the wavevectors $\boldsymbol{k}_1$, $\boldsymbol{k}_2$, etc., represent the wavevector of just the carrier (not of the carrier + phonon system). Methods using the density matrix apply even to non-equilibrium situations. They would be, on the other hand, only of academic interest if there were no methods to determine ρ_{mn} or $\rho_{\boldsymbol{k}_1\boldsymbol{k}_2}$, avoiding the necessity of finding ρ_{tot}. Fortunately, such methods do exist (see [2] and papers cited therein). (Here, for concreteness, we again treat just the case of a conduction-band electron. For the hole or exciton, only the corresponding particle creation and annihilation operators are substituted.) Let us notice, however, that getting rid of all the information about phonons does *not* mean that we ignore their existence and influence on the carrier state.

The diagonal elements ρ_{mm} and $\rho_{\boldsymbol{kk}}$ of the single-carrier density matrix $\rho = \sum_{mn} \rho_{mn} A_m^\dagger A_n = \sum_{\boldsymbol{k}_1\boldsymbol{k}_2} \rho_{\boldsymbol{k}_1\boldsymbol{k}_2} A_{\boldsymbol{k}_1}^\dagger A_{\boldsymbol{k}_2}$ represent the mean number of carriers in the given localized or plane-wave state, respectively. (As we have just one carrier, they also represent the corresponding probabilities of finding the carrier in such states.) With that, we come to a crucial question: If the single-carrier density matrix can be given in both (localized as well as extended) representations, which is the most proper way to describe the carriers? Localized or extended? So, which indices should be preferred: site indices or wavevectors? In still other words: Is the carrier state or proper physical picture of the carrier rather localized or extended?

To answer these questions, let us remind ourselves again that (due to the competing carrier tendencies to localization and propagation which are due to the fourth and second terms on the right-hand side of (7.40), respectively) neither the site indices nor the phonon occupation numbers (i.e., the carrier wavevector in periodic OMC – the case we now focus our attention to) are good quantum numbers representing quantities preserved during the time evolution. On the other hand, because of the transformation between the localized and extended pictures in (7.50), both bases (as well as any other basis in the carrier Hilbert space, as already stressed above) is admissible. One can tell only that the basis is perhaps more appropriate in a given physical situation in which the off-diagonal elements of ρ are, in a sense, smaller. This criterion seems to prefer, at first sight and in the case of very narrow bare carrier bands in OMC at least, the localized picture. That the off-diagonal elements ρ_{mn}, $m \neq n$, become small follows physically from the assumption of independent phases of the carrier wavefunctions localized at two different molecules (sites). If we, however, set $\rho_{mn} \propto \delta_{mn}$ and, as a consequence of equivalence of all elementary cells, independent of m, from (7.50) we find that $\rho_{\boldsymbol{k}_1\boldsymbol{k}_2}$ becomes diagonal and independent of the wavevectors as well. This means full equivalence of the localized and extended pictures in the sense of the above criterion. On the other hand, because $\rho_{\boldsymbol{kk}}$ determines the probability of finding the carrier in the plane-wave state created by $A_{\boldsymbol{k}}^\dagger$ irrespective of the state of the phonon system, this situation may take place only at temperatures $k_{\mathrm{B}}T$ (in energy units) much higher than the width of the corresponding effective polaron band. Owing to very narrow bare-carrier band widths determined by hopping (transfer or resonance) integrals J_{mn} (usually <0.1 eV) and owing to further appreciable shrinking of the band width provided by the Debye–Waller renormalization (7.47), this is perhaps the usual situation in periodic OMC at not extremely low temperatures. In the opposite case, at temperatures so low that $k_{\mathrm{B}}T$ becomes appreciably less than the respective shrunk polaron band-width, one may expect that $\rho_{\boldsymbol{k}_1\boldsymbol{k}_2}$ is still diagonal, but its diagonal elements $\rho_{\boldsymbol{kk}}$ become appreciably $\boldsymbol{k}$-dependent. Then, still in periodic OMC with just one molecule per elementary cell, one obtains from the reciprocal transformation of (7.50), i.e.,

$$\rho_{mn} = \frac{1}{N} \sum_{\boldsymbol{k}_1 \boldsymbol{k}_2} \mathrm{e}^{-i\boldsymbol{k}_2 \boldsymbol{R}_n^{(0)} + i\boldsymbol{k}_1 \boldsymbol{R}_m^{(0)}} \rho_{\boldsymbol{k}_1 \boldsymbol{k}_2} , \tag{7.51}$$

that off-diagonal elements of ρ_{mn} appear. So, in the above sense, the extended picture becomes more appropriate. This, *inter alia*, is also the reason why ground-state trial wavefunctions in variational methods aiming at determining properties of carriers in interacting carrier-phonon systems must be rather extended [11, 12]. In the case of periodic OMC, with more molecules per elementary cell, the above localized versus extended picture transformations become more complex because of the necessity of including the index of the molecule in the elementary cell. Let us designate their number as M. We then get M small-polaron bands. Then, the above statements of equivalence and non-equivalence of the localized and extended pictures apply only to the appropriate extended polaron states mixing contributions from all the M molecules in each elementary cell. Otherwise, the localized picture might remain more appropriate, even at very low temperatures. At the end, however, let us notice that the carriers may be in the above sense rather localized, but can simultaneously be found everywhere within the OMC in question, as all the diagonal elements ρ_{mm} determining the site-occupation probabilities of the carrier may be non-zero. So, this notion of localization is certainly more general than in, for example, soliton theories, where the carrier surrounded by a cloud of the virtual phonons (lattice polarization) moves in space but can be, at any time, found only in a limited area within bulk of the material. Arguing by, for example, the variational principle, one can find that such states of even standing solitons always yield a higher mean value of energy than the corresponding extended states, in terms of which we speak here [13, 14]. At the end of the section, let us warn that the above polaron states are not necessarily the only states existing in OMC. It seems that at higher energies and from the same band of bare-carrier states, coexisting polaron and non-polaron states could be found. Experimental arguments on grounds of optical properties of OMC and theoretical evidence exist. The reader is referred to, for example, [2], and the more specialized literature cited therein.

7.8 Establishing the Electronic States in OMC

In the previous section, we distinguished six typical regimes in interacting carrier-phonon systems and argued that regimes (e) and (f) defining the small-polaron regime are likely to represent carrier states in typical OMC. In both these regimes, the condition $\hbar\omega_0 \gg J$ implies that the phonon system has enough time to start to respond to the carrier presence at any site before the carrier hops to a neighboring one. In principle, this still does not mean, however, that the phonon cloud around the carrier (lattice and molecular skeleton polarization) has enough time to fully relax around the carrier up to a locally stationary state corresponding to a prescribed carrier position. As usual, on the other hand, we have tacitly assumed above that this favorable situation really takes place. Arguments exist mainly in terms of the phonon dephasing time t_d, given as reciprocal value of the phonon dispersion $\Delta\omega$ in individual phonon bands; as far as $t_\mathrm{d} \ll \hbar/J$ (i.e. $J \ll \hbar \cdot \Delta\omega$) the polaron cloud has enough time to become well developed and almost stationary before the ensuing carrier hop. Hence, we can see the importance of the hierarchy of relevant times characterizing general as well as individual OMC. This observation is, in particular, important for bands of intramolecular phonon modes where $\Delta\omega$ might be relatively small irrespective of values of typical frequencies ω_0.

This question becomes even more important for the above electron-polaron problem around conduction-band electrons or holes. In this case, inclusion of the possibility that the central carrier can move (as above for the case of the interaction with phonons) is technically more difficult, and the problem has not so far been discussed in detail. The importance of the problem stems from the observation that the exciton bands are very narrow, i.e., the exciton dephasing times are very long. Thus, the cloud of the virtual excitons around the central charge in the center of the electron polaron requires a longer time to reach a locally stationary situation before the next carrier hop. Recent calculations and analysis [15] show that the time reserve is not very big in real OMC. Nevertheless, the electronic polarization (cloud of virtual excitons) around the central carrier, i.e., the electron polaron, does have time enough to become well developed before the next carrier hop within the bulk of OMC [15]. Thus, the above discussion of the electronic polarization around standing carriers corresponds to reality in typical OMC.

7.9 Prospects for Theoretical Work

The above text presents the contemporary understanding of electronic (i.e., electron, hole, exciton, etc.) states in OMC. As the whole problem given by (7.18–7.25) is unusually complicated, further progress will most likely be rather slow. This situation is even more complicated by the fact that real OMC contain many imperfections which have been so far ignored here, such as that structure of real phonon modes (in particular their local structure in, for example, the vicinity of imperfections) might be relatively complicated etc. Imperfections of any type can lead to many new effects, such as the appearance of real localized carrier eigenstates, dynamic trapping and detrapping, etc. Some of these effects are briefly discussed below. For perfectly periodic OMC, we expect progress in especially the following directions:

- Clarification of the internal structure of elementary excitations in terms of so-called quasiparticles (most of the above types of excitations preserve, due to other types of coupling, their structure as above for just a limited time interval – hence the importance of finite quasiparticle lifetime effects).
- Inclusion of more (and sometimes even competing) mechanisms of interaction of the elementary excitations with their surroundings.
- Discussion of the optimal type of description for given experiments (polaron, i.e. dressed versus bare excitation, localized versus extended states, etc.). This is irrespective of the fact that, as stressed above, all the bases in the corresponding Hilbert space are, in principle, admissible.
- Role of (quasi)stationary states in dynamic processes (transport etc.).

One should stress that inclusion of more mechanisms may lead not only to, for example, a greater intensity of perturbation of the local structure of excitations as above. Instead, under special situations, even cooperative (self-organizing) phenomena may appear in even quasistationary situations, which may change the structure and dynamic behavior of elementary excitations appreciably [16].

7.10 Charge-Carrier-Trapping States in OMC

In organic solid state, values of microscopic mobility of charge carriers are low, of the order of $\mu_0 < 1\ \mathrm{cm}^2/\mathrm{Vs}$. This consequently leads to a marked tendency for localization of charge carriers on individual molecules of the crystal. As the most important, one should mention the phenomenon of electronic polarization of the molecular subsystems surrounding the charge carriers. We should again emphasize that electronic polarization is an essentially many-electron interaction phenomenon [2, 6, 17]. Charge carriers become strongly localized on individual molecules and their motion should be described in the framework of a hopping model. A typical charge-carrier localization time on a lattice site in polyacene crystals is several orders of magnitude greater than the relaxation time for electronic polarization of the surrounding lattice molecules. It is thus appropriate to regard the localized charge carrier as "static", relative to the very fast process of electronic polarization. Subsequently, quasifree charge carriers move in molecular crystals together with their electronic polarization "cloud". Such a quasifree charge carrier may be then described as the electronic polaron.

The localized charge carriers mainly affect the π orbitals of conjugated molecules, which have high polarizability. The characteristic relaxation time τ_e is of the order of $\tau_e \approx 10^{-16}$–10^{-15} s. Under such conditions, the charge carriers carry the polarization field with them, forming the above-discussed quasiparticle – the electronic polaron (EP) (see Fig. 7.1).

The mean charge-carrier localization time τ_h on a separate lattice site describes the mean time interval between ensuing hopping events of the carrier

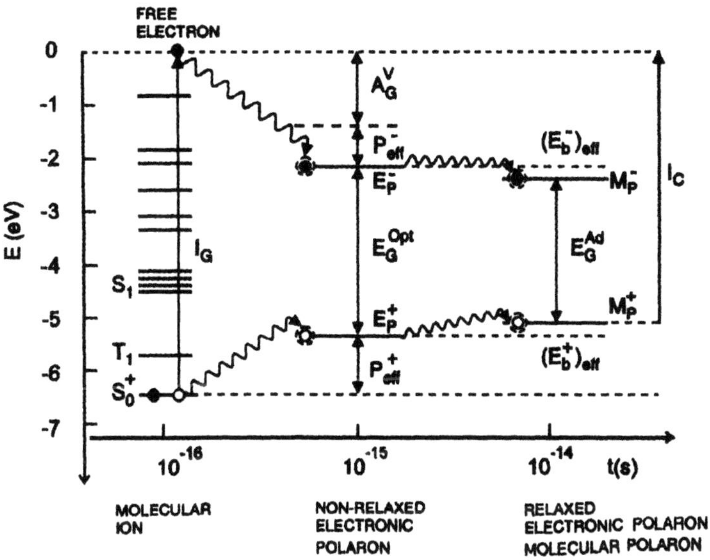

Fig. 7.1. Schematic of the formation of polaron states in organic molecular crystals. S_0^+ is the energy level of a molecular ion; T_1 and S_1 are the energy levels of neutral excited states; $E_G^{\rm Opt}$ and $E_G^{\rm Ad}$ are the optical and adiabatic energy gap of the crystal. $P_{\rm eff}^-$ and $P_{\rm eff}^+$ are the effective electronic polarization energies of the positive and negative charge carriers. A_G^V is the vertical electronic affinity; E_P^- and E_P^+ are the non-relaxed electronic polaron states. $(E_b^+)_{\rm eff}$ and $(E_b^-)_{\rm eff}$ are the effective formation energies of a molecular polaron; M_P^- and M_P^+ are the molecular polaron conductivity levels; I_G and I_C are the ionization energies of the molecule and the crystal, respectively. The horizontal axis shows the time scale corresponding to the polaron formation. From [18]

from one localized state to another. In anthracene-type crystals, the value of τ_h is of the order of $\tau_h \approx 10^{-14}$s. The time τ_h depends on distance between neighboring molecules and, consequently, the values of τ_h are different in different crystallographic directions. In anthracene-type crystals, the mean hopping time of τ_h of the charge carrier is two orders of magnitude larger than the time τ_e necessary to form the polarization.

In the course of vibronic relaxation of a molecule, changes in bond lengths and vibrational frequencies take place as a result of the redistribution of a localized surrounding charge. The molecule passes, during the charge localization time, from an equilibrium configuration of its nuclei in the neutral unpolarized state into an equilibrium configuration of the ionized or polarized state. According to this model, a nearly small molecular polaron (MP) is formed as a result of the interaction of the charge carrier with intramolecular vibrations of the molecule on which it is localized during the localization, as well as those in the close vicinity. The value of the mean vibronic relaxation time τ_v is of the order of $\tau_v = 10^{-14}$–10^{-15} s, which is faster than the hopping of the carrier onto a neighboring molecule. The plausibility of formation of the MP is confirmed by the ionized-state energy-structure studies in polyacene crystals.

The relaxation time of intermolecular vibrations under the influence of the charge (interaction with lattice optical and acoustic phonons) is in the range of the charge-carrier localization time in the anthracene crystal ($\tau_l \geq 10^{-13}$ s) [2]. In the ab-plane of the anthracene crystal, the process of formation of the lattice polaron is slow, and, most likely, the polaron does not manage to form.

In real molecular crystals, the charge carriers may be trapped in shallow traps of structural origin. The mean capture time τ_t of the carrier in such traps of depth $E_t \geq 0.03$ eV is larger than 10^{-12} s. Therefore, the capture time in shallow traps is larger than the lattice relaxation time, and the local polarization of the lattice becomes possible. Hence, the local lattice deformation appears due to accommodation by the molecules of the trapped-carrier state (molecular polaron). Formation of these polaron states through time is schematically illustrated for polyacene crystals in Fig. 7.1. If free electrons and holes are created as a result of ionization of a neutral molecule, they polarize, within time τ_e, the electronic subsystem of the surrounding molecules of the solid. The positions of the EP states E_P^- and E_P^+ in the energy diagram of the crystal are determined by the corresponding effective electronic polarization energies P_{eff}^- and P_{eff}^+. It is shown that the charge-induced dipole P_{ind} and charge-permanent quadrupole interactions provide the main contribution to P_{eff}. The optical energy gap characterizes the EP in OMC and separates the EP states, i.e., $E_G^{\text{Opt}} = E_P^- - E_P^+$ [2, 6, 18, 19].

At the next stage of the polarization dynamics, the localized charge carriers produce vibronic polarization of the skeleton of the atomic nuclei of the molecule, on which the charge carrier is localized, as well as that of the nuclei of the nearest molecules. As a result, the relaxed EP states M_P^+ and M_P^- are formed, and are called MP. The adiabatic energy gap E_G^{Ad} separates the MP states, i.e., $E_G^{\text{Ad}} = M_P^- - M_P^+$. The relaxation energy of the electronic polaron is determined by the energies of vibrational origin, viz.

$$(E_b^+)_{\text{eff}} + (E_b^-)_{\text{eff}} = E_G^{\text{Opt}} - E_G^{\text{Ad}}. \tag{7.52}$$

The energy levels M_P^+ and M_P^- are in fact the conduction levels of the negative and positive charges in the crystal. It is shown that transfer of the MP is determined by a non-activated tunnelling mechanism.

The data obtained on the energy gaps allow the full energy level diagram of the ionized (polaron) states in the polyacene crystals to be drawn. For this purpose one traditionally uses experimental values of the ionization energy of the molecule I_G and of the crystal I_C, the electron affinity of the molecule A_G, and the electronic polarization energy P_{eff}. I_G and A_G characterize LUMO (the lowest unoccupied molecular orbital) and HOMO (the highest occupied molecular orbital) by the energy which is needed to remove the first electron to infinity and which is gained upon adding one electron from infinity, respectively. In the framework of this model, these levels give rise to a narrow band when equal molecules interact in their crystal lattice. The solid-state I_{C} is reduced and A_{C} is increased with respect to those of the free gas molecules by P_{eff} (see Fig. 7.1). Since the polarization energy is essentially determined by some sort of average polarizability of the surrounding molecules, it can be expected, in a first approximation, to provide a fairly universal constant for organic solids.

The total balance of the energy terms in the diagram is determined by the following expressions:

$$P_{\mathrm{eff}}^{+} + (E_{\mathrm{b}}^{+})_{\mathrm{eff}} + E_{G}^{\mathrm{Ad}} + (E_{\mathrm{b}}^{-})_{\mathrm{eff}} + P_{\mathrm{eff}}^{-} + A_{\mathrm{G}}^{\mathrm{V}} = (I_{\mathrm{G}})_{\mathrm{calc}} , \tag{7.53}$$

$$P_{\mathrm{eff}}^{+} = P_{\mathrm{id}}^{+} + W_{Q_0}^{+} , \tag{7.54}$$

$$P_{\mathrm{eff}}^{-} = P_{\mathrm{id}}^{-} - W_{Q_0}^{-} , \tag{7.55}$$

$$|W_{Q_0}^{+}| = |W_{Q_0}^{-}| = W_{Q_0} , \tag{7.56}$$

$$(E_{\mathrm{b}}^{+})_{\mathrm{eff}} = (E_{\mathrm{b}}^{-})_{\mathrm{eff}} = \frac{E_{\mathrm{G}}^{\mathrm{Opt}} - E_{\mathrm{G}}^{\mathrm{Ad}}}{2} \tag{7.57}$$

where W_{Q_0} is the charge-quadrupole moment interaction and P_{id} is the self-consistent electronic polarization energy. The levels of the EP states are determined by the corresponding energies of the electronic polarization. Positions of the levels are obtained with the respect to level S_0^{+}. The polaron level E_{P}^{+} lies above the level S_0^{+} by the value of P_{eff}^{+}. The polaron level E_{P}^{-}, in its turn, lies below the vacuum level by a value of $P_{\mathrm{eff}}^{-} + A_{\mathrm{G}}^{\mathrm{V}}$. The sum $P_{\mathrm{eff}}^{-} + A_{\mathrm{G}}^{\mathrm{V}}$ thus characterizes the electronic component of the affinity of the crystal for a negative charge (electron) $A_{\mathrm{V}}^{\mathrm{C}}$.

The parameter of the crystal ionization energy I_{C} is determined by the value of the terms I_{G}, P_{eff}^{+} and $(E_{\mathrm{b}}^{+})_{\mathrm{eff}}$, according to the expression

$$I_{\mathrm{C}} = I_{\mathrm{G}} - P_{\mathrm{eff}}^{+} - (E_{\mathrm{b}}^{+})_{\mathrm{eff}} . \tag{7.58}$$

At present, this model of four electronic conduction levels is generally accepted as a basis of energy structure of ionized states in OMC, and it has been confirmed both experimentally and by calculation data [2,6].

However, any ideal crystal with perfect lattice structure is just an abstraction. Real crystals possess different kinds of structural defects and contains impurities as well. The structural defects of the crystal may form local centers for both excitons and charge carriers, i.e., they can create both neutral and ionized local states. It has been shown, that these defects really create local states in OMC: trapping states (traps) and scattering states (anti-traps) for both charge carriers and excitons (physical traps) [6,17]. However, the forces which give rise to these local states, their physical nature, and the energy values differ in both cases. Local trapping centers for excitons are also created through local change of dispersion and resonance interaction between the excited molecule and surrounding molecules in the region of the structural defect. On the other hand, local trapping centers for charge carriers are also formed as a result of a local change in the electron polarization by the excess charge carrier located in the structural defect of the lattice. Further on, we shall deal with the charge-carrier-trapping states some more. The structural irregularities and lattice defects create local states for electron polarons. A detailed phenomenological model describing the formation and nature of the local electronic states of the structural origin in OMC was proposed by E.A. Silinsh [20] and J. Sworakowski [21]. According to this model, local charge-carrier trapping states with quasicontinuous energy spectra are due to the local electronic polarization variations for the charge carriers located in the regions of structural irregularities of the crystal. Physical lattice defects are possible, such as unoccupied lattice sites (vacancies), lattice sites occupied by orientationally misaligned molecules, interstitial molecules, dislocation lines, and grain boundaries. It can be shown that the most feasible approximation of energy spectra for such local states is the Gaussian distribution model [6,17,22]. The shallow, quasicontinuous local trapping states, with $\Delta P < 0$ at the conduction level edge, and their symmetrical counterpart, antitrapping states, with $\Delta P > 0$, may be approximated by their density $h_e(E)$ (see Fig. 7.2), centered symmetrically around the conduction levels M_P^+ and M_P^-.

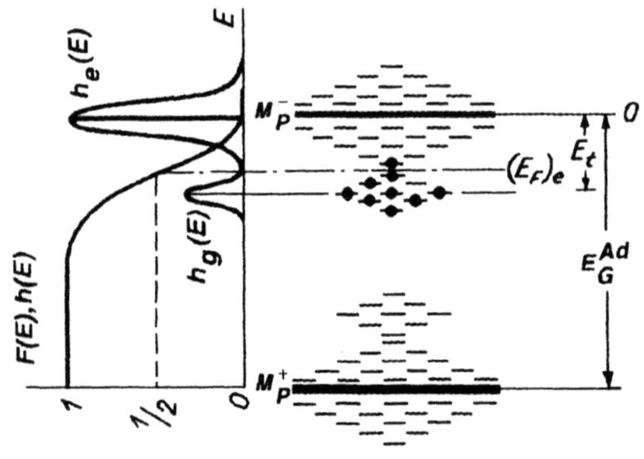

Fig. 7.2. Schematic energy diagram for a molecular crystal with Gaussian distribution of local states of structural origin. $h_e(E)$: Gaussian distribution of local states centered at conductivity level; $h_g(E)$: Gaussian distribution of local states situated in energy gap. (From [6])

An electron trap at E_t is said to be shallow if the Fermi level E_F lies below E_t, i.e., $(E_t - E_F)/k_BT > 1$. This type of distribution may be caused by, for example, edge dislocations, having both structural deformation counterparts, namely, a compressed lattice region above and a dilated lattice region below the dislocation line. On the other hand, trapping states, located in the energy gap, may be approximated by the $h_g(E)$ type of the distribution centred at E_t in the energy gap. This type of distribution may be formed in compressed lattice regions of various more complex extended structural defects, such as stacking fault ribbons, dislocation aggregates, second-phase inclusions, etc.

Both kinds of the Gaussian distribution can be described by the formula [6,20,21]

$$h(E) = \frac{N_t}{\sigma_G\sqrt{2\pi}} \exp[-(E - E_t)^2/(2\sigma_G^2)] , \tag{7.59}$$

where E_t is the position of the Gaussian distribution peak inside the energy gap, N_t is the total density of the local trapping states of the corresponding distribution, and σ_G is the distribution parameter (see Fig. 7.3).

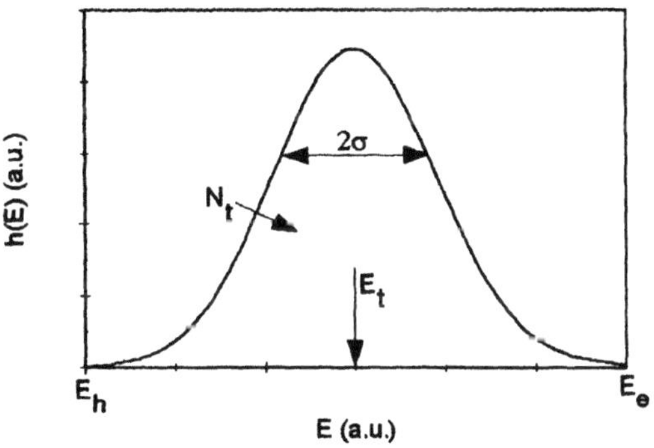

Fig. 7.3. Gaussian type distribution $h(E)$ of local electronic state. E_t is the position of the Gaussian distribution peak inside energy gap, N_t is the total density of local trapping states of the corresponding distribution and σ_G is the distribution parameter

Such a generalized Gaussian model includes in the common description limiting cases of the local state distribution. Thus, in the limit when the parameter $\sigma_G \rightarrow 0$, the distribution $h_e(E)$ gives an idealized picture of discrete conduction levels M_P^+ and M_P^- for a perfect crystal. On the other hand, the $h_g(E)$ type of distribution, in the limit when $\sigma_G \rightarrow 0$, describes discrete trapping states at E_t which may be considered as a δ-function.

Lattice vacancies are dominant point defects in OMC. As a charge carrier approaches a vacancy, the absolute value of polarization energy P decreases. This means that the vacancy would act as a scattering center. I. Eisenstein and R. Munn [23] included in their treatment the charge-carrier–quadrupole interaction term. Since the sign of W_{q-Q_0} is dependent on both the sign of the charge carrier and the quadrupole moment of molecule, the variation of the polarization energy may have positive as well as negative values. The calculations also predict asymmetry of the hole and electron trapping. According to these calculations, in pentacene crystals, the traps of the charge–quadrupole origin for electrons are 0.209 eV deep and for holes −0.066 eV deep. This trapping asymmetry has actually been observed experimentally in pentacene crystals [24].

Guest (impurity) molecules in the host molecular crystals form local states for charge carriers and excitons. Their energy levels can be described well within the framework of the polarization model. These trapping states are often called "chemical traps" to distinguish them from the structural (physical) ones.

In the zero-order approximation, the local states formed in a host lattice by the guest molecules can be regarded as discrete, determined only by molecular parameters of the guest and host molecules. The guest molecule forms, in a host crystal, an energetically unaltered energy which the guest molecule would form in its own crystal lattice. Thus, the local states for electrons $E_{\mathrm{g}}^{\mathrm{e}}$ formed by the guest molecules are determined by the difference between the electron affinities of the guest and host molecules. On the other hand, the local states for holes $E_{\mathrm{g}}^{\mathrm{h}}$ are determined by the difference between the ionization energies of the guest and host molecules:

$$E_{\mathrm{g}}^{\mathrm{e}} = (A_{\mathrm{G}})_{\mathrm{guest}} - (A_{\mathrm{G}})_{\mathrm{host}} \tag{7.60}$$

and

$$E_{\mathrm{g}}^{\mathrm{h}} = (I_{\mathrm{G}})_{\mathrm{guest}} - (I_{\mathrm{G}})_{\mathrm{host}} \, . \tag{7.61}$$

Several experiments have demonstrated that this model is correct for at least unsubstituted aromatic and heteroatomic hydrocarbons [25, 26]. For example, tetracene molecules in the anthracene host lattice form traps for both electrons and holes, but anthracene molecules in the tetracene host lattice form anti-traps for both electrons and holes. Typical electron acceptors (e.g., anthraquinone) create only trapping states for electrons, while typical donors create trapping states for just holes. One should also mention that any guest molecule produces some distortion of the surrounding host lattice. The displaced host molecules surrounding the guest one should form local states of the structural origin, i.e., physical traps by inducing compressive strain among host neighbors so that they form a cluster of trap levels (so-called X-traps). The guest molecule causing the X-trap formation may be a chemical trap at the same time. If the chemical trap depth E_{t} is smaller than that of the deepest X-trap, then the captured electron will most likely occupy the deepest X-trap. Conversely, a chemical trap may be surrounded itself by shallower X-traps which act as a trapping "funnel" and thus increase the trapping probability of the chemical trap. Alternatively, it may lead to a locally dilated lattice, in which the host molecules then act as structural anti-traps and reflect charge carriers, reducing the trapping probability [27].

Another case concerning chemical traps is when a molecular system consisting of non-polar molecules is doped with polar guest molecules. If the values of I_G and/or A_G of the guest molecules are sufficient to form the chemical trap, then detrapping of charge carriers from such traps may be field-enhanced due to a charge-permanent dipole interaction [28]. The permanent dipole moment of the guest molecule locally modifies the polarization energy because of a contribution to the local field acting on the localized carrier. Consequently, the local states appear in the vicinity of the guest molecule [28]. As follows from calculations [28], values up to 10 Debye of the permanent dipole moment of the guest molecule create the local states as deep as a few tenths of an eV. These results are in good agreement with results for doped anthracene crystals obtained from low-temperature stimulated luminescence experiments [29].

Local perturbations of the polarization energy because of presence of polar impurities may extend over several lattice constants and create a considerable number of shallower local states, which are spatially connected. These extended domains consist of local traps with energy distributed in space. This set of traps may act as a "macrotrap", whose presence was predicted by J. Kalinowski and J. Godlewski [30, 31]. These traps may be more realistic in molecular random systems such as amorphous organic films.

7.11 Experimental Methods for Trapping States in OMC

Different methods of charge-carrier injection are used for investigation of local electronic states in organic molecular solid state. Experimental papers dealing with local electronic states do not allow the physical and chemical origins of the observed traps to be clearly distinguished.

The method of space-charge-limited current (SCLC) is most widely used. It was significant that M. A. Mark and P. Lampert dedicated their excellent monograph to current injection in solids [32]. The method is based on a monopolar injection of charge carriers from an ohmic contact into bulk. The structure and shape of the current–voltage (IU) characteristics and current-thickness characteristics contain useful information on the local states, their energy spectrum and average density. As well as the SCLC method, modified SCLC methods have also been developed and introduced. Consequently, general methods based on monopolar charge-carrier injection influenced by an electric field can be classified as follows:

- method of SCLC [6, 32]
- differential SCLC (DM-SCLC) [22, 33] and

- thermally modulated SCLC (TM-SCLC) [34–38].

Several methods for the determination of charge-carrier-trapping states are based on trap thermal or optical release processes. Some of these methods are the following:

- time-of-flight(TOF) method [25, 26, 39]
- thermally stimulated current (TSC) method [40–42] and
- optically stimulated currents (OSC) [25].

All these methods have been used in the investigation of energy spectra parameters of the local trapping states in organic materials both in OMC and in multilayer assemblies of LB films. Further brief descriptions of all these methods are given in the following sections.

Space Charge Limited Current Methods(SCLC)

SCLC methods utilize mutual dependencies of current (I), voltage (U) and thickness of sample (L). Since there is no possibility to describe these curves analytically, different approximate formulae have been developed according to different trap distribution functions. The following assumptions are usually used in developing different phenomenological theories:

- the distribution of the electric field E throughout the sample is uniform
- the diffusion of the charge carrier is negligible
- only a monopolar charge-carrier injection from electrodes to OMC occurs, and therefore the recombination process is negligible and
- the density of thermal equilibrium charge carriers is negligible.

It should be to taken into account that at low electric fields and close to the electrodes, the diffusion current may influence the total value of the current and application of SCLC methods is therefore limited. A characteristic feature of SCLC is a superlinear dependence of the dark current I on voltage U as well as a superlinear dependence of I on thickness of the sample L [32]:

$$I \propto U^n / L^l \tag{7.62}$$

where $n \geq 2$ and $l \geq 3$.

The structure and shape of the $I = f(U)$ and $I = f(L)$ curves contain useful information on the local states, their energy spectra, and the total density of traps. The IU characteristics of a trap-free insulator can be described by Child's law [32]:

$$j = \frac{9}{8} \varepsilon \mu \frac{U^2}{L^3} \tag{7.63}$$

where j is the current density, ε is the static permittivity, and μ is the free-electron drift mobility. One usually assumes a mono-energetic trap distribution occurring in single crystal materials. However, SCLC theory with an exponential approximation for the trap density has been widely used for investigation of trapping states with a quasi continuous energy spectrum, there being a dominant wide dispersion, as a rule, in amorphous and polycrystalline samples. In real molecular crystals, the energy distribution of trapping states has a presumably Gaussian character. The advantage of the Gaussian model is that energy spectra of the local states of the structural origin are able to be described, both from the physical and the statistical standpoints, as done in [6,17].

At low electric fields which correspond to filling traps with charge carriers in the tail part of the Gaussian distribution and with the quasi-Fermi level being situated far away from the maximum of the distribution E_t, the slope of the *IU* characteristics n $[n = \mathrm{d}(\ln j)/\mathrm{d}(\ln U)]$ is close to $n \approx 2$.

The medium high-voltage region may be adequately described by the following approximation proposed by S. Nešpůrek and co-authors [17,37]:

$$j = \frac{e\mu N_e U}{L} \frac{(\varepsilon U)^{\alpha} \exp\left[-E_t/(k_B T)\right]}{(eL^2 N_t - \varepsilon U)^{\alpha}} \tag{7.64}$$

where $\alpha = \sqrt{(2\pi\sigma_G^2)/(16k_B^2 T^2) + 1}$ and N_e is density of the electronic states at the conductivity level. This region includes the transition of the quasi-Fermi level through the maximum E_t of the Gaussian curve and the post-maximum region. This corresponds to the slope $n \to \infty$. The approximation is valid up to the trap-filled-limit voltage U_{TFL}. It does not, however, include U_{TFL} itself since at $U \to U_{TFL}$, the denominator of approximation (7.64) exhibits singularity, according to the expression for U_{TFL}

$$U_{TFL} = \frac{eL^2 N_t}{\varepsilon} \, . \tag{7.65}$$

The *IU* characteristics in the neighborhood of U_{TFL} is more adequately approximated by the set of equations

$$j = e\mu n_d \frac{U}{L} \, , \tag{7.66}$$

$$U = \frac{eL^2}{\varepsilon} \left(n_d + \frac{N_t}{1 + \left[\frac{N_e}{n_d} \exp\left(-\frac{E_t}{k_B T} \right) \right]^{1/\alpha}} \right) , \tag{7.67}$$

where n_d is the concentration of the free charge carriers at the collecting electrode. If the total trap density is not too high, U_{TFL} may be experimentally reached under SCLC conditions. In such a case, the approximate set (7.66–7.67) should be used.

At higher electric fields, where the trap-free SCLC region lies, the *IU* characteristics can be described by the conventional Child's law (7.63).

The *IU* characteristics for a single set of traps is S-shaped, but in the case of several sets of traps, the curve is step-like. Then, every step is connected to the filling of a definite set of traps by the charge carriers.

Space-Charge-Limited Current Differential Method

The SCLC differential method (DM-SCLC) was proposed by S. Nešpůrek and J. Sworakowski in the 1970s [21, 33, 37, 43]. It permits one to obtain, directly from the IU characteristics, not only parameters N_t, σ_G and E_t but also the profile of the distribution of the local-state energy spectra. However, the method was shown to give correct results only for discrete or narrow trap distributions with σ_G values not exceeding k_BT ($\sigma_G \leq k_BT$). The quasi-Fermi level can be determined from the equation

$$E_F(U) = k_BT \ln \frac{\vartheta_1 N_e e\mu}{L} + k_BT \ln \frac{U}{j} , \tag{7.68}$$

where ϑ_1 is a parameter ($1 \leq \vartheta_1 \leq 2$). The energy profile of the energy distribution of the local states can be described, as shown in [22], by

$$h(E) = \gamma \frac{1}{n-1} \exp\left(-\frac{1}{k_BT} \int_{E_0}^{E} \frac{1}{n-1} \, dE_F \right) , \tag{7.69}$$

where n is the value which characterizes the slope of the IU characteristic in the *log-log* scale, E_0 is the energetically deepest trap level, and γ is a numerical constant. It is shown that the dispersion parameter of the distribution influences the value of U_{TFL}. This has to be taken into account when determining the value of N_t.

Thermally Modulated Space-Charge-Limited Current Method

The thermally modulated SCLC method (TM-SCLC) is one of the modified methods of SCLC provided by S. Nešpůrek and co-authors [34, 35], and recently developed by F. Schauer and co-authors [36, 37]. The TM-SCLC method investigates dependence of current I on temperature T in the so-called "fractional cooling" regime [35, 38]. Dependence of the activation energy E_a on the applied voltage U is determined. At a given voltage $U =$ const., the sample is slowly cooled (~0.05 K/s) and the corresponding current-temperature [$I = f(T)$] dependence is measured in the SCLC regime under monopolar injection of holes from the positive electrode. The temperature range in the cooling cycle should not exceed ~10 K. The activation energy E_a at the given voltage U is determined as

$$E_a = k_B \frac{\Delta \ln(I)}{\Delta(1/T)} , \tag{7.70}$$

where k_B is the Boltzmann constant. As the quasi-Fermi level passes, with growing U, through the given trap level, the values of E_a remain constant. This allows one to obtain values of E_t of the given trap distribution. The step-like dependence of $E_a(U)$ thus directly gives the energy spectra of the whole set of trap distributions [35, 38].

After the cooling cycle, the temperature is slowly increased (~ 0.1 K/s), reaching several degrees above the initial one (in order to depopulate the filled traps). Then, the cycle is repeated again at an another value of U, etc. As a result, one obtains the $E_a(U)$ dependence in the form of the above step-like structure which directly reflects the energy spectrum of the trapping states. The activation energy E_a in this case characterizes the position of the quasi-Fermi level E_F, which, in turn, is determined by the level of filling of the traps as a function of the injecting voltage U. It can be shown that at the voltage $U = 0.5U_{TFL}$ (where U_{TFL} is the trap-filled-limit voltage), the value of E_a becomes equal to E_t ($E_a = E_t$) and remains constant until all traps of the given set are filled ($U = U_{TFL}$). After that, it jumps to a zero value or to the next step of the $E_a(U)$ dependence. As well as the $E_t(U)$ spectrum, the method allows, under certain conditions, the parameters σ_G and N_t to be evaluated from the $E_a(U)$ dependence also.

It should be emphasized that the value of the trapping-state density N_t obtained by the SCLC methods mentioned above does not include all local states, i.e., the number of displaced molecules around the structural defect, e.g., edge dislocation. Such a kind of structural defect includes about one hundred or more molecules, but experimentally only the deepest trapping state of the local-state assembly emerges [2,38].

Time-of-Flight Method (TOF)

The TOF method has been widely used for the determination of charge carrier transport. It was introduced into the study of organic materials by R. G. Kepler [44] and O. H. LeBlanc [45]. The TOF technique determines the flight time of pulse-excited charge carriers across a sample. In the case when charge carriers are excited at one electrode and move at constant average drift velocity, a nearly rectangular TOF pulse is observed. The shape of the pulse may be distorted by finite flight pulse duration, a finite light penetration depth, and charge carrier traps as well. In the case of shallow traps, a decrease of current and an increase of the thermally activated transit time of the TOF pulse are observed. These changes of the TOF pulse are caused by sequential trapping and thermal release events during the charge-carrier transport through the sample. The initial part of the TOF pulse is influenced by charge carriers which move freely without trapping. Due to the presence of the deep traps, the ideal TOF pulse is distorted. This leads to a rapid decay of the current pulse before charge carriers arrive at the other electrode [25,46]. Only in the case of shallow traps or, if the density of the deep traps is sufficiently small to allow for charge-carrier transport over macroscopic distances without very large losses, the transit time, and hence charge-carrier drift mobility, can be measured by the TOF method.

The effective drift mobility μ_{eff} is determined by the time of flight t_f of the pulse-excited charge carrier across a macroscopic distance L at given applied voltage U as [25]

$$\mu_{\text{eff}} = \frac{L^2}{Ut_{\text{f}}} \,. \tag{7.71}$$

Here we assume that the value of the effective drift mobility is independent of the electric field.

The measurements of the temperature dependence of the effective drift mobility have been accepted as a technique to study parameters of traps (see [25] and references therein). In the case of shallow trapping, temperature dependence of μ_{eff} can be described as [46]

$$\mu_{\text{eff}} = \mu^0(T) \left[1 + \frac{N_{\text{t}}}{N_{\text{e}}} \exp\left(\frac{E_{\text{t}}}{k_{\text{B}}T}\right)\right]^{-1} , \tag{7.72}$$

where μ^0 is the mobility of a pure ideal crystal. The energy depth E_{t} and the density of traps N_{t} can be determined by plotting $\log[(\mu^0/\mu_{\text{eff}}) - 1]$ versus $(1/T)$.

Thermally Stimulated Current (TSC) Method

Thermally activated spectroscopy is especially convenient for studying deep quasi-discrete trapping states with E_{t} values exceeding 0.6–0.7 eV, not accessible for SCLC techniques.

In the TSC method, filling of the given set of traps with charge carriers is done by photoinjection with UV light or injection with electric field at low temperatures. After that, the temperature of the sample is raised according to a definite law and the charge carriers which have been thermally released are observed by means of the peaks of the thermally stimulated current $I = f(T)$ in an external electric field. For OMC, the TSC method was developed by J. G. Simmons [40] and M. Samoć with co-authors [41, 42]. The TSC spectra measurements can determine the energy distribution of a given set of traps $h(E)$ as well as energy parameters E_{t}, N_{t}, and σ_{G} from

$$j(T) = 0.6\, e\, L\, \beta\, \frac{E}{T}\, f(E)\, h(E) \,. \tag{7.73}$$

Here $f(E)$ is the probability function of the trap population and β is the linear heating rate of the sample. The energy depth of a given set of traps E_{t} is determined by the temperature T at which the release of the traps occurs and a peak of the thermally stimulated current appears. Even if the depths of two different sets of traps are the same, the maximum of the TSC spectra can occur at different temperatures because of the pre-exponential factor n and the heating rate β. The energy depth E_{t} can be described as [40]

$$\frac{\beta(M_{\text{P}}^- - E_{\text{t}})}{k_{\text{B}}T_{\text{m}}^2} = \nu \exp\left(\frac{M_{\text{P}}^- - E_{\text{t}}}{k_{\text{B}}T_{\text{m}}}\right) \,. \tag{7.74}$$

It should be noted that the degree of trap filling and the magnitude of the collecting electric field influence the determination of the trapping state parameters.

In some cases decreasing the temperature influences the structure of the film and, consequently, the energy parameters of the sets of traps. For example, in tetracene thin films, a phase transition at $T = 160$ K [47] occurs, and this process must be taken into account in the analysis of TSC spectra.

Optically Stimulated Current (OSC)

Optical trap excitation with subsequent auto-ionization leads directly, via relaxation, into the molecular electron and hole polaron states or direct optical trap to conductivity-level transition. These processes are also characterized by energetic parameters of the trapping states and, experimentally, can be investigated by a method developed by N. Karl [25]. The OSC spectra show an increase of photocurrent at wavelengths suitable for exciting transitions of the ions of the guest molecules. It is necessary to take into account that the light intensity and duration used for the OSC spectra should be sufficiently small in order to avoid noticeable bleaching of the traps during measurement. The wavelength of the appropriate maximum gives the value of the local trap radical ion transition of the guest molecules E_t. Such electronic transitions of different organic radicals (tetracene, biphenylene, pentacene) in naphthalene and anthracene crystals are shown in [25]. The experimentally obtained transitions in OSC spectra give values which are close to those observed in photoelectron spectroscopy.

In conclusion, it can be said that comparison of the data from the several methods which can be applied leads to complete and plausible values of the energy parameters of the local trapping states. In Table 7.1, values obtained for parameters of sets of hole and electron traps of pentacene thin films are shown [38]. In summing-up all the results obtained by different methods (SCLC, DM-SCLC, TM-SCLC), it is argued that in pentacene thin films five sets of hole traps exist at energy depths $E_t = 0.08$, 0.21–0.23, 0.28, 0.35, and 0.47 eV. Besides these traps, two electron traps at energy depths $E_t = 0.11$ and 0.68 eV are determined.

The conventional SCLC method is most appropriate for studying shallow trap distributions. The TM-SCLC method yields parameters E_t, N_t, and σ_G and provides a higher resolution of energy spectra in comparison with the conventional SCLC method. Thus, in pentacene thin films, only two sets of shallow traps with a wide dispersion parameter were obtained by the conventional SCLC method. In contrast, four sets of traps were obtained in the same energy range by the TM-SCLC method. At low electric fields, sometimes the conditions of applicability of the SCLC method regime may be violated. Consequently, values of the activation energy obtained with this method may not characterize the organic material itself.

Table 7.1. Parameters of hole and electron trap distribution in pentacene thin films obtained by different methods and by different techniques of trap population. Charge carriers are injected by electric field (F) or by photoinjection ($h\nu$)

E_t (eV)	σ_G (eV)	N_t (cm^{-3})	Type of charge carrier	Technique of trap population	Method of determination
0.06 ± 0.02		4×10^{15}			TM-SCLC
0.08 ± 0.05	$0.029 - 0.093$	$(1.6 - 5) \times 10^{15}$	h	F	SCLC
0.08 ± 0.03	0.05 ± 0.01	1×10^{15}			DM-SLCL
0.11 ± 0.02	0.009 ± 0.002	$(5.3 \pm 2.0) \times 10^{15}$	e	$h\nu$	TSC
0.21 ± 0.03		1×10^{15}	h		TM-LSCL
0.23 ± 0.02	0.014 ± 0.02	$(4.1 \pm 3) \times 10^{14}$		F	TSC
0.28 ± 0.02	0.014 ± 0.03	$(1 - 8) \times 10^{13}$	h	F	TSC
0.28 ± 0.05		4×10^{14}			TM-SCLC
$0.35 - 0.4$	$0.045 - 0.162$	$(0.7 - 3) \times 10^{14}$			SCLC
0.35 ± 0.04	0.05 ± 0.03	$(2 \pm 1) \times 10^{14}$	h	F	DM-SCLC
0.35 ± 0.05	0.014 ± 0.02	3×10^{14}			TM-SCLC
0.47 ± 0.02		$(0.4 - 1) \times 10^{14}$	h	F	TSC
0.68 ± 0.02	0.022 ± 0.02	$(2 - 5) \times 10^{14}$	e	F	TSC

Comparison of the experimental results obtained with the TOF, TSC and OSC methods [25] gives information on the presence of impurities which create chemical traps. Thus, for the example, the traps of tetracene (10^{-6} mole/mole anthracene) doped anthracene crystal were investigated by the above method. The electron trap energy depth $E_t = 0.42$ eV was obtained by the TOF method and the same value followed from the TSC peak between 150 and 160 K. The photocurrent excitation spectrum obtained by the OSC method showed peak at 863.8 nm. This peak characterizes doublet–doublet D0–D1 electronic transition of the tetracene radical generated by the trapping of a hole on a tetracene molecule. Combining both the TSC and OSC measurements, it is shown that the energy depth at $E_t = 0.42$ eV is caused by a tetracene molecule in the anthracene crystal.

References

1. M. Pope, C.E. Swenberg, *Electronic Processes in Organic Crystals* (Clarendon Press, Oxford 1982)
2. E.A. Silinsh, V. Čápek, *Organic Molecular Crystals. Interaction, Localization, and Transport Phenomena* (AIP, New York 1994)
3. T. Förster, Ann. Phys. (Leipzig) B **2**, 55 (1948)
4. S. Nakajima, Y. Toyozawa, R. Abe, *The Physics of Elementary Excitations.* Springer Series in Solid-State Sciences, Vol. 12 (Springer-Verlag, Berlin, Heidelberg, New York 1980)
5. W. Jones, N. March, *Theoretical Solid State Physics*, Vol. I-II (John Wiley & Sons, London, New York, Sydney, Toronto 1973)

6. E.A. Silinsh, *Organic Molecular Crystals. Their Electronic States.* Springer Series in Solid-State Sciences, Vol. 16 (Springer-Verlag, Berlin, Heidelberg, New York 1980)
7. Y. Toyozawa, Prog. Theor. Phys. (Kyoto) **12**, 421 (1954)
8. V. Čápek, Czech. J. Phys. B **28**, 567 (1978)
9. E.A. Silinsh, V. Čápek, L. Nedbal, Phys. Stat. Sol. B **102**, K149 (1980)
10. E.I. Rashba, In: *Excitons*, ed. by E.I. Rashba, M. D. Sturge (North-Holland, New York 1987), p. 273
11. Y. Zhao, D.W. Brown, K. Lindenberg, J. Chem. Phys. **107**, 3159 (1997)
12. D.W. Brown, K. Lindenberg, Y. Zhao, J. Chem. Phys. **107**, 3179 (1997)
13. V. Čápek, D. Krausová, Czech. J. Phys. B **37**, 1201 (1987)
14. A. Scott, Phys. Rep. **217**, 1 (1992)
15. V. Čápek, E.A. Silinsh, Chem. Phys. **200**, 309 (1995)
16. V. Čápek, H. Tributsch, J. Phys. Chem. B **103**, 3711 (1999)
17. S. Nešpůrek, E.A. Silinsh, Phys. Stat. Sol. A **34**, 747 (1976)
18. E.A. Silinsh, M. Bouvet, J. Simon, Molec. Mat. **5**, 1 (1995)
19. M. Bouvet, E.A. Silinsh, J. Simon, Molec. Mat. **5**, 255 (1995)
20. E.A. Silinsh, Phys. Stat. Sol. A **3**, 817 (1970)
21. J. Sworakowski, Mol. Cryst. Liq. Cryst. **11**, 1 (1970)
22. S. Nešpůrek, J. Sworakowski, Phys. Stat. Sol. A **41**, 619 (1977)
23. I. Eisenstein, R. Munn, Chem. Phys. **77**, 47 (1983)
24. E.A. Silinsh, I.J. Muzikante, A.J. Rampans, L.F. Taure, Chem. Phys. Lett. **105**, 617 (1984)
25. N. Karl, In: *Defect Control in Semiconductors*, Vol. II, ed. by K. Sumino (Elsevier Sci. Publ., North Holland, Amsterdam 1990), p. 1725
26. K.-H. Probst, N. Karl, Phys. Stat. Sol. A **27**, 499 (1975); erratum *ibid.* **31**, 793 (1975)
27. N. Emoto, M. Kotani, Chem. Phys. Lett. **101**, 386 (1984)
28. J. Sworakowski, IEEE Trans. Diel. Electr. Insul. **7**, 531 (2000)
29. A.K. Kadashchuk, N. I. Ostapenko, Yu. A. Skryshevskii, V. I. Sugakov, M.P. Shpak, JETP Lett. **46**, 207 (1987)
30. J. Kalinowski, J. Godlewski, P. Monalski, Mol. Cryst. Liq. Cryst. **175**, 67 (1989)
31. J. Kalinowski, J. Godlewski, P.G. Di Marco, V. Fattori, Jpn. J. Appl. Phys. **31**, 818 (1992)
32. M.A. Lampert, P. Mark, *Current Injection in Solids* (Academic Press, New York, London 1970)
33. I. Muzikante, L. Taure, Izv. AN Latv SSR, Ser. Fiz. Techn. Nauk No. 6, 31 (1987)
34. S. Nešpůrek, O. Zmeškal, F. Schauer, Phys. Stat. Sol. A **85**, 619 (1984)
35. F. Schauer, S. Nešpůrek, O. Zmeškal, J. Phys. C: Solid State Phys. **19**, 7231 (1986)
36. F. Schauer, S. Nešpůrek, H. Valerian, J. Appl. Phys. **80**, 880 (1996)
37. S. Nešpůrek, P. Smejtek, Czech. J. Phys. B **22**, 160 (1972)
38. I. Muzikante, E.A. Silinsh, Acta Physica Polonica A **88**, 389 (1995)
39. K. Oyama, I. Nakada, J. Phys. Soc. Jpn. **24**, 792 (1968)
40. J.G. Simmons, G.W. Taylor, M.C. Tamm, Phys. Rev. B **7**, 3714 (1973)
41. A. Samoć, M. Samoć, J. Sworakowski, Phys. Stat. Sol. A **36**, 735 (1976)
42. A. Samoć, M. Samoć, J. Sworakowski, J.M. Thomas, J. O. Williams, Phys. Stat. Sol. A **37**, 271 (1976)

43. J. Sworakowski, S. Nešpůrek, Vacuum **39**, 7 (1989)
44. R.G. Kepler, Phys. Rev. **119**, 1226 (1960)
45. O.H. LeBlanc, Jr., Chem. Phys. **33**, 626 (1960)
46. D.C. Hoesterey, G.M. Letson, J. Phys. Chem. Solids **24**, 1609 (1963)
47. H. Baessler, H. Müller, Chem. Phys. Letters **36**, 312 (1975)

8 Charge-Carrier Mobility in Organic Crystals

N. Karl

8.1 Introduction

Organic crystals are usually lacking main valence bonds between their constituent molecular repeat units. Therefore, in organic crystals there is usually no substantial electronic overlap between the molecules[1]; interactions are based on mere van der Waals forces in the very large class of neutral molecular crystals (or on a combination of van der Waals and local [closed shell] Coulombic interactions in the barely studied class of ionic organic crystals). As a consequence, exchange of a possibly present conduction electron or hole between neighboring molecules is not a very efficient process and, hence, electronic transport is generally slow in organic solids. Charge carriers are to be considered as rather localized (on individual molecules), with the consequence, specific for this class of materials, of considerable local changes of nuclear positions, vibrational frequencies, and electronic wavefunctions by polarization interactions (see, e.g., [1–3]). This situation is usually described by introducing the concept of a polaron [1,4] as an appropriate quasiparticle, comprising the electronic charge and the induced surrounding polarization based on electronic, vibronic and phononic relaxation. Due to strong intra- and intermolecular vibrational fluctuations, no coherent propagation of this quasiparticle is usually possible around room temperature and above, not even in a pure and structurally perfect crystal. The strong localization is reflected macroscopically by strong inertial resistance against acceleration (e.g. by an applied electric field), which can formally be ascribed to a high "effective" mass of the "polaronic" charge carrier [4].

As we will see below, however, gradual freezing out of uncorrelated vibrations upon cooling allows the charge carriers to move faster at lower temperatures, with the consequence of a reduction of the local interaction times with the polarizable environment, which in turn leads to a gradual reduction of the effective masses in addition. It is therefore not too much of a surprise to find that charge-carrier transport in extremely purified and highly perfect organic crystals at sufficiently low temperature can be nearly as fast as, for example, in silicon at room temperature [5,6]. A description of

[1] Except for a few well-defined polymer crystals with a one-dimensionally extended system of main valence bonds–mainly from the class of the polydiacetylenes.

transport by coherent Bloch waves – instead of stochastic polaron hopping – can therefore be more adequate at the lowest temperatures (e.g., at 4 K) [5]. The fundamental theoretical concepts of possible competition and apparent transition between band and hopping transfer in this class of materials have been elaborated long ago (see [1]; see also [7]). More recent investigations have treated the different basic theoretical concepts in a rather exhaustive, albeit general, manner [2, 3, 8]. A more detailed understanding, based on three-dimensional microscopic structural, energetic and dynamic parameters, which could predict absolute values of mobilities, their temperature dependencies, anisotropies and eventual non-linearities, has still remained a very challenging unsolved task in transport theory for this class of materials.

Although rather high mobilities can be obtained with organic crystals, the practical relevance may be impaired by severe experimental difficulties with ultrapurification, with growth of perfect single crystals, with cutting and polishing of samples without introduction of lattice defects, and by the necessity of operation at low temperature. (For purification and crystal growth see Chap. 6 as well as [9–12] and the literature cited therein).

To gain more profound insight, reliable experimental data on the fundamental material property describing electrical transport, *the charge-carrier mobility* μ, are required for a wide variety of different classes of molecules. The magnitude and anisotropy of μ should be followed over wide temperature ranges, and attention paid to eventual non-linearities. A (primitive) understanding of mobility by decomposition into an average effective mass and an average scattering rate of the ensemble of charge carriers is highly desirable as a first step. However, it is important to note that this concept is problematic for hopping transport (see below), deserves special attention in the presence of trapping impurities, and breaks down for dispersive transport, which is a characteristic of disordered organic solids as usually encountered with vapor-deposited or spin-coated thin films and with polymers.

As will be shown, a single (shallow) impurity level is still compatible with the concept of non-dispersive transport. However, for disordered solids where a wide range of site energies and intermolecular interactions lead to dispersive transport [13], mobility is no longer a well-defined unambiguous material property, but rather has to be replaced by a wide distribution of mobilities of different subclasses of charge carriers, and, more severely, with an average mobility of the ensemble of mobile charge carriers decreasing with time (see, e.g., [14]). These features obviously are an important obstacle to the development of fast and low-loss organic electronic devices on the basis of disordered organic solids, a fact which, however, is becoming recognized in the literature only gradually. Nevertheless, there may be niches where special applications or the cost/performance relation can tolerate slow and dispersive transport.

For these reasons this chapter will focus on transport in well-defined, periodically ordered and high-purity organic crystals and epitaxially ordered thin films, which, though often difficult to make, can at least contribute to a reliable fundamental understanding of charge-carrier transport in this class of materials, and point out directions to go in the development of devices.

In Sect. 8.2 the basic concept of the fundamental transport property, charge carrier mobility, is analyzed. Then the most successful experimental method for determining this property, the time-of-flight (TOF) method, is presented with a choice of its different realizations and with selected aspects one has to be aware of under non-ideal conditions. Other methods which can be used to measure charge-carrier mobilities are summarized. Special emphasis is laid on the surface acousto-electric wave (SAW) technique, which deserves more attention than given until now, because it can determine charge-carrier mobilities in dark-conductive samples in a rather straightforward manner and is thus complementary to the TOF method, which has been successfully used with organic photoconductors only, so far.

8.2 The Concept of Charge-Carrier Mobility

8.2.1 Basic Concept

For a free particle a constant accelerating force exerted by a field leads to a linear increase of momentum with time, associated with a linear temporal increase of velocity (in the non-relativistic case). A general definition of a field is by "the force per unit of that property on which the field acts" (charge, mass, etc.). For a constantly accelerated particle, which experiences interactions with its surroundings, velocity is usually not continuously increasing, but assumes a constant average value. In many different fields of physics (mass motion in a viscous medium, ion motion in an electrolyte, electron or hole motion in a semiconductor or photoconductor) this average velocity is found to be proportional to the field, but this is not at all trivial. A steady-state linear (average) velocity–field relation can be traced back to an average momentum loss per unit time proportional to the average distance travelled per unit time, i.e., proportional to the average velocity.

$$-\langle\frac{\mathrm{d}p}{\mathrm{d}t}\rangle = \beta\,\frac{\mathrm{d}\langle\Lambda\rangle}{\mathrm{d}t} = m\,\frac{\langle v\rangle}{\tau}\ , \tag{8.1}$$

which is counterbalanced by momentum gain transferred from the force exerted by the field. The constant β in (8.1) can formally be expressed as the ratio of a mass m and a relaxation time τ. In the case of a charged particle momentum change in an electric field E is $\langle \mathrm{d}p/\mathrm{d}t\rangle = eE$; hence, for constant mass,

$$\langle v\rangle = \frac{e}{m}\,\tau\,E\ . \tag{8.2}$$

The average velocity gained per unit field, $\langle v\rangle/E = (e/m)\tau$, is called the "mobility" μ.

$$\langle\mu\rangle = \frac{e}{m}\tau \ . \tag{8.3}$$

It is clear from this derivation that a material property of a (constant) "mobility" is a very fundamental and general concept, and not a specific feature of a semiconductor with band transport, not even of any specific electrically conducting material.

One should be aware, however, that on a microscopic scale it is not immediately obvious that the picture of a continuous momentum gain (acceleration) and momentum loss ("friction") should be adequate. In a band-type description the microscopically discontinuous (accelerated) transport of a quasifree charged particle with occasional (stochastic) scattering events at phonons can indeed be fitted to this concept. However, for hopping transport, consisting in an uncorrelated sequence of waiting times, jumps, and energy dissipation, this picture is questionable, because the concept of a quasi-continuous acceleration and deceleration brakes down here.

For a more general treatment of the problem of the electric field dependence of velocity, it is useful to start from diffusional motion, characterized by a diffusion constant, D, which we assume to describe a thermally activated random motion in an isotropic medium with sufficiently small energetic diagonal disorder. While the same (thermal) energy is involved (for every site, on average) in forward and in backward jumps in diffusion, leading to a sort of "bipolar" motion with the characteristic that the range extends as the square root of time, the additional energy taken up from an electric potential gradient changes its sign for the two directions, giving rise to a "polar" contribution to motion. This has not only fundamental importance for the functional difference of the time dependence of the range a charge carrier travels according to one or the other process (diffusion or drift), but also leads to a *linear* drift-velocity–electric-field relation for the superimposed polar motion in a good approximation, as can be seen from the following most elementary approach: The probability of a hop is proportional to $\exp(-\mathcal{E}_{\text{act}}/k_{\text{B}}T \pm eEa/2k_{\text{B}}T)$, where $\mathcal{E}_{\text{act}}$ is the thermal activation energy needed for a charge-carrier to hop across the intermolecular potential barrier (assumed here to be symmetric) in the absence of an electric field to a neighboring molecule at distance a or $-a$; the $\pm eEa/2k_{\text{B}}T$ term decreases or increases this probability due to the additional energy available from the potential difference to the maximum of the barrier at $+a/2$ or $-a/2$, respectively. The difference of the probabilities of two average jumps in opposite directions determines the net directional movement to be proportional to $\exp(-\mathcal{E}_{\text{act}}/k_{\text{B}}T)\cdot 2\sinh(eaE/k_{\text{B}}T)$. Series expansion of the hyperbolic sine for $eaE \ll k_{\text{B}}T$ leads to

$$j \propto \frac{eEa}{2k_{\text{B}}T} + \frac{1}{3!}\left(\frac{eEa}{2k_{\text{B}}T}\right)^3 + \dots , \tag{8.4}$$

i.e., to a linear current/field relation in good approximation, even if eaE is not very much smaller than k_BT.

The linear velocity–field relation (8.2), albeit fundamental for small field strength for both, band and hopping transport, is, however, by no means generally valid, neither for hopping transport, as already seen from Eq. (8.4), nor for band transport, for the following reason: at higher particle velocity (energy) the relaxation probability $1/\tau$ may increase or decrease and/or the (effective) mass m may change (see below). Thus, the $\mathcal{E}(p)$ energy–momentum relation is a very important source of information in semiconductor physics. Moreover, m and τ may also comprise anisotropies of the interactions, leading to different resulting (average) velocities and field dependencies for different orientations of a given field. Last, but not least, elastic and inelastic momentum relaxation should be distinguished and their relative contributions taken into account as a function of velocity.

It should be noted that, in a (sufficiently rigid) periodic lattice, with periodicity interval a, $\mathcal{E}(p) = \mathcal{E}(\hbar k)$ is a uniquely defined multivalued function for the first Brillouin zone already, with k, and thus p, proportional to a limited range of reciprocal wavelengths, $k = 2\pi/\lambda$ ($-\pi/a < k < \pi/a$, where λ is the de Broglie wavelength of the interacting particle considered). However, in summary, it is not necessary for the concept of a mobility that a band scheme $\mathcal{E}(k)$ can be established and Bloch waves are well-defined quantum states. There can be, nevertheless, allowed energy bands in which transport can occur (namely by hopping) and energetic gaps in which a mobile charge-carrier cannot exist. Only in the case of dispersive transport, as mentioned above, is charge-carrier mobility not really a material property.

8.2.2 Temperature Dependence

In typical hopping models mobility is associated with an exponential thermal activation term (see, e.g., [2]). For band model descriptions let us limit the consideration to the field-independent constant mass/constant relaxation time approximation of mobility (8.3) for the moment. These quantities, however, and hence charge-carrier mobility, depend on temperature. In classical semiconductor physics (see, e.g., [15]) the model of acoustic deformation potential scattering of quasifree charge-carriers moving at temperature T through a vibrationally fluctuating (non-ionic) crystal lattice has been shown to be a very broadly valid approximation. As a consequence of increasing occupation of acoustic phonon modes, the mean free path Λ of a charge-carrier decreases with rising temperature as $\Lambda_{ac} \propto 1/T$. In addition, the encounter probability between a charge-carrier and an acoustic lattice deformation is proportional to the thermal charge carrier velocity $v_{th} \propto \sqrt{k_BT}$; hence, the scattering probability $1/\tau \propto v_{th}/\Lambda_{ac}$ is proportional to $T^{3/2}$, and therefore (assuming a temperature-independent effective mass) $\mu(T)$ should follow a power law,

$$\mu(T) \propto T^n, \quad n = -3/2 \; . \tag{8.5}$$

Eventual deviations from this law can – at least phenomenologically – be explained by a temperature-dependent mass and/or by additional scattering at optical phonons. It is thus characteristic of all band model descriptions that mobility is not thermally activated as in hopping models, but a sort of "deactivated", instead. This is a very important distinction between band and hopping transport.

8.2.3 Ohmic and High-Field Non-linear Transport

Ohmic transport is microscopically defined by a *linear* relation between the average field-induced directional drift velocity of the charge-carrier (superimposed on its stochastic thermal motion) and the accelerating field, i.e., by a constant, velocity- and energy-independent ratio of mass and relaxation time (8.2)); for a given field a constant average drift velocity results from an average momentum loss proportional to the velocity, as discussed before. With *non-linear* transport, the velocity can either be *superlinear* with the field (e.g., if field-assisted detrapping or surmounting of potential barriers is involved, a situation often found in impure and/or disordered organic solids), or *sublinear*, if the velocity tends to saturate with increasing field (usually as a consequence of an increased scattering rate as the field-induced velocity, and hence the energy of the charge-carrier, increases). The latter sublinear case, observable in high-purity, well-ordered materials, is of fundamental importance because it exhibits effects of "warm" or "hot" charge-carriers, and thus, whenever observed, strongly supports a band model type of transport description under the respective experimental parameters. It can yield information on the kinetic energy of the charge-carriers and hence, with an independently measured velocity, on the effective mass [5, 16]. Alternatively, band structure effects can manifest themselves in non-linear transport when hot carriers are involved with an energy-dependent effective mass.[2] These are very important aspects for a correct description of high mobility transport, which was indeed recently found in organic crystals at low temperatures [5]. Sublinear velocity–field dependencies have been known for typical inorganic semiconductors such as Ge, Si and GaAs, for a long time (see, e.g., [18–23]; a compilation of experimental results may be found in [24]).

The most simple model, which was shown to describe the non-linear transport observed, for example, in naphthalene crystals at low temperatures [5], shall be briefly described here (see Shockley 1951 [18]; see also [25]).

The model starts from the assumption that charge-carriers during the process of energy take up from a sufficiently strong electric field decouple from the initial equipartition of the energy between the electrons/holes and

[2] In what follows we assume that the $\mathcal{E}(k)$ relation is parabolic to a good approximation up to the energy of the lowest optical phonon (typically around 5 meV for materials like naphthalene and anthracene; see, e.g., [17]) so that the effective mass can be taken to be constant.

the lattice, as a consequence of a finite energy relaxation time $\tau_{\mathcal{E}}$. With the further assumption that among the charge-carriers energy and momentum are quickly randomly distributed to follow a Maxwell–Boltzmann distribution function still within $\tau_{\mathcal{E}}$, a separate charge-carrier *temperature*, T_e, which is higher than that of the lattice, can be defined.

A standard perturbation theory description of (directional) momentum relaxation by (elastic) acoustic deformation potential scattering leads to

$$\langle \tau_{m,\mathrm{ac}} \rangle = \frac{4}{3\sqrt{\pi}}\, l_{\mathrm{ac}} \left(\frac{2k_{\mathrm{B}}T_{\mathrm{e}}}{m^*} \right)^{-1/2} , \tag{8.6}$$

with the carrier mean free path $l_{\mathrm{ac}} = \pi^{3/2}\hbar^4 \rho u_{\mathrm{L}}^2/(2m^{*2}\mathcal{E}_{\mathrm{ac}}^2 k_{\mathrm{B}}T)$; u_{L} is the longitudinal sound velocity, ρ the mass density, and $\mathcal{E}_{\mathrm{ac}}$ the deformation potential constant, i.e., the energetic shift of the band edge per unit of relative lattice dilatation (see, e.g., [25]). If now with increasing field the charge-carrier temperature T_e begins to rise above the lattice equilibrium value T, the acoustic momentum relaxation time decreases continuously, $\langle \tau_{m,\mathrm{ac}} \rangle \propto T_{\mathrm{e}}^{-1/2}$, according to (8.6); thus $\mu(E)/\mu_0 = \sqrt{T/T_{\mathrm{e}}(E)}$. The final charge-carrier temperature T_e is reached for a given field when the rate of energy gain from the field, $e\mu(T_e)E^2$, is balanced by the rate of energy loss to phonons, and $\mu(T_e)$ is obtained from $\langle \tau_{m,\mathrm{ac}} \rangle$ at T_e. Finally a field-dependent mobility $\mu(E)$ is obtained from the field-dependent drift velocity at $T_e(E)$, $v(E)$, as

$$\mu(E) = \mu_0 \sqrt{2} \left\{ 1 + \left[1 + \frac{3\pi}{8} \left(\frac{\mu_0 E}{u_{\mathrm{L}}} \right)^2 \right]^{1/2} \right\}^{-1/2} . \tag{8.7}$$

In this model the deviation from linearity is determined by the ratio of the equivalent Ohmic velocity $v_{\mathrm{d,lin}} = \mu_0 E$ (defined by the low field mobility μ_0 which can be measured) and the longitudinal sound velocity u_{L}, which can be obtained, for example, from sound-pulse propagation or from inelastic neutron-scattering experiments. For the field regime where $\mu_0 E$ is negligible in comparison to u_{L} we have $\mu \approx \mu_0$, whereas for the extreme case of large fields where $\mu_0 E \gg u_{\mathrm{L}}$ we obtain the "sub-Ohmic" limiting behavior

$$\mu_0 \propto \frac{1}{\sqrt{E}} \quad \text{and} \quad v_{\mathrm{d}} \equiv \mu(E)E \propto \sqrt{E} . \tag{8.8}$$

In the range of first deviations from linearity, where $\mu_0 E$ is substantially smaller than sound velocity, we obtain

$$\mu = \mu_0 \left[1 - \sqrt{\frac{3\pi}{64}} \left(\frac{\mu_0 E}{u_{\mathrm{L}}} \right)^2 \right] . \tag{8.9}$$

Attention should be paid to the fact that (8.7) does not contain any adjustable parameters, which means that the experiment can directly prove the applicability of the model for a given case.

While the above model can explain sublinear deviations from Ohm's law, it does not finally lead into a velocity saturation, as has been observed experimentally for very high fields. An appropriate model to include the latter case is based on the assumption that for sufficiently high charge-carrier velocity, and hence energy, an additional scattering channel opens up which allows for efficient inelastic losses [18]. This channel can be attributed to generative optical phonon scattering. Its onset is associated with a kinetic energy of the charge-carriers reaching the lowest optical phonon energy. The charge-carrier is assumed to be subjected to a free acceleration, $dv/dt = (e/m^*)E$, which increases its momentum linearly with time, $\hbar k(t) = eEt$, until after time τ it has gained enough energy from the field to generate an optical phonon of energy $\hbar\omega = [\hbar k(\tau)]^2/2m^*$. A cyclic repetition of this interrupted acceleration process leads to an average field independent drift velocity

$$v_{\mathrm{d,sat}} = \sqrt{\frac{\hbar\omega}{2m^*}} \; . \tag{8.10}$$

It is important to notice that for a measured saturated drift velocity, and with known optical phonon energy, the effective mass m^* of the charge-carrier can be obtained (or at least estimated) in a most elegant way. For more details [5], and [18] may be consulted. It is worth mentioning that in organic crystals high mobilities with tendency towards saturation in high fields have also been observed for (thermalized) positrons [26].

In the derivations given above the effective mass m^* has been taken temperature-independent. There are reasons, however, why a temperature-independent mass is a crude approximation only for organic crystals.

- The band width is governed by the intermolecular $\pi-\pi$ interactions, and these change with lattice contraction/expansion upon cooling/heating. For this reason the band width is – at least to some extent – temperature-dependent.
- The assumption of a parabolic band structure $\mathcal{E}(k) = \hbar^2k^2/2m^*$ which is associated with an energy-independent mass, i.e. $1/m^* = (1/\hbar^2)\delta^2\mathcal{E}/\delta k^2$ = const. is also an approximation. When electrons or holes reach higher band states with increasing temperature eventual non-parabolicity will lead to a change of the effective mass.
- Last, but not least, as discussed in the Introduction, the charge-carriers are expected to become gradually more delocalized as, upon cooling, the occupation of the phonons decreases; increasing delocalization – in conjunction with increasing mobility/diffusion constant – gradually reduces the phononic, vibronic and electronic contributions to the localizing polarization interactions, and hence the effective mass may drastically decrease upon cooling to low temperatures, and therefore the bandwidth strongly increases [4].

A strong temperature dependence of the effective mass and of the bandwidth appears to be the most important characteristic that distinguishes organic molecular semiconductor crystals from classical semiconductors such as, for example, silicon or GaAs. The exponent $n < -1.5$, frequently observed for the temperature-dependence of mobility $\mu = (e/m^*)\tau \propto T^n$ for fast directions, could thus at least phenomenologically be deconvoluted as $n = -(1.5 + p)$, where $\tau \propto T^{-1.5}$ as derived above, and $1/m^* \propto T^{-p}$, i.e. a mass increasing with temperature in a power-law dependence; for further discussion on temperature-dependent effective mass $m^*(T)$, see also [3,4,27].

8.2.4 The Mobility Tensor

The material property *mobility*, μ, linearly relates a (polar) vectorial response, the average drift velocity $\boldsymbol{v}$ with Cartesian components v_i, to a (polar) vectorial cause, the local electric field $\boldsymbol{E}$ inside the sample, with components E_j. This relation is, in general, anisotropic; for a given field strength $|\boldsymbol{E}|$ the velocity depends on the orientation of $\boldsymbol{E}$ relative to the crystallographic axes, which in abbreviated matrix notation (using Einstein's summation convention) reads

$$v_i = \mu_{ij} E_j \quad (i,j = 1 \dots 3) \ . \tag{8.11}$$

Therefore, μ is a (polar) tensor of the second rank. For thermodynamic reasons the matrix μ_{ij} is symmetric, $\mu_{ij} = \mu_{ji}$, which means that in the most general case six mobility components have to be measured for a given material, temperature and charge-carrier sign. For a suitable special choice of the coordinate system ($'$), the tensor matrix is diagonal. The mobility tensor can be visualized either by a tensor ellipsoid whose 3 principal radii represent the 3 principal mobilities, μ'_{11}, μ'_{22}, and μ'_{33}, in this coordinate system, or, for a more convenient presentation of the actually measured "mobilities parallel to given directions" (see below) and their error bars, by the tensor representation quadric $\mu'_{ij} x'_i x'_j = 1$, with semiaxes $1/\sqrt{\mu'_{ii}}$ [28]; an example may be found in [17], Fig. 18. The material property mobility, and hence its representation surfaces, have to conform with crystal symmetry. Thus, for cubic symmetry all three principal mobilities are equal; there is no transport anisotropy (this is the case for silicon). However, the great majority of organic crystals is of lower than cubic symmetry; for these, transport is necessarily anisotropic (and it can be highly anisotropic [29]); in these cases more than one principal mobility component must be given to describe the full tensor property. For trigonal, tetragonal and hexagonal crystals $\mu'_{11} = \mu'_{22} \neq \mu'_{33}$, and for orthorhombic symmetry $\mu'_{11} \neq \mu'_{22} \neq \mu'_{33}$, but with the principal axes parallel to the crystallographic axes a, b, c. For monoclinic symmetry, frequently encountered among organic crystals, $\mu'_{11} \neq \mu'_{22} \neq \mu_{33}$, and only one tensor principal axis is fixed by symmetry ($\mu_{bb} = \mu_{22}$, where μ_{22} is parallel to the twofold rotation or screw axis, or perpendicular to the mirror plane), while the orientation of the other two principal axes cannot be predicted from first principles, but must be determined experimentally; these can even rotate with temperature [17,30]. Finally a triclinic crystal requires measurement of 6 tensor components for each temperature (and, as always, separately for electrons and holes); an example of this non-trivial task may be found in [29].

It is important to mention that for an arbitrarily oriented sample and a given direction of the electric field vector $\boldsymbol{E} = (E_1, E_2, E_3) = |\boldsymbol{E}| \cdot (\ell_1, \ell_2, \ell_3)$, the transport quantity which is actually measured experimentally is often (e.g., in a sandwich configuration type of TOF experiment as described below), the *mobility parallel to the given direction* $\boldsymbol{L}/|\boldsymbol{L}| = (\ell_1, \ell_2, \ell_3)$,

$$\mu_{\|} = \mu_{\ell\ell} = \mu_{ij}\ell_i\ell_j \ , \tag{8.12}$$

i.e., a certain linear combination of all six independent tensor components for the most general, i.e., lowest symmetry, triclinic case. From six such measurements in sufficiently different directions the tensor (matrix) can be evaluated. If more measurements are made, a best tensor can be fitted; this procedure is advantageous, because it allows the amount of deviation of each individual measurement point from the best tensor to be determined, and thus it allows one to judge if among the TOF measurements or sample orientations one (or more) might have been erroneous (see, e.g., [28,29]).

8.3 Measurement Techniques

8.3.1 The Time-of-Flight (TOF) Method

Basic Principle

The TOF method is a rather direct and general measurement technique for determining mobilities of charge carriers in that it directly visualizes their field-induced transit motion through a sample from the start at one electrode to the arrival at the opposite one. In early inorganic semiconductor physics a number of variants were developed to measure charge-carrier mobilities in Si, Ge, CdS, CdTe, GaAs, S, Se, Te, and J_2 crystals. In the Haynes and Shockley type of experiments [31–37] minority carriers were electric-pulse injected into *n*- or *p*-doped semiconductor samples, or photogenerated near a surface [36,37], and their arrival selectively detected; depending on the duration of the travel time relative to the dielectric relaxation time, the neat mobilities, or ambipolar mobilities, were so obtained. In a second type of experiments no selective minority-carrier detection is necessary. In these experiments electron-beam pulses [19,20,23,38–44], α or β particles, or γ quanta have been used for charge-carrier generation, the latter mainly with photoconductors such as AgCl and AgBr (crystal counter technique [45–47]), and, very broadly, excitation of charge-carrier pairs by short light pulses of appropriate photon energy has been applied [38,48,49]. A noteworthy combination of the Haynes–Shockley and TOF methods was demonstrated with

Si [22]. For dark-conductive materials often a pulsed drift field was applied to avoid excessive heating. Field switching, however, may require compensation of the switching transients [50]. An overview on the advantages and drawbacks of different excitation methods may be found in [51].

For organic crystals the variant of the TOF method, exclusively used so far, consists of measuring in a plane-parallel "sandwich" electrode and sample geometry the travel time τ_0 of a thin "sheet" of extra carriers (i.e., of a plus or minus space charge) dragged by an externally applied constant electric field, E_0 (of appropriately chosen polarity), from the front electrode at $x = 0$ across the sample of thickness L to the rear electrode at $x = L$ (typically 0.1–1 mm) (see Figs. 8.1 and 8.2). Let us assume for the moment that the

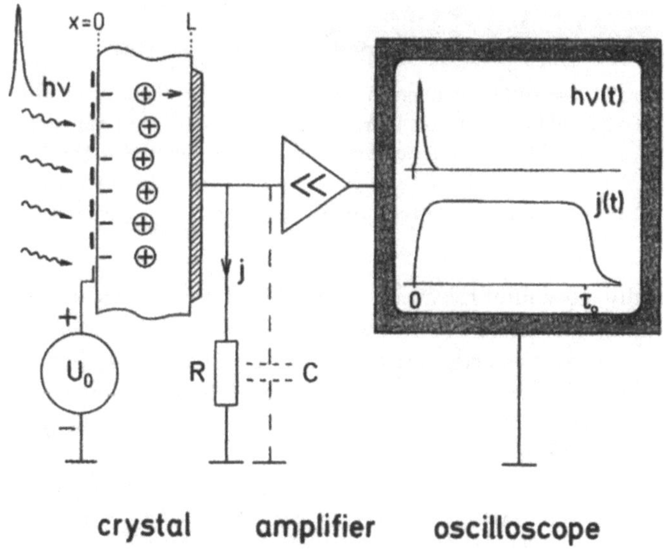

Fig. 8.1. Principle of the time-of-flight (TOF) method which allows the electron and hole mobilities in a photoconductor to be separately determined. Mobile charge-carriers are generated near one electrode by a sufficiently short, strongly absorbed excitation pulse, $h\nu(t)$. For the drawn polarity of the applied electric field $E_0 = -U_0/L$, a rectangular current pulse, $j(t)$, of duration τ_0 results which reflects the average drift velocity of the holes, $v = L/\tau_0$, and their mobility $\mu = v/E_0$. By reversing the sign of the applied voltage U_0, the electron mobility can be measured. The pulse edges are rounded by a finite charge-carrier generation time and a non-negligible penetration depth of the exciting radiation, as well as by the RC time constant and amplifier/display bandwidth limitations; the trailing edge in addition reflects Coulombic, diffusional and detrapping delay broadening of the arriving charge-carrier distribution. (From [6])

Fig. 8.2. Cross section at two different magnifications (originally $0.555 \times 11\ \text{mm}^2$) of a typical crystal slice as used for TOF measurements; a high degree of parallelism of both surfaces is necessary. (Anthracene, crystallographic *ab* plane – slightly distorted during the cleavage; due to glide systems the $\boldsymbol{a}$ and $\boldsymbol{b}$ directions are indicated by faint steps; sample photographed between crossed polarizers)

sample has essentially no equilibrium concentration of free-charge carriers (no dark conductivity); this is typical of a wide-band-gap photoconductor[3].

Ideally, the sheet of extra charge-carriers of number density N_0 per unit area is generated infinitesimally close to the front electrode by a short excitation pulse as a delta function $N_0\delta(x-0)$ at time $t=0$. To achieve this, most authors have used strongly absorbed short light pulses of sufficient photon energy, but the early methods of excitation mentioned above, by individual α or β particles, by pulsed particle beams, or by pulses of electrons, impinging (under vacuum) during a fast deflection of a dc electron beam of an opened cathode ray tube across the sample, can also be applied in principle. Subsequent spatial separation of the light- or impact-generated free electron–hole pairs occurs through the action of the applied field, in competition with bimolecular recombination. Optically induced unipolar injection from the illuminated electrode or from sensitizer molecules at the sample surface (see [53], Fig. 8.7, [54–56]) is also possible. To allow a clear evaluation, the charge-generation time should be substantially shorter than the travel time τ_0, and the amount of space charge should be small enough that the field emerging from it does not noticeably disturb the externally applied field

[3] In dark-conductive semiconductors with a sufficiently short dielectric relaxation time, the propagation of charge-neutral (ambipolar) density inhomogeneities can be monitored in a TOF experiment if Ohmic contacts can be provided (see, e.g., [22, 52]); this kind of experiment has not yet successfully been applied to organic semiconductors.

E_0 (other situations require more sophisticated evaluations). During the time $t < \tau_0$ while such a carrier sheet moves with constant velocity $v = \mu E_0$, it is associated with the constant current density

$$j(t) = (1/L)\, e\, \mu\, N_0\, E_0 \tag{8.13}$$

that can be registered in an external low-resistance circuit–coupled by the displacement current (or, equivalently, caused by the redistribution of the counter charge on the plates of the capacitor formed by the electrodes, when

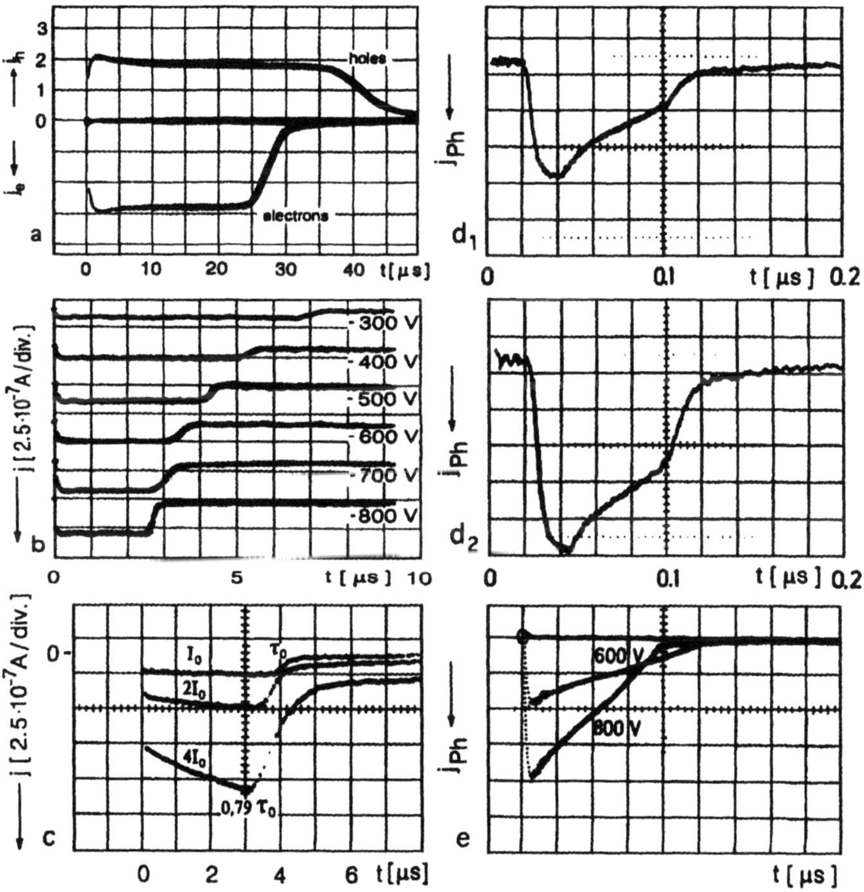

Fig. 8.3. Typical TOF mobility pulses obtained in the short-circuit current mode for samples typically several tenths of a mm thick. (**a**) For one polarity of the field, hole pulses (positive) are obtained, whereas for the opposite polarity electron pulses (negative) – with generally different flight times – are measured; equal pulse areas

Fig. 8.3 (continued) indicate generation of an equal number of electrons and holes. (T = 300 K; $\boldsymbol{E} \parallel \boldsymbol{a}$; the evaluation yields μ_{aa} = 1.2 cm^2/Vs. (**b**) Electron TOF pulses for different field strengths $\boldsymbol{E}_i \parallel \boldsymbol{b}$. The transit times $\tau_{0,i}$ obey a relation $\tau_{0,i}E_i$ = const., which proves that $v \propto E$, i.e., Ohm's law is valid. (**c**) Space-charge-limited/distorted electron TOF pulses. Distortions originating from the (additional) field, which emerges from a non-negligible amount of moving space charge, and the long tail, caused by afterinjection, can be eliminated by reducing the exciting light intensity I, and, hence, the yield. (a-c: Anthracene, room temperature; original single shot oscilloscope photographs, mounted into one frame [62, 63].) (**d**) At sufficiently low temperatures/high fields TOF pulse durations may no longer decrease with increasing field, as in (**b**), but remain unaltered, indicating a saturation of the charge-carrier drift velocity, which constitutes an unexpected non-linearity for van der Waals bonded solids. (Ultrapurified naphthalene crystal, T = 3.8 K, electron pulses with fractional trapping, field approximately parallel to $\boldsymbol{a}$, E = 13.5 V/cm in (**d**$_1$), 23.1 kV/cm in (**d**$_2$); the pulse rise time is given by the 16 ns FWHM XeCl excimer laser pulses used for excitation; original single-shot photographs of the screen of a 1 GHz oscilloscope with screen intensifier [63].) (**e**) Triangular TOF (short-circuit current) pulses obtained for homogeneous generation of charge carriers in the bulk. (Naphthalene, excited by ca. 1 ns light pulses of a nitrogen laser, λ = 337 nm, which is only weakly absorbed at room temperature and generates charge-carriers via an exciton–exciton annihilation process [63])

the sheet moves).[4] The current drops to zero when, at time $\tau_0 = L/v$, the drifting charge-carriers reach the rear electrode where they are discharged or, at least, stopped. Thus, the current pulse is *rectangular* under ideal circumstances, has a duration corresponding to the flight time τ_0 across the sample, an area proportional to the moved charge, and has a sign reflecting the charge-carriers selected by the polarity of the field (see Fig. 8.3a and b). If, in contrast, charge carriers are *homogeneously* generated in the bulk by weakly absorbed light or via a two-photon-excitation direct or indirect process (for the latter see, for example, [57]) – and thus have different travel distances/times – ideally a *triangular* TOF pulse of duration τ_0 occurs (see Fig. 8.3e) which can also be evaluated. Care has to be taken, however, in praxis to distinguish a non-ideal triangular TOF pulse, which has been shaped by recombination or by exponential trapping losses, from all trivial decaying signals which do not bear any arrival time information at all. It

[4] The electrodes can be blocking or covered by an insulating layer in principle; there is no a priori need for Ohmic contacts, which is a great advantage of the method described. Any sort of charge exchange with the electrodes, however, may be welcome, because it supports the reaching of the initial condition of space-charge neutrality before the next light pulse starts the next transit. Applying several light pulses in between, with the external voltage switched off and the electrodes short-circuited, may help to reestablish space-charge neutrality.

is an indispensable characteristic of all variations of the TOF method that at least a discernible fraction of the charge-carriers must arrive at the rear electrode and is not stopped anywhere in between the electrodes by trapping. Otherwise, in the best case, only a quantum yield times mobility product can be obtained.

So far unipolar charge-carrier transport has been treated in this description of the TOF method. Simultaneous pulsed injection of electrons from one and holes from the opposite electrode is also possible. This can not only be used to measure the mobilities of both signs in one single (stepped) pulse but also to study bimolecular electron-hole recombination during the time interval of mutual interpenetration of the charge clouds (see [58], Fig. 31 and [53], Fig. 13). Charge-carrier recombination often leads to recombination-luminescence ($S_1 \to S_0$ fluorescence, see Chap. 6). Recombination-luminescence near one electrode can be used to detect flight arrival times optically, if suitable structures are used; delayed recombination-luminescence, however, due to a preceding space-charge accumulation-phase, may cause ambiguities [59, 60].

Historical

The TOF method was introduced in 1960 into the organic field by Kepler, who measured electron and hole mobilities in anthracene crystals by recording the short circuit, i.e., small external resistance, transit current density $j(t)$ as a voltage signal $U(t) = RAj(t)$ [61] (where A is the sample area), and independently by LeBlanc, who recorded the time evolution

$$q(t) = \int_0^t j(t')\,\mathrm{d}t' , \quad \text{with} \quad q(\tau_0) = \int_0^{\tau_0} j(t')\,\mathrm{d}t' = eN_0 , \tag{8.14}$$

of charge q (delivered per unit sample area) as the open circuit voltage on an integrating capacitor $U(t) = q(t)A/C$ [64]. These two methods are equivalent, but the former has the advantage that the shape of the pulse is more intuitively instructive, especially under non-ideal conditions, where the arrival time often is manifested by a weak kink in the current pulse only and barely visible in the integral. After the pioneering work by Kepler and LeBlanc, a number of authors applied the TOF method to a variety of organic crystals and determined temperature dependencies and anisotropies. For a few extremely-high-purity and perfect samples with large mobilities non-linear (non-Ohmic) transport with velocity saturation was discovered at low temperatures [5, 6, 30, 65]. The literature data available up to 1984 were compiled in [17].

Non-ideal TOF Pulses

For describing more general cases of one-dimensional current flow, it is useful to start with the full detailed equation for the local total current density (in the absence of a magnetic field):

$$j(x,t) = e\mu\, n(x,t)E(x,t) - eD\frac{\mathrm{d}n(x,t)}{\mathrm{d}x} + \varepsilon\varepsilon_0\frac{\mathrm{d}E(x,t)}{\mathrm{d}t}\ , \tag{8.15}$$

where the first term describes the conduction current, which is based on the number density distribution $n(x,t)$ of charge-carriers per interval $\mathrm{d}x$ (and unit area in the real three-dimensional situation), which are moved by the local field, the second the diffusion current, based on the concentration gradient of mobile charge-carriers, and the third the displacement current, based on changes of the local electric field; the latter originates not only from the externally applied voltage difference, but may also have non-negligible contributions from the field emerging from the actual (mobile and immobile) space-charge distribution; D is the charge-carrier diffusion constant and the other symbols have their usual meanings. In this equation the respective elementary charge e and the mobility μ are to be taken as positive for holes and negative for electrons, while D has the same sign for both sorts of carriers; if there are contributions of both signs, their respective terms are to be added and bimolecular recombination has to be taken into account. As the total current density $j(x,t)$ as defined by (8.10) is the same everywhere, (8.15) can be averaged over L for further evaluation, and the total current density measured externally is

$$j(t) = \frac{1}{L}\int_0^L j(x,t)\,\mathrm{d}x\ . \tag{8.16}$$

For negligible contributions by diffusion ($\mathrm{d}n/\mathrm{d}x \approx 0$) and with the field solely given by the external voltage difference $U = -\int_0^L E_0\,\mathrm{d}x$ (i.e., with negligible field contributions from space charge), and held constant, and further, with

$$\int_{x=0}^{x=L} n(x,t)\,\mathrm{d}x = N_0\int_{x=0}^{x=L} \delta(x-vt)\,\mathrm{d}x = N_0\,, \tag{8.17}$$

the total number of charge-carriers (per unit area), an ideal *rectangular* TOF drift current pulse results (provided the time constants of the amplifier and oscilloscope are appropriately chosen):

$$\begin{aligned} j(t) &= \frac{e\mu N_0 E_0}{L} = -\frac{e\mu N_0 U}{L^2} \quad (0 < t < \tau_0)\ , \\ j(t) &= 0 \quad (t > \tau_0)\ , \end{aligned} \tag{8.18}$$

where $\tau_0 = L/v_0 = L/\mu E_0$.

However, in real situations the charge-carrier distribution is not a delta function, but has a finite width due to a finite duration and penetration depth of the excitation; the local field has contributions from the mobile and possibly immobile (trapped) space charge (e.g., left behind from previous experiments or injected from the electrodes in the dark pause, in the worst case with contributions of both signs), which leads to space-charge-perturbed or limited current pulses. Charge-carrier concentration gradients cannot always be neglected and cause diffusional contributions to the current, as well as a non-negligible broadening of the drifting distribution, as does Coulombic repulsion among the charge-carriers [66]. Moreover, the integration length L may not be uniform across the sample as a consequence of non-parallel or corrugated surfaces (see Fig. 8.2). In the initial overlapping region of electrons and holes substantial recombination losses may occur [67], and last but not least, trapping losses may occur underway. All of these effects lead to characteristic deviations from the near ideal rectangular shape of the TOF drift current pulse. A number of these cases, which can occur superimposed on each other, have been treated analytically (see [68]). We here only wish to present some examples by brief qualitative argumentation, without going into the details. For the experimentalist it is, however, important to know about these effects for a suitable choice of experimental parameters and for an appropriate evaluation of the results.

Trapping

Hoesterey and Letson [69] have described TOF pulses obtained with weakly tetracene-doped anthracene crystals by multiple *shallow trapping*/thermal release-limited transport. Basically, the transported charge-carriers freely move in the band with their typical lattice (or "microscopic") mobility, but transport is occasionally delayed by trapping with subsequent thermal release (i.e., by so-called "shallow traps"). Thus, the effective total flight time τ_{eff} across a distance L is the sum of the total free flight time τ_0 and of M trap reside times of average duration τ_{tr},

$$\tau_{\text{eff}} = \tau_0 + M\tau_{\text{tr}} , \tag{8.19}$$

and an effective mobility can be defined as $\mu_{\text{eff}} = L/\tau_{\text{eff}}E$. Assuming a discrete trapping level, separated by an energy $\mathcal{E}_{\text{tr}}$ from the band, and using Boltzmann statistics with sufficiently many shallow trapping events, a time-independent (non-dispersive) average ensemble mobility μ_{eff} is obtained as (see [70])

$$\mu_{\text{eff}}(T) = \frac{\mu_0(T)}{1 + (N_{\text{tr}}/N_{\text{b}})\exp(\mathcal{E}_{\text{tr}}/k_{\text{B}}T)} . \tag{8.20}$$

Trapping and detrapping statistics lead to a symmetric Gaussian-type broadening of an initially delta function shaped sheet of carriers, with a half width which is usually much smaller than the travelled distance, giving rise to a fairly narrow trailing edge of an otherwise still approximately rectangular TOF pulse. Thus, a unique average travel time and hence average velocity can be read from the oscilloscope display. (Trapping-delayed arrival and broadening of the spatial distribution function is, by the way, very similar to that which occurs in gas chromatography.)

Upon plotting experimental results as $\ln(\mu_0/\mu_{\text{eff}} - 1) = \mathcal{E}_{\text{tr}}/k_{\text{B}}T + \ln(N_{\text{tr}}/N_{\text{b}})$ versus $1/T$, a linear dependence should be obtained, from which the detrapping activation energy ("trap depth") and the relative trap concentration $N_{\text{tr}}/N_{\text{b}}$ can be read (see Fig. 8.12). Notice that for lower temperatures the mobility decreases exponentially upon cooling, whereas for sufficiently high temperature/small trap depth and small trap concentration a microscopic mobility μ_0 often can be obtained to a reasonable accuracy, even if no pure or undoped crystal is available. More details and the expected temperature dependencies of effective mobilities calculated for a number of parameters may be found in [70].

Optical detrapping at appropriate wavelengths can be applied to selectively remove charge-carriers from traps [70]. Intense laser pulses of a photon energy insufficient for charge-carrier production, but large enough for optical detrapping, applied during a TOF transit, might help to come closer to microscopic mobilities.

Deep trapping describes situations where during a reasonable experimental time span no charge-carrier, once trapped, is released again. A free-carrier concentration decays exponentially with time. In typical TOF experiments then only a fraction of the generated charge-carriers arrive at the rear electrode (see Fig. 8.3d) and frequently even none. Depending on temperature and trap depth, trap residence times can extend to hours or months. This may not only cause severe problems with *space charge*, but also can be used for thermally and optically stimulated trap release spectroscopy, usually abbreviated as TSC and OSC, respectively, or even a combination of both [70]. The value of these kinds of electrically detected spectroscopy for characterizing – and possibly improving – purity and, concomitantly, transport properties has not yet been broadly recognized.

Space-Charge-Limited TOF Pulses

For transport a violation of space-charge neutrality has primarily the effect that these charges are sources of an additional (internally generated) electric field. Only when the space charge distribution is precisely known can the influences of the additional (generally inhomogeneous) field on transport adequately be taken into account. It is also important to know if a given space charge is mobile or trapped in deep or in shallow traps, or in both. Only two limiting cases are simple enough to be of practical importance: a field-driven TOF pulse as well as a dc current may be fully space charge limited, i.e., fed by a reservoir of charge-carriers (near one electrode) with sufficient concentration to support a space-charge-limited current without exhaustion. This is equivalent to saying that all electric field lines end at charges in the bulk of the sample and the reservoir is field-free. As soon as a charge-carrier arrives at and leaves through the opposite electrode, another one can be picked up from the reservoir by the field line that became free and a space-charge-limited constant current may be established (see Sect. 8.3.4); alternatively, if the charge-carrier concentration in the reservoir has already decayed before the arrival of the fastest charge-carriers of the rear electrode, a space-charge-limited TOF pulse will be registered. For instantaneous unipolar current injection the duration of the TOF pulse as defined by the kink is reduced to $0.786\tau_0$, and the current is increasing during this time (see, for example, [71,72]), as can be seen from Fig. 8.3c. More general cases of space-charge distributions and the possibility of partial compensation of effects of (mobile) space-charge effects and trapping in the TOF pulse shape have been discussed in [68].

8.3.2 The Acousto-electric Travelling Wave (SAW) Method

The acousto-electric travelling wave (SAW) method for measuring charge-carrier mobilities [73, 74] requires a certain (dark) conductivity; in this respect it is somehow complementary to the TOF method, which is optimally applicable on photoconductors with vanishing dark conductivity, as described in Section 8.3.1. The method is based on a running surface acoustic wave (SAW) resonantly generated on top of a piezoelectric wafer (e.g., $LiNbO_3$) by applying a high-frequency (e.g., 20 MHz) driving voltage to a suitably spaced interdigital electrode (see Fig. 8.4). Because of the piezoelectric property of the wafer, the mechanical Rayleigh wave is accompanied by an (alternating) electric field wave which decays exponentially into the outer space. In a thin semiconductor slice placed parallel to the surface of the wafer, the charge-carriers therefore experience an alternating electrical field force whose z-component drags them periodically along and against the propagation direction, z, of the wave. Because the wave (usually) runs across the charge-carriers faster than they can follow, an unsymmetric forced motion results, which contains a dc component. This dc field force component can be measured by the dc current component in the semiconductor slab, if the latter is short circuited between its ends. In the open circuit case a corresponding dc (counter) voltage builds up. These signals are a direct measure of charge-carrier mobility if conductivity is based on a single sign. (If there are mobile electrons *and* mobile holes, their signals will partially cancel each other, however.) Let us treat the open circuit case in more detail (the short circuit current follows as $I_{sc} = U_{oc}/R$, where R is the sample resistance–if there are Ohmic contacts).

To obtain more insight into the cause of the non-linearity of this effect, consider an electron in a thin semiconducting film of thickness d placed parallel to the surface of the piezoelectric substrate at a separation distance h, and choose for the moment $d \ll h$. The motion of the electron along with the negative half wave of average z-amplitude $-|E_w(h)|$ lasts longer than its counter motion when the positive half wave of average amplitude $+|E_w(h)|$

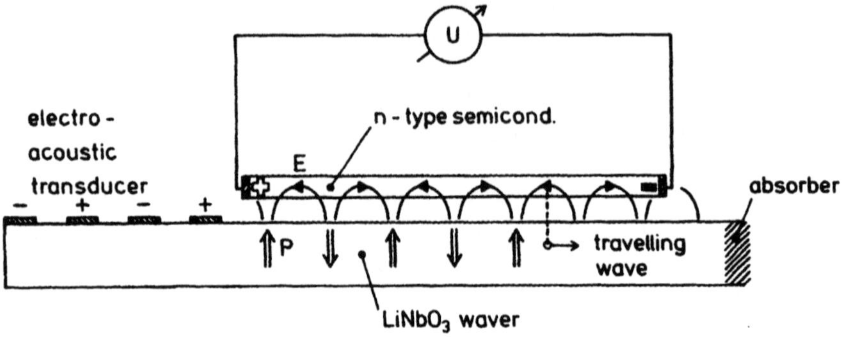

Fig. 8.4. Principle of the SAW method for charge-carrier mobility measurements. A piezoelectrically driven electric field wave drags the charge-carriers in a semiconductor slice or thin film – due to a net nonlinear interaction – preferentially into one direction, as explained in the text. The open circuit voltage difference so established is a rather direct measure of mobility. (From [65])

passes.[5] This unbalance can be calculated as an *effective* constant driving field $E_L(h) < 0$ acting on the electron on average. In an open circuit configuration, therefore, a voltage develops due to the (in the example negative) charge flowing to and accumulating at the (right) end of the sample of length L; a counter field is thus built up which ultimately stops a net current flow when it reaches the magnitude E_L. In this equilibrium situation the electron drift velocity in the negative and in the positive half wave, respectively is as follows:

$$v^- = \mu|E_\mathrm{w}| - \mu E_L \qquad (\mu\,\text{taken} > 0) \tag{8.21}$$

and

$$v^+ = -\mu|E_\mathrm{w}| - \mu E_L \ . \tag{8.22}$$

The interaction times with the negative and with the positive half wave, $\tau^-/2$ and $\tau^+/2$, are obtained as the Doppler-like expressions

$$\frac{\tau^-}{2} = \frac{\tau_\mathrm{R}}{2} \frac{1}{1 - \frac{\mu|E_\mathrm{w}|}{v_\mathrm{R}} + \frac{\mu E_L}{v_\mathrm{R}}} \tag{8.23}$$

and

$$\frac{\tau^+}{2} = \frac{\tau_\mathrm{R}}{2} \frac{1}{1 + \frac{\mu|E_\mathrm{w}|}{v_\mathrm{R}} + \frac{\mu E_L}{v_\mathrm{R}}} \tag{8.24}$$

[5] For tutorial simplicity we approximate a sine wave by a rectangular wave with an average amplitude E_w of the former.

where $\tau_R = 2\pi/\omega$ is the period and v_R the propagation velocity of the high-frequency (HF) wave; μ is taken to be positive. The condition for a stationary state, namely a displacement $\Delta s = 0$ in a full period τ_R, i.e.,

$$\Delta s = v^+ \frac{\tau^+}{2} + v^- \frac{\tau^-}{2} = 0 \tag{8.25}$$

(together with the realistic approximation[6] $v^+, v^- \ll v_R$), leads to the following simple expression for the generated balancing field E_L:

$$\frac{E_L}{|E_w|} = \frac{\mu |E_w|}{v_R} = \frac{v_d}{v_R}, \quad \text{hence } \mu = \frac{E_L v_R}{|E_w|^2} \ , \tag{8.26}$$

where $v_d = \mu |E_w|$ is the drift velocity of the electron in the semiconducting thin-film sample gained from the z-amplitude E_w of the traveling wave field; the latter can be calculated for a given piezoelectric wafer material and orientation, HF amplitude and distance of the semiconducting thin-film sample (see, for example, [52, 77] and the references given therein); v_R, the propagation velocity of the surface wave is known for common transducer materials and orientations. The (open circuit) counter voltage across the sample length L, which can be measured by an electrometer, is (for the n-type sample considered)

$$U_L' = -\frac{\mu E_w^2 L}{v_R} \ . \tag{8.27}$$

A positive sign is obtained for the holes of a p-type sample. Notice that despite the application of a high-frequency ac driving voltage a dc open circuit signal is obtained. (A good check for whether a measured dc signal is genuine consists of tuning the HF to observe a typical resonance curve of the SAW transducer in the dc signal, and reversing the travel direction of the SAW, which should reverse the sign of the dc voltage; moreover, the square dependence on the wave and thus on the HF amplitude is a characteristic of this non-linear process.) A typical signal as obtained in a genuine mobility measurement by the SAW method is displayed in Fig. 8.5.

Let us now briefly consider the short-circuit case; here a build-up of a counterfield E_L' is prevented. A similar derivation as given above, but with $E_L' = 0$ it yields for the short-circuit current density in the sample

$$j_{sc} = -\frac{\sigma \mu E_w^2}{v_R} = \sigma E_L = -\frac{\sigma U_L}{L} = \frac{\sigma U_L'}{L} \ , \tag{8.28}$$

i.e., j_{sc} is just driven by that acousto-electrically generated dc electromotive force in the semiconducting thin film, which can be measured as U_L'/L in the

[6] For a yz-cut $LiNbO_3$ wafer $v_{R,z} = 3.488 \times 10^5$ cm/s [75]. For $\omega/2\pi = 20$ MHz, the corresponding wavelength is $\lambda = 0.174$ mm; (y is the normal to the wafer).

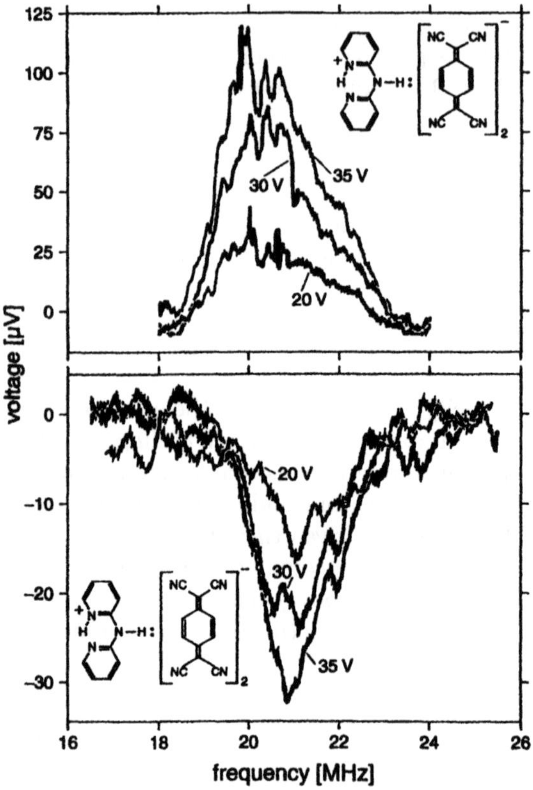

Fig. 8.5. Typical SAW-induced dc voltage signals between the ends of the semiconductor slice or thin film (see Fig. 8.4) as obtained upon tuning the frequency of the ac driving high voltage across the resonance of the SAW transducer. (51 nm vapor-deposited films of the dark-conductive organic radical ion salt bipyridylammonium:TCNQ; L=5 mm in the upper and L=1.3 mm in the lower part of the figure) [65]

open-circuit configuration at zero current. An open-circuit voltage measurement has the advantage that contact barrier problems are avoided.

The crucial quantity for the evaluation is the actual magnitude of the exponentially decaying field $E_w(y) = E_w(y = 0)\exp(-y/\lambda)$ above the surface of the piezoelectric wafer after the semiconductor film has been placed at $y = h$, which in praxis needs consideration not only of dielectric and Debye length space-charge shielding (see, for example, [52, 76] and the references given therein), but also of possible non-ideal coupling of the HF amplitude to the Rayleigh wave amplitude. All of these effects tend to reduce the field, and neglecting them may thus lead to an underestimation of the mobility.

The SAW method can still be used for mobility measurements even if the sample is thicker than a fraction of λ. In this case the z-component of the electric field inside the sample is depth dependent; the sample can be considered as subdivided into a set of sufficiently thin parallel sheets. While for the short circuit case these sheets contribute individually to the short-circuit current, the open-circuit case requires consideration of parallel wired resistors with individual internal electromotive forces [52]. A number of different cases with different Debye screening lengths have been treated analytically in [76]. More details and references may be found in [77]. The method has been successfully applied to measure charge-carrier mobilities in organic thin films [77] as well as in organic single crystals [52].

8.3.3 The Field Effect Transistor (FET/TFT) Method

The thin-film (field effect) transistor (TFT or FET) method determines the (parallel) mobility of charge carriers in a semiconducting (organic) thin-film in a space-charge accumulation layer, (capacitively induced by a perpendicular gate field applied across an insulating layer) (see Fig. 8.6) by measuring the extra contribution of these charge-carriers to the conductivity. The gate field is directly varied by applying a gate voltage U_G to the gate electrode G, but it is indirectly counterbalanced to a larger or lesser extent by the local value of the drain to source voltage drop and hence the induced space charge is position-dependent. The space-charge enhanced drain to source-current can be calculated as a function of the drain voltage, U_D, and the gate voltage U_G, and the quantity of interest, the mobility can be obtained from the increment of conductivity (the "transconductance") (8.29) in the linear regime of the drain (D) to source (S) current–drain voltage characteristics, $I_D(U_D)$, or from the saturation current $I_{D,sat}$ in the saturation regime (8.30) (see Fig. 8.15):

$$\left(\frac{\partial I_D}{\partial U_G}\right)_{U_D=\text{const}} = \pm\frac{Z}{X}\,\mu C_i U_D \tag{8.29}$$

and

$$I_{D,sat} = \pm\frac{Z}{2X}\,\mu C_i (U_G - U_0)^2 \ , \tag{8.30}$$

where X is the separation of the source and drain electrodes ("channel length"), Z the channel width, and C_i the capacitance per unit area across the insulating layer between the gate electrode and the semiconductive film. The plus and minus sign are for n- and for p-type conduction, respectively. The threshold voltage U_0 takes into account the field free-carrier concentration as well as details of the internal charge distribution and trapping at the interface and at the contacts.

In addition to the simplifications made in deriving (8.29) and (8.30) (see, for example, [78, 79]), there are the following important problems, but also new challenges:

- The source and drain electrodes should be Ohmic to prevent space-charge-limited currents, and thus to avoid the contact properties (e.g. barriers) being influenced by the gate field, and also to eliminate the problem of unknown field distributions within the sample.

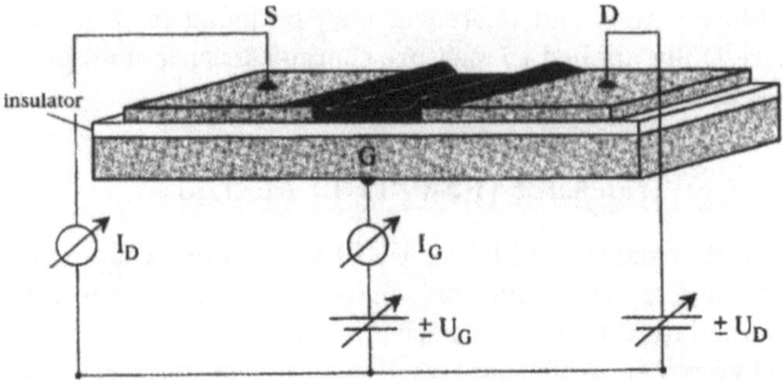

Fig. 8.6. Schematic of an organic thin-film field-effect transistor (OFET). In a weakly semiconducting film (black) between two narrow-spaced electrodes (S: source; D: drain) the equilibrium charge-carrier density and hence the conductivity is altered by a field which emerges from the gate electrode G and induces space charge. In praxis SiO_2 and Si_3N_4 layers on a silicon wafer have frequently been used as the insulting gate spacer, but organic insulating films have also been applied successfully

- Charge-carrier traps should be essentially absent or carefully taken into account; otherwise, an unknown part of the induced charge may be intermittently trapped, with the consequence that a mobility that is too low (a so-called effective mobility) would be found for the residual free charge-carriers. (In addition, the field distribution would be changed and the response-time constants increased.)
- As the locally induced charge density can be very high, considerable Fermi level shifts may occur with variation of the gate voltage, which has the consequence of gradually including/excluding trap levels from the transport dynamics; this leads to the fact that the effective mobility is no longer a constant, but instead gate-voltage dependent (see [78,79]). Charge-carrier densities exceeding the trap-filled limit, however, can assume their intrinsic mobility even for impure samples. This presents a great challenge to measure high intrinsic mobilities in *single crystals* at low temperature in field-effect transistor geometry, where extremely high gate fields can be applied. Indeed, FET-mobilities of up to 10^5 cm^2/Vs were recently reported for tetracene and pentacene crystals at liquid He temperature [80,81], and even transitions to superconductivity were observed for anthracene, tetracene, pentacene [82], and C_{60} [83]. These exciting results are extremely remarkable!

8.3.4 Other Direct and Indirect Methods to Determine Mobilities

There exists a number of other transport effects which allow one to determine or at least estimate charge-carrier mobilities. A brief overview will be given here (see also [51]).

The Xerographic Discharge Method

This method is based on the principle of electrophotography as used in most copiers and laser printers and is a variant of the TOF method. A dark-insulating thin sample, usually a thin film in contact with a metallic support, is negatively corona-charged on its exposed surface. A light flash of sufficient intensity creates free holes on the opposite side (the interface to the metal, usually carrying a sensitizer layer) and initiates a space-charge-limited transport pulse. Ideally, the film is left fully discharged afterwards. The duration of the current pulse can be evaluated to yield the hole mobility [55].

The Equilibrium Charge-Carrier Extraction Method

In a weakly conductive sample of sufficiently long dielectric relaxation time equilibrium charge carriers can be extracted by a field ramp; the resulting current pulse contains information on the mobility of the majority carriers [84–87].

Drift-Current Methods Under Limited Range Conditions

If a definite quantity of mobile charge-carriers can be generated quasi-instantaneously inside a dark-insulating sample, e.g., by stopping high energy particles such as, e.g., MeV electrons, the increase of conductivity monitored by time-resolved microwave absorption [88–90], or the height of a current pulse (measured in an externally applied field), can be used to evaluate an effective average mobility, even if the range in which charge carriers move freely before they are trapped is not known ("pulse radiolysis method"). The only condition is that the measurement must be completed before a substantial fraction of the free carriers becomes localized.

Estimation from Conductivity

In a semiconductive sample intrinsic conductivity is a measure of an average mobility, if the density of mobile charge-carriers is uniform and known. For organic radical ion salts the latter can be estimated from stoichiometry for the metallic case, or from a known energy gap if the material is intrinsic, for a semiconductive phase [91,92].

Space-Charge-Limited Currents

Given a plane-parallel electrode configuration and a sample sandwiched in between, it is easy to calculate the maximum space charge which is compatible with a given applied voltage (it is approximately determined by the capacitance of the so-formed plate capacitor). For ideal (unipolar) space-charge-limited (see Chap. 7, Sect. 11) currents ("SCLC") injected at one electrode, a field distribution results which gives rise to an externally measured current density that depends on the amount of injected charge *and* on the driving field, and hence displays a square dependence on voltage and an inverse-cube dependence on sample thickness:

$$j = \frac{9}{8}\,\mu\,\varepsilon\,\varepsilon_0\,\frac{U^2}{L^3} \tag{8.31}$$

(see, for example, [93]), where ε and ε_0 are the relative and absolute dielectric permittivities, respectively. A measurement of SCLC thus yields the mobility. However, for the formula to be valid, the current should by far exceed the "trap-filled" limit, and follow a $j \propto U^2$ dependence; to make sure that this is the case, a current–voltage dependence should regularly be recorded over a wide voltage range. In addition, the $j \propto L^{-3}$ dependence should be checked, before relying on mobility results.

If space-charge conduction is balanced partly by double injection, or if the current is injection-limited by contact barriers, extraction of reliable mobility data is a rather hopeless enterprise.

Charge-Carrier Diffusion

Charge-carrier diffusion is related to mobility by the Einstein relation

$$D = \frac{\mu k_{\mathrm{B}} T}{e}\,; \tag{8.32}$$

it is difficult to study under well-controlled conditions, because the effects are small ($k_{\mathrm{B}}T/e$ corresponds to a potential of 25 mV at room temperature) and easily disturbed by, or even hidden under, other effects. In [66] an attempt has been made to evaluate the diffusional broadening of the drifting carrier sheet in TOF experiments (which should be $\propto \sqrt{t}$, whereas mutual Coulombic repulsion of the charges leads to a $\propto t$ broadening).

The Dember effect (see [94]) consists of the development of a diffusion potential (or current) in a concentration gradient of electrons and holes with different diffusion constants or trapping kinetics. A concentration gradient comes about by a spatially inhomogeneous generation of charge-carrier pairs, e.g., by (pulsed) illumination with light of a certain absorption length. The Dember effect may be difficult to interpret because of possible unknown surface states and surface traps. However, for organic materials, which allow surface cleaning by evaporation without decomposition, pulsed Dember effect measurements under a good high vacuum allow at least the determination of the faster charge-carrier sign. In highly purified anthracene crystals, for example, the electrons are more mobile in certain crystallographic directions than the holes, and vice versa for other directions. Hence, a sign reversal of the Dember effect can be observed and the zero-signal directions ($\mu^+ = \mu^-$) determined [68]. In C_{60} thin films, electrons were found as the more mobile carriers [95].

Spin diffusion can more easily be measured than charge diffusion, using electron spin resonance in a magnetic field gradient. Under favorable conditions spin diffusion is based on free charge-carrier motion and thus allows the determination of mobilities [96, 97].

Hall Effect and Cyclotron Resonance

The Hall effect is based on a (small) deflection of drifting charge-carriers by a perpendicular component of a magnetic field, which can be described by an internal electromotive force that acts in a direction different from that of the externally impressed current. These carriers can be equilibrium, photogenerated, or injected charge-carriers. The (average) deflection angle can be determined by a Hall angle measurement, the deflected component as a magnetic field-induced transverse short-circuit current, or as a counter voltage, preventing a Hall current flow under open-circuit conditions. These signals are often small and difficult to detect reliably for materials with a small charge-carrier concentration and/or low mobility, and thus easily mixed up with thermovoltages and other spurious signals. It is therefore mandatory to check the sign reversal upon magnetic field reversal and to uniquely attribute a signal to the required linear response to a magnetic field. In all cases the primary information obtained for a material with cubic symmetry (like Si, Ge, GaAs) is a Hall (specific) resistance which relates the generated electromotive force per unit magnetic field to the impressed (imposed) current, and which is proportional to the inverse charge-carrier density. For anisotropic materials, however, the material property describing the Hall effect response is an axial tensor of the third rank, relating the simultaneous action of a current and a magnetic field vector of given directions with respect to the crystallographic axes to a generated Hall field vector. For triclinic symmetry the Hall tensor has 9 independent components.

Cyclotron resonance requires that a charge-carrier experiences so little scattering that it by far exceeds the constraint of small angular deflection, allowing it to perform at least one full orbit about a magnetic field line, thus defining a cyclotron resonance frequency $\omega_c = eB/m^*c$ (in the most simple case of isotropic m^*; for more general cases the tensorial nature of $1/m^*$ has to be taken into account). Together with the scattering time, reflected by the line width, the mobility can be calculated.

Hall effect, cyclotron resonance and other related magnetotransport phenomena can, in principle, give valuable transport information (charge-carrier mobilities, effective masses, scattering times), but in the organic field these methods have essentially only been used successfully with some *quasimetallic* or *highly semiconductive* organic radical ion salts, where planar aromatic molecules form linear segregated stacks (see, for example, [98–100]). In these structures high mobilities have been found even at room temperature; however, because transport is extremely anisotropic, often quasi-one-dimensional, an unambiguous evaluation of the primary experimental data is not trivial in view of the complicated tensorial nature of the Hall effect. Hall effect measurements published for a few organic *photoconductors*, such as, for example, anthracene, gave either contradictory results among different authors or lacked a sign reversal check, and thus have not allowed reliable final conclusions to be drawn [93]. The signals attributed to cyclotron resonance of holes in the only cyclotron resonance experiment reported for a wide band gap organic crystal, anthracene, performed at 2 K [101, 102], were probably genuine, but could never be confirmed with other samples, probably because the exceptional crystal quality needed for such measurements was not reproducible. The high mobility derivable from these data, however, is not unreasonable for unrelaxed charge-carriers [103], and not in contradiction with low-temperature TOF mobility data obtained for anthracene [16, 30] and naphthalene [5, 16], and with recent field effect transistor mobility data [80, 81].

Photo-Electromagnetic Effects

Charge-carrier diffusion in the absence of an external electric field can be modified by a magnetic field. A number of such effects have been reported in the literature and their applicability to organic materials might be worth checking (e.g., changes of a Dember effect signal in the presence of a magnetic field, or generation of a pulsed circular current and a related magnetic field, caused by carriers diffusing off a light-pulse-excited point source).

8.4 Experimental Results

8.4.1 Magnitude, Anisotropy, and Temperature Dependence of Mobilities

In Figs. 8.7, 8.8, and 8.9 typical TOF mobility results are presented, plotted as $\log \mu$ versus $\log T$, where a power law temperature dependence should lead to a straight line.

While Figs. 8.7 and 8.9 display primary results ("mobilities parallel to the field direction", see Sect. 8.2.4), Fig. 8.8 gives the principal electron mobilities in naphthalene as the result of a tensor evaluation of mobilities measured in more than 10 different directions. Below ca. 100 K all principal electron mobility components exhibit a $\mu \propto T^{-3/2}$ temperature dependence, which in inorganic semiconductor physics usually is taken as a clear indication of acoustic deformation potential scattering (see (8.5)). Above 100 K there is a weaker temperature dependence superimposed, a fact which has briefly been mentioned already in Sect. 8.1 as being indicative of a transition to a more efficient polaronic hopping transport at higher temperatures. More about this subject may be found in [4, 30]. At low temperature/high electric field the hole mobilities in naphthalene (Fig. 8.7) are field-dependent, reflecting a sublinear velocity/field relation, i.e., non-Ohmic transport, with the highest

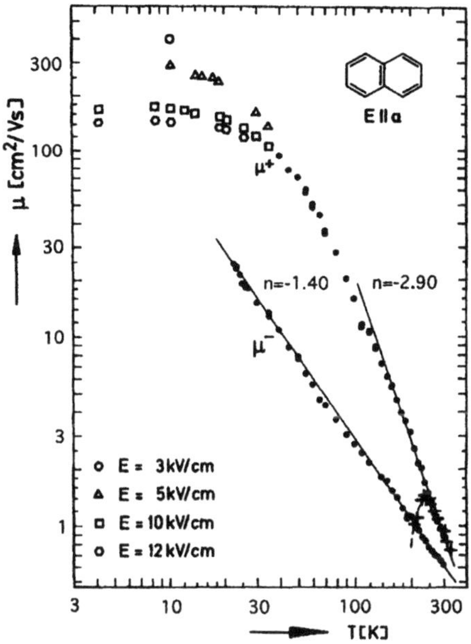

Fig. 8.7. TOF electron and hole mobilities measured in an ultrapurified naphthalene crystal between room temperature and 4 K (• and open symbols). For comparison, previously available data, obtained with less purified material (+) [104] are inserted (TOF measurements). At low temperature/high electric field mobilities are field-dependent with the highest mobilities measured at the lowest practicable fields, reflecting a sublinear velocity/field relation, i.e., non-Ohmic transport. (From [6])

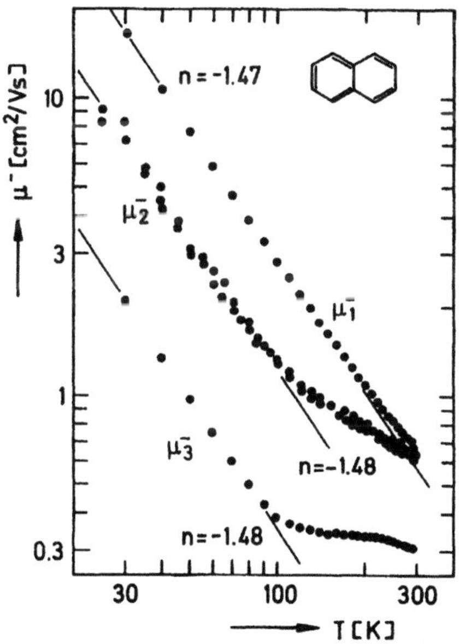

Fig. 8.8. The three principal electron mobility tensor components in naphthalene between 30 and 300 K (TOF measurements). While below 100 K the temperature dependence is $\mu^- \propto T^{-1.5}$ in all directions; much weaker dependences are obtained at higher temperatures for the principal directions 2 and 3, the latter being approximately perpendicular to the *ab* plane, i.e., parallel to c'. In direction 3 the electron mobility is surprisingly nearly temperature-independent over two hundred Kelvin; a weaker temperature dependence is also observed in the orthogonal *b* direction (μ_2) [62,63]

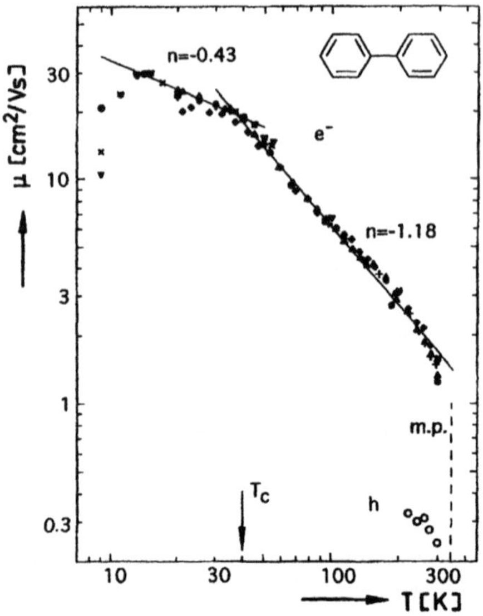

Fig. 8.9. The electron and the hole mobility in a single crystal of highly purified biphenyl (with a lifetime of triplet excitons of 560 ms [9]) increase with decreasing temperature. At the soft mode phase transition to an incommensurately modulated phase at 40 K (see [105, 106]), the exponent of the power-law temperature dependence of the electron mobility changes from −1.18 to −0.43 (5 series of TOF measurements approximately for $\boldsymbol{E} \parallel [010]$). – A microscopic hole transport could only be followed down to 220 K; below it became multiple shallow trapping limited with a trap activation energy of 370 meV in the sample available [65]

mobilities measured at the lowest practicable fields. Such non-linearities have also been found at sufficiently low temperature for the other sign, and for a few other ultrapure organic crystals (biphenyl, anthracene, perylene), see, for example, Fig. 8.13 and the references [5,6,10,30,56]. In Fig. 8.10 typical examples of the anisotropy of charge-carrier mobility tensors are displayed for the *ac* mirror plane of these monoclinic crystals. From these representations the mobility μ in a given direction can be read, although the field direction necessary to induce this velocity direction is not directly obtained. The respective perpendicular principal mobility components μ_{bb} may be read from Table 8.1.

A polaronic hopping-type temperature dependence of transport, as mentioned in connection with Fig. 8.8 for naphthalene has also been measured for the electron mobility in pyrene and for the hole mobility in perylene (see Fig. 8.11a,b) whereas for the respective other charge-carrier sign a $\mu \propto T^n$ $(-3 < n < -1)$ dependence has been obtained for these materials down to low temperatures, which can be taken as proof of successful purification. Notice that a $\log \mu$ versus inverse temperature Arrhenius type of plot has been used in Fig. 8.11, which indicates thermally activated mobilities, albeit not with a single activation energy; more seriously, if one tries to nevertheless fit the data with a straight line, an activation energy around $k_{\mathrm{B}}T$ is obtained, which would need the assumption of a (very shallow) trap of a concentration near to 100%; one would thus exclude a multiple shallow trapping model. Notice also that in Figs. 8.8a,b the absolute mobility values and the temperature dependencies are surprisingly similar. It is worth mentioning that both materials, pyrene and perylene crystallize in a sandwich pair configuration of the molecules. This fact leads to the speculation that these abnormal temper-

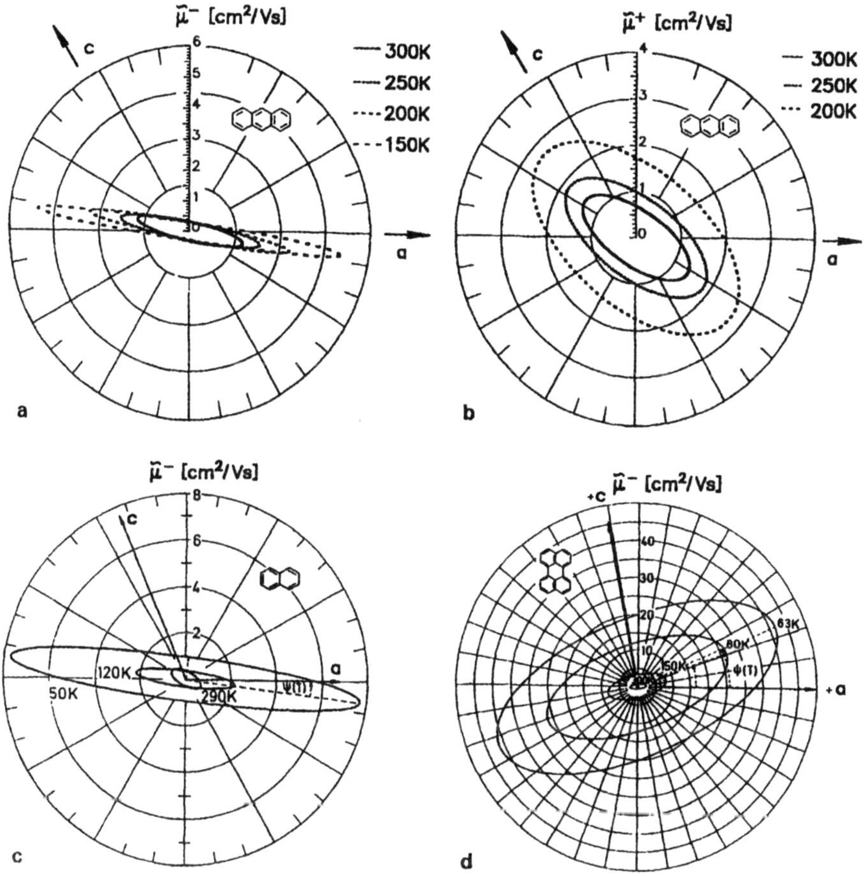

Fig. 8.10. Anisotropy and temperature dependencies of TOF charge-carrier mobilities in the crystallographic *ac* plane, presented by the μ_1-μ_3 principal axis cross-section of the tensor ellipsoids. (**a**) Electron mobilities in anthracene [30]; (**b**) hole mobilities in anthracene [30]; (**c**) electron mobilities in naphthalene [17,63], (**d**) electron mobilities in perylene [17, 107]. The respective mobilities μ_2 along the third principal axis, oriented in the [010] = *b* direction of these monoclinic crystals are given in Table 8.1

ature dependencies of the mobility have to do with the question of how the molecular pair interacts in the monopositive or in the mononegative radical ion state. For one sign the charge-induced additional interaction could be attractive, for one sign repulsive, depending on the details of the wavefunctions. In the attractive case the "band-type" mobility may be suppressed and only a rather unspecific hopping across intermolecular atom–atom contacts may survive, while in the repulsive case charge-carrier localization may be substantially smaller, which would support a dominance of a larger "band-like" contribution.

Table 8.1. The measured *b*-axis principal mobilities of the materials presented in the Fig. 8.10a–d (in cm^2/Vs). *Especially pure *perdeuterated* sample $C_{14}D_{10}$. **Slightly less pure/less perfect $C_{14}H_{10}$ sample at 7.5 kV/cm. ***For μ_{bb}^{+} of perylene (see Fig. 8.11)

	T(K)	μ_{bb}^{-}	μ_{bb}^{+}
(a,b) Anthracene [30]	300	1.16	2.93
	250	1.39	3.74
	200	1.74	5.05
	150	2.36	7.4
	120	2.96	11.3
	80	4.7	22
	50	7.7	39
	35	11.0	54
(c) Naphthalene [17,63]	290	0.64	1.47
	120	1.07	13.0
	50	3.00	115*
	30	7.7	140**
	20	–	205**
(d) Perylene [17,107]	300	5.53	***
	150	18.3	
	80	53.5	
	63	80.0	
	50	115	
	40	150	

8.4.2 Mobilities Across Phase Transitions

Following mobilities across structural phase transitions provides a unique possibility to study structure–transport interrelations on a microscopic scale and to put theoretical concepts on a broader basis. This field has not yet been exploited to a large extent. An experimental example, phenanthrene, has been given in [10] and in [65]: the hole mobility, which amounts to $\mu_{aa} = 6.0\ cm^2/Vs$, $\mu_{bb} = 8.1\ cm^2/Vs$, $\mu_{c'c'} = 2.1\ cm^2/Vs$ at 230 K, displays a rather abrupt steep decrease by more than one order of magnitude as temperature upon rising approaches the order/disorder phase transition at $T_c = 345$ K [10], where the dipolar molecules begin to rotate in the initially acentric crystal, giving rise to a pseudo-centrosymmetric structure, as was proven by the loss of second-order optical non-linearity [108] (see also [109]). Surprisingly, the electrons do not see much of the phase transition [65].

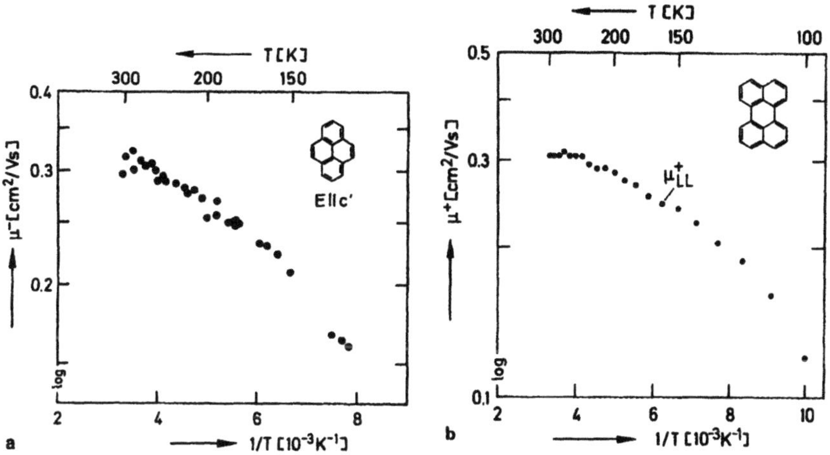

Fig. 8.11. The TOF electron mobility $\mu^-_{c'c'}$ in pyrene (**a**) and the hole mobility μ^+_{LL} in perylene (**b**) follow neither a $\mu \propto T^n$ nor a uniformly thermally activated temperature dependence, as can be seen from the $\log \mu$ versus $1/T$ Arrhenius plot. As the crystals were made from highly purified material, it is assumed that electron transport in pyrene and hole transport in perylene are based on polaron hopping – in contrast to that of the opposite sign. For the perylene measurement the orientation of the unit vector L was $L = (0.581;\ 0;\ 0.814)$ in the crystallographic a, b, c' coordinate system, but the hole mobilities were only weakly anisotropic [107]

Biphenyl, which in some respects may be considered a model substance for poly-*p*-phenylene, undergoes, upon cooling, a soft mode structural phase transition to an incommensurately modulated phase at 40 K [105, 106]. At this transition temperature the exponent of the power law temperature dependence of the electron mobility changes from $n = -1.18$ to -0.43 (see Fig. 8.9); incommensurability "disorder" thus drastically slows down the further increase of the mobility upon further cooling (below 15 K the influence of very shallow residual traps becomes dominant).

8.4.3 The Role of Physical and Chemical Defects

Chemical defects are ubiquitous in organic crystals and usually dominate the trapping-governed transport behavior. Thus, the influences of physical defects on transport can only be studied in ultrapurified material.

Physical defects such as "zero-dimensional" empty lattice sites or interstitial molecules as well as "one-dimensional" dislocations and two or three-dimensionally strained lattice regions increase or decrease the local polarization energies in organic crystals and thus can form traps or antitraps for charge carriers. Little is experimentally known on a systematic basis about the influences on charge-carrier transport caused by such physical defects in organic crystals. While point defects are difficult to assess, dislocations can be made visible by X-ray topography [110]. Under favorable conditions large cross sections of organic crystals can be free of dislocations (see Chapter 6).

Physical defects can also be introduced by dopants which do not trap charge-carriers themselves ("antitraps" [111]), but distort the lattice environment to some extent. Such physical defects, as well as lattice strain, often form large concentrations of *very shallow* traps, which therefore noticeably affect TOF mobilities only in chemically pure crystals at sufficiently low temperatures. An example (carbazole in anthracene) has been given in [112]. Minor effects at higher temperatures (see calculations in [70]) are often overlooked and the absolute accuracy of TOF mobilities is overestimated, even if apparently a power-law temperature dependence is approximately obeyed. Therefore, in retrospect, reported effects of isotopic substitution (H by D) on the mobilities have remained questionable under more realistic estimates of the experimental uncertainties.

The properties of *disordered organic solids* are dominated by physical defects; these compete with chemical defects or may even govern transport. However, this is not the subject of this chapter. It should be nevertheless mentioned that there are unique model systems to study disorder: dipolarly disordered crystals such as 2,3-dimethyl-naphthalene and 2,3-dimethyl-anthracene, which combine a translationally symmetric lattice with a very simple dipole moment up/dipole moment down molecular disorder on the lattice sites [113,114]. Charge-carrier mobilities are polaron-hopping-like (see Fig. 9 of [65]), with a temperature dependence very similar to that found for the electrons in pyrene, and for the holes in perylene (see Fig. 8.11). A study of *chemical defects* under well-controlled conditions requires the availability of high purity material as a reference. Early reliable investigations performed with the system anthracene and anthracene doped with tetracene have been communicated in [69, 115–117]. An example of multiple shallow-trapping-limited transport, measured by TOF pulses, is displayed in Fig. 8.12 [12,118], see also Fig. 6.3. It is important to realize that even ppm concentrations of impurities can reduce the charge-carrier mobilities by orders of magnitude.

8.4.4 Non-linear Transport

Non-linear transport is expected when the uptake of kinetic energy of the charge-carrier from the field is fast enough (high mobility) to compete with inelastic scattering losses, and the charge-carrier therefore heats up to more than $k_B T_{\text{lattice}}$. Fig. 8.13a and b display velocity–field relations as measured by TOF experiments for electrons in perylene and for holes in (perdeuterated) naphthalene. A distinct sublinear non-Ohmic behavior is obtained for velocities exceeding ca. 0.5×10^6 cm/s, with a tendency to saturate at several 10^6 cm/s for the electrons in perylene and a very clear saturation at 0.9 $\times 10^6$ cm/s for the holes in naphthalene, for which TOF pulses could be obtained down to the temperature of liquid He. Saturated electron velocities in

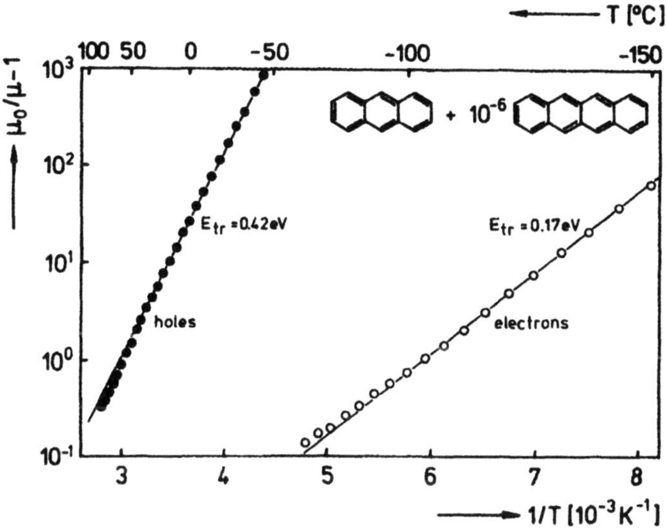

Fig. 8.12. The c' electron and hole mobility in an anthracene single crystal, doped with ca. 10^{-6} mol/mol tetracene (see Fig. 6.3) turn out to be multiple shallow-trapping-limited "effective" mobilities. A Hoesterey–Letson plot of $\log(\mu_0 - \mu_{\text{eff}})/\mu_{\text{eff}}$ versus $1/T$ yields straight lines over 3 orders of magnitude; from the slopes the electron and hole detrapping activation energies ("trap depths") can be read (TOF measurements) [70, 119]. Extrapolation to $1/T \to 0$ allows an estimation of the effective trap concentration to be made

naphthalene are demonstrated by the field-independent TOF pulse durations displayed in Figs. 8.3d$_1$,d$_2$.

These results can be used to estimate the effective mass of the charge carriers (see Sect. 8.2.3). In favorable cases (lowest attainable temperature, fastest crystallographic direction) these masses come out to be on the order of a free electron mass [5, 16]. Thus, a band-type transport is proven for the low-temperature regime for these ultrapure and high-perfection single crystals. It is clear that physical and chemical defects can only deteriorate these properties.

8.4.5 Mobilities in Organic Thin Films

SAW Mobilities

As an example of a mobility measurement by the SAW method, Fig. 8.14 displays the temperature dependence of the electron mobility in 50 nm films of the radical ion salt $\text{BIPA}^+:(\text{TCNQ})_2^-$, vacuum vapor deposited onto a planar glass substrate. SAW mobilities could also be measured in thin films of Y^{+3} bis-phthalocyanine and Lu^{+3} bis-phthalocyanine [120] and in single crystals of the dark-conductive charge-transfer complexes tetrathiafulvalene:chloranil

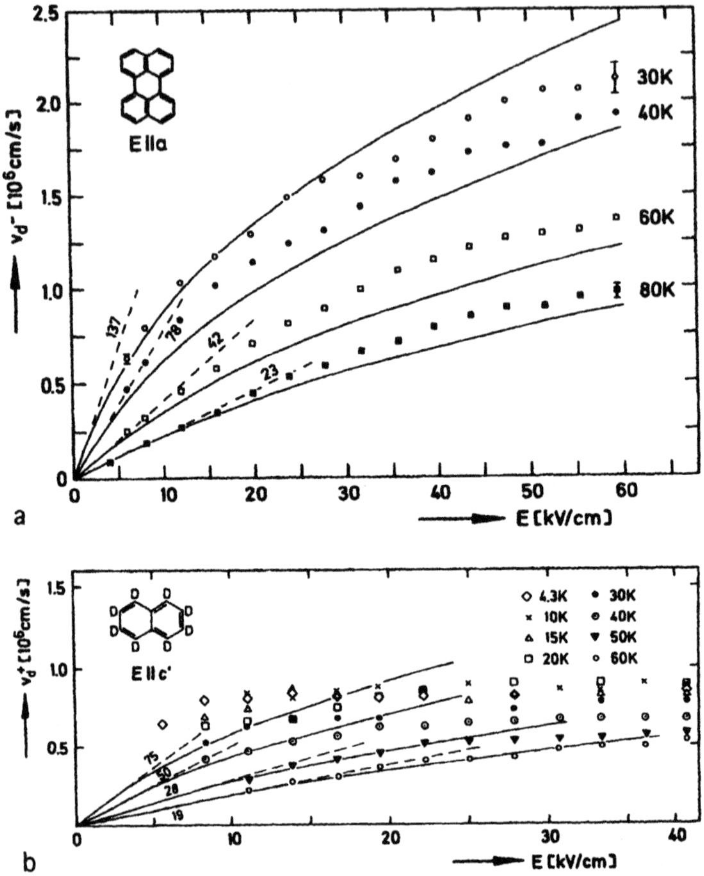

Fig. 8.13. The low-temperature TOF charge-carrier velocities in several organic crystals exhibit a gradual transition from a linear to a sublinear electric field dependence as temperature is lowered and/or the field intensity increased, with a tendency towards saturation. The low field slopes represent the Ohmic mobilities. As examples (**a**) the electron drift velocity v_a in perylene [16] and (**b**) the hole drift velocity $v_{c'}$ in perdeuterated naphthalene [16] are presented with the low field mobilities indicated. The full lines are fits by the acoustic deformation potential scattering model (see Eq. (8.7)). (The results for normal, i.e., hydrogenated naphthalene, $C_{10}H_8$, were essentially the same within the experimental uncertainty limits)

(TTF:CA) and phenothiazine:tetracyanoquinodimethane (PTZ:TCNQ) [52]. More details may be found in [77].

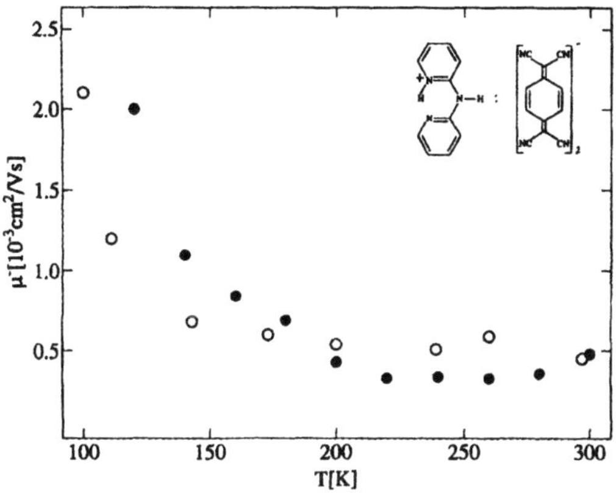

Fig. 8.14. Electron mobility in dark-conductive thin films (○ 50 nm, • 51 nm) of the radical ion salt $BIPA^+{:}(TCNQ)_2^-$, measured by the surface acousto-electric travelling wave (SAW) method between 100 and 300 K [65]. The films were prepared by vacuum vapor deposition of the radical ion salt onto a planar glass substrate

FET Mobilities

Early trials to measure an electric field effect with organic materials were made by Bauser and Pernisz [121] (see also the references given therein). Recent examples of FET mobility measurements are given in Fig. 8.15a and b for a 52 nm pentacene film and for a 59 nm diindeno(1,2,3-cd,1′, 2′, 3′-ℓm)perylene (DIP) film on a 270 nm SiO_2 gate insulation layer on a Si gate electrode. From the observed sign asymmetry of the characteristics, *p*-type conduction is concluded for both materials. The hole mobility evaluated from the saturation regime from a $I_{D,sat}^{1/2}$ versus U_G plot according to (8.30) amounts to 1.1×10^{-2} cm^2/Vs for the pentacene sample and to 0.9×10^{-2} cm^2/Vs for the DIP sample, whereas from the linear regime 3.5×10^{-2} cm^2/Vs was obtained for the latter.

In the literature FET hole mobilities of around 0.1 cm^2/Vs were also reported for di-hexyl-sexithiophene [122–126, 132] and even up to 0.6 cm^2/Vs for the extreme donor type (i.e., very low ionization energy) material pentacene [127–129]. It was recognized that – besides purity [79, 129] – the kind and degree of ordering of the organic semiconductor film is very important for achieving high FET mobilities; a good choice of the insulating layer on the gate electrode is also crucial (see, for example, [126, 129]), which demonstrates the importance of details of the layer to substrate interactions and of interface states. Eloquent examples of the importance of structural order for obtaining high mobilities are given in Fig. 8.16. Eight transistors were made simultaneously during one thin-film deposition run; the points in one vertical row show the scatter of the mobilities of different transistors of one batch.

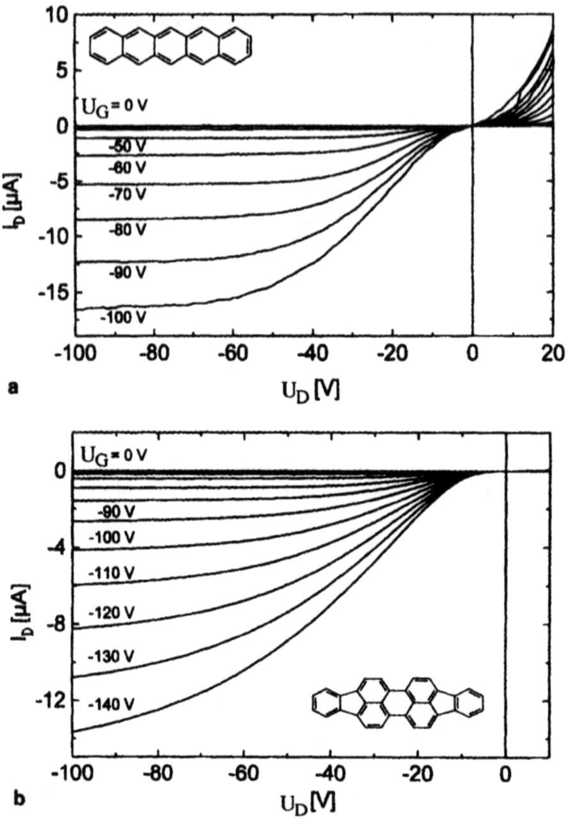

Fig. 8.15. Characteristics of 2 mm channel-width organic thin-film field-effect transistors (OFETs) made on a SiO_2 gate insulation layer on a highly doped *p*-Si (111) wafer (with Ohmic contact), used as the gate electrode. (**a**) Pentacene, 52 nm film on 270 nm SiO_2, vacuum vapor deposited at a substrate temperature T_s of 363 K at a rate of 0.09 nm/s (base pressure of 2×10^{-8} torr); channel length 35 μm; in situ measurements. (**b**) Diindeno(1,2,3-cd, 1′, 2′, 3′-ℓm)perylene (DIP), 59 nm film on 430 nm SiO_2, vacuum vapor deposited at a T_s of 423 K at a rate of 0.09 nm/s (base pressure of 10^{-7} torr); channel length 45 μm; in situ measurements [130,131]

The different vertical rows display batches of different X-ray rocking width, which was deliberately varied by depositing the organic thin film at different substrate temperatures.

Two facts deserve special attention: although the substrate was a disordered SiO_2 layer on a silicon wafer, the organic layers displayed remarkably narrow X-ray rocking widths, in the best cases down to 0.16° for pentacene and 0.02° (the experimental resolution limit) for DIP, which is attributed to a tendency for self-ordering (structural self-organization) of the molecules. The charge-carrier mobilities varied over 6 orders of magnitude within a small in-

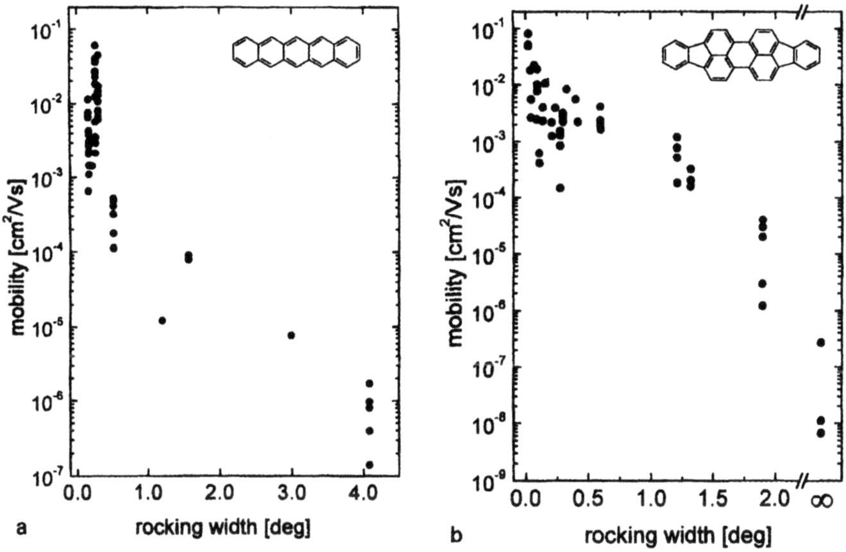

Fig. 8.16. Organic field-effect transistors (see caption of Fig. 8.15) with different structural order display widely varying field-effect mobilities. Structural order was varied by the substrate temperature used during deposition, and characterized by the angular width of X-ray rocking curves [130, 131]

terval of rocking widths, and the highest mobilities approached 0.1 cm^2/Vs, albeit with one order of magnitude of scatter. A correlation between X-ray rocking width and magnitude of mobilities has also been observed for TOF mobilities through thin films of PTCDA [133]. A lot of work is left to understand the *microscopic* correlates of the highest mobilities in terms of size and azimuthal orientation of single crystal grains, grain interfaces (see [134, 135]), sample purity, incorporation and segregation of impurities and dopants [135], contact properties, etc.

For organic *thin films* mobilities around 10^{-1} cm^2/Vs are usually classified as "high"; however, it can be expected that for samples with higher chemical purity and more well-defined, uniform (hopefully epitaxial) order between the source and drain electrode, mobility values comparable to those obtained with single crystals, as presented in this chapter, will be achievable.

8.5 Concluding Remarks

I have attempted to give a broad overview on charge-carrier transport in well-ordered organic crystals, focusing on the fundamental transport property, *charge-carrier mobility*, starting from a presentation of the theoretical foundaments, to a discussion of suitable experimental measurement techniques, and to a basic understanding of the results. It has been shown that mobilities in well-ordered, high-purity organic crystals can be rather large, approaching and even exceeding values at *low temperature* which are typical for silicon at room temperature, a fact which supports a conventional band-type description for the low-temperature regime. The smooth continuation of a band-type $\mu \propto T^n$ $(-3 < n < -1)$ power-law decrease of mobility with increasing temperature up to room-temperature and above, observed for quite a number of organic crystals, is not well understood.

It has been shown that *room temperature* mobilities around 1–5 cm^2/Vs can be obtained for typical aromatic molecular crystals. Mobilities in well ordered organic thin films can approach such values, an important fact for potential applications. The broad and much more intricate field of low-mobility dispersive transport in (more or less) disordered organic solids, which in the eyes of the author will have, at best, only limited practical importance for future electronic applications, has been omitted.

For further details on the topics touched in this chapter and for a broader overview of the entire field of organic photoconductors and semiconductors, the books by Kao and Hwang [136], Silinsh and Čápek [3], and Pope and Swenberg [93] should be consulted, and the citations of further books and papers given in [17] followed.

References

1. T. Holstein, Ann. Phys. (N.Y.) **8**, 325 and 343 (1959)
2. G.G. Robertson, N. Aspley, and R.W. Munn, Physics Reports **60**, 59–150 (1980)
3. E.A. Silinsh, V. Čápek, *Organic Molecular Crystals* (American Inst. of Phys., New York 1994)
4. E.A. Silinsh, A. Klimkāns, S. Larsson, V. Čápek, Chem. Phys. **198**, 311 (1995)
5. W. Warta, N. Karl, Phys. Rev. B **32**, 1172 (1985)
6. W. Warta, R. Stehle, N. Karl, Appl. Phys. A **36**, 163 (1985)
7. R.W. Munn, Inst. Phys. Conf. Ser. **58**, 98 (1981)
8. V. Čápek, R.W. Munn, Phys. Stat. Sol. B **108**, 521 (1981); **109**, 245 (1982)
9. N. Karl, In: *Crystals, Growth, Properties and Applications*, ed. by H.C. Freyhardt (Springer-Verlag, Berlin 1980) pp. 1–100
10. N. Karl, Mol. Cryst. Liq. Cryst. **171**, 31 (1989)
11. N. Karl, In: *Materials for Nonlinear and Electrooptics*, ed. by M.H. Lyons, Institute of Physics Conference Series N° 103 (Institute of Physics, Bristol 1989), pp.107–118
12. N. Karl, J. Cryst. Growth **99**, 1009 (1990)
13. H. Scher, E.W. Montroll, Phys. Rev. B **12**, 2455 (1975)
14. H. Scher, In: *Photoconductivity and Related Phenomena*, ed. by J. Mort, D.M. Pai (Elsevier, Amsterdam 1976) pp. 71–115
15. W. Shockley, *Electrons and Holes in Semiconductors* (van Nostrand, Princeton, NJ 1950)
16. N. Karl, J. Marktanner, R. Stehle, W. Warta, Synth. Met **42–43**, 2473 (1991)

17. N. Karl, "*Organic Semiconductors*", In: Landolt-Boernstein (New Series), Group III, Vol. 17 *Semiconductors*, ed. by O. Madelung, M. Schulz, H. Weiss (Springer-Verlag Berlin, Heidelberg, New York, Tokyo 1985) Subvolume 17i, pp. 106–218. Updated edition: Vol. 41E (2000) (Update of "Organic semiconductor" on CD)
18. W. Shockley, Bell Syst. Tech. J. **4**, 990 (1951)
19. J.G. Ruch, G.S. Kino, Appl. Phys. Letters **10**, 40 (1967)
20. J.L. Su, Y. Nishi, J.L. Moll, A. Neukermans, Solid State Elect. **13**, 1115 (1970)
21. P.A. Tove, G. Andersson, G. Ericsson, R. Lidholt, IEEE Trans. Electron Dev. **ED17**, 407 (1970)
22. A. Neukermans, G.S. Kino, Phys. Rev. B **7**, 2693 (1973)
23. C. Canali, C. Jacoboni, F. Nava, G. Ottaviani, A. Alberigi-Quaranta, Phys. Rev. B **12**, 2265 (1975)
24. H. Kahlert, "*Hot Electrons*", In: Landolt-Boernstein (New Series), Group III, Vol. 17 *Semiconductors*, ed. by O. Madelung, M. Schulz, H. Weiss (Springer-Verlag, Berlin, Heidelberg, New York, Tokyo 1985) Subvolume 17i, pp. 106–218
25. K. Seeger, *Semiconductor Physics* (Springer-Verlag, Berlin 1985)
26. A.P. Mills, N. Karl, Phys. Rev. B **48**, 7050 (1993)
27. C.E. Swenberg, M. Pope, Chem. Phys. Lett. **287**, 535 (1998)
28. J.F. Nye, *Physical Properties of Crystals* (Clarendon Press, Oxford, 1957, 1972)
29. D. Massa, N. Karl, Mol. Cryst. Liq. Cryst. **175**, 93 (1989)
30. N. Karl, J. Martanner, Mol. Cryst. Liq. Cryst. **357**, 1 (2000)
31. R. Bray, Phys. Rev. **76**, 152 (1949)
32. J.R. Haynes, W. Shockley, Phys. Rev. **75**, 691 (1949)
33. F.S. Goucher, Phys. Rev. **81**, 475 (1951); Bell System Tech. J. **30**, 990 (1951)
34. J.R. Haynes, W. Shockley, Phys. Rev. **81**, 835 (1951)
35. J.R. Haynes, W.C. Westphal, Phys. Rev. **85**, 680 (1952)
36. R. Lawrance, A.F. Gibson, Proc. Phys. Soc. (Lond.) **65B**, 994 (1952)
37. W. Shockley, W.T. Read Jr., Phys. Rev. **87**, 835 (1952)
38. W.E. Spear, Proc. Phys. Soc. (Lond.) **B 70**, 669 (1957)
39. W.E. Spear, Proc. Phys. Soc. **76**, 826 (1960)
40. W.E. Spear, J. Phys. Chem. Solids **21**, 110 (1961)
41. W.E. Spear, J. Mort, Proc. Phys. Soc. **81**, 130 (1963)
42. A.R. Adams, W.E. Spear, J. Phys. Chem. Solids **25**, 1113 (1964)
43. A.R. Adams, D.J. Gibbons, W.E. Spear, Solid State Comm. **2**, 387 (1964)
44. C. Canali, G. Ottaviani, M. Martini, K. Zaino, Appl. Phys. Lett. **19**, 51 (1971)
45. K.A. Yamakawa, Phys. Rev. **82**, 522 (1951)
46. Ch. Allemand, J. Rossel, Helvetica Phys. Acta **27**, 519 (1954)
47. F.C. Brown, Phys. Rev. **97**, 355 (1955)
48. A. Many, M. Simhony, S.Z. Weisz, J. Levinson, J. Phys. Chem. Solids **22**, 285 (1961)
49. W.E. Spear, A.R. Adams, J. Phys. Chem. Solids **27**, 281 (1966)
50. W. Helfrich, P. Mark, Z. Physik **166**, 370 (1962)
51. F.K. Dolezalek, In: *Photoconductivity and Related Phenomena*, ed. by J. Mort, D.M. Pai (Elsevier, Amsterdam, 1976) pp. 27–69
52. K.-H. Kraft, PhD thesis dissertation, University of Stuttgart (1993)
53. N. Karl, In: *Festkörperprobleme/Adv. Solid State Phys.* XIV, ed. by H.J. Queisser (Pergamon/Oxford, Vieweg/Braunschweig 1974) pp. 261–290

54. M. Eichhorn, F. Willig, K.-P. Charlé, K. Bitterling, J. Chem. Phys. **76**, 4648 (1982)
55. J. Mort, G. Pfister, In: *Electronic Properties of Polymers* ed. by J. Mort, G. Pfister (J. Wiley, New York 1982) pp. 215–265
56. K. Bitterling, F. Willig, Phys. Rev. B **35**, 7973 (1987)
57. K. Nakagawa, M. Kotani, H. Tanaka, Phys. Stat. Sol. B **102**, 403 (1980)
58. C. Bogus, PhD thesis dissertation, University of Giessen (1967)
59. D.F. Williams, M. Schadt, J. Chem. Phys. **53**, 3480 (1970)
60. G. Sommer, N. Karl, unpublished results; G. Sommer, PhD thesis dissertation, University of Stuttgart (1974)
61. R.G. Kepler, Phys. Rev. **119**, 1226 (1960)
62. N. Karl, In: *Spektroskopie amorpher und kristalliner Festkoerper*, ed. by D. Haarer and H.W. Spiess (Steinkopff Verlag, Darmstadt 1995), pp. 453–481
63. W. Warta, Diplomarbeit, University of Stuttgart (1978) and PhD thesis dissertation, University of Stuttgart (1985)
64. O.H. LeBlanc Jr., J. Chem. Phys. **33**, 626 (1960)
65. N. Karl, K.-H. Kraft, J. Marktanner, M. Muench, F. Schatz, R. Stehle, H.-M. Uhde, J. Vac. Sci. Technol. A **17**, 2318 (1999)
66. N. Karl, E. Schmid, M. Seeger, Z. Naturforsch. **25a**, 382 (1970)
67. N. Karl, G. Sommer, Phys. Stat. Sol. A **6**, 231 (1971)
68. N. Karl, PhD thesis dissertation, University of Freiburg/Brg. (1968)
69. D.C. Hoesterey, G.M. Letson, J. Phys. Chem. Solids, **24**, 1609 (1963)
70. N. Karl, In: *Defect Control in Semiconductors*, Vol. II ed. by K. Sumino (North Holland, Amsterdam 1990) pp. 1725–1746
71. R.H. Tredgold, *Space Charge Conduction in Solids* (Elsevier, Amsterdam 1966)
72. A.C. Papadakis, J. Phys. Chem. Solids **28**, 641 (1967)
73. R. Adler, D. Janes, B.J. Hunsinger, S. Datta, Appl. Phys. Lett. **38**, 102 (1981)
74. D. Janes, S. Datta, R. Adler, B.J. Hunsinger, In: *Topical Conference on Tetrahedrally Bonded Amorphous Semiconductors*, ed. by R.A. Street, D.F. Biegelsen, J.C. Knights (AIP, New York 1981) pp. 222–226
75. H. Matthews, *Surface Wave Filters* (J. Wiley, New York, 1977)
76. J. Marktanner, Ph.D. thesis dissertation, University of Stuttgart (1995)
77. N. Karl, K.-H. Kraft, J. Marktanner, Synth. Met. **109**, 181 (2000)
78. G. Horowitz, R. Hajlaoui, H. Bouchriha, R. Bourguiga, M. Hajlaoui, Adv. Mater. **10**, 923 (1998)
79. G. Horowitz, R. Hajlaoui, D. Fichou, A. El Kassmi, J. Appl. Phys. **85**, 3202 (1999)
80. J.H. Schön, S. Berg, Ch. Kloc, B. Batlogg, Science **287**, 1022 (2000)
81. J.H. Schön, Ch. Kloc, B. Batlogg, Science **288**, 2338 (2000)
82. J.H. Schön, Ch. Kloc, B. Batlogg, Nature **406**, 702 (2000)
83. J.H. Schön, Ch. Kloc, R.C. Haddon, B. Batlogg, Science **288**, 656 (2000) and J.H. Schön, Ch. Kloc, B. Batlogg, Nature **408**, 549 (2000)
84. J.B. Arthur, W. Bardsley, M.A.C.S. Brown, A.F. Gibson, Proc. Phys. Soc. B **68**, 43 (1955)
85. R. Bray, Phys. Rev. **100**, 1047 (1955)
86. K. Arlauskas, V. Gaidelis, K. Genevičius, G. Juška, Synth. Met. **109**, 101 (2000); G. Juška, K. Arlauskas, R. Österbacka, H. Stubb, Synth. Met. **109**, 173 (2000)

87. G. Juška, K. Arlauskas, M. Viliūnas, J. Kočka, Phys. Rev. Lett. **84**, 4946 (2000)
88. A.M. van de Graats, J.M. Warman, M.P. de Haas, D. Adam, J. Simerer, D. Haarer, P. Schuhmacher, Adv. Mater. **8**, 823 (1996)
89. G.H. Gelnick, J.M. Warman, J. Phys. Chem. **100**, 20035 (1996)
90. R.J. Hoofman, L.D. Sibbeles, M.P. de Haas, A. Hummel, D. Bloor, J. Chem. Phys. **109**, 1885 (1998)
91. P.H. Nguyen, G. Paasch, W. Brütting, W. Riess, Phys. Rev. B **49**, 5172 (1994)
92. W. Brütting, P.H. Nguyen, W. Riess, G. Paasch, Phys. Rev. B **51**, 9533 (1995)
93. M. Pope, C.E. Swenberg, *Electronic Processes in Organic Crystals and Polymers*, 2nd edn. (Oxford Sci. Publ., New York, Oxford 1999)
94. T.S. Moss, L. Pincherle, A.M. Woodward, Proc. Phys. Soc. (Lond.) B**66**, 743 (1953)
95. D. Sarkar, N.J. Halas, Sol. State Comm. **90**, 261 (1994)
96. G.G. Maresch, A. Grupp, M. Mehring, J.U. von Schütz, Synth. Met. **16**, 161 (1986)
97. G.G. Maresch, M. Mehring, J.U. von Schütz, H.P. Werner, K. Gökelmann, V. Enkelmann, K. Müllen, K.U. Klabunde, J. Chem. Phys. **91**, 4543 (1989)
98. B. Korin-Hamzic, L. Forro, J.R. Cooper, Phys. Rev. B **41**, 11646 (1990)
99. D. Schweitzer, E. Balthes, S. Kahlich, I. Heinen, H.J. Keller, W. Strunz, W. Biberacher, A.G.M. Jansen, E. Steep, Synth. Met. **70**, 857 (1995)
100. J. Wosnitza, *Fermi Surfaces of Low-Dimensional Organic Metals and Superconductors*, Springer Tracts in Modern Physics, Vol. 134 (Springer-Verlag, Berlin 1996)
101. D.M. Burland, Phys. Rev. Lett. **33**, 833 (1974)
102. D.M. Burland, U. Konzelmann, J. Chem. Phys. **67**, 319 (1977)
103. B. Munn, Chem. Phys. Lett. **52**, 168 (1977)
104. J. Berréhar, M. Schott, Mol. Cryst. Liq. Cryst. **46**, 223 (1978)
105. H. Cailleau, A. Girard, F. Moussa, C.M. E. Zeyen, Solid State Comm. **29**, 259 (1979)
106. H. Cailleau, J.C. Messager, F. Moussa, F. Bugaut, C.M.E. Zeyen, C. Vettier, Ferroelectrics **67**, 3 (1986)
107. R. Stehle, PhD thesis dissertation, University of Stuttgart (1984)
108. J.P. Dougherty, S.K. Kurtz, J. Appl. Cryst. **9**, 145 (1976)
109. J. Kroupa, J. Fousek, N.R. Ivanov, B. Březina, M. Pavel, A. Fousková, V. Lhotská, V. Petříček, I. Císařová, Solid State Comm. **66**, 1003 (1988)
110. H. Klapper, In: *Crystals, Growth, Properties, and Applications*, Vol. 13, ed. by N. Karl (Springer-Verlag, Berlin 1991) pp. 109–162
111. N. Emoto, M. Kotani, Chem. Phys. Lett. **101**, 386 (1983)
112. N. Karl, Chem. Scripta **17**, 201 (1981)
113. N. Karl, H. Heym, J.J. Stezowski, Mol. Cryst. Liq. Cryst. **131**, 163 (1985)
114. M. Dörr, H. Gerlach, J. Kalus, N. Karl, M. Monkenbusch, I. Natkaniec, U. Schmelzer, W. Schmidt, J.J. Stezowski, P. Vonderwisch, G. Voss, M. Warth, J. Phys. Cond. Matt. **10**, 10879 (1998)
115. K. Bogus, Z. Physik **184**, 219 (1965)
116. K. Oyama, I. Nakada, J. Phys. Soc. Jpn. **24**, 792 (1968)
117. A. Schmillen, W. Falter, Z. Physik **218**, 401 (1969)
118. K.-H. Probst, N. Karl, Phys. Stat. Sol. A **27**, 499 (1975); erratum: *ibid.* **31**, 793 (1975)

119. K.-H Probst, Dissertation (PhD thesis), University of Stuttgart (1984)
120. H. Holdik, N. Karl, unpublished results (1991)
121. H. Bauser, U. Pernisz, Chem. Phys. Lett. **11**, 213 (1971)
122. F. Garnier, A. Yassar, R. Hajloui, G. Horowitz, F. Deloffre, B. Severt, S. Ries, P. Alnot, J. Am. Chem. Soc. **115**, 8716 (1993)
123. A. Dodabalapur, L. Torsi, H.E. Katz, Science **268**, 270 (1995)
124. L. Torsi, A. Dodabalapur, A.J. Lovinger, H.E. Katz, R. Ruel, D. D. Davis, K.W. Baldwin, Chem. Mater. **7**, 2247 (1995)
125. H.E. Katz, A. Dodabalapur, L. Torsi, D. Elder, Chem. Mater. **7**, 2238 (1995)
126. C.D. Dimitrakopoulos, B.K. Furman, F. Graham, S. Hedge, S. Purushothaman, Synth. Met. **92**, 47 (1998)
127. D.J. Gundlach, Y.-Y. Lin, T.N. Jackson, S.F. Nelson, D.G. Schlom, IEEE Electron Dev. Lett. **18**, 87 (1997)
128. Y.-Y. Lin, D.J. Gundlach, S.F. Nelson, T.N. Jackson, IEEE Electron Dev. Lett. **18**, 607 (1997)
129. C.D. Dimitrakopoulos, I. Kymissis, S. Purushothaman, D.A. Neumayer, P.R. Duncombe, R.B. Laibowitz, Adv. Mater. **11**, 1372 (1999)
130. M. Münch, PhD thesis dissertation, University of Stuttgart (2000)
131. M. Münch, N. Karl, to be published
132. B. Servet, G. Horowitz, S. Ries, O. Lagorsse, P. Alnot, A. Yassar, F. Deloffre, P. Srivastava, R.Hajlaoui, P. Lang, F. Garnier, Chem. Mater. **6**, 1809 (1994)
133. N. Karl, J. Marktanner, Mol. Cryst. Liq. Cryst. **315**, 163 (1998)
134. J.H. Schön, B. Batlogg, Appl. Phys. Lett. **74**, 260 (1999)
135. J.H. Schön, Ch. Kloc, B. Batlogg, Appl. Phys. Lett. **75**, 1 (1999)
136. K.C. Kao, W. Hwang, *Electrical Transport in Solids, With Particular Reference to Organic Semiconductors* (Pergamon Press, Oxford, New York 1981) (Contains about 2600 citations until 1978)

9 Dielectric and Charge-Transfer Approaches to Excitons

R.W. Munn and P. Petelenz

9.1 Introduction

Excitons are the characteristic elementary excitations of poorly conducting molecular crystals. They are commonly treated in what may be termed the *Hamiltonian approach.* A Hamiltonian is constructed for the crystal with separate terms for the isolated molecules and for the interactions between them. Because the interactions are weak, the crystal states are expressed in terms of the molecular states, and the energy eigenvalues are deduced. The eigenstates are then Frenkel excitons tracing their parentage to different isolated molecular excited states, possibly split and coupled by the crystal interactions. In this chapter, we explore an alternative to the Hamiltonian approach and an extension to it.

The alternative is the *dielectric approach.* This uses the molecular polarizability to construct the response of the crystal charge density to an electric potential that varies in time and space (or the response of the crystal polarization to an electric field). Divergences or poles in the response function as a function of frequency indicate responses that are self-sustaining in the absence of any stimulus and therefore correspond to the elementary excitations of the crystal. Evidently, the dielectric approach should offer advantages in treating the dielectric and optical properties of excitons. Although it appears very different from the Hamiltonian approach, the dielectric approach has an obvious connection through the Hohenberg–Kohn theorem that the energy is a functional of the electron density.

Both approaches rely on properties of the isolated molecule. Given weak interactions between molecules, this may be adequate for many purposes. However, the observation of photoconductivity with an action spectrum that follows the exciton absorption spectrum suggests the need for an approach that is extended to include charge-separated states. Such states clearly cannot exist in the isolated molecule, but the crystal admits the existence of *charge-transfer excitons.* These are constructed like Frenkel excitons from neutral localized molecular configurations, but now the configurations consist of anion–cation pairs. Such configurations have large electric dipole moments that should interact strongly with an external electric field, and hence they should also play an important role in the electro-absorption spectra of molecular crystals.

Extending the Hamiltonian approach to include charge-transfer excitons requires the energies of the localized anion–cation pairs in the crystal environment. Although they could be calculated from first principles, this involves calculating the molecular polarizability (much as calculating the interaction between two polar molecules from first principles yields the induction energy). It is more efficient to recognize the role of the polarizability from the start, and hence a dielectric approach starts to inform the Hamiltonian approach. There is also a question of how the dielectric approach can be extended to include charge-transfer contributions to the response function that will correctly yield poles corresponding to charge-transfer excitons.

The structure of the chapter is as follows. We first introduce the dielectric approach. We then show how the dielectric approach can be applied to nonlinear optics. Next we treat charge-transfer excitons and show the importance of their mixing with Frenkel excitons, drawing on the dielectric approach to support the Hamiltonian approach where appropriate. Finally, we illustrate how the mixed exciton manifold of states can be used to understand electroabsorption spectra and charge-carrier photogeneration, with particular reference to the C_{60} fullerene crystal and the aromatic hydrocarbon crystals as examples that have been intensively studied by many researchers.

9.2 Dielectric Theory of Excitons

The dielectric function plays an important unifying role in theories of the dielectric, optical and dynamical properties of solids. Its conceptual value is well recognized, but its practical value is often limited by the difficulty of obtaining explicit results. In molecular crystals this is not a problem: the natural approach is to describe the properties of the crystal in terms of the molecular properties, the molecular arrangement, and the molecular interactions [1]. Such an approach corresponds to the tight-binding approximation, which is known to help in obtaining explicit results for the inverse dielectric function [2].

The dielectric function approach usually starts with scalar quantities, which it is then convenient to develop into an equivalent treatment of vector and tensor quantities. A set of external charges produces a potential $V_0(\boldsymbol{r}')$ at point $\boldsymbol{r}'$. This potential produces in the crystal a change of charge density $\Delta\rho(\boldsymbol{r})$ at point $\boldsymbol{r}$, given by

$$\Delta\rho(\boldsymbol{r}) = \varepsilon_0 \int \mathrm{d}^3\boldsymbol{r}' \chi(\boldsymbol{r},\boldsymbol{r}') V_0(\boldsymbol{r}') \,. \tag{9.1}$$

Here $\chi(\boldsymbol{r},\boldsymbol{r}')$ is the *electron density response function*, which in general is non-local, as shown, i.e., the charge density changes at point $\boldsymbol{r}$ in response to a potential at points $\boldsymbol{r}'$, which do not have to coincide with $\boldsymbol{r}$. Since the charge density changes, the total potential in the crystal changes to $V(\boldsymbol{r})$, and the change of charge density is then related to the total potential by

$$\Delta\rho(\boldsymbol{r}) = \varepsilon_0 \int \mathrm{d}^3\boldsymbol{r}' \tilde{\chi}(\boldsymbol{r},\boldsymbol{r}') V(\boldsymbol{r}') \;, \tag{9.2}$$

where $\tilde{\chi}(\boldsymbol{r},\boldsymbol{r}')$ is the *polarizability function.*

Equivalent results follow on expressing the change of charge density in terms of a polarization $\boldsymbol{P}(\boldsymbol{r})$ through

$$\Delta\rho(\boldsymbol{r}) = -\boldsymbol{\nabla} \cdot \boldsymbol{P}(\boldsymbol{r}). \tag{9.3}$$

Then for linear response the polarization can be expressed as

$$\boldsymbol{P}(\boldsymbol{r}) = \varepsilon_0 \int \mathrm{d}^3\boldsymbol{r}' \tilde{\boldsymbol{\chi}}(\boldsymbol{r},\boldsymbol{r}') \cdot \boldsymbol{E}(\boldsymbol{r}') \;, \tag{9.4}$$

where $\boldsymbol{E}(\boldsymbol{r}') = -\boldsymbol{\nabla} V(\boldsymbol{r}')$ is the total electric field after charge density rearrangements, and $\tilde{\boldsymbol{\chi}}(\boldsymbol{r},\boldsymbol{r}')$ is the *tensor polarizability function.*

As crystal excitations, excitons can be classified by their wavevector. This leads us to consider the Fourier transforms of the various response functions from direct space into reciprocal space, for example, $\chi(\boldsymbol{Q},\boldsymbol{Q}')$. (To reduce proliferation of symbols, the same symbol is frequently used in direct and reciprocal space.) Then translational symmetry requires the wavevectors $\boldsymbol{Q}$ and $\boldsymbol{Q}'$ to differ by a reciprocal lattice vector, for example, by each being of the form $\boldsymbol{q}+\boldsymbol{G}$, where $\boldsymbol{q}$ is restricted to the first Brillouin zone and $\boldsymbol{G}$ is a reciprocal lattice vector. In this form, the restriction of $\boldsymbol{q}$ to the first Brillouin zone means that it corresponds to real-space behavior that varies on a scale larger than unit-cell dimensions, while $\boldsymbol{G}$ corresponds to real-space behavior that varies on a scale inside one cell.

Properties characteristic of the whole crystal can be ascribed to those of one unit cell, from which the crystal can be assembled. Variations on a scale smaller than the unit cell are then of no significance. This leads to the idea of the *macroscopic* polarization

$$\boldsymbol{P}(\boldsymbol{q}) = \lim_{\boldsymbol{G}\to 0} \boldsymbol{P}(\boldsymbol{q}+\boldsymbol{G}) \tag{9.5}$$

and the corresponding macroscopic electric field $\boldsymbol{E}(\boldsymbol{q})$, which both describe behavior for wavevectors inside the first Brillouin zone corresponding to real-space behavior that varies on a scale larger than unit-cell dimensions. These macroscopic quantities are related according to

$$\boldsymbol{P}(\boldsymbol{q}) = \varepsilon_0 \boldsymbol{\chi}(\boldsymbol{q}) \cdot \boldsymbol{E}(\boldsymbol{q}) \;, \tag{9.6}$$

where $\boldsymbol{\chi}(\boldsymbol{q})$ is the *linear electric susceptibility.*

So far, time dependence has been ignored, but in general the external potential and the changes it produces in the charge density depend on time. This means that both stimulus and response depend on both position and time, and the response functions depend on two pairs of variables. Under most circumstances, a crystal will exhibit ergodic behavior such that its response at time t' to a stimulus at time t depends only on the difference $t' - t$, and not on t and t' separately. This means that on Fourier transformation to examine the frequency dependence of the response, the two times t and t' give rise to a single frequency dependence that originates from $t' - t$, rather than two separate frequency dependences. Hence generalization of (9.6) to incorporate time dependence leads in the frequency domain to

$$\boldsymbol{P}(\boldsymbol{q},\omega) = \varepsilon_0 \boldsymbol{\chi}(\boldsymbol{q},\omega) \cdot \boldsymbol{E}(\boldsymbol{q},\omega) \ . \qquad (9.7)$$

At this stage, the arguments indicated in Sect. 9.1 are invoked. If we choose a value of wavevector $\boldsymbol{q}$ and explore the frequency dependence of $\boldsymbol{\chi}(\boldsymbol{q},\omega)$, poles indicate resonant frequencies at which a finite polarization can result from a vanishingly small macroscopic electric field. In direct space this polarization radiates at its own frequency, in accordance with standard electromagnetic theory, and therefore excites a further polarization, whereby the excitation propagates through the crystal. These considerations allow one to identify the resonant excitations as the excitons (which may be pure electronic, vibronic or pure vibrational in nature). Hence, in order to study the excitons, we need to construct a susceptibility tensor appropriate to a molecular crystal.

In a molecular crystal, the molecules retain their separate identities, as shown, for example, by diffraction studies of electron densities. Indeed, such studies are routinely used to determine molecular structure, although details of the molecular geometry such as bond angles may be changed from those of the free molecule by the intermolecular forces. We can therefore think of the crystal volume as being divided up into a set of disjoint molecular volumes, v_{lk} for molecule k in unit cell l. A mathematically well-defined and physically meaningful way of defining such volumes is provided by topological analysis of the electron density; extending the atoms-in-molecules approach to molecules-in-crystals [3]. It is then natural to think of each molecule having its own properties within its own volume. However, these properties may differ from those of the free molecule, because the molecule may be distorted as already mentioned, and because its electron density is restricted to lie within its own volume (orbital confinement or valence compression). The properties of the molecule in the crystal environment are referred to as *effective* properties, such as the effective polarizability, which may be regarded as the properties of a molecule "dressed" by the interactions with its neighbors.

From this point of view, the tensor polarizability function for an insulating molecular crystal should be a sum of terms, one for each molecule, that describe response confined to the molecular volume [4]:

$$\varepsilon_0 \tilde{\boldsymbol{\chi}}(\boldsymbol{r},\boldsymbol{r}') = \sum_{lk} \boldsymbol{\sigma}_k(\boldsymbol{r} - \boldsymbol{r}_{lk}, \boldsymbol{r}' - \boldsymbol{r}_{lk}) \ . \qquad (9.8)$$

Here the molecular polarizability function $\boldsymbol{\sigma}_k(\boldsymbol{r} - \boldsymbol{r}_{lk}, \boldsymbol{r}' - \boldsymbol{r}_{lk})$ depends on the molecule k in the unit cell but not on the unit cell itself. It depends on the positions $\boldsymbol{r}$ and $\boldsymbol{r}'$ relative to an origin $\boldsymbol{r}_{lk}$ that labels each individual molecule, and is understood to be zero unless both $\boldsymbol{r}$ and $\boldsymbol{r}'$ lie within the molecular volume. Thus, the polarization in each molecular volume depends only on the electric field within that volume v_{lk}. This polarizing electric field in turn arises from the macroscopic electric field and the additional field that originates from the polarization in all the other molecular volumes.

Given a tensor polarizability function of the form (9.8), a general but rather formal solution can be deduced for the self-consistent response of a molecular crystal [4]. More explicit solutions can be deduced for special forms of the molecular polarizability function [5]. In particular, the essential features are displayed by the point-dipole model, in which

$$\boldsymbol{\sigma}_k(\boldsymbol{r} - \boldsymbol{r}_{lk}, \boldsymbol{r}' - \boldsymbol{r}_{lk}) = \boldsymbol{\alpha}_k \delta(\boldsymbol{r} - \boldsymbol{r}')\delta(\boldsymbol{r} - \boldsymbol{r}_{lk}) \; , \tag{9.9}$$

where $\boldsymbol{\alpha}_k$ is the frequency-dependent polarizability of molecule k. Then the susceptibility is given by

$$\chi(\boldsymbol{q}, \omega) = \sum_{k\,k'} \chi_{kk'}(\boldsymbol{q}, \omega) \tag{9.10}$$

$$\chi_{kk'}(\boldsymbol{q}, \omega) = \left[\boldsymbol{a}(\omega)^{-1} - \boldsymbol{L}(\boldsymbol{q})\right]^{-1}_{kk'} \; . \tag{9.11}$$

Here for Z molecules per unit cell, the matrices on the RHS of (9.11) are of order $3Z$: $\boldsymbol{a}(\omega)$ is the dimensionless matrix of polarizabilities $\boldsymbol{a}_{kk'}(\omega) = \boldsymbol{\alpha}_k(\omega)\delta_{kk'}/\varepsilon_0 v$, where v is the unit-cell volume, and $\boldsymbol{L}(\boldsymbol{q})$ is the matrix of Lorentz-factor tensors (dimensionless wavevector modulated dipole sums [6]) $\boldsymbol{L}_{kk'}(\boldsymbol{q})$.

The poles of the susceptibility, and hence the exciton frequencies, can now be deduced given $\boldsymbol{\alpha}_k(\omega)$. Near a pole ω_0 of the polarizability, it takes the form

$$\boldsymbol{\alpha}_k(\omega) = \frac{\boldsymbol{n}_k}{\omega_0 - \omega} \; , \tag{9.12}$$

where complications from degeneracy are ignored; $\boldsymbol{n}_k$ is a tensor amplitude proportional to the oscillator strength of the molecular excitation at ω_0. Simplifying to one molecule per unit cell, $Z = 1$, and scalar response, we obtain [7]

$$\chi(\boldsymbol{q}, \omega) = \frac{n}{[(\omega_0 - nL_q) - \omega]} \; , \tag{9.13}$$

which evidently has a pole at a frequency $\Omega_q = \omega_0 - nL_q$. This corresponds to the usual primary result of exciton theory: the molecular excitation is broadened into a band of excitations depending on transition dipoles (contained in n) interacting through L_q. The other main result is the site shift of ω_0 from the isolated molecular excitation frequency. This is understood to arise from the change in interactions with the rest of the crystal when a molecule is excited. In the dielectric approach, this shift can be understood to be incorporated within the change from the free molecular polarizability to the effective polarizability in the crystal environment.

Once there is more than one molecule per unit cell, the dimension of the matrix to be inverted on the RHS of (9.11) increases correspondingly, and so does the number of poles that arise from the single molecular pole at ω_0. Thus, we obtain the final main result of exciton theory, the Davydov or factor-group splitting, with branches of differing polarization because of the differing interactions between translationally equivalent and inequivalent molecules in the unit cell [8].

9.3 Non-linear Optics

Introductory accounts of excitons in molecular crystals seldom use the dielectric approach, whereas introductory accounts of non-linear optics in molecular crystals seldom use anything else. Perhaps as a consequence, exciton effects are not often mentioned in the context of non-linear optics in molecular crystals, even though they are the elementary excitations of the crystal in terms of which it must be possible to express the non-linear response. In this section we outline how to extend the dielectric approach to non-linear optics and show how it can be reconciled with exciton theory.

In Sect. 9.2, we worked from (9.4), relating the polarization to the total electric field. However, as specified there, that expression related to linear response. More generally, response is non-linear, and the polarization is a series of terms in different powers of the electric field. (Further generalizations, and complications, arise on including response to electric field gradients and magnetic fields, and on including electric quadrupole and magnetic dipole contributions to the polarization [9], but all these will be ignored here.) Then the leading correction to (9.7) is

$$\boldsymbol{P}^{(2)}(\omega_\sigma) = \varepsilon_0 \boldsymbol{\chi}^{(2)}(-\omega_\sigma;\omega_1,\omega_2) : \boldsymbol{E}(\omega_1)\boldsymbol{E}(\omega_2) \ . \tag{9.14}$$

Here ω_σ is the output frequency, which must equal the sum of the input frequencies ω_1 and ω_2, and $\boldsymbol{\chi}^{(2)}$ is the quadratic susceptibility. The wavevector dependence has been suppressed for clarity, and in accordance with common practice the factor 1/2 that would be required in a Taylor series has been omitted.

Clearly, in a molecular crystal we expect the quadratic susceptibility to arise from a corresponding non-linear molecular response. In the point-dipole approximation this response is described by the molecular first hyperpolarizability $\boldsymbol{\beta}(-\omega_\sigma;\omega_1,\omega_2)$. The procedures that relate the linear susceptibility to the polarizability $\boldsymbol{\alpha}$ can then be adapted to relate the quadratic susceptibility to $\boldsymbol{\beta}$. The result is [10,11]

$$\chi^{(2)}(-\omega_\sigma;\omega_1,\omega_2) = \sum_k \boldsymbol{b}_k(-\omega_\sigma;\omega_1,\omega_2) \vdots \boldsymbol{d}_k(\omega_\sigma)\boldsymbol{d}_k(\omega_1)\,\boldsymbol{d}_k(\omega_2)\ , \qquad (9.15)$$

where $\boldsymbol{b}_k$ is the reduced hyperpolarizability $\boldsymbol{\beta}_k/\varepsilon_0 Zv$ and $\boldsymbol{d}_k$ is the local-field tensor at molecule k for linear response. This tensor gives the local polarizing electric field $\boldsymbol{F}_k$ at molecule k as $\boldsymbol{d}_k \cdot \boldsymbol{E}$, where $\boldsymbol{E}$ is the macroscopic electric field. It can also be used to express the linear susceptibility as

$$\chi^{(1)}(\omega) = \sum_k \boldsymbol{a}_k(\omega) \cdot \boldsymbol{d}_k(\omega)\ , \qquad (9.16)$$

where the superscript (1) has been added to avoid any confusion. This expression allows us to extend the previous treatment of exciton effects into non-linear response.

As we saw in the simple case of one molecule per unit cell, the molecular excitation frequency ω_0 became the exciton frequency Ω_0 given by $\omega_0 - nL$, ignoring wavevector dependence. This occurs because near the excitation the local-field tensor takes the form [7,12]

$$d(\omega) = \frac{\omega_0 - \omega}{\Omega_0 - \omega}\ . \qquad (9.17)$$

This suppresses the pole at ω_0 in the isolated molecule, and replaces it by one at Ω_0. The situation in $\chi^{(2)}$ is more complicated, because the hyperpolarizability $\boldsymbol{\beta}$, and hence $\boldsymbol{b}$, has poles at various combinations of the frequencies ω_1 and ω_2. However, when one of these frequencies or their sum ω_σ (say ω_i) lies near the molecular excitation frequency ω_0, the reduced hyperpolarizability takes the form $m_i/(\omega_0 - \omega_i)$, like (9.12), where m_i is an amplitude that depends on products of dipole matrix elements and the remaining two frequencies. In the quadratic susceptibility, the local-field tensor (9.17) near $\omega = \omega_i$ then suppresses the pole at ω_0 and again replaces it by one at the exciton frequency Ω_0 [7]. We therefore confirm that, despite its more complicated frequency dependence, the quadratic susceptibility has poles at the same exciton frequencies as deduced from the poles of the linear susceptibility.

For more than one molecule per unit cell, the local field tensor has poles at the frequencies of the various Davydov components of the exciton, as required for (9.16) to yield multiple poles from a single molecular excitation frequency. Essentially, the same arguments apply to the next non-linear response coefficient, the cubic susceptibility $\chi^{(3)}$. This is a sum of a *direct* term that depends on the second hyperpolarizability γ and a *cascading* term that depends quadratically on $\boldsymbol{\beta}$. The same mechanism as before guarantees that each term separately has its poles shifted from the molecular excitation frequencies to the crystal exciton frequencies. Each term also appears to have poles where the sum of two input frequencies equals a molecular excitation frequency, which would not accord with the expectation that only the exciton frequencies are relevant in the crystal; however, detailed analysis [7] shows that the poles cancel between the two terms.

This treatment confirms that the conventional treatment of non-linear optics in molecular crystals does incorporate exciton effects properly, through the local-field tensors. Studies of the frequency dependence of non-linear optical effects in solid C_{60} confirm the need for such reassurance [13]. Hence, analysis and prediction of frequency dependence from molecular response in non-linear optics of molecular crystals can be fully consistent with linear optics and spectroscopy, but only if the local field is treated as a function of frequency rather than as a numerical factor.

9.4 Non-local Response: Charge-Transfer Excitons

As noted earlier, dielectric response is usually non-local: the polarization at a given point is in general affected by the electric field at all points. Even in the point-dipole treatment adopted for illustration, the crystal response is non-local, because the dipoles at all points contribute to the electric field at a given point, even though the polarizability is then strictly local. This non-localness is incorporated through the wavevector dependence of the Lorentz-factor tensors. Further non-localness is incorporated when the tensor polarizability function is local to molecular volumes rather than individual points, as in (9.8). However, in general response in one molecular volume is also affected by electric fields in other molecular volumes. The tensor polarizability function must then be written as a double sum over lattice points of molecular polarizability functions that describe response between or within molecular volumes [14]. Here we develop this idea to provide a perspective on charge-transfer (CT) excitons.

Previously, we had local molecular polarizability responses consistent with the tight-binding approximation. This was associated with localized molecular excitations from which conventional Frenkel exciton states develop in the crystal. If we now admit non-local molecular polarizability response, we go beyond the tight-binding approximation. We therefore admit excitations that are associated with more than one molecule, and from these excitations new exciton states develop in the crystal. The process by which these excitons develop in the dielectric approach is similar to that by which the Davydov components develop. Increasing the size of the irreducible polarizable unit increases the dimension of the susceptibility matrix in (9.11) that determines the linear susceptibility $\boldsymbol{\chi}(\boldsymbol{q},\omega)$, and this in turn increases the number of poles of $\boldsymbol{\chi}(\boldsymbol{q},\omega)$. Note that the additional exciton states arise from excitations that exist only in the crystal and have no counterpart in an isolated molecule.

The phenomenon of photoconductivity in molecular crystals (discussed in more detail below) indicates that CT states must have at least a transient existence. In the spirit of the tight-binding approximation, a CT configuration is constructed from a molecular cation and a molecular anion. The *configuration* is the local entity (albeit associated with two molecules) from which the CT exciton *state* is constructed. Such a configuration has a large dipole moment and so is expected to be important in electro-absorption spectra. It can also contribute to the optical spectrum, either directly or through the interaction with Frenkel excitons, discussed in detail later.

From the point of view of the dielectric theory, a CT configuration gives rise to a dipole moment associated with a pair of molecules. As a result, there is an additional non-local CT polarizability, which is in principle accessible via the usual sum-over-states expression, suitably generalized. This suffices to expand the polarizability matrix and hence to generate a susceptibility that has additional poles near the energies of the CT configurations [15]. This approach also shows how CT and Frenkel states interact and provides an alternative means of explaining how this interaction can greatly enhance the polarizability in states of predominantly Frenkel parentage through the large dipole moment of CT configurations. This result, previously explained from energetic considerations [16] serves to justify values obtained from Stark shifts in electro-absorption spectra that are much larger than those deduced for isolated molecules. This process of generalization is comparable with that involved in developing the concept of distributed polarizability within a molecule [17]. Within each atomic basin, a dipole moment can develop in response to an electric field. However, in general, charge also transfers between atoms, yielding an extra contribution to the overall polarizability from CT dipole moments that are not local to one atomic basin.

We now consider in more detail the energy of a CT configuration in a crystal, relative to that of the isolated ions (which we can relate separately to that of the isolated ground-state neutral molecule). The various contributions are basically electrostatic in origin, and can be summarized as follows [18]:

- The Coulomb energy of interaction between the two ions, W_{C}. This is readily calculated from the crystal geometry.
- The energy of interaction between each ion and the permanent multipole moments of all other molecules, W_{M}. This is zero to a good approximation, since most terms are equal and opposite for the two ions.
- The polarization energy P of distortion of the charge distribution of the other molecules by the ions and vice versa. This can be calculated by the self-consistent polarization field method [19], which is essentially numerical, or by the Fourier-transform method [20, 21], which evaluates an explicit algebraic expression. The former method has the advantage of being applicable even in the absence of translational symmetry, while the latter has the advantage of yielding macroscopic dielectric results in suitable limits. In particular, the sum of Coulomb energy W_{C} and the polarization energy P is the screened Coulomb energy, which at large sep-

arations of the ions takes the standard form with the screening described by an apparent dielectric constant that depends on direction [20,21].

- Finally, these interaction energies all depend on distance and hence produce forces and torques under which the crystal structure adjusts to reach a new equilibrium arrangement with a relaxation energy W_{R}. Direct calculations show that for CT states this energy is negligible compared with the other terms [22].

In Sect. 9.7 we discuss electro-absorption spectra, which depend on the energy of a CT state in an electric field. Various assumptions have been made about the form of this energy, in particular whether the quadratic Stark effect depends on the macroscopic electric field $\boldsymbol{E}$ or the local field $\boldsymbol{F}$. Detailed analysis [23] shows that the form of the energy depends on whether the CT state is a delocalized combination of local CT configurations (as in a pure crystal) or is itself localized (as in doped crystals under suitable conditions). For the pure crystals considered later in this chapter, the energy is bilinear in the fields $\boldsymbol{E}$ and $\boldsymbol{F}$.

9.5 Hamiltonian Description of Excitons

The dielectric description discussed above provides an all-encompassing, consistent framework in which to treat the electronic excitations of condensed matter. It is very well suited for general considerations on the one hand, and for simple few-parameter phenomenological fits on the other. However, apart from the calculation of electrostatic energies, it is less practical for actual derivation of the properties of crystal exciton states from the fundamental microscopic characteristics, entailed in the wavefunctions of the constituent entities (such as molecules or atoms). In this latter context, the more conventional Hamiltonian approach is the recommended practical alternative.

At the level of the total untruncated many-particle Hamiltonian, the two approaches are equivalent [7] and equally impractical in application. The advantage of the Hamiltonian formulation is its ability to provide a set of well-defined approximations and models. These models depend on parameters that are amenable to independent microscopic evaluation, being expressed in terms of molecular or atomic wavefunctions which are the primary substrate of quantum theory, and are obtainable from quantum-chemical calculations.

In the most straightforward formulation, the Hamiltonian is denoted in terms of the creation and annihilation operators of electrons in specified atomic orbitals, localized at specified sites of the crystal. In order to make the problem tractable, the set of orbitals has to be restricted in size (typically to the $2p_z$ orbitals of constituent atoms [24–29]), and only the leading matrix elements of the Hamiltonian can be taken into account. These necessarily comprise in Ising- or Hückel-type approximations the diagonal energies of electrons and terms describing their transfer, usually only between the neighboring sites. In the more refined Hubbard- or PPP-type versions, the interaction between particles located at the same site, and in most neighboring [24, 25, 28] or even more distant sites [26, 27, 29], is also explicitly included. The intra- and intermolecular interactions are treated on the same footing, taking no advantage of the weakness of intermolecular interactions in molecular crystals.

In the spectroscopic context where low-energy excitations are of main interest, the formulation described above is not efficient numerically. At reasonably low excitation densities, most electrons in the crystal form only a background for the few that are directly engaged in the optical transitions, and only these latter ones need to be explicitly considered. This allows one to reduce considerably the size of the necessary basis set, saving computational effort for more realistic handling of the relevant interactions. Computational complexity can be reduced further by making proper use of the fact that in molecular solids intermolecular overlap is small and intermolecular interactions are much weaker than those within one molecule.

In this alternative formulation, sometimes referred to as the *molecular-exciton approach* [26], the ground state of the crystal serves as the vacuum state for the electronic excitations. The excited-state Hamiltonian is denoted in terms of the creation and annihilation operators of localized electron–hole pairs, each characterized by a specific relative position of the two particles in the crystal lattice. In effect, the excited states are represented as linear combinations of the localized Frenkel and CT excitations and are obtained through the process of configuration interaction (CI).

Usually, only the excitations from a relatively narrow energy range are of primary interest. Then the Hamiltonian can be further reduced to the single-excitation subspace by neglecting terms that do not conserve the number of excitations, and (in the limit of low excitation density) those non-linear in exciton number operators. However, in some non-linear optics applications more general forms of the Hamiltonian need to be considered [30].

Since the nearest-neighbor intermolecular overlap integrals (IOI) are normally small in molecular crystals, they constitute a convenient expansion parameter for classifying the various parts of the effective Hamiltonian. The zeroth-order terms are the diagonal energies of the localized excitations and the dipole–dipole resonance terms that govern the transfer of Frenkel excitons. With the ground state taken as the zero of energy, the former are of the order of several eV, their spacings typically ranging from 0.1 to 1 eV (although even quasi-degeneracies may sometimes occur), while the latter range from 0.01 eV to 0.5 eV. In this approximation, only Frenkel excitons corresponding to allowed molecular excited states can propagate in the crystal; the CT states are strictly localized and may be identified with those calculated within the dielectric approach described in Sect. 9.2.

The CT states are coupled to one another by the first-order terms (0.01–0.1 eV), referred to as the charge-transfer integrals, and to the Frenkel states by the so-called dissociation integrals. At this stage, the CT states can propagate stepwise, by transfer of a single charge at a time.

The new terms that appear in the second order mediate the propagation of the CT states by transfer of both charges together, some new couplings in the CT manifold, and the propagation of optically forbidden Frenkel excitons mediated by exchange terms. However, a consistent formulation of the Hamiltonian correct to the second order in nearest-neighbor IOI is essentially straightforward but inadequate, since some terms linear in second-nearest-neighbor overlaps are comparable to those quadratic in the nearest-neighbor IOI and should also be included, leading to new complications. As the second-order matrix elements are altogether quite small (10^{-4}–10^{-2} eV), the effective Hamiltonian is normally truncated after the linear terms. It then assumes the form

$$H = \sum_{\boldsymbol{m},\boldsymbol{n}} E(\boldsymbol{n}) B^{+}_{\boldsymbol{m},\boldsymbol{m}+\boldsymbol{n}} B_{\boldsymbol{m},\boldsymbol{m}+\boldsymbol{n}} + \sum_{\boldsymbol{m},\boldsymbol{\lambda}} M(\boldsymbol{\lambda}) B^{+}_{\boldsymbol{m},\boldsymbol{m}} B_{\boldsymbol{m}+\boldsymbol{\lambda},\boldsymbol{m}+\boldsymbol{\lambda}} +$$

$$\sum_{\boldsymbol{m},\boldsymbol{n},\boldsymbol{\kappa}} J_{\mathrm{h}}(\boldsymbol{n},\boldsymbol{\kappa}) B^{+}_{\boldsymbol{m},\boldsymbol{m}+\boldsymbol{n}} B_{\boldsymbol{m}+\boldsymbol{\kappa},\boldsymbol{m}+\boldsymbol{n}} + \sum_{\boldsymbol{m},\boldsymbol{n},\boldsymbol{\kappa}} J_{\mathrm{e}}(\boldsymbol{n},\boldsymbol{\kappa}) B^{+}_{\boldsymbol{m},\boldsymbol{m}+\boldsymbol{n}} B_{\boldsymbol{m},\boldsymbol{m}+\boldsymbol{n}+\boldsymbol{\kappa}} \quad (9.18)$$

where the operator $B_{\boldsymbol{m},\boldsymbol{m}+\boldsymbol{n}}$ annihilates the excited state consisting of a hole located at site $\boldsymbol{m}$ and of an electron located at site $\boldsymbol{m}+\boldsymbol{n}$, which corresponds to a Frenkel state for $\boldsymbol{n}=0$ and to a CT state otherwise. The diagonal matrix element $E(0)$ denotes the energy of a molecular excited state, corrected for the gas-to-crystal shift, while for $\boldsymbol{n} \neq 0$ the diagonal elements $E(\boldsymbol{n}) = I - A + P + C(\boldsymbol{n})$ denote the energies of localized CT states, with I, A and P standing for the molecular ionization potential, molecular electron affinity and the polarization energy of the crystal by an infinitely spaced electron–hole pair. $C(\boldsymbol{n}) = e^2/4\pi\varepsilon_0\varepsilon_{\mathrm{eff}}R(\boldsymbol{n}) = \Delta P(\boldsymbol{n}) - e^2/4\pi\varepsilon_0 R(\boldsymbol{n})$ is the screened electrostatic interaction between the charges [20,21]. Here $\varepsilon_{\mathrm{eff}}$ is the effective dielectric constant, $\Delta P(\boldsymbol{n})$ stands for the change of the polarization energy resulting from bringing the two charges to the finite distance $R(\boldsymbol{n})$, and the remaining term represents their unscreened Coulombic interaction at that distance. The off-diagonal matrix elements are:

- $M(\boldsymbol{\lambda})$, denoting the Frenkel exciton transfer integrals, and
- $J_{\mathrm{e}}(\boldsymbol{n},\boldsymbol{\kappa})$ and $J_{\mathrm{h}}(\boldsymbol{n},\boldsymbol{\kappa})$, denoting the electron and hole transfer integrals for $\boldsymbol{n} \neq 0$ and the Frenkel exciton dissociation integral for $\boldsymbol{n} = 0$.

This Hamiltonian was originally proposed by Merrifield for a linear molecular crystal [31]. In the above form, it takes into account only one Frenkel state per molecule and only one series of CT states, originating from a given donating and a given accepting orbital on each molecule (most commonly, HOMO and LUMO [highest occupied and lowest unoccupied molecular orbitals, respectively]). Its generalization to include several such series is straightforward, but in actual implementation produces a numerical problem of considerable complexity. Fortunately, even if several series of CT states are experimentally observable and therefore interesting (as is the case in C_{60}, for example [32–34]), they are well separated in energy and may be treated one by one.

On the other hand, it is relatively common for several Frenkel states to be located in the vicinity of the same CT manifold. In this case we label the members of the set of Frenkel states by an extra index α. They are then easily incorporated into the basis set by introducing for $\boldsymbol{n} = 0$ the corresponding creation operators $B^{+}_{\alpha,\boldsymbol{m},\boldsymbol{m}}$. The Hamiltonian retains the same general structure and hence is not given explicitly here. It merely acquires additional terms [35, 36] describing the diagonal energies and the propagation and dissociation of the new Frenkel excitons. All the intramolecular excitations are then treated on the same footing; it is worth noting that they need not have the same orbital parentage as the CT states included in the model, although in some implementations of the molecular-exciton approach they in fact do, since the selection of the Frenkel states is based solely on the criterion of orbital parentage [29].

The Hamiltonian (9.18) conforms to the requirement stated at the beginning: all of its matrix elements are accessible to independent theoretical evaluation, based on the wavefunctions of individual molecules calculated by quantum-chemical methods.

Technically, the most demanding task is the calculation of the CT integrals as matrix elements of the intermolecular interaction operator. (It is implicitly assumed in this formulation that the wavefunctions of different molecules have been orthogonalized prior to the evaluation of the matrix elements, or that their non-orthogonality is neglected.) The main difficulty is that reproducing the exponentially small intermolecular overlap terms correctly requires an extended representation of atomic orbitals [37–39]. The calculation of the dissociation integrals is very similar, but the difference between these integrals and the CT integrals is usually small [40] and for many applications may be neglected.

Routine MO-type calculations yield the molecular ionization potentials, electron affinities, diagonal energies of intramolecular excitations and transition dipole moments from the ground to the excited states of individual molecules. Then the Frenkel exciton resonance integrals may be obtained directly in the dipole approximation; if needed, more accurate values are obtainable from the detailed form of the excited-state wavefunctions [41].

Quantum chemical methodology can also provide the molecular polarizabilities. These are used for calculating the dispersion-induced gas-to-crystal shift and the electrostatic stabilization energies of localized CT states. The latter calculations are performed either by the self-consistent polarization field method [19] or by the more recent Fourier-transform approach [20, 21], as discussed in the Sect. 9.2.

As described above, the Hamiltonian approach satisfies the stated requirements in principle: all the necessary parameters can be evaluated from fundamental theory, via quantum-chemical methodology. In practice, a complete calculation of this kind has never been performed. The normal objective of a theoretical approach is to interpret experimental data; the problem is the accuracy obtainable for the quantum-chemical results, which is often insufficient for actual applications. For instance, a discrepancy of several tenths of an eV, typical for calculated molecular excitation energies, ionization potentials and electron affinities, is still far from the standard necessary for interpreting crystal electro-absorption spectra. Calculated molecular polarizabilities seldom agree well with those of larger free molecules, let alone those in the crystal environment that are appropriate to calculations of polarization energies (although calculations for benzene and urea taking account of environmental effects can now reproduce the linear response quite well [42, 43]).

For this reason, in actual applications a somewhat less ambitious strategy is usually adopted: the properties of an individual molecule are taken for granted, and the objective of the theory is to derive the spectral properties of the crystal from these data as the starting point. Similarly, effective molecular polarizabilities for the molecules in the crystal environment deduced from the crystal linear susceptibility are normally used.

9.6 Mixing of Frenkel and Charge-Transfer States

It is readily seen that, owing to the overlap-dependent CT and dissociation integrals J, the eigenstates of the Hamiltonian of (9.18) are no longer pure Frenkel or pure CT excitons. Once the solutions have been obtained, by identifying the leading CI coefficients in the wavefunction, some dominant parentage can be attributed to each eigenstate in order to interpret the results, but there are also contributions from all other basis states. This phenomenon is referred to as "mixing between the Frenkel and CT states".

This phenomenon was probably first recognized in the case of triplet exciton Davydov splitting [39, 44]. When the terms of the Hamiltonian (9.18) that are linear in intermolecular overlap integrals are treated by perturbation theory, in the second order there appears a super-exchange type contribution from the following sequence of virtual processes.

- A Frenkel exciton localized at molecule A dissociates by electron transfer to another molecule B to give the CT state A^+B^- (or by hole transfer to give A^-B^+).
- This CT state recombines by subsequent hole transfer from the molecular ion A^+ to molecular ion B^- (or by electron transfer from A^- to B^+).

The net result is equivalent to exciton transfer from molecule A to molecule B.

If the molecules are nearest neighbors in translationally inequivalent positions, the corresponding eigenstates of the crystal are the in-phase and out-of-phase combinations of the wavefunctions A*B and AB* (where the asterisk denotes the excited molecule), and the usual energies of the Davydov components are split by the exciton-transfer interaction. For dipole-allowed Frenkel excitons the Davydov splitting is dominated by the strong resonance interactions (zeroth-order in IOI) and the small indirect contribution mediated by the CT states is negligible. However, for triplet excitons whose transfer is governed by very small electron-exchange terms (second order in IOI), both contributions are of the same order of magnitude [39]. In effect, the experimentally observed Davydov splitting of triplet excitons in polyacenes cannot be reproduced properly without taking due account of the Frenkel-CT exciton mixing [39]. A similar argument allows one to rationalize some peculiarities of triplet exciton transfer in donor–acceptor complexes [45].

The same mechanism as for triplets may also apply for any other forbidden Frenkel state. As shown in several recent papers [46–48], this is the case for the lowest singlet excited state of the C_{60} crystal . In the archetypal case of the lowest singlet state in the naphthalene crystal, which is also optically forbidden, the CT-type overlap-dependent terms apparently dominate the Davydov splitting and contribute significantly to the observed gas-to-crystal shift. This manifests itself in the pressure dependence of these quantities [49–51]; it seems that even in anthracene, where the lowest excited state is optically allowed but the dipole-controlled part of the Davydov splitting happens to be small, the CT terms strongly affect the pressure dependences.

CT effects in the energetics of the eigenstates that derive from forbidden intramolecular excitations have been demonstrated for naphthalene crystals doped with benzene derivatives [52–54]. The dopant molecule perturbs the energies of the host excitons in its vicinity. In favorable cases local levels are formed slightly below the host exciton band, and can be detected spectroscopically [52–54]. There may be one or more such states for a given dopant, depending on its kind. It has been shown experimentally that the stabilization energies of the local levels deriving from the lowest singlet state of the naphthalene host crystal correlate only with the ionization potential, and with no other parameter characterizing the guest molecule (providing that the guest is small enough to substitute just one molecule in the host lattice). This is a strong indication of the involvement of ionized molecules, and hence of CT states.

Qualitatively, the effect is readily rationalized [52–54]. The naphthalene-to-naphthalene CT states are located several tenths of an eV higher than the lowest Frenkel state of the crystal. Hence, the coupling with the CT manifold tends to stabilize the eigenstates of Frenkel parentage. If naphthalene is replaced by a molecule with a lower ionization potential, the diagonal energy of the guest-to-host CT state is lower than that of the host-to-host CT state, so that for the naphthalene molecules surrounding the impurity the stabilizing effect is stronger, giving rise to local levels.

A simple generalization of the Hamiltonian (9.18), making allowance for the presence of dopant molecules, affords a reasonably good quantitative description of the observed dependence of local-state energies on the guest ionization potential [55]. For technical reasons, in the actual calculations a finite cluster must be used instead of the infinite crystal, but the results are stable with respect to changes of cluster size. Because the naphthalene crystal structure consists of a herringbone packing within well-separated *ab* planes, the model may be safely limited to two dimensions. The model allows one to rationalize the number of local levels induced by a given dopant, and the level structures at high impurity concentrations [56–58].

The notion of "mixing" introduced at the beginning of this section implies that the Frenkel states acquire some admixture of CT character and vice versa. The former effect increases the polarizability of the eigenstates of Frenkel origin deduced from electro-absorption experiments. The latter effect increases the inherently weak optical absorption of the eigenstates of CT origin and is crucial for understanding both the electro-absorption and photocurrent spectra, as will be discussed in the following sections.

9.7 Electro-Absorption

Electro-absorption (EA) spectroscopy consists of measuring the change of the absorption spectrum induced by a slowly varying external electric field. Traditionally, EA spectra are interpreted by resolving them into first- and second-derivative signals [59, 60]. The first-derivative signals, governed by the polarizability, are attributed to Frenkel excitons where the field-induced level shift is quadratic in electric field strength, while the second-derivative signals are commonly associated with entities that exhibit a permanent dipole moment leading to a linear Stark shift, namely the CT states. In view of the mixing between the two groups of states, this classification can only be approximately valid.

The polarizabilities of the lowest Frenkel state of polyacene crystals, as evaluated from such resolutions, in some cases substantially exceed the corresponding molecular values [59,60]. Even a simple two-state model [16] allows one to interpret the excess in terms of a CT admixture, embodying the non-local polarizability contribution predicted previously [14]. However, once the $J(0,1)$ dissociation integrals linear in IOI that mediate the above effect are admitted into the Hamiltonian, the CT integrals $J(1,1)$ which are of the same order must be included as well.

This has profound conceptual consequences. The CT states are no longer localized ion pairs, characterized by a well-defined relative position of the charges; instead, the new eigenstates comprise contributions corresponding to different electron–hole distances. The new interaction, off-diagonal in the localized basis, has a particularly strong effect on the symmetry-equivalent CT states that in the first order in IOI were degenerate: they split into different symmetry-adapted eigenstates, in some respects analogous to Davydov components [21]. This splitting, whose size depends on the CT integrals and on the lattice co-ordination number, may attain values comparable to the Coulombic energy gaps between different localized CT states. Moreover, for symmetry reasons, in a centrosymmetric crystal the localized (+−) and (−+) CT states must enter the eigenvectors with equal weights – in effect, the non-polar crystal structure does not support solutions with a permanent dipole moment, which are essential for the traditional interpretation of second-derivative EA signals.

Consequently, the second-derivative signals are now attributed not to inherently polar states, but to pairs of quasi-degenerate non-polar states, coupled by the interaction with the modulating external electric field. The coupling lifts the degeneracy, shifting one of the levels to lower and the other to higher energy, and giving rise to one simple and one inverted first-derivative signal, respectively. When the zero-field gap between the states is smaller than the spectral width of the corresponding absorption bands, the two first-derivative signals fuse into one second-derivative shape.

The above argument explains the existence of second-derivative EA signals. It also shows (a) that such signals might emerge also for quasi-degenerate states of Frenkel rather than CT parentage, and (b) that, if the splitting of the zero-field eigenstates of CT parentage by the off-diagonal interaction linear in IOI is sufficiently large, these states can produce first-derivative signals. Hence, the simple scheme where the shape of the EA signal identifies the origin (Frenkel or CT) of the corresponding exciton is not generally valid.

In these circumstances, a detailed analysis of the EA spectrum must necessarily be supported by quantum-mechanical calculations; they may be based on the Hamiltonian (9.18) which, as has already been mentioned, may be parameterized either in a completely non-empirical way or with some assistance from isolated-molecule experimental data. As only low-energy CT excitations are practically observable, the set of CT basis functions may be limited to those where the electron–hole distance does not exceed one lattice period of the crystal. After diagonalizing this Hamiltonian, one can calculate the absorption spectra at zero and non-zero electric field and generate the theoretical EA spectrum as their difference. So far, this has been done for several crystals [61, 62], and the results exhibited remarkably good agreement with experiment, both in the shape of the signal and in its absolute scale. This latter aspect is especially worth noting in view of notorious difficulties with reproducing absolute intensities in any spectroscopy, strengthened for EA by the unforgiving differential shape of the spectra. In this regard, taking proper account of the dependence of the energies on the product of the macroscopic and local electric fields [23] is essential.

The results may be summarized as follows. The CT admixture amplifies the EA signal of Frenkel excitons, as predicted by simple models [16]. On the other hand, this same mixing lends some Frenkel-derived absorption intensity to the eigenstates of CT parentage, which also increases their EA signal. Note that, in contrast to absorption, the EA signal is not positive definite, and no sum rule is known that conserves the global EA "intensity". Note also that, contrary to early expectations [35], the CT states also carry some inherent transition dipole moment from the ground state [21]. Being linear in IOI, it is of the same order as that borrowed from the Frenkel states, although it is somewhat smaller numerically.

For polyacenes, treated in the two-dimensional model already mentioned (but no longer restricted to a cluster of finite size), the absorption intensity is lent primarily by the lowest Frenkel state [63], but the contributions borrowed from higher allowed states are not negligible. Forbidden Frenkel states, if energetically close (as in anthracene [63]), may also considerably perturb the CT excitons by affecting their energies and drawing some of their intensity. Owing to this coupling and to the splitting by the off-diagonal interactions within the CT manifold, the CT part of the EA spectrum is complex, especially for anthracene. Another complication results from the zigzag herringbone molecular packing, which makes the intermolecular overlap integrals non-negligible between the origin molecule and those at both the (1/2, 1/2, 0) and the (0, 1, 0) positions in the crystal. The electrostatic energies of the corresponding localized CT states then differ only by about 0.1 eV, making the mixing between the two manifolds substantial. Further problems emerge from the fact that the transitions to the eigenstates of CT parentage strongly overlap with those due to the vibronic replicas of the lowest Frenkel transition. Accordingly, the coupling with the 0.17 eV intramolecular stretching mode has to be included in the theoretical model, which can be done in the approximation of strong vibronic coupling by appropriate scaling of the Hamiltonian matrix elements [61]. By and large, while the microscopic and essentially non-empirical model based on the Hamiltonian (9.18) provides a successful reproduction of the EA signal of linear polyacenes, the complexities described above suggest that the results of purely phenomenological numerical analyses of these EA spectra have to be interpreted with great caution.

In C_{60}, the EA spectrum shows several series of CT states , deriving from the HOMO-to-LUMO (at about 2.5 eV), HOMO−1-to-LUMO (~3.5 eV) and HOMO-to-LUMO+1 (~3.5 eV) orbital promotions [32–34]. As the excitation energies of the fullerene cation and anion are similar, the combined EA signal of the latter two series is quite strong. The CT signals generally dominate the spectrum, in contrast to the situation in polyacenes where the Frenkel and CT contributions are comparable. This is not surprising, since in C_{60} both HOMO and LUMO are degenerate (5 and 3 components, respectively), so that the total number of equivalent primary CT states is multiplied by 15 [32]. For the nearest-neighbor (1/2, 1/2, 0) CT states (where the notation indicates the position of the electron, with the hole at the origin), this is further increased by the large co-ordination number (12) of the lattice.

Therefore, although individual intermolecular overlap integrals in fullerene are as small as in other molecular crystals, the total absorption intensity of the CT states is substantial. Accordingly, also the off-diagonal splitting within the (1/2, 1/2, 0) CT manifold is anomalously large (0.2–0.3 eV), the latter effect being compounded by the interaction with nearby absorption-inactive Frenkel states [32, 62, 64]. As a result, the (1/2, 1/2, 0) family of CT states gives rise to two distinct EA features at about 2.5 eV and 2.75 eV, of which the first one has a shape strongly resembling a first derivative [65], a phenomenon impossible to rationalize within a localized model of CT excitations.

Originally, EA was introduced as a convenient tool to investigate the low-energy CT states, where the distinct second-derivative shape of the signals and their sensitivity to electric field could compensate their small absorption intensity [59, 60]. While the advantages of EA spectroscopy still remain valid, it is now apparent that in most cases interpretation must be assisted by calculations of exciton energies. It should be noted that, owing to the sensitivity of EA spectra to the shortcomings of the model, these calculations must be reasonably realistic energetically. This gives the molecular-exciton approach some advantage over even sophisticated approaches that treat the intra- and intermolecular interactions on equal footing.

Detailed eigenstate calculations are indispensable to interpret an EA spectrum; once the results are available, it is possible to assign individual EA features to specific groups of eigenstates. Then a procedure analogous to quantum-chemical population analysis allows one to attribute to each eigenstate some approximate parentage expressed in terms of localized states characterized by the relative position of the two charges with respect to each other. It is worth noting that, although the off-diagonal splitting may be substantial, the centre of gravity of the set of split eigenstates that derive from the same localized CT state is usually close to this state's diagonal electrostatic energy, calculated within the dielectric approach. Hence, if excitons deriving from several different CT configurations are experimentally observable and properly assigned, one can use the Coulomb formula to extrapolate their energies to infinite electron–hole distance; this gives a direct bridge to the dielectric formulation. The energy of a localized infinitely spaced electron–hole pair, often referred to as the optical band gap, is an important ingredient in phenomenological analyses of EA spectra [19, 59, 60]. However, in actual evaluations of the optical gap [59, 60], the energy of one of the split components was used instead of the corresponding centre of gravity of the eigenstates deriving from a given localized CT state.

9.8 Photoconductivity

As photocurrent is produced by thermal dissociation of electron–hole pairs, the CT states relevant for photoconductivity are not the low-energy ones observed in EA. In order to produce free charge carriers, a bound electron–hole pair must be supplied with thermal energy at least equal to the Coulombic stabilization energy; in effect, prohibitively small Boltzmann factors preclude substantial contributions from small-radius CT states.

This calls for a slightly different theoretical description. It is still based on the Hamiltonian (9.18), but for practical reasons it has to be less detailed than that used in the context of EA spectroscopy. In fact, with increasing electron–hole radius, the number of CT states grows very rapidly, and one arrives at a formidable numerical problem if all of them are included in the basis set. However, by the same token, the density of higher eigenstates of CT parentage is large and the details of the energy spectrum are not really needed, as they would be hidden in the relatively poor spectral resolution typical of photocurrent measurements.

Therefore, the pertinent eigenvalue problem was actually solved [20] only for a hypothetical one-dimensional crystal where exact analytical solutions are available [31]. The calculations revealed that the off-diagonal splittings of the eigenstates decrease rapidly with eigenstate energy (i.e., with the expectation value r_{eh} of the electron–hole distance). Hence, in the states relevant for the photocurrent, the Coulomb formula, valid for localized CT states, is also a very reasonable approximation for the delocalized eigenstates. At the same time, the absorption intensities of consecutive eigenstates decrease with increasing r_{eh}, approximately following an exponential law.

The above relationships should remain valid also for a three-dimensional crystal. This substantiated a simple semi-empirical description of the observed photocurrent spectrum [20,66,67]; the parameters of the fit were found to be in reasonably good agreement with theoretical expectations based on the one-dimensional model. The conclusions are based on the case of anthracene crystal, so far the only one where at the same time sufficiently reliable experimental data and theoretical calculations are available.

The findings for anthracene lead to the following view: In the photocurrent threshold region (about 3.8–4.4 eV for anthracene) the eigenstates of CT parentage are formed by direct optical transitions from the ground state. Their energies follow the Coulomb law to a good approximation, and their absorption intensity is almost exclusively borrowed from the Frenkel states according to the Hamiltonian (9.18). The intensities depend approximately exponentially on distance because the eigenstate deriving predominantly from the CT state with an electron–hole distance of n nearest-neighbor spacings borrows its absorption intensity from the Frenkel state via $n-1$ CT states, and each of these couplings introduces a factor linear in the nearest-neighbor overlap integral. The inherent CT transition moments (*vide supra*) would also decrease exponentially, but more rapidly.

Subsequently, the primary CT states formed optically either recombine or dissociate thermally into free charge carriers, the latter process being approximately described by the Onsager model. The contribution from a given state to the photocurrent is a compromise between this state's absorption intensity (decreasing with increasing r_{eh}) and its Boltzmann factor (increasing with r_{eh}); the largest contributions come from the CT states characterized by an electron–hole distance in the range 20–30 Å. The effect of Boltzmann factors is reflected in the observed decrease of the photocurrent activation energy E_a with increasing photon frequency. On heuristic grounds, extrapolation of the corresponding experimental plot to $E_a = 0$ is expected to yield the minimum energy necessary to produce an unbound electron–hole pair, identified with the electrical band gap of the crystal.

It should be emphasized that at the bottom of the band each charge carrier is completely delocalized. In contrast, the optical band gap estimated from EA experiments, as mentioned in Sect. 9.7, corresponds to a hypothetical pair of completely localized charges (impossible to produce in reality, by virtue of the uncertainty principle). The difference between the two band gaps is the charge delocalization energy [68].

In the unbound electron–hole continuum each of the charges moves independently; in effect, the splitting of these states in the solid is governed by terms linear in IOI. On the other hand, the band dispersion of CT excitons is governed by terms quadratic in IOI, since in order to move a CT state in the lattice it is necessary to transfer two charges simultaneously. As the intermolecular overlap integrals are small, the bandwidth resulting from delocalization is much larger for the unbound states than for the bound CT excitons. Hence, the energy separation between the bottom of the exciton band and the bottom of the unbound pair continuum, obtainable by comparing the EA and photocurrent data, is actually smaller than the separation between the centres of the corresponding bands, equal to the electrostatic binding energy of the CT state. It is irrelevant in this context whether the band gap is direct or indirect; the effect occurs regardless. However, in crystals with an indirect gap an intriguing possibility can be envisaged: the energy of a weakly bound CT state, generated optically at the centre of the Brillouin zone, may be higher than the bottom of the conduction band, located at the edge of the Brillouin zone. Then the optically accessible CT state is in fact immersed in the (absorption-inactive) electron–hole continuum into which it can decay. The decay may be mediated by the interaction with phonons, which would also cover the deficit of crystal momentum. It is reasonable to expect that such activationless decay should be easier than the normal Onsager-type CT state dissociation and could lead to especially efficient charge-carrier photogeneration.

The fact that the difference between the electric and optical band gap is a consequence of the charge delocalization energy [68] seems to have passed unnoticed for a long time. The discrepancy was attributed to vibrational relaxation [19] until recent calculations [22] demonstrated the relaxation energy to be an order of magnitude smaller, and so proved the orthodox explanation to be untenable.

Views concerning the mechanism by which the primary CT states are created have also gradually changed with time. Originally, there was a tendency in the literature to attribute the generation of photocarriers to the auto-ionization of Frenkel excitons. This process is certainly feasible for incident photons of relatively high frequency that create Frenkel excitons whose energies exceed the band gap of the crystal [67]. These intramolecular excitations are immersed in the continuum of unbound electron–hole states and its vibronic replicas, into which they can decay. Accordingly, during its short lifetime, such a Frenkel state may either relax to lower intramolecular states, or auto-ionize into an unbound electron–hole pair. Although the branching ratio favors the former process, the probability of the latter is not negligible [67]. Most of the primary charge carriers produced in this way have some excess kinetic energy with respect to the bottom of the continuum, but not enough to escape the Coulombic force, by which they are recaptured after travelling some distance. This average distance (thermalization length) determines the radius of the created bound CT states and can be estimated by considering the rate of energy loss due to carrier scattering on low-frequency phonons [67]. This radius turns out to be substantial, in anthracene of the order 30–90 Å (depending on the spectral range). Subsequently, the generated CT states may either recombine (geminate recombination), or thermally dissociate by the Onsager mechanism into free charge carriers, monitored in photocurrent measurements.

Contrary to early interpretations [19], energy conservation prevents the auto-ionization mechanism from operating at low exciting photon frequencies, where the energy of the generated Frenkel states is lower than that of an unbound electron–hole pair (band gap). Still, some vibronic replicas of the lowest allowed Frenkel state S_1 are almost degenerate with low CT states, so that at least from the point of view of energy conservation a radiationless transition to the CT manifold could be conceivable. (This is not auto-ionization, as the final state is a discrete CT state, not the free electron–hole continuum.) However, such processes are very inefficient, since the electron transfer has to compete with very fast vibrational relaxation within the S_1 state. The vibrational excitations of reasonably high frequency (which could efficiently fill the energy separation between the Frenkel and the CT states) are too short-lived to be able to transfer the charge to a sufficiently large distance before they relax. For the transitions from highly excited vibrational states of the S_1 exciton to a vibronically unexcited CT state, the vibrational overlap integrals very strongly suppress the value of the inducing Hamiltonian matrix element. On the other hand, lower vibronic states of S_1 have enough energy to produce only small-radius CT states whose large electrostatic stabilization and consequently negligibly small Boltzmann factors eliminate them as efficient contributors to the photocurrent.

In effect, in the photocurrent threshold region direct optical creation of CT excitons (described at the beginning of this section) is the prevailing mechanism. In this energy range, Frenkel excitons contribute only through their virtual admixture to the wavefunctions of the eigenstates of CT parentage (intensity borrowing), not as states that are actually populated. On the other hand, the photocurrent at high energies is dominated by CT states created by auto-ionization of Frenkel excitations, optically generated as real states. Additional evidence for this latter interpretation is provided by two-color two-step excitation studies of higher excited states [69], where the yield of photocarriers is insensitive to the energy of the second excitation, apparently because the states initially created all relax to the same auto-ionizing state.

9.9 Concluding Remarks

The dielectric and Hamiltonian approaches may be used as alternative descriptions of the excitons in molecular crystals. The two descriptions are equivalent in theoretical status, and are mostly complementary in applications: they support and supplement each other. The dielectric approach is unequaled in the efficient handling of electrostatic interactions, and is also extensively used in the theory of non-linear optical effects. Although both approaches can describe the mixing of Frenkel and CT-type excitations, the Hamiltonian approach is more practical in detailed calculations of this kind, and more easily reducible to first principles of quantum mechanics.

References

1. R.W. Munn, In: *Molecular Electronics: Materials and Methods*, ed. by P.I. Lazarev (Kluwer, Dordrecht 1991)
2. S.K. Sinha, Phys. Rev. **177**, 1256 (1969)
3. R.F.W. Bader, T.T. Nguyen-Dang, Y. Tal, Rept. Prog. Phys. **44**, 893 (1981)
4. L.M. Hafkenscheid, J. Vlieger, Physica **75**, 57 (1974)
5. R.W. Munn, Chem. Phys. **50**, 119 (1980)
6. T. Luty, R.W. Munn, Chem. Phys. **43**, 295 (1979)
7. R.W. Munn, J. Chem. Phys. **106**, 3870 (1997)
8. V.M. Agranovich, Sov. Phys.–Usp. **17**, 103 (1974)
9. R.W. Munn, Mol. Phys. **89**, 555 (1996)
10. J.A. Armstrong, N. Bloembergen, J. Ducuing, P.S. Pershan, Phys. Rev. **127**, 1918 (1962)
11. M. Hurst, R.W. Munn, J. Mol. Electron. **2**, 35 (1986)
12. S. Mukamel, In: *Molecular Non-linear Optics – Materials, Physics and Devices*, ed. by J. Zyss (Academic Press, San Diego 1993)
13. R.W. Munn, Z. Shuai, J.-L. Brédas, J. Chem. Phys. **108**, 5975 (1998)
14. R.W. Munn, T. Luty, Chem. Phys. **81**, 41 (1983)
15. R.W. Munn, Chem. Phys. **215**, 301 (1997)
16. P. Petelenz, Chem. Phys. Lett. **215**, 607 (1993)
17. A. J. Stone, Mol. Phys. **56**, 1065 (1985)

18. R. W. Munn, In: *Electrical and Related Properties of Organic Solids*, ed. by R.W. Munn, A. Miniewicz, B. Kuchta (Kluwer, Dordrecht 1997)
19. E.A. Silinsh, *Organic Molecular Crystals: Their Electronic States* (Springer-Verlag, Berlin 1980)
20. P.J. Bounds, P. Petelenz, W. Siebrand, Chem. Phys. **63**, 303 (1981)
21. P.J. Bounds, W. Siebrand, I. Eisenstein, R.W. Munn, P. Petelenz, Chem. Phys. **95**, 197 (1985)
22. I.V. Brovchenko, Chem. Phys. Lett. **278**, 355 (1997)
23. R.W. Munn, Chem. Phys. **236**, 151 (1998)
24. T. Tsubo, K. Nasu, J. Phys. Soc. Jpn. **63**, 2401 (1994)
25. T. Tsubo, K. Nasu, Solid State Comm. **91**, 907 (1994)
26. D. Mukhopadhyay, G.W. Hayden, Z.G. Soos, Phys. Rev. B **51**, 9476 (1995)
27. Z.G. Soos, M.H. Hennessy, G. Wen, Chem. Phys. **227**, 19 (1998)
28. H. Nitta, M. Suzuki, T. Iida, Solid State Comm. **109**, 283 (1999)
29. E.L. Shirley, L.X. Benedict, S.G. Louie, Phys. Rev. B **54**, 10970 (1996)
30. O. Dubovsky, S. Mukamel, J. Chem. Phys. **96**, 9201 (1992)
31. R.E. Merrifield, J. Chem. Phys. **34**, 1835 (1961)
32. S. Kazaoui, N. Minami, Y. Tanabe, H.J. Byrne, A. Eilmes, P. Petelenz, Phys. Rev. B **58**, 7689 (1998)
33. S. Kazaoui, R. Ross, N. Minami, Phys. Rev. B **52**, R11665 (1995)
34. S.A. Jagliński, PhD thesis, University of Utah (1996)
35. J.P. Hernandez, S.I. Choi, J. Chem. Phys **50**, 1524 (1969)
36. W.L. Pollans, S.I. Choi, J. Chem. Phys. **52**, 3691 (1970)
37. R. Silbey, J. Jortner, S.A. Rice, M.T. Vala, J. Chem. Phys. **42**, 733 (1965)
38. R. Silbey, J. Jortner, S. A. Rice, M. T. Vala, J. Chem. Phys. **43**, 2925 (1966)
39. A. Tiberghien, G. Delacôte, J. Phys. (Paris) **31**, 644 (1970)
40. R.W. Munn, P. Petelenz, W. Siebrand, Chem. Phys. **111**, 207 (1987)
41. D.W. Schlosser, M.R. Philpott, Chem. Phys. **49**, 181 (1980)
42. H. Reis, S. Raptis, M.G. Papadopoulos, R.H.C. Janssen, D.N. Theodorou, R.W. Munn, Theor. Chem. Accounts **99**, 384 (1998)
43. H. Reis, M.G. Papadopoulos, R.W. Munn, J. Chem. Phys. **109**, 6826 (1998)
44. S.I. Choi, J. Jortner, S.A. Rice, R. Silbey, J. Chem. Phys. **41**, 3294 (1964)
45. P. Petelenz, V.H. Smith Jr, Chem. Phys. Lett. **82**, 430 (1981)
46. A.-M. Janner, R. Eder, B. Koopmans, H.T. Jonkman, G.A. Sawatzky, Phys. Rev. B **52**, 17158 (1995)
47. R. Eder, A.-M. Janner, G.A. Sawatzky, Phys. Rev. B **53**, 12768 (1996)
48. R.W. Munn, B. Pac, P. Petelenz, Phys. Rev. B **57**, 1328 (1998)
49. J.A. Gisby, S.H. Walmsley, Chem. Phys. Lett. **135**, 275 (1987)
50. J.A. Gisby, S.H. Walmsley, Chem. Phys. **122**, 271 (1988)
51. P. Petelenz, Chem. Phys. **138**, 35 (1989)
52. I.V. Brovchenko, N.I. Ostapenko, V.I. Sugakov, M.T. Shpak, Izv. Akad. Nauk USSR **52**, 659 (1988)
53. I.V. Brovchenko, N.I. Ostapenko, M.T. Shpak, Fiz. Tverd. Tela **30**, 261 (1988)
54. P. Petelenz, I.V. Brovchenko, D. Mucha, M. Slawik, Chem. Phys. **143**, 415 (1990)
55. P. Petelenz, A. Eilmes, Chem. Phys. **155**, 247 (1991)
56. I.V. Brovchenko, A. Eilmes, P. Petelenz, J. Chem. Phys. **98**, 3737 (1993)
57. A. Eilmes, P. Petelenz, Mol. Cryst. Liq. Cryst. **257**, 251 (1994)
58. I.V. Brovchenko, A. Eilmes, P. Petelenz, J. Chem. Phys. **102**, 3099 (1995)

59. L. Sebastian, G. Weiser, H. Bässler, Chem. Phys. **61**, 125 (1981)
60. L. Sebastian, G. Weiser, G. Peter, H. Bässler, Chem. Phys. **75**, 103 (1983)
61. P. Petelenz, M. Slawik, K. Yokoi, M.Z. Zgierski, J. Chem. Phys. **105**, 4427 (1996)
62. B. Pac, P. Petelenz, M. Slawik, R.W. Munn, J. Chem. Phys. **109**, 7932 (1998)
63. M. Slawik, P. Petelenz, J. Chem. Phys. **107**, 7114 (1997)
64. B. Pac, P. Petelenz, A. Eilmes, R.W. Munn, J. Chem. Phys. **109**, 7923 (1998)
65. P. Petelenz, M. Slawik, B. Pac, Synth. Met. **64**, 335 (1994)
66. P. Petelenz, D. Mucha, Chem. Phys. **154**, 145 (1991)
67. P. Petelenz, D. Mucha, J. Chem. Phys. **100**, 4607 (1994)
68. P. Petelenz, G. Mazur, Chem. Phys. Lett. **301**, 223 (1999)
69. R. Katoh, M. Kotani, Chem. Phys. Lett. **201**, 141 (1993)

10 Excitons in Organic-Based Multilayers

V.M. Agranovich, G.C. La Rocca and F. Bassani

10.1 Introduction

A lot of effort has been devoted in the last few years to the growth of organic crystalline layered structures (including organic quantum wells [QWs] and superlattices): much improvement in the technique of organic molecular beam deposition has led to a variety of good-quality heterostructures based on molecular solids as well as combinations of organic and inorganic semiconductors [1]. The weak van der Waals bonding between organic epilayers and substrates facilitates the epitaxial growth even in the presence of lattice mismatch. The possibility of growing tailor-made systems incorporating different organic crystalline materials with even more flexibility than for multiple quantum wells based on inorganic semiconductors alone opens up a promising field of research from the point of view of fundamental as well as applied physics. It is only natural that the advent of such a new class of organic crystalline materials prompted scientists to investigate their optical properties, and in particular their non-linear and electro-optical properties, in order to understand their possible advantages in comparison with the usual organic or inorganic materials. In this chapter, we review the results of theoretical studies for two types of heterostructures.

In Sect. 10.2, we discuss the optical properties of interacting organic and inorganic nanostructures with resonating Frenkel excitons (FEs) in the organic material and Wannier–Mott excitons (WEs) in the inorganic semiconductor. In a number of cases, rather strong modifications of the optical properties take place that are interesting from the point of view of both fundamental and applied physics. We show that in such heterostructures two regimes can be realized: the strong coupling regime and the weak coupling one. The former can be achieved in structures where the level splitting due to the resonant interaction is large compared to the exciton widths in the inorganic and organic materials. In such a case, new hybrid exciton states appear in which the electronic excitation spends part of its time as an FE and part as a WE. On the one hand, FEs can have a large oscillator strength, which is typical of some organic molecules. On the other hand, WEs can have strong optical non-linearities due to their large Bohr radii and small saturation densities. Hybrid exciton states have properties of both FEs (large oscillator strength) and WEs (small saturation density), which enhances their optical non-linearities. The latter regime (weak coupling) occurs in structures where the width of the exciton states in one or both of the subsystems (organic and inorganic) is large in comparison with the energy splitting. In such a case, the hybrid excitons do not arise, but the energy transfer from one subsystem to the other can be very fast. This process could be particularly interesting for certain applications. Usually, the FE width in organic materials is large compared to that of the WEs in inorganic semiconductors; thus, the case of energy transfer from the inorganic subsystem to the organic one is of relevance. The WEs in the inorganic semiconductors can be pumped electrically (as in light emitting diodes, LEDs) and, consequently, the fast energy transfer to the organics can be used to pump the latter material. Such a mechanism could be used, for example, to realize bright organic LEDs bypassing the problems hindering the electrical pumping of the organic material itself (small carrier mobilities, space-charge formation, charge-injection difficulties, etc.). We consider different resonance phenomena in hybrid heterostructures and discuss the theoretical results for quantum-well, microcavity and quantum-dot configurations.

In Sect. 10.3, we summarize the studies of charge-transfer excitons (CTEs) at donor–acceptor (D–A) interfaces. The repulsion among CTEs, arising from their static dipole moments, can be responsible at high excitation densities for a new kind of phase transition to a conducting state (cold photoconductivity). Below the threshold for such transition, the CTEs dipole interaction leads to strong optical non-linearities and, in asymmetric superlattices, to a new photovoltaic mechanism.

10.2 Excitons in Organic-Inorganic Heterostructures

In this section, we discuss the possibility of obtaining qualitatively new physical effects, potentially useful also for technological applications, by ingeniously combining organic with inorganic materials in one and the same hybrid structure. Excitons play a fundamental role in the optical properties of dielectric solids [2,3]; two models are conventionally used to classify excitons: the Frenkel exciton model (appropriate mainly for molecular solids) and the Wannier–Mott exciton model (appropriate mainly for semiconductors). The mean electron–hole distance in a WE is typically large (in comparison with the lattice constant). On the other hand, the FE is represented as an electronic state of a crystal in which electrons and holes are placed on the same molecule. We can say that FEs in organic crystals have radii a_F, comparable to the lattice constant $a_F \sim a \sim 5$ Å. In contrast, weakly bound WEs in semiconductor QWs have large Bohr radii ($a_B \sim 100$ Å in III-V materials and $a_B \sim 30$ Å in II-VI ones; in both cases $a_B \gg a$). The oscillator strength of an FE is close to a molecular oscillator strength F, whereas the oscillator strength f of a WE is usually much weaker: in a QW $f \sim a^3 a_B^{-2} L^{-1} F$, where L is the QW width ($a_B > L > a$). Both types of excitons interact with lattice vibrations through exciton–phonon coupling and can also be inhomogeneously broadened by structural imperfections. In good quality inorganic semiconductor QWs, the WE linewidth is of the order of 1 meV (usually limited by inhomogeneous broadening). FEs in organic materials typically have a much larger linewidth (often due to strong electron–phonon coupling); for instance, about 200 meV in thin films of 3,4,9,10-perilenetetracarboxylic-dianhydride (PTCDA) [1]. However, it is possible to choose resonating organic materials with sharp FEs, such as coronene (exciton linewidth $\simeq$ 4 meV [4]) or the surface exciton of anthracene (linewidth of about 2 meV [5]). It is important to notice that the dipole–dipole interaction coupling the FEs and WEs at an organic–inorganic heterojunction can be of order of 10 meV [6,7]. Therefore, the case of strong coupling (Sect. 10.2.1), in which the exciton linewidths are smaller than the anticrossing energy splitting and hybrid excitons (HEs) exhibiting pronounced optical non-linearities are formed must be distinguished from the case of weak coupling (Sect. 10.2.2) in which the FEs are much broader and the dipole–dipole coupling gives rise to an irreversible energy transfer from the inorganic to the organic material.

10.2.1 Strong Coupling: Hybrid Excitons

In high-quality semiconductors as well as in organic crystalline materials, the optical properties near and below the band gap are dominated by the exciton transitions and the same situation takes place also for organic and inorganic QWs (or wires or dots). The excitonic optical non-linearities in semiconductor QWs can be large because the ideal bosonic approximation for WEs breaks down as soon as they start to overlap with each other, i.e., when their two-dimensional (2D) density n becomes comparable to the saturation density $n_{\mathrm{S}} \sim 1/(\pi\, a_{\mathrm{B}}^2)$ (n_{S} is, typically, 10^{12} cm^{-2}). Then, due to phase-space filling (PSF), exchange and collisional broadening, the exciton resonance is bleached. However, a generic figure of merit for the optical non-linearities scales as follows: $I_{\mathrm{P}}^{-1}\,(\Delta\chi/\chi)$, where $\Delta\chi$ is the non-linear change in the susceptibility in the presence of a pump of intensity I_{P}. As $\Delta\chi/\chi \sim n/n_{\mathrm{S}} \sim n\, a_{\mathrm{B}}^2$, but also $n \propto f\, I_{\mathrm{P}} \propto a_{\mathrm{B}}^{-2}\, I_{\mathrm{P}}$, such a figure of merit is nearly independent of the exciton Bohr radius [8]. As for the FEs in organic crystals, just because they have small radii, they have very large saturation density. Thus, pronounced PSF non-linearities of the exciton resonance in molecular crystals are practically impossible to achieve as very high excitonic concentrations are needed. Of course, other mechanisms may effectively enhance the optical non-linearities of organic materials, but the discussion of these falls outside the scope of this chapter.

Here we will consider hybrid structures in which FEs and WEs are in resonance with each other and coupled through their dipole–dipole interaction at the interface and through cavity photons in a microcavity. The basic idea is to realize the formation of new eigenstates given by appropriate coherent linear combinations of large radius exciton states in the inorganic material and small radius exciton states in the organic one. We can expect that these hybrid electronic excitations will be characterized by a radius dominated by their Wannier component and by an oscillator strength dominated by their Frenkel component. Thus, they can have at the same time a large oscillator strength and a small saturation density n_S. In this way, the desirable properties of both the inorganic and organic materials conspire to overcome the basic limitation mentioned above for the figure of merit of the exciton resonance non-linearities. One of the most natural choices to implement this idea is a layered structure with an interface between a covalent semiconductor and a crystalline molecular semiconductor. In the following, we will discuss at length the electronic excitation spectra arising from the Frenkel–Wannier exciton hybridization; the non-linear optical properties of hybrid excitons will be considered in detail. We predict a large enhancement of the excitonic resonant non-linearities, in some cases of two orders of magnitude compared to traditional systems. We also consider the resonant interaction between FEs in the organic QW and WEs in the semiconductor QW in a microcavity where organic and inorganic QWs are separated. In this case the resonant interaction appears mainly through the cavity photons and can be very strong if the cutoff frequency of the cavity photon is close to the excitonic resonances. We demonstrate that in this case new hybrid Frenkel–Wannier–Mott exciton–cavity photon states can be tailored to engineer the fluorescence efficiency and relaxation processes.

Two-Dimensional Hybrid Excitons and Linear Optics

Here we study the dipole–dipole interaction between an organic quantum well (OQW) and an inorganic one (IQW) and demonstrate how new hybrid states arise [6]. The configuration we consider is given in the following. A plane semiconductor IQW of thickness L_w occupies the region $|z| < L_w/2$, the z axis being chosen along the growth direction. All the space with $z > L_w/2$ is filled by the barrier material and that with $z < -L_w/2$ by the organic material in which the OQW is placed (Fig. 10.1). For simplicity, we treat the organic molecules in the dipole approximation, neglecting the contribution of higher multipoles to the interaction, and we consider the OQW as a single monolayer, i.e., as a 2D lattice of molecules at discrete sites $\boldsymbol{n}$, placed at $z = -z_0 < -L_w/2$ (the generalization to the case of several monolayers is easy). All the semiconductor well-barrier structure ($z > -L_w/2$) is assumed to have the same background dielectric constant ε, while the organic half-space ($z < -L_w/2$) the dielectric constant $\tilde{\varepsilon}$ (corresponding to the organic substrate).

Due to the different electronic structure of the two QWs under consideration, one may neglect the single-particle wavefunctions mixing; in other words, the OQW and the IQW states are assumed to have zero wavefunction overlap. Assuming a perfect 2D translational invariance of the system, we classify the excitons by their in-plane wavevector $\boldsymbol{k}$. Suppose that for some

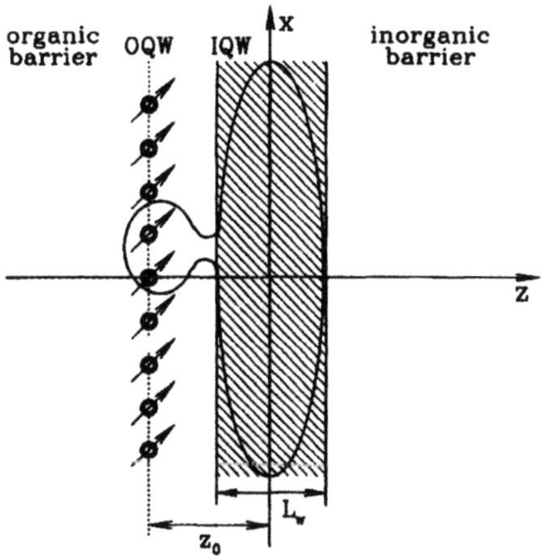

Fig. 10.1. The physical configuration under study: a hybrid exciton state is sketched

bands of FEs in the OQW and WEs in the IQW the energy separation is much less than the distance to other exciton bands. Then we take into account only the mixing between these two bands. We choose as a basis set the "pure" Frenkel and Wannier states, i.e., the state when the OQW is excited, while the IQW is in its ground state (denoted by $|F,\boldsymbol{k}\rangle$), and vice versa (denoted by $|W,\boldsymbol{k}\rangle$), their energies being $E_{\mathrm{F}}(\boldsymbol{k})$ and $E_{\mathrm{W}}(\boldsymbol{k})$. We seek the new hybrid states in the form

$$|\alpha,\boldsymbol{k}\rangle = A_\alpha(\boldsymbol{k})\,|F,\boldsymbol{k}\rangle + B_\alpha(\boldsymbol{k})\,|W,\boldsymbol{k}\rangle \ , \tag{10.1}$$

where $\alpha =$ "u", "l" labels the two resulting states (upper and lower branches). The Schrödinger equation for the coefficients A, B is then written as:

$$[E_{\mathrm{F}}(\boldsymbol{k}) - E]\,A(\boldsymbol{k}) + \langle F,\boldsymbol{k}|\,\hat{H}_{\mathrm{int}}\,|W,\boldsymbol{k}\rangle\,B(\boldsymbol{k}) = 0\,,$$

$$\langle W,\boldsymbol{k}|\,\hat{H}_{\mathrm{int}}\,|F,\boldsymbol{k}\rangle\,A(\boldsymbol{k}) + [E_{\mathrm{W}}(\boldsymbol{k}) - E]\,B(\boldsymbol{k}) = 0\,, \tag{10.2}$$

where $\hat{H}_{\mathrm{int}}$ is the Hamiltonian of the dipole–dipole interaction between the QWs. Solution of (10.2) gives the energies of the upper and lower branches and the splitting $\Delta(\boldsymbol{k})$:

$$E_{\mathrm{u,l}}(\boldsymbol{k}) = \frac{E_{\mathrm{F}}(\boldsymbol{k})+E_{\mathrm{W}}(\boldsymbol{k})\pm\Delta(\boldsymbol{k})}{2}\,,$$

$$\Delta(\boldsymbol{k}) \equiv \sqrt{[E_{\mathrm{F}}(\boldsymbol{k})-E_{\mathrm{W}}(\boldsymbol{k})]^2+4\Gamma^2(\boldsymbol{k})}\,, \tag{10.3}$$

where we use the notation $\Gamma(\boldsymbol{k}) \equiv |\langle W,\boldsymbol{k}|\,\hat{H}_{\mathrm{int}}\,|F,\boldsymbol{k}\rangle|$ for the coupling matrix element. For the orthonormalized new states the weighting coefficients are given by

$$|A_{\mathrm{u}}(\boldsymbol{k})|^2 = |B_{\mathrm{l}}(\boldsymbol{k})|^2 = \frac{1}{2}\left(1 + \frac{E_{\mathrm{F}}(\boldsymbol{k}) - E_{\mathrm{W}}(\boldsymbol{k})}{\Delta(\boldsymbol{k})}\right)\,, \tag{10.4}$$

$$|A_{\mathrm{l}}(\boldsymbol{k})|^2 = |B_{\mathrm{u}}(\boldsymbol{k})|^2 = \frac{1}{2}\left(1 - \frac{E_{\mathrm{F}}(\boldsymbol{k}) - E_{\mathrm{W}}(\boldsymbol{k})}{\Delta(\boldsymbol{k})}\right)\,. \tag{10.5}$$

To evaluate the matrix element $\Gamma_{\boldsymbol{k}}$, we write down the interaction Hamiltonian as

$$\hat{H}_{\text{int}} = -\sum_{\boldsymbol{n}} \hat{\boldsymbol{p}}^{\text{F}}(\boldsymbol{n}) \cdot \hat{\boldsymbol{\mathcal{E}}}(\boldsymbol{n}) \,, \tag{10.6}$$

where $\hat{\boldsymbol{p}}^{\text{F}}(\boldsymbol{n})$ is the operator of the dipole moment of the organic molecule situated at the lattice site $\boldsymbol{n}$, and $\hat{\boldsymbol{\mathcal{E}}}(\boldsymbol{n})$ is the operator of the electric field at the point $\boldsymbol{n}$, produced by the IQW exciton. If we introduce the operator of the IQW polarization $\hat{\boldsymbol{P}}^{\text{W}}(\boldsymbol{r})$, then the operators $\hat{\boldsymbol{\mathcal{E}}}(\boldsymbol{n})$ and $\hat{\boldsymbol{P}}^{\text{W}}(\boldsymbol{r})$ are related to each other exactly in the same way as the corresponding classical quantities in electrostatics [6,7]. The matrix element of the IQW polarization between the ground state $|0\rangle$ and $|W,\boldsymbol{k}\rangle$ for a 1s-exciton with the Bohr radius a_{B} is equal to [6,9,10]

$$\langle 0|\, \hat{\boldsymbol{P}}^{\text{W}}(\boldsymbol{r})\, |W,\boldsymbol{k}\rangle = \sqrt{\frac{2}{\pi}}\, \frac{\boldsymbol{d}^{\text{vc}}}{a_{\text{B}}}\, \frac{\mathrm{e}^{i\boldsymbol{k}\boldsymbol{r}_{\parallel}}}{\sqrt{S}}\, \chi^{\text{e}}(z)\, \chi^{\text{h}}(z) \,, \tag{10.7}$$

where $\sqrt{2/(\pi a_{\text{B}}^2)}$ is the value of the 1s-wavefunction of the relative motion of the electron and hole, taken at $r_{\parallel} = 0$; $\chi^{\text{e}}(z)$, $\chi^{\text{h}}(z)$ are the envelope functions for the electron and hole in the IQW confinement potential (we assume the IQW to be thin, so that the transverse and relative in-plane motions of the electron and hole are decoupled), and S is the in-plane normalization area. Finally,

$$\boldsymbol{d}^{\text{vc}} = \int_{\text{u.c.}} u_{\text{v}}^{*}(\boldsymbol{r})\, (-e\boldsymbol{r})\, u_{\text{c}}(\boldsymbol{r})\, \mathrm{d}^3\boldsymbol{r} \tag{10.8}$$

is the matrix element of the electric dipole moment between the conduction and valence bands ($\boldsymbol{d}^{\text{vc}}$ is taken to be independent of $\boldsymbol{k}$, $u_{\text{c(v)}}$ are the Bloch functions for the conduction [valence] band extremum and the integration in (10.8) is performed over the unit cell). For the FE the dipole moment matrix element is given by

$$\langle F,\boldsymbol{k}|\, \hat{\boldsymbol{p}}(\boldsymbol{n})\, |0\rangle = \frac{\mathrm{e}^{-i\boldsymbol{k}\boldsymbol{n}}}{\sqrt{N}}\, \boldsymbol{d}^{\text{F}^*} = \frac{\mathrm{e}^{-i\boldsymbol{k}\boldsymbol{n}}}{\sqrt{S}}\, a_{\text{F}}\, \boldsymbol{d}^{\text{F}^*} \,, \tag{10.9}$$

where $\boldsymbol{d}^{\text{F}}$ is the transition dipole moment for a single organic molecule (analogous to $\boldsymbol{d}^{\text{vc}}$ in the semiconductor), N is the total number of sites in the lattice, a_{F} is the lattice constant, which may be considered as the radius of the FE. The final expression for the coupling matrix element is [7]:

$$\langle F,\boldsymbol{k}|\, \hat{H}_{\text{int}}\, |W,\boldsymbol{k}\rangle = -\sqrt{\frac{2}{\pi}}\, \frac{d_i^{\text{F}^*}}{a_{\text{F}}}\, \frac{d_j^{\text{vc}}}{a_{\text{B}}} \int \mathrm{d}z\, \mathcal{D}_{ij}(\boldsymbol{k}, -z_0, z)\, \chi^{\text{e}}(z)\, \chi^{\text{h}}(z) \,, \tag{10.10}$$

with

$$\mathcal{D}_{ij}(\boldsymbol{k}, z, z') = \frac{4\pi}{\varepsilon + \tilde{\varepsilon}} k e^{k(z-z')} \left(\frac{ik_i}{k} + \delta_{i,z} \right) \left(\frac{ik_j}{k} + \delta_{j,z} \right) . \tag{10.11}$$

The only contributing polarizations for the semiconductor are those along $\boldsymbol{k}$ (L-modes) and along the growth direction z (Z-modes, only for the light holes). For simplicity, we take the electron and hole confinement wavefunctions for the lowest subbands in the approximation of an infinitely deep IQW,

$$\chi^{\mathrm{e}}(z)\, \chi^{\mathrm{h}}(z) = \frac{2}{L_{\mathrm{w}}} \cos^2 \left(\frac{\pi z}{L_{\mathrm{w}}} \right) , \tag{10.12}$$

and assume the transition dipole moment in the organics $\boldsymbol{d}^{\mathrm{F}}$ to be real (which is always possible with an appropriate choice of molecular wavefunctions). Without loss of generality we may take the vector $\boldsymbol{k}$ along the x axis. Evaluating the integral in (10.10), we obtain the interaction parameter $\Gamma_{\mathrm{L(Z)}}$ for the L- and Z-modes:

$$\Gamma_{\mathrm{L(Z)}}(k) = \frac{8\sqrt{2\pi}}{\epsilon + \tilde{\epsilon}} \frac{e^{-kz_0} \sinh(kL_{\mathrm{w}}/2)}{1 + (kL_{\mathrm{w}}/2\pi)^2} \frac{|d^{\mathrm{vc}}_{x(z)}| \sqrt{(d^{\mathrm{F}}_x)^2 + (d^{\mathrm{F}}_z)^2}}{a_{\mathrm{F}} a_{\mathrm{B}} L_{\mathrm{w}}} . \tag{10.13}$$

It is seen that $\Gamma(k)$ has a maximum $\Gamma_{\max}$ at $k = k_{\max}$. The value of $k_{\max}$ for arbitrary z_0 and L_{w} may be found numerically; for $z_0 - L_{\mathrm{w}} > 0.1\, L_{\mathrm{w}}$ it is well described by the formula

$$k_{\max} \simeq \frac{1}{L_{\mathrm{w}}} \ln \left(\frac{2z_0 + L_{\mathrm{w}}}{2z_0 - L_{\mathrm{w}}} \right) , \tag{10.14}$$

while in the limit $z_0 \simeq L_{\mathrm{w}}/2$ we have $k_{\max} \simeq 2.4/L_{\mathrm{w}}$.

We approximate the WE energy by a parabola, with the in-plane effective mass $m_{\mathrm{W}} = m_{\mathrm{e}} + m_{\mathrm{h}}$, $m_{\mathrm{e(h)}}$ being the electron (hole) mass, and neglect the FE dispersion since the typical masses are $(5\text{–}100)m_0$:

$$E_{\mathrm{W}}(\boldsymbol{k}) = E_{\mathrm{W}}(0) + \frac{\hbar^2 k^2}{2m_{\mathrm{W}}} , \quad E_{\mathrm{F}}(\boldsymbol{k}) = E_{\mathrm{F}}(0) , \quad E_{\mathrm{F}}(0) - E_{\mathrm{W}}(0) \equiv \delta . \tag{10.15}$$

We will measure all energies with respect to $E_{\mathrm{W}}(0)$. The dispersion of the hybrid states (10.3) can be written as

$$E_{\mathrm{u,l}}(\boldsymbol{k}) - E_{\mathrm{W}}(0) = \frac{\delta}{2} + \frac{\hbar^2 k^2}{4m_{\mathrm{W}}} \pm \sqrt{\left(\frac{\delta}{2} - \frac{\hbar^2 k^2}{4m_{\mathrm{W}}} \right)^2 + \Gamma^2(\boldsymbol{k})} . \tag{10.16}$$

To perform numerical estimates we choose the following values of parameters. For the IQW those representative of II-VI semiconductor (e.g.,

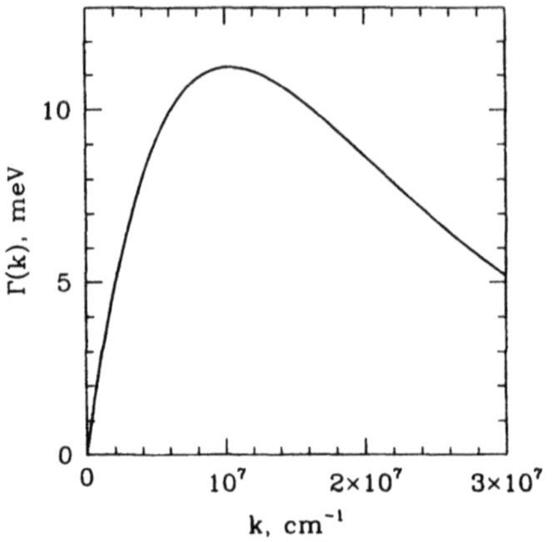

Fig. 10.2. The interaction parameter $\Gamma(k)$ for $d^{\mathrm{vc}} = 12$ Debye, $d^{\mathrm{F}} = 5$ Debye, $a_{\mathrm{B}} = 25$ Å, $a_{\mathrm{F}} = 5$ Å, $L_{\mathrm{w}} = 10$ Å, $z_0 = 10$ Å, $\epsilon_\infty = 6$, $\tilde{\epsilon}_\infty = 4$. (From [7])

ZnSe/ZnCdSe) QWs are taken [11]: $\epsilon = \epsilon_\infty = 6$, $d^{\mathrm{vc}}/a_{\mathrm{B}} \approx 0.1\,e$ (which corresponds to $d^{\mathrm{vc}} \simeq 12$ Debye and a Bohr radius of 25 Å), the exciton mass $m_{\mathrm{W}} = 0.7 m_0$ and the well width $L_{\mathrm{w}} = 10$ Å. For the organic part of the structure, we take parameters typical of such media (e.g., see [1,12,13]): $\tilde{\epsilon} = \tilde{\epsilon}_\infty = 4$, the transition dipole for the molecules in the monolayer $d^{\mathrm{F}} = 5$ Debye, $a_{\mathrm{F}} = 5$ Å and $z_0 = 10$ Å. We plot $\Gamma(k)$ for these values of parameters in Fig. 10.2. We see that $\Gamma_{\max} \simeq 11$ meV. The dispersion curves $E_{\mathrm{u,l}}(k)$ along with the FE weight in the lower branch $|A_{\mathrm{l}}(k)|^2$ for a detuning $\delta = 10$ meV are plotted in Fig. 10.3.

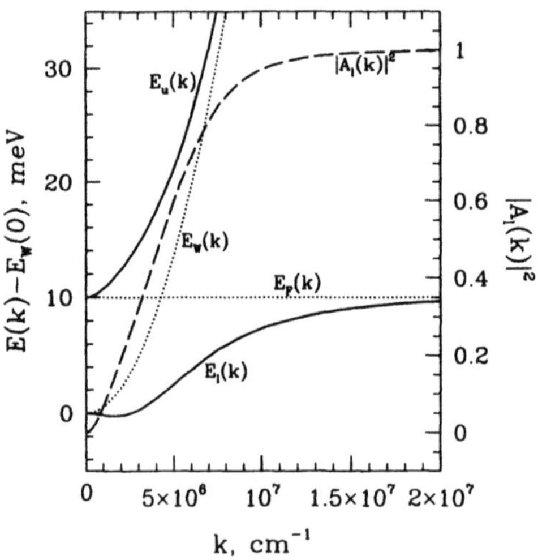

Fig. 10.3. The dispersion $E_{\mathrm{u,l}}(k)$ of the upper and lower hybrid exciton branches (*solid lines*) and that of the unperturbed FE and WE (*dotted lines*). The "weight" of the FE component in the lower branch $|A_{\mathrm{l}}(k)|^2$ is shown by the *dashed line*. The parameters are the same as in Fig. 10.2 ($m_{\mathrm{W}} = 0.7\,m_0$); the detuning $\delta = 10$ meV. (From [7])

A non-trivial feature of the lower branch dispersion is a minimum away from $k = 0$, which is always present for $\delta \leq 0$, as well as for small positive values of δ, and is the deepest for $\delta = 0$ [14]. As a consequence, at low temperatures and under optical pumping at frequencies above the excitonic resonance, excitons will accumulate in this minimum, which can be detected, for example, by pump–probe experiments. The fluorescence from these states should increase with temperature since states with small k become populated.

If an incident electromagnetic wave with the electric field $\boldsymbol{\mathcal{E}}(\boldsymbol{r}) = \boldsymbol{\mathcal{E}}_0 \, e^{i\boldsymbol{Q}\boldsymbol{r}}$ is present, then the interaction with the hybrid structure is described by the Hamiltonian (neglecting the local field corrections)

$$\hat{H}_{\rm em} = -\boldsymbol{\mathcal{E}}_0 \cdot \left(\sum_{\boldsymbol{n}} \hat{\boldsymbol{p}}^{\rm F}(\boldsymbol{n}) \, e^{i\boldsymbol{Q}_{\parallel}\boldsymbol{n}} + \int {\rm d}z \int {\rm d}^2\boldsymbol{r}_{\parallel} \, \hat{\boldsymbol{P}}^{\rm W}(\boldsymbol{r}) \, e^{i\boldsymbol{Q}_{\parallel}\boldsymbol{r}_{\parallel}} \right) , \tag{10.17}$$

where we have neglected the z-dependence of the incident field since the thickness of our structure is much less than the light wavelength. The corresponding matrix element is different from zero only if $\boldsymbol{k} = \boldsymbol{Q}_{\parallel}$ and in this case is equal to

$$\langle \alpha, \boldsymbol{k} | \, \hat{H}_{\rm em} \, | 0 \rangle \equiv -\boldsymbol{\mathcal{E}}_0 \cdot \boldsymbol{M}_{\boldsymbol{k}}^{\alpha} = -\boldsymbol{\mathcal{E}}_0 \cdot \left(A_{\alpha}^{*}(\boldsymbol{k}) \, \boldsymbol{M}^{\rm F} + B_{\alpha}^{*}(\boldsymbol{k}) \, \boldsymbol{M}_{\boldsymbol{k}}^{\rm W} \right) , \tag{10.18}$$

where

$$\boldsymbol{M}^{\rm F} = \sqrt{N} \, \boldsymbol{d}^{\rm F^*} = \frac{\sqrt{S}}{a_{\rm F}} \, \boldsymbol{d}^{\rm F^*} \tag{10.19}$$

and

$$\boldsymbol{M}_{\boldsymbol{k}}^{\rm W} = \sqrt{\frac{2}{\pi}} \, \frac{\sqrt{S}}{a_{\rm B}} \, \boldsymbol{d}^{\rm vc^*} \int \chi^{\rm e^*}(z) \, \chi^{\rm h^*}(z) \, {\rm d}z \tag{10.20}$$

are the optical matrix elements for the isolated OQW and IQW respectively, which are independent of $\boldsymbol{k}$. Usually, we have $M^{\rm F} \gg M^{\rm W}$, since $a_{\rm F} \ll a_{\rm B}$, and in the region of strong mixing the oscillator strengths f^{α} of a hybrid state are determined by its FE component:

$$f^{\alpha}(\boldsymbol{k}) \simeq |A_{\alpha}(\boldsymbol{k})|^2 \, F \, . \tag{10.21}$$

At the crossing point $k = k_0$ (for $\delta > 0$), we have $|A_{\alpha}(k_0)|^2 = 1/2$, and the FE oscillator strength is equally distributed between the two hybrid states. For the hybrid exciton radii the opposite relation holds (as sketched in Fig. 10.1). Calculating the expectation value of the exciton radius squared $\hat{r}^2$ in the state $|\alpha, \boldsymbol{k}\rangle$, we obtain

$$\begin{aligned} \langle \alpha, \boldsymbol{k} | \hat{r}^2 | \alpha, \boldsymbol{k} \rangle &= |A_{\alpha}(\boldsymbol{k})|^2 \langle F, \boldsymbol{k} | \hat{r}^2 | F, \boldsymbol{k} \rangle + |B_{\alpha}(\boldsymbol{k})|^2 \langle W, \boldsymbol{k} | \hat{r}^2 | W, \boldsymbol{k} \rangle \\ &\simeq |B_{\alpha}(\boldsymbol{k})|^2 a_{\rm B}^2 \, , \end{aligned} \tag{10.22}$$

since $a_B \gg a_F$ and we neglect the latter. Cross terms do not appear since we neglect the single-particle wavefunction mixing between the two QWs.

We see that the new states can possess both large oscillator strengths and exciton radii. This effect is especially pronounced if the crossing of the FE and WE dispersion curves occurs for the value of the wavevector close to that of the maximum of the coupling strength: $k_0 \simeq k_{max}$. Since k_0 is determined by the detuning δ, and k_{max}, in turn, depends on L_w and z_0 (10.14), a special choice of these parameters should be made to maximize the effect. Also, in order to take advantage of the hybrid states in optics, the wavevector of light in the medium $q = n_\infty \omega / c$ (n_∞ being the background refraction index) should not be far from k_0. Usually near-excitonic resonances, $q < k_0$, and special care should be taken to overcome this difficulty (e.g., using a coupled diffraction grating with period $2\pi/k_0$ [15] or a prism). We mention, however, that even in the region of small wavevectors in which the 2D excitons are radiative, the hybridization may be realized not due to the instantaneous dipole–dipole interaction, but due to the retarded interaction stemming from the exchange of photons. Such a situation has been analysed (even in the non-linear regime) with an appropriate transfer matrix approach, which is equivalent to the solution of the full Maxwell equations [16].

Concerning the choice of materials for the implementation of the system considered here, as examples of molecular substances having small-radius ($\leq$5 Å) excitons with energies of a few eV and small linewidths, which may be grown as crystalline layers on a variety of inorganic (including semiconductor) crystals, we mention tetracene (2 eV), pentacene (1.5 eV), and coronene (2.9 eV). Semiconductors with large-radius excitons with matching energies are, for instance, the III-V and II-VI ternary solid solutions such as GaAlAs, ZnCdSe and ZnSSe [17]; as well as a judicious choice of alloy composition and well thickness, a fine tuning of the resonance condition can be achieved by applying an external static electric field along the growth direction (quantum confined Stark effect [18]; for HEs it has been considered in [19]). A major experimental problem, besides the choice of the organic material, is the control of the interface quality: the inhomogeneous broadening should remain small, and the in-plane wavevector $\boldsymbol{k}$ a (sufficiently) good quantum number; organic superlattices with high-quality interfaces have been demonstrated [1]. The necessary condition for the hybrid states to be observable is that the exciton linewidths must be smaller then the splitting $\Delta(k)$. This is the case in the present calculations, where for $k_0 = k_{max}$ we have $\Delta(k_0) = 2\Gamma_{max} \simeq 20$ meV, while in inorganic QWs the homogeneous linewidth at low temperatures is $\sim$1 meV [20, 21]. As already mentioned, the non-radiative linewidth of a 2D FE in an OQW can also be small. For example, in the case of a 2D exciton in the external monolayer of anthracene, the linewidth at low temperatures is $\sim$2 meV [5]; for the exciton in coronene it is $\sim$4 meV [4]. In principle, apart from the resonance condition and the large difference in excitonic radii, the present model demands no specific requisite, and the rapid progress in the growth of organic crystalline multilayers justifies some optimism about its concrete realization.

We also wish to mention the work [22] in which the effects of the exciton–phonon interaction in hybrid systems were studied. In this work the resonant Raman spectroscopy is also suggested as a tool for studying hybrid organic–inorganic QWs.

Non-linear Optics of Two-Dimensional Hybrid Excitons

From the results of the previous subsection we may expect that the exciton hybridization should strongly modify the non-linear optical properties of the structure under consideration. Indeed, hybrid excitons can combine both a large oscillator strength, which makes it easy to produce large populations, and a large radius, which, in turn, leads to low saturation densities. In this subsection we analyze the situation quantitatively [23], calculating the response of the interband polarization $\boldsymbol{P} = \boldsymbol{P}^{\mathrm{W}} + \boldsymbol{P}^{\mathrm{F}}$ to the external driving electric field (corresponding to a cw experiment) $\boldsymbol{\mathcal{E}}(\boldsymbol{r},t) = \boldsymbol{\mathcal{E}}_0\, \mathrm{e}^{i\boldsymbol{Q}\boldsymbol{r}-i\omega t}$ + c.c., in the presence of a high density of excitations using the standard technique of semiconductor Bloch equations [18, 24]. Since we are considering a cw experiment, the populations are stationary and may be treated as parameters in the equation for the time-dependent interband polarization.

First, we express the operator of the electron–hole interband polarization $\hat{\boldsymbol{P}}^{\mathrm{W}}(\boldsymbol{r})$ in terms of the electron and hole creation and annihilation operators in the envelope function approximation, following the standard procedure [9, 24]:

$$\hat{\boldsymbol{P}}^{\mathrm{W}}(\boldsymbol{r}) = \frac{\boldsymbol{d}^{\mathrm{vc}}}{S}\, \chi^{\mathrm{e}}(z)\, \chi^{\mathrm{h}}(z) \sum_{\boldsymbol{k},\boldsymbol{q}} \mathrm{e}^{i\boldsymbol{k}\boldsymbol{r}_\parallel}\, \hat{h}_{-\boldsymbol{q}}\, \hat{c}_{\boldsymbol{k}+\boldsymbol{q}} + \mathrm{h.\,c.}\ . \tag{10.23}$$

Here $\chi^{\mathrm{e}}(z)$ and $\chi^{\mathrm{h}}(z)$ are electron and hole wavefunctions in the given IQW subbands (resonant with the FE), $\hat{c}_{\boldsymbol{k}}$ and $\hat{h}_{\boldsymbol{k}}$ are annihilation operators for an electron and hole with the in-plane wavevector $\boldsymbol{k}$ in the subbands under consideration, S is the in-plane normalization area and $\boldsymbol{d}^{\mathrm{vc}}$ is the matrix element (10.8). We do not take into account the spin degeneracy, considering thus the polarization produced by electrons and holes with a given spin (thus, the final expression for the susceptibility should be multiplied by two). An analogous expression for the OQW polarization is

$$\hat{\boldsymbol{P}}^{\mathrm{F}}(\boldsymbol{r}) = \frac{\boldsymbol{d}^{\mathrm{F}}}{a_{\mathrm{F}}\sqrt{S}}\, \delta(z+z_0) \sum_{\boldsymbol{k}} \mathrm{e}^{i\boldsymbol{k}\boldsymbol{r}_\parallel}\, \hat{B}_{\boldsymbol{k}} + \mathrm{h.\,c.}\ , \tag{10.24}$$

where $\hat{B}_{\boldsymbol{k}}$ is the annihilation operator for the FE, which is assumed to be tightly bound.

As well as the term of the Hamiltonian describing free FEs and free electron–hole pairs (with the single-particle energies $E_{\mathrm{F}}(\boldsymbol{k})$, $\epsilon_{\mathrm{e}}(\boldsymbol{k})$ and $\epsilon_{\mathrm{h}}(\boldsymbol{k})$ correspondingly), the Hamiltonian we consider here includes the following:

(a) The Coulomb interaction between electrons and holes

$$\hat{H}_{\text{coul}} = \frac{1}{2S}\sum_{\boldsymbol{q}\neq 0} v(\boldsymbol{q}) \sum_{\boldsymbol{k},\boldsymbol{k}'} \Big(\hat{c}^{\dagger}_{\boldsymbol{k}+\boldsymbol{q}} \hat{c}^{\dagger}_{\boldsymbol{k}'-\boldsymbol{q}} \hat{c}_{\boldsymbol{k}'} \hat{c}_{\boldsymbol{k}} + \hat{h}^{\dagger}_{\boldsymbol{k}+\boldsymbol{q}} \hat{h}^{\dagger}_{\boldsymbol{k}'-\boldsymbol{q}} \hat{h}_{\boldsymbol{k}'} \hat{h}_{\boldsymbol{k}} - 2\hat{c}^{\dagger}_{\boldsymbol{k}+\boldsymbol{q}} \hat{h}^{\dagger}_{\boldsymbol{k}'-\boldsymbol{q}} \hat{h}_{\boldsymbol{k}'} \hat{c}_{\boldsymbol{k}} \Big) \tag{10.25}$$

where

$$v(\boldsymbol{q}) = \frac{2\pi e^2}{\varepsilon_0\, q}\,, \tag{10.26}$$

ε_0 being the static dielectric constant of the IQW.

(b) The dipole–dipole interaction between the QWs

$$\hat{H}_{\text{hyb}} = \sum_{\boldsymbol{k}} V_{\text{hyb}}(\boldsymbol{k})\, \hat{B}^{\dagger}_{\boldsymbol{k}} \sum_{\boldsymbol{q}} \hat{h}_{-\boldsymbol{q}} \hat{c}_{\boldsymbol{k}+\boldsymbol{q}} + \text{h. c.}\,, \tag{10.27}$$

where

$$V_{\text{hyb}}(\boldsymbol{k}) = -\frac{d_i^{\text{F}^*}\, d_j^{\text{vc}}}{a_{\text{F}}\sqrt{S}} \int \mathrm{d}z\, \mathcal{D}_{ij}(\boldsymbol{k}, -z_0, z)\, \chi^{\text{e}}(z)\, \chi^{\text{h}}(z)\,, \tag{10.28}$$

which corresponds to (10.10) with $\sqrt{2/(\pi a_{\text{B}}^2)}$ replaced by $1/\sqrt{S}$, since we use plane waves as the basis for the semiconductor states. Of course, this interaction is also of a Coulomb nature, but since we treat the OQW and the IQW as completely different systems and neglect all effects of electronic exchange between them, these pieces of the Hamiltonian come separately. We notice that $\hat{H}_{\text{coul}}$ gives rise to *intra* band scattering, while $\hat{H}_{\text{hyb}}$ gives rise to *inter* band scattering [23].

(c) The interaction with the driving electric field

$$\hat{H}_{\text{dr}} = -(\boldsymbol{\mathcal{E}}_0 \cdot \boldsymbol{M}^{\text{F}})\mathrm{e}^{-i\omega t} \hat{B}^{\dagger}_{\boldsymbol{Q}_\parallel} - (\boldsymbol{\mathcal{E}}_0 \cdot \boldsymbol{M}^{\text{eh}})\mathrm{e}^{-i\omega t} \sum_{\boldsymbol{q}} \hat{c}^{\dagger}_{\boldsymbol{Q}_\parallel+\boldsymbol{q}} \hat{h}^{\dagger}_{-\boldsymbol{q}} + \text{h. c.}\,, \tag{10.29}$$

and

$$\boldsymbol{M}^{\text{eh}} = \boldsymbol{d}^{\text{vc}^*} \int \chi^{\text{e}^*}(z) \chi^{\text{h}^*}(z)\, \mathrm{d}z\,, \quad \text{and} \quad \boldsymbol{M}^{\text{F}} = \frac{\sqrt{S}}{a_{\text{F}}}\, \boldsymbol{d}^{\text{F}^*}\,, \tag{10.30}$$

where we again neglect the z-dependence of the field and the wavevector dependence of $\boldsymbol{M}^{\text{eh}}$.

Given the Hamiltonian, the generalized Bloch equations for the polarization functions can be written, projected on the resonant exciton states and solved perturbatively to the first order in the excitation density. Finally, we obtain [23] for the 2D susceptibility (not forgetting the factor of 2 originating from spin degeneracy as mentioned at the beginning of this subsection):

$$\chi_{ij}(\omega, \boldsymbol{k}) = \frac{(2\, d_i^{\text{F}^*} d_j^{\text{F}}/a_{\text{F}}^2)(E_{\text{W}}(\boldsymbol{k}) + \delta E_{\text{W}}(\boldsymbol{k}) - \hbar\omega)}{(E_{\text{W}}(\boldsymbol{k}) + \delta E_{\text{W}}(\boldsymbol{k}) - \hbar\omega)\,(E_{\text{F}}(\boldsymbol{k}) - \hbar\omega) - \beta_{\boldsymbol{k}} |V_{\text{FW}}(\boldsymbol{k}) + \delta V_{\text{FW}}(\boldsymbol{k})|^2}\,. \tag{10.31}$$

In (10.31) the non-linearities appear through the blue shift δE_{W}, the Pauli blocking factor β and the modification of the hybridization δV_{FW}; all these effects are typical of WEs [18], but here they belong to the HEs which also have a large oscillator strength characteristic of FEs. When only excitons are present (i.e., under resonant excitation at low temperature), the non-linear corrections can be calculated to the first order in the total exciton density n_{T}, assuming equal populations of resonating FEs and WEs. The blue shift δE_{W} reduces to

$$\delta E_{\mathrm{W}} \simeq 0.48\, E_{\mathrm{b}}\, \pi a_{\mathrm{B}}^2\, n_{\mathrm{T}}\,, \tag{10.32}$$

where E_{b} is the binding energy of a 2D WE. The blocking factor turns out to be

$$\beta_{\boldsymbol{k}} \simeq 1 - 0.57\, \pi a_{\mathrm{B}}^2 n_{\mathrm{T}}\,. \tag{10.33}$$

The effect of δV_{FW} can be roughly estimated as

$$|V_{\mathrm{FW}} + \delta V_{\mathrm{FW}}|^2 \simeq (1 - 0.48\, \pi a_{\mathrm{B}}^2 n_{\mathrm{T}})\, |V_{\mathrm{FW}}|^2\,. \tag{10.34}$$

The characteristic feature of (10.31) is the presence of the factor $(d^{\mathrm{F}}/a_{\mathrm{F}})^2$ instead of $(d^{\mathrm{vc}}/a_{\mathrm{B}})^2$ in the analogous expression for an isolated IQW. This leads to the enhancement of absorption, determined by $\mathrm{Im}\,\chi$. Thus, while the saturation density is comparable to that of WEs ($n_{\mathrm{S}} \sim 1/a_{\mathrm{B}}^2$), the density of photogenerated electron–hole pairs, for a given light intensity, can be two orders of magnitude larger (by a factor $\sim(a_{\mathrm{B}}/a_{\mathrm{F}})^2$); for the same reason, also the linear susceptibility $\chi(n_{\mathrm{T}} = 0)$ can be two orders of magnitude larger. Therefore, the present theory substantiates the intuitive expectation of very pronounced non-linear optical properties of the HEs.

Numerical examples of the predictions of (10.31) have been obtained using the values of semiconductor parameters representative of III-V semiconductor (e.g., GaAs/AlGaAs) QWs, having a smaller linewidths with respect to excitons in II-VI QWs. Namely, we set $\epsilon_\infty = 11$, $d^{\mathrm{vc}} = 20$ Debye, the Bohr radius $a_{\mathrm{B}} = 60$ Å and the binding energy is $E_{\mathrm{b}} \simeq 20\,\mathrm{meV}$; all of the others are the same as in the previous subsection. This gives $|V_{\mathrm{FW}}| \simeq 4$ meV at $k = 10^7\ \mathrm{cm}^{-1}$. Assuming a phenomenological linewidth $\hbar\gamma_{\mathrm{W}} = \hbar\gamma_{\mathrm{F}} = 2$ meV for both excitons, Fig. 10.4 shows the split resonance of the HEs at different excitation densities (linear regime, $n_{\mathrm{T}} = 10^{11}\ \mathrm{cm}^{-2}$, and $n_{\mathrm{T}} = 2\times10^{11}\ \mathrm{cm}^{-2}$); it is noticeable, in particular, that for vanishing excitation density the mixing is complete and the oscillator strength is equally shared by the two peaks (at $\Delta E = \pm 5$ meV), whereas for high excitation density, due to the small blue shift of the WE, the stronger line corresponds to the lowest (more Frenkel-like) HE. From numerical results such as those shown in Fig. 10.4, we can estimate the relative non-linear change in the absorption coefficient close to resonance $|\Delta\alpha|/\alpha \sim 10^{-11}\ \mathrm{cm}^2\, n_{\mathrm{T}}$, which is analogous to the case of a semiconductor multiple QW. However, for a given pump intensity the 2D density

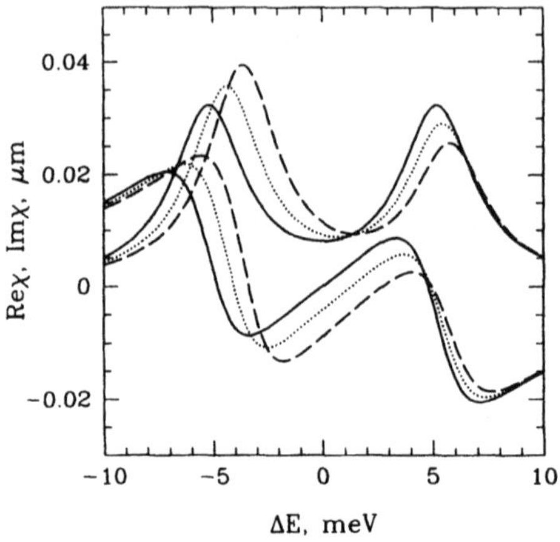

Fig. 10.4. Real and imaginary (positive definite) parts of the 2D susceptibility χ near the HE resonances in the linear regime (*solid lines*), medium excitation density ($n_T = 10^{11}$ cm^{-2}, *dotted lines*) and high excitation density ($n_T = 2 \times 10^{11}$ cm^{-2}, *dashed lines*). Other parameters are $d^{vc} = 20$ Debye, $a_B = 60$ Å, $E_b = 20$ meV; $\epsilon_\infty = 11$, all of the other are the same as in the previous subsection. The linewidths are $\hbar\gamma_W = \hbar\gamma_F = 2$ meV. (From [7])

of photogenerated excitons n_T in our case of HEs is about two orders of magnitude larger because the oscillator strength of HEs is comparable to that of FEs rather than WEs.

A similar theoretical approach can be used to calculate the dynamical Stark effect for HEs, which shows qualitative and quantitative differences with respect to the case of the usual inorganic semiconductor QWs [25].

Hybrid Excitons in Microcavities and Other Confined Structures

The structure described at the start of Sect 10.2.1 raises the technologically challenging problem of growing high-quality organic–inorganic heterojunctions only a few nanometers apart. A more promising way of realizing an HE system is to couple FEs and WEs through a microcavity (MC) electromagnetic field [26]. Strong exciton-radiation interactions are observed in MCs [27] and we can expect hybridization to arise not due to the Coulombic short-range interaction, but due to the strong long-range interaction stemming from virtual cavity photon exchange. For cavity embedded QWs, the fabrication problems would be much alleviated as their separation can be of the order of an optical wavelength. For the sake of simplicity, however, in the following discussion we assume that both QWs lie at the center ($z \simeq 0$) of a single MC at a distance $d \ll \lambda$ from each other. This situation is qualitatively equivalent to that of two coupled MCs for which the growth conditions could be separately optimized for the organic and inorganic QWs [26].

MC-embedded organic QWs in the weak-coupling regime are well known [28], and effects such as spectral narrowing and increased directionality of light emission have been demonstrated. To achieve the strong-coupling regime, as observed for inorganic QWs [27], with organic materials, we need molecular compounds combining a large oscillator strength of the lowest energy electronic transition with an absorption linewidth smaller than the cavity-mode splitting. Good candidates for such structures are thin-film crystals of aromatic molecules such as anthracene, tetracene, terrylene and many others. For example, five monolayers of terrylene ($d \simeq 50$ Å) exhibit an oscillator strength per unit area as large as $10^{15}\,\mathrm{cm}^{-2}$, more than a hundred times that of a GaAs QW exciton. In order to illustrate the results obtained for such a system [26], we use for the material parameter data from available experiments or realistic estimates. We assume that $E_{\mathrm{F}}(k) = E_{\mathrm{C}}(k=0)$ and $E_{\mathrm{W}}(k=0) = E_{\mathrm{C}}(k=0)\,(1+\eta)$, i.e., an FE resonant with the cavity mode E_{C} (we neglect the dispersion of the FE) and a WE with a fractional detuning of η at $k=0$. Using the reduced variable $\bar{k} = k/k_{\mathrm{cav}}$ with $k_{\mathrm{cav}} = \pi/L$, we have for this case $E_{\mathrm{C}}(k)/E_{\mathrm{C}}(0) = \sqrt{1+\bar{k}^2}$ and $E_{\mathrm{W}}(k)/E_{\mathrm{W}}(0) = 1+\eta+a\,\bar{k}^2$ with $a = \hbar^2 k_{\mathrm{cav}}^2/2m_W E_{\mathrm{C}}(0)$. For resonance at $E_{\mathrm{C}}(0) = 1.5$ eV and $\epsilon \simeq 10$, we have $k_{\mathrm{cav}} = 2.4 \times 10^5\,\mathrm{cm}^{-1}$ and, using an exciton mass $m_W = 0.3\,m_0$ (m_0 is the free electron mass), $a = 10^{-5}$. The inorganic QW Rabi splitting Δ_1 is taken to be 3 meV; then, assuming a ratio of the organic to inorganic QW oscillator strength $F/f \simeq 60$, we have for the organic QW Rabi splitting $\Delta_2 \simeq 23$ meV. The ratio $\Delta_2/\Delta_1 \simeq 8$ is by no means unusually large and, as a matter of fact, even larger oscillator strengths can easily be attained with many organic materials. For example, from the standard longitudinal-transverse (LT) splittings of 0.08 meV in GaAs ($\epsilon \simeq 12$) and $\simeq$50 meV for the lowest singlet exciton in tetracene ($\epsilon \simeq 9$) [29], their oscillator strength ratio is about 500. The large splitting $\Delta_2 \approx 100$ meV expected from such estimates gave reasonable hope for reaching the strong coupling regime even at room temperature, since the absorption linewidth can be as low as a few tens of meV in selected organic systems. As a matter of fact, the strong coupling regime in an MC with an organic active layer has very recently been observed [30], employing several organic compounds. For example, using the organic semiconductor tetra-(2,6-*tert*-butyl)phenol-porphyrin zinc (4TBPPZn), a splitting $\Delta_2 \simeq 100$ meV has been found. In this case, the thickness L_2 of the active layer was about 1000 Å, and as $\Delta_2 \propto \sqrt{L_2}$, such an experimental value [30] corresponds very well to the estimate of $\Delta_2 \simeq$ 25 meV [26] made for $L_2 \simeq 50$ Å. In these experiments [30], different dyes were blended in a polymer matrix to realize active thin films. In order to suppress the inhomogeneous broadening and also to further enhance the mode splitting, it is very important to employ crystalline organic semiconductors. We are confident that in this same way it will be possible to observe the Frenkel–Wannier–Mott exciton hybridization in an organic MC.

We assume such a situation in our demonstrative calculations and neglect dissipation for both bare excitonic states. The dispersion of cavity polaritons

$E_j(k)$ and weighting coefficients $N_j^{\mathrm{F,W,C}}(k)$ (analogous to A and B of (10.1)) are shown in Fig. 10.5.

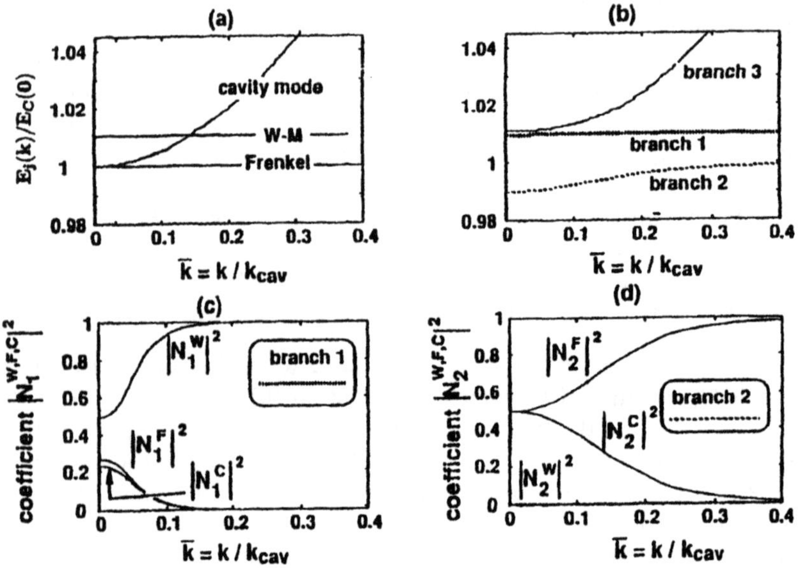

Fig. 10.5. (**a**) Bare dispersion curves of cavity photon, WE and FE normalized to the cavity mode at $k = 0$. The FE is resonant with the cavity mode and the WE has a positive detuning. (**b**) Cavity polariton dispersion curves: for large wavevectors, branches 1, 2 and 3 turn into WE, FE and cavity photon, respectively. (**c**) Weighting coefficients of branch 1. (**d**) Weighting coefficients of branch 2. (From [7])

From Fig. 10.5c it is seen that branch 1 (which at large wavevectors turns into a pure WE) contains a large part of the FE state ($|N_1^{\mathrm{F}}|^2$) for $\bar{k} < 0.1$. As seen from Fig. 10.5d branch 2 (which at large wavevectors turns into a pure FE) for $\bar{k} < 0.25$ also retains a large part of the FE state ($|N_1^{\mathrm{F}}|^2$) while exhibiting a large cavity photon component. The FE component is crucial in assisting the inelastic relaxation that will be considered, whereas the cavity photon component obviously has a large radiative width. For $\bar{k} \ll 1$, even for high mirror reflectivities ($1 - R = 10^{-3}$), the cavity-mode radiative lifetime is of the order of $\tau \simeq 1$ ps. The better mixing of branch 2 with the cavity photon means faster radiative decay in a larger phase space. Such a short lifetime is only effective in a very narrow region of phase space ($k < 0.05\, k_{\mathrm{cav}}$) in the case of typical inorganic QW splittings; such a region can only be reached in about 100 ps due to slowed-down relaxation [31] in the flat part of the dispersion curve, poorly coupled to the cavity mode. In our case, to populate the states of branch 2 with a large radiative width (i.e., those with $k < 0.2\, k_{\mathrm{cav}}$), we can assume that the parameters of the MC with two QWs are such that for $k, k' < 0.2\,k_{\mathrm{cav}}$ an inelastic resonance condition is realized, i.e., that the energy difference $E_1(k) - E_2(k')$ is close to the energy of some intramolecular optical phonon strongly coupled to excitons. For this case, the relaxation rate can be of order 10 ps or less [26], i.e., at least one order of magnitude faster than for MC with an inorganic QW.

Hybrid states of FEs and WEs have also been considered for other geometries, such as quasi-1D parallel organic and inorganic quantum wires [32] and a spherical system (quantum dot) [33]. We do not analyze the experimental possibilities of constructing such systems and mention only some essential points which are different from those in the plane geometry studied above. An important feature of the hybrid states in quasi-1D systems is the fact that the matrix element of the resonant dipole–dipole coupling between the quantum wires is different from zero even at zero wavevector (which is not the case for 2D systems). This makes it possible to excite these states directly without using any special methods (such as coupled grating or attenuated total reflection). Evidently, the analogous situation takes place in the case of quantum dots, where the states cannot be described by the wavevector at all.

In [33] the third-order non-linear susceptibility $\chi^{(3)}$ for a semiconductor quantum dot covered with organic material was found. In this work it was assumed that the inhomogeneous broadening in the organic material is absent and it is possible to consider FE states. A strong enhancement of $\chi^{(3)}$ near the HE resonance was predicted; however the relevant physics is quite different from that described above.

10.2.2 Weak Coupling: Förster Energy Transfer

A large effort has recently been devoted to the study of organic LEDs (see [34] and references therein) and lasers [35]. Förster-like energy transfer between different dye molecules in solid solutions has already been used to achieve light amplification in optically pumped organic thin films [36]. However, light-emitting organic materials have poor transport properties compared to inorganic semiconductors, and to achieve efficient electrical pumping of such devices is a challenging problem. We consider here a novel hybrid configuration in which both inorganic semiconductors and organic materials are present: the basic idea is to pump the optically active organic molecules via electronic energy transfer from the 2D WEs of a semiconductor QW, or the 0D excitons of a semiconductor quantum dot (possibly sitting on a wetting layer). Hopefully, in this hybrid system it would be possible to use the efficient electrical pumping of the inorganic semiconductor material to produce luminescence in the organic one. While the former have electronic excitations with spatially coherent wavefunctions extending over many unit cells, those of the latter are, in general, strongly scattered by phonons (incoherent excitons). Usually, the relaxation of the final excited states in the organics is much faster then the back transfer rate (of course, the interesting case is when the luminescence quantum yield of the organics is high); in this respect, our considerations conceptually correspond to the Förster picture of energy transfer between donor and acceptor molecules [37]. Employing realistic material parameters and different experimental arrangements, the energy-transfer mechanism we consider will be shown to be fast enough to efficiently quench the WE luminescence and to turn on the organic-molecule light emission.

General Theory

The configuration we consider consists of a confined semiconductor system embedded in a bulk-like organic material (for the sake of simplicity, we choose a symmetric quantum well or dot and consider the organic material to be isotropic). The background dielectric constant of the semiconductor material is taken to be real, whereas the total dielectric constant of the organic material has both a real and an imaginary part in the frequency region of interest. In fact, we are interested in an organic material having a broad absorption band in the optical range overlapping with the semiconductor WE sharp resonance, as sketched in Fig. 10.6 appropriate for a QW.

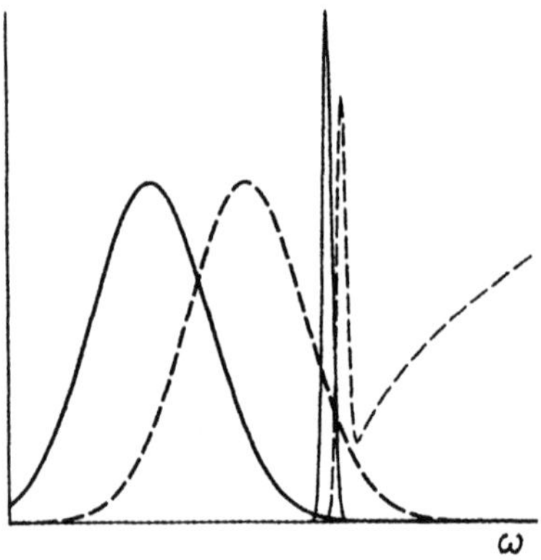

Fig. 10.6. Schematic inorganic semiconductor QW (*thin lines*) and organic medium (*thick lines*) absorption (*dashed lines*) and luminescence (*solid lines*) spectra. Notice the overlap of the QW exciton luminescence peak and the broad organic medium absorption. (From [14])

The Förster-like energy-transfer rate due to the dipole–dipole interaction can be calculated simply from the Joule losses [14,38] in the organic material. These are proportional to the imaginary part of the dielectric function and are produced by the penetration into the organics of the electric field generated by the semiconductor exciton polarization (which has to be explicitly taken into account as a source term in the macroscopic electrostatic equations). From such considerations, it follows that the corresponding decay rate $1/\tau$ can be written as

$$\frac{1}{\tau} = \frac{1}{2\pi\hbar} \int \mathrm{d}^3\boldsymbol{R}\, \mathrm{Im}\,\varepsilon_{ij}(\boldsymbol{R},\omega)\, \mathcal{E}_i^{\mathrm{vc}}(\boldsymbol{R})\, \left(\mathcal{E}_j^{\mathrm{vc}}(\boldsymbol{R})\right)^* , \tag{10.35}$$

where $\varepsilon_{ij}(\boldsymbol{R},\omega)$ is the dielectric function of the absorbing organic medium. Note that (10.35) multiplied by the energy of the excitation $\hbar\omega$ coincides with the expression for the power dissipated in the medium in the presence of a classical external electric field of frequency ω and an amplitude numerically equal to $\boldsymbol{\mathcal{E}}^{\mathrm{vc}}(\boldsymbol{R})$ [39]. In this respect the quantum-mechanical problem is equivalent to that of a classical oscillating polarization inside the semiconductor.

Summarizing, we give the following recipe for calculating the energy transfer rate. Suppose that inside the semiconductor we have the classical macroscopic quasistationary polarization, oscillating with the frequency ω:

$$\boldsymbol{P}(\boldsymbol{r},t) = \boldsymbol{d}^{\mathrm{vc}}\,\psi(\boldsymbol{r},\boldsymbol{r})\,\mathrm{e}^{-i\omega t} + \text{c.c.}\,. \tag{10.36}$$

Then, solve the electrostatical problem and find the corresponding electric field

$$\boldsymbol{\mathcal{E}}(\boldsymbol{R},t) = \boldsymbol{\mathcal{E}}(\boldsymbol{R})\,\mathrm{e}^{-i\omega t} + \text{c.c.}\,. \tag{10.37}$$

The latter, substituted into (10.35), will give the correct quantum-mechanical decay rate if the complex dielectric function of the organics is known (independently of its microscopic structure). Both the microscopic dipole approximation and the macroscopic description of the organic medium are valid as long as the obtained electric field is slowly varying in space on the molecular scale. This condition is fulfilled in all cases considered below, since the typical wavevectors of excitons in the QW are small compared to the inverse lattice constant and the localization length or the dot radius are larger than the lattice constant. The expression (10.35) will be the starting point for our further considerations.

Quantum Well Free and Localized Excitons

We consider a symmetric structure, consisting of a semiconductor QW of thickness L_{w} between two barriers of thickness L_{b} each, the whole semiconductor structure being surrounded by thick slabs of an organic material (actually, we assume each slab to be semi-infinite); the z-axis is chosen to be along the growth direction, $z = 0$ corresponding to the center of the QW. We assume that in the frequency region here considered the semiconductor background dielectric constant ϵ_{b} is real and the same for the well and the barrier, while that of the organic material $\tilde{\epsilon}$ is complex.

We adopt the same simplified microscopic quantum mechanical model of a 2D WE as in Sect. 10.2.1. However, we choose as x the direction of the in-plane component of the exciton dipole moment $\boldsymbol{d}^{\mathrm{vc}}$, preferring to consider the polarization not with respect to the wavevector $\boldsymbol{k}$, but to some fixed frame. This little complication is justified since next to the free exciton we intend to study the case of the localized exciton, i.e., a system with broken 2D translational symmetry. Evidently, we need to consider two cases: $\boldsymbol{d}^{\mathrm{vc}}$ being parallel and perpendicular to the QW plane. We will refer to them as X and Z polarizations respectively. When dealing only with free excitons in a single well, three modes of different symmetry would be identified: longitudinal (L), transverse (T) and perpendicular (Z). The L and Z modes correspond to the X and Z polarizations above (their energies are split by the depolarization shift, but this is immaterial for the following). For the T mode the dipole–dipole interaction here considered vanishes [6].

The corresponding electric field can again be obtained from the solution of the electrostatics problem [6,14,38]. The electric field penetrating the organic material is given by

$$\mathcal{E}(\boldsymbol{r}) = [-i\boldsymbol{k} + k\boldsymbol{e}_z]\phi(z)\,\mathrm{e}^{i\boldsymbol{k}\boldsymbol{r}_\parallel}\,, \tag{10.38}$$

with

$$\phi(z) = \rho_0\, C_{\boldsymbol{k}}\, \mathrm{e}^{-k(z-L_\mathrm{b}-L_\mathrm{w}/2)}\,, \tag{10.39}$$

$$C_{\boldsymbol{k}}^{(\mathrm{X})} = -\frac{ik_x}{k}\,\frac{8\pi^2 q}{k\,(k^2+q^2)}\,\frac{\sinh\,(kL_\mathrm{w}/2)}{\epsilon_\mathrm{b}\sinh\,(kL_\mathrm{b}+kL_\mathrm{w}/2)+\tilde{\epsilon}\cosh\,(kL_\mathrm{b}+kL_\mathrm{w}/2)}\,, \tag{10.40}$$

$$C_{\boldsymbol{k}}^{(\mathrm{Z})} = \frac{8\pi^2 q}{k\,(k^2+q^2)}\,\frac{\sinh\,(kL_\mathrm{w}/2)}{\epsilon_\mathrm{b}\cosh\,(kL_\mathrm{b}+kL_\mathrm{w}/2)+\tilde{\epsilon}\sinh\,(kL_\mathrm{b}+kL_\mathrm{w}/2)}\,, \tag{10.41}$$

and

$$\rho_0 = \sqrt{\frac{2}{\pi a_\mathrm{B}^2}}\,;\quad q = \frac{2\pi}{L_\mathrm{w}}\,. \tag{10.42}$$

Now we simply substitute this electric field into (10.35) and get the decay rate:

$$\frac{1}{\tau} = \frac{S}{2\pi\hbar}\,\mathrm{Im}\,\tilde{\epsilon}\int_{L_\mathrm{b}+L_\mathrm{w}/2}^{+\infty} 2k^2\,|\phi(z)|^2\,\mathrm{d}z = \frac{\mathrm{Im}\,\tilde{\epsilon}}{\pi^2\hbar}\,\frac{|d^{\mathrm{vc}}|^2}{a_\mathrm{B}^2}\,\frac{k\,|C_{\boldsymbol{k}}|^2}{L_\mathrm{w}^4}\,, \tag{10.43}$$

where we have considered the absorption only at $z > L_\mathrm{w}/2 + L_\mathrm{b}$ (considering the organic material in $z < -L_\mathrm{w}/2 - L_\mathrm{b}$ as well, τ would be twice shorter).

We evaluate τ from (10.43) for parameters representative of II-VI semiconductor (e.g., ZnSe/ZnCdSe) QWs [11] ($\epsilon_\mathrm{b} \approx 6$, $d^{\mathrm{vc}} \approx 0.1\,ea_\mathrm{B}$ [about 12 Debye]; the Bohr radius is taken to be 25 Å) and of organic condensed media such as metallophthalocyanines [12] or fullerenes [13] ($\tilde{\epsilon} \approx 4+3i$). This value is not even the most optimistic one, e.g., for PTCDA, widely used in experiments with organic nanostructures [1], one has $\sqrt{\tilde{\varepsilon}} = n+i\kappa = 2.16+i\,1.04$ [40], which gives an even larger $\mathrm{Im}\,\tilde{\varepsilon}$. The lifetime does not depend drastically on the polarization or on the real parts of dielectric constants. Fig. 10.7 shows that the dependence on L_w is also weak, while L_b (Fig. 10.8), when it is

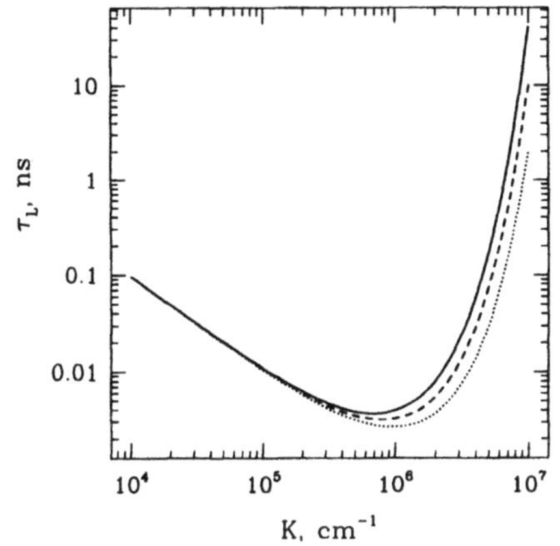

Fig. 10.7. Free L-exciton lifetime τ (ns) versus the in-plane wavevector k (cm^{-1}) for three well widths: $L_w = 20$ Å (*dotted line*), $L_w = 40$ Å (*dashed line*), and $L_w = 60$ Å (*solid line*). Other parameters are $L_b = 40$ Å, $\epsilon_b = 6$, $\tilde{\epsilon} = 4 + 3i$. (From [14])

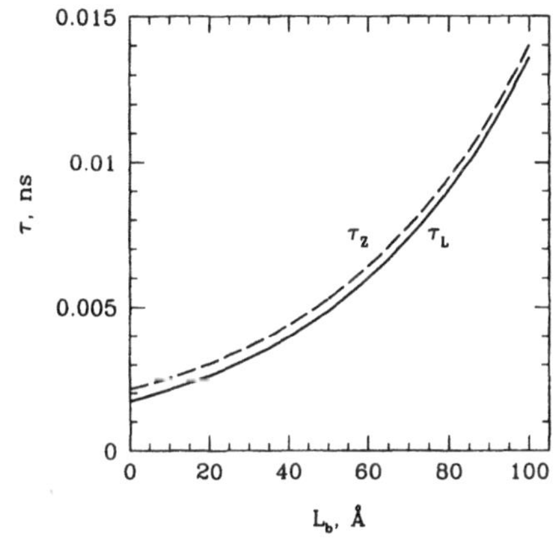

Fig. 10.8. Free L-exciton (solid line) and Z-exciton (dashed line) lifetime τ (ns) versus the barrier width L_b (Å). $K = 10^6$ cm^{-1}, $L_w = 60$ Å, $\epsilon_b = 6$, $\tilde{\epsilon} = 4 + 3i$. (From [14])

increased, shows an obvious exponential factor (clearly seen from the hyperbolic functions in the denominators of (10.40) and (10.41)). The most interesting dependence is that on k. We see, that τ exhibits a minimum at $k_{min} \sim 1/L_b$. This dependence may be easily understood if one recalls that the dipole–dipole interaction between two planes behaves like

$$V(\boldsymbol{k}, z) \sim k\,\mathrm{e}^{-kz}\,, \tag{10.44}$$

which, when substituted into the Fermi Golden Rule, gives the correct asymptotics $\tau \sim 1/k$ at $k \to 0$ and exponential growth at $k \to \infty$. Such behavior corresponds to that shown in Fig. 10.2, where the dipole–dipole interaction was responsible for the hybridization of FEs and WEs with the same wavevector.

Considering at first a quasi-thermalized exciton distribution, typical values of k at a temperature of ~100 K are ~3 $\times 10^6$ cm^{-1}. The corresponding energy-transfer lifetime (tens of picoseconds) is much less than the exciton recombination lifetime, which is about 200 ps in II-VI semiconductor QWs, as reported by different authors (see [11] and references therein, [41]). We remark that for the case of free excitons in a QW, the effective radiative lifetime (which, assuming a thermal distribution, increases linearly with temperature) is determined by the population transfer from non-radiative excitons with large k to small k excitons undergoing a fast radiative decay [42]. Thus, the dipole–dipole energy-transfer mechanism considered here proves to be efficient enough to quench a large fraction of the semiconductor excitons, thereby activating the organic-medium luminescence. Moreover, the intraband relaxation of excitons due to the acoustic phonon scattering occurs at time scales of the order of 20–30 ps at 10 K [41], which is larger than the minimal transfer lifetime, obtained here (less then 10 ps for $k_{\min} \sim 10^6\ \mathrm{cm}^{-1}$). This makes it possible to excite the QW so as to produce the initial non-equilibrium distribution of excitons with $k = k_{\min}$, tuning the frequency of the excitation pulse to exceed the energy $\hbar\omega_{\mathrm{exc}}(k_{\min})$ of the exciton with $k = k_{\min}$ by one LO-phonon frequency Ω_{LO} (since in II-VI semiconductors the free-carrier-to-exciton relaxation is governed mainly by LO-phonon scattering and happens at times of about 1 ps [41,43–45]) or an integer multiple of Ω_{LO}, if the exciton binding energy is larger than $\hbar\Omega_{\mathrm{LO}}$. A numerical estimate for ZnSe gives $\hbar\omega_{\mathrm{exc}}(k_{\min}) - \hbar\omega_{\mathrm{exc}}(k = 0) \sim 1$ meV, while $\hbar\Omega_{\mathrm{LO}} \approx 31$ meV [41], so the following kinetics of excitons at $k \sim k_{\min}$ is governed mainly by the acoustic phonons. Finally, another possibility would be to resonantly pump excitons with the appropriate k by using a coupling grating configuration [15].

Analogous calculations may be performed for the case of III-V semiconductor materials [14,38]. The lifetime is longer compared to that in Fig. 10.7 by about an order of magnitude, which is due to the larger values of a_{B} and ϵ_{b}. However, the energy transfer discussed here is still efficient enough because the effective exciton recombination time in III-V materials is also larger (about 1 ns [46]).

Now we turn to the situation in which the QW width fluctuations, alloy disorder or impurities localize the 2D exciton (such a situation is more frequent for II-VI semiconductor QWs than for III-V ones). Then, the wavefunction of the center-of-mass exciton motion $\Phi(\boldsymbol{r}_\parallel)$ is no longer just a plane wave, and the corresponding polarization is given by

$$\boldsymbol{P}(\boldsymbol{r}) = \boldsymbol{d}^{\mathrm{vc}} \sqrt{\frac{2}{\pi a_{\mathrm{B}}^2} \frac{2}{L_{\mathrm{w}}}} \cos^2\left(\frac{\pi z}{L_{\mathrm{w}}}\right) \Phi(\boldsymbol{r}_\parallel) , \tag{10.45}$$

which implies that $\Phi(\boldsymbol{r}_\parallel)$ is normalized according to

$$\int \mathrm{d}^2\boldsymbol{r}_\parallel\, |\Phi(\boldsymbol{r}_\parallel)|^2 = 1 . \tag{10.46}$$

The solution of the Schrödinger equation for a particle in the random potential, caused by the QW width fluctuations and the alloy disorder, is beyond the scope of the present paper (much work has been done in this field, e.g., see [47] and references therein). We can mention only some general properties that $\Phi(\boldsymbol{r}_\parallel)$ should have: (a) it should be localized within some distance $L \gtrsim a_B \gtrsim L_\mathrm{w}$, (b) it should be smooth and without nodes. As a consequence, its spatial Fourier expansion should contain mainly the components with wavevectors $k \lesssim 1/L$.

From an expansion of the wavefunction $\Phi(\boldsymbol{r}_\parallel)$ into plane waves,

$$\Phi(\boldsymbol{r}_\parallel) = \int \frac{\mathrm{d}^2\boldsymbol{k}}{(2\pi)^2}\, \Phi_{\boldsymbol{k}}\, \mathrm{e}^{i\boldsymbol{k}\boldsymbol{r}_\parallel}\,, \tag{10.47}$$

we obtain [14] the following for the decay rate:

$$\frac{1}{\tau} = \frac{\mathrm{Im}\,\tilde{\epsilon}\,|d^{\mathrm{vc}}|^2}{\pi^2\hbar\, a_\mathrm{B}^2}\frac{1}{L_\mathrm{w}^4}\int \frac{\mathrm{d}^2\boldsymbol{k}}{(2\pi)^2}\, k\,|\Phi_{\boldsymbol{k}}|^2\,|C_{\boldsymbol{k}}|^2\,, \tag{10.48}$$

with the same $C_{\boldsymbol{k}}$ as given by (10.40) and (10.41).

It is possible to get some information about the decay rate (10.48) based only on general properties of the wavefunction mentioned above. We have three length scales in our problem: L_w, L_b and L. First, we have the condition $L_\mathrm{w} \lesssim L$. Since wavevectors with $kL \gtrsim 1$, being cut off by $|\Phi_{\boldsymbol{k}}|^2$, do not contribute to the integral, we may set $kL_\mathrm{w} \to 0$. The subsequent analysis depends on the relation between L_b and L.

If $L_\mathrm{b} \ll L$, we may put $kL_\mathrm{b} \to 0$ as well. Then we have

$$\frac{C_{\boldsymbol{k}}^{(\mathrm{X})}}{L_\mathrm{w}^2} \longrightarrow -\frac{2\pi i}{\tilde{\epsilon}}\frac{k_x}{k} \quad \text{and} \quad \frac{C_{\boldsymbol{k}}^{(\mathrm{Z})}}{L_\mathrm{w}^2} \longrightarrow \frac{2\pi}{\epsilon_\mathrm{b}}\,, \tag{10.49}$$

and the integral may be estimated as

$$\frac{1}{\tau_\mathrm{X}} = A\,\frac{2}{\hbar}\frac{\mathrm{Im}\,\tilde{\epsilon}}{|\tilde{\epsilon}|^2}\frac{|d^{\mathrm{vc}}|^2}{a_\mathrm{B}^2}\frac{1}{L} \quad \text{and} \quad \frac{1}{\tau_\mathrm{Z}} = A\,\frac{4}{\hbar}\frac{\mathrm{Im}\,\tilde{\epsilon}}{\epsilon_\mathrm{b}^2}\frac{|d^{\mathrm{vc}}|^2}{a_\mathrm{B}^2}\frac{1}{L}\,, \tag{10.50}$$

up to a numerical factor $A \sim 1$, determined by the detailed shape of $\Phi_{\boldsymbol{k}}$. We have set the average value of k over the wavefunction $\Phi_{\boldsymbol{k}}$ to be A/L, and, if for the X polarization we assume $\Phi_{\boldsymbol{k}}$ to be cylindrically symmetric (which may be considered as the average over the realizations of disorder), then the numerical factor A is the same for both cases.

In the opposite limit, $L_\mathrm{b} \gg L$ (which also implies $L_\mathrm{b} \gg L_\mathrm{w}$), and we may set $\Phi_{\boldsymbol{k}} = \Phi_{\boldsymbol{k}=0}$, since the values of $\boldsymbol{k}$, contributing to the integral, are determined by $C_{\boldsymbol{k}}$ (namely, $k \lesssim 1/L_\mathrm{b}$), which in this limit takes the form

$$\frac{C_{\boldsymbol{k}}^{(\mathrm{X})}}{L_\mathrm{w}^2} \longrightarrow -\frac{ik_x}{k}\frac{4\pi}{(\tilde{\epsilon}+\epsilon_\mathrm{b})\,\mathrm{e}^{kL_\mathrm{b}} + (\tilde{\epsilon}-\epsilon_\mathrm{b})\,\mathrm{e}^{-kL_\mathrm{b}}} \tag{10.51}$$

and

$$\frac{C_{\boldsymbol{k}}^{(\mathrm{Z})}}{L_\mathrm{w}^2} \longrightarrow \frac{4\pi}{(\tilde{\epsilon}+\epsilon_\mathrm{b})\,\mathrm{e}^{kL_\mathrm{b}} - (\tilde{\epsilon}-\epsilon_\mathrm{b})\,\mathrm{e}^{-kL_\mathrm{b}}}\,. \tag{10.52}$$

Estimating the integral, we have

$$\frac{1}{\tau_X} = A\frac{1}{\pi\hbar}\frac{\mathrm{Im}\,\tilde{\epsilon}}{|\tilde{\epsilon}+\epsilon_b|^2}\frac{|d^{vc}|^2}{a_B^2}\frac{|\Phi_{k=0}|^2}{L_b^3}, \tag{10.53}$$

where $|\Phi_{k=0}|^2 \sim L^2$, which follows from the normalization condition. The expression for $1/\tau_Z$ differs from this by an additional factor of 2 and the factor A may be different in the two cases. It is determined by the values of $\tilde{\epsilon}$ and ϵ_b is bounded by

$$A = \int_0^\infty \frac{\xi^2\,d\xi}{\left|\frac{\epsilon_b}{|\epsilon_b+\tilde{\epsilon}|}\sinh\xi + \frac{\tilde{\epsilon}}{|\epsilon_b+\tilde{\epsilon}|}\cosh\xi\right|^2}, \tag{10.54}$$

$$\frac{\pi^2}{12} = \int_0^\infty \frac{\xi^2\,d\xi}{\cosh^2\xi} < A < \int_0^\infty \frac{\xi^2\,d\xi}{\sinh^2\xi} = \frac{\pi^2}{6}. \tag{10.55}$$

So we see that at $L \ll L_b$ the decay rate is proportional to L^2, and at $L \gg L_b$ to L^{-1}; therefore, it has a maximum at some $L \sim L_b$. This is in agreement with the results of the previous section, since the plane waves, giving the largest contribution to the wavefunction $\Phi(\boldsymbol{r}_\parallel)$ and thus determining the decay rate, have values of the wavevector of the order of $k \sim 1/L$, and we have seen that wavevectors, corresponding to the shortest lifetimes, were $k_{min} \sim 1/L_b$.

To illustrate these considerations, we choose a specific example of the localized wavefunction – that of the ground state in the isotropic parabolic potential:

$$\Phi_k = \sqrt{4\pi}\,L\,e^{-k^2L^2/2}, \tag{10.56}$$

which obviously has all the necessary features mentioned in the beginning of this section. For this wavefunction the integral in (10.48) may be evaluated numerically for the arbitrary parameters L, L_w, and L_b (recall that only $L \gtrsim L_w$ are physically relevant). The results of the calculation (τ versus L) are plotted in Fig. 10.9 along with the asymptotic dependencies for $L_b = 40$ Å, $\epsilon_b = 6$ and $\tilde{\epsilon} = 4 + 3i$ (we have set $L_w \to 0$ for the plots in Fig. 10.9). In the limit $L \gg L_b$, the coefficient $A = \sqrt{\pi}/2$, the coefficient for $L \ll L_b$, given by (10.54) was calculated numerically.

The transfer rate from an electron–hole gas (uncorrelated free carriers) has also been considered [14]: according to our results, the kinetics of the initial free-carrier population (produced, for example, by the electrical pumping) is not significantly changed by the presence of the organic medium, since the energy transfer from free carriers turns out to be slower than the process of exciton formation (unless excitation density and temperature are very high). On the other hand, the subsequent evolution of free or localized excitons is strongly affected by the presence of the organic medium. In an isolated QW

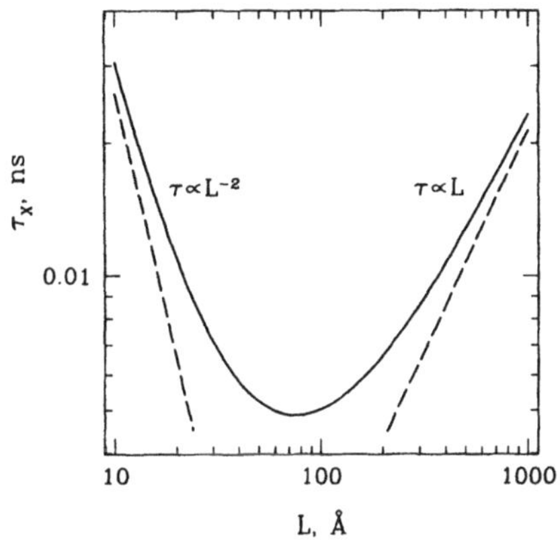

Fig. 10.9. Localized X-exciton lifetime τ (ns) versus the localization length L (Å) (*solid line*) along with the limiting cases $L \ll L_b$ and $L \gg L_b$ (*dashed lines*). $L_w \ll L$, $L_b = 40$ Å, $\epsilon_b = 6$, $\tilde{\epsilon} = 4 + 3i$. (From [14])

the effective lifetime of the exciton distribution may be several hundred ps. However, excitons coupled to the organic medium efficiently transfer their energy to the organic molecules before they can recombine inside the QW. In fact, for QWs based on the II-VI semiconductors and in a realistic configuration, we have seen that such transfer may occur at time scales of the order of 10 ps. This simple physical picture shows that the system studied here may be promising for optical devices based on hybrid organic–inorganic structures, combining the good transport properties of inorganic semiconductors and the large oscillator strengths of organic materials. Finally, we wish to stress that our model calculations can be easily applied to any hybrid multilayer structure and could be generalized to the case in which the structure is embedded in an MC.

Quantum-Dot Excitons

One family of semiconductor nanostructures, widely studied in recent years, is that of quantum dots (see [48] and references therein). Förster energy transfer from a semiconductor quantum dot to the surrounding organic medium has also been predicted [49], considering only the transfer rate from the lowest excited state of the dot and only the limiting cases of strong and weak confinement (in other words, the dot size much smaller or much larger than the exciton Bohr radius). However (see [50] and references therein), due to the discreteness of states, carrier relaxation in quantum dots happens in a way qualitatively different from that in higher-dimensionality nanostructures. The role and efficiency of different relaxation mechanisms is not yet clear [48, 50], so we consider the problem of the energy transfer from higher levels of the dot to be also relevant. We show that the carrier relaxation itself may be affected by Förster transfer and, moreover, present the results for intermediate dot sizes obtained from a variational calculation [51].

We consider for simplicity a nanostructure consisting of a spherical quantum dot of radius R with dielectric constant ε, surrounded by a concentric semiconductor barrier with thickness L_b and the same dielectric constant. The barrier is assumed to be infinitely high, thus carriers cannot penetrate it. The space outside the barrier is filled with the organic substance with complex dielectric constant $\tilde{\varepsilon}$. Let the electron–hole pair in the dot be described by the envelope wavefunction $\psi(\boldsymbol{r}_e, \boldsymbol{r}_h)$ with the corresponding energy $\hbar\omega$; such an approach implies that the size of the dot should be significantly larger than the lattice constant. The exciton polarization in the quantum dot is given by (10.36) and, from the corresponding electric field in the organics, the transfer rate can be obtained as given by (10.35).

In the spherical geometry the states of the electron–hole pair may be classified by the total angular momentum quantum numbers l, m, which determines the angular dependence of the electron–hole pair wavefunction

$$\psi(\boldsymbol{r}, \boldsymbol{r}) \propto Y_{lm}(\theta, \varphi) , \tag{10.57}$$

where $Y_{lm}(\theta, \varphi)$ are the spherical harmonic functions. Putting the z axis along the vector $\boldsymbol{d}^{\mathrm{vc}}$, for the polarization we may write

$$P_z(\boldsymbol{r}) = P^{(r)}(r)\, Y_{lm}(\theta, \varphi) ; \tag{10.58}$$

the corresponding electrostatic potential can also be expressed in a multiple expansion and, from the usual boundary conditions, can be calculated [51] in terms of the bare 2^{l+1}-pole moment of the exciton polarization:

$$Q_{l+1}^{\mathrm{exc}} = -i\sqrt{(l+1)^2 - m^2}\sqrt{\frac{4\pi}{2l+1}} \int_0^{R_2} r^l\, P^{(r)}(r)\, r^2\, \mathrm{d}r . \tag{10.59}$$

From the potential, we find the decay rate

$$\begin{aligned} \frac{1}{\tau} &= \frac{\operatorname{Im}\tilde{\varepsilon}}{2\pi\hbar} \int_{|\boldsymbol{r}|>R+L_b} (\nabla\phi^* \cdot \nabla\phi)\, \mathrm{d}^3\boldsymbol{r} \\ &= \frac{2l+4}{2l+3} \frac{\operatorname{Im}\tilde{\varepsilon}}{\hbar} \frac{|Q_{l+1}^{\mathrm{eff}}|^2}{(R+L_b)^{2l+3}} , \end{aligned} \tag{10.60}$$

with

$$Q_{l+1}^{\mathrm{eff}} = \frac{2l+3}{(l+1)\varepsilon + (l+2)\tilde{\varepsilon}} Q_{l+1}^{\mathrm{exc}} . \tag{10.61}$$

One may want to average (10.60) over m, as the energy does not depend on the magnetic quantum number m. This is certainly relevant for the case when the dot is pumped electrically, while for the case of excitation by polarized light one should choose the state corresponding to the given polarization. Such averaging corresponds just to the substitution in (10.59):

$$\sqrt{(l+1)^2 - m^2} \;\rightarrow\; \sqrt{(l+1)(2l/3+1)} , \tag{10.62}$$

since the radial part of the polarization cannot depend on the quantum number m.

The pair wavefunction is determined by two interactions: (a) the confinement potential, and (b) the Coulomb attraction of the electron and the hole, with the corresponding characteristic length scales R and a_{B}. We do not solve the Schrödinger equation for arbitrary R and a_{B}; we focus instead on two limiting cases.

If $R \ll a_{\mathrm{B}}$ (strong confinement), Coulomb interaction may be neglected and the electron–hole pair wavefunction will be the product of two one-particle wavefunctions. Each one-particle state is labeled by three quantum numbers: the orbital quantum number $l_{\mathrm{e,h}} = 0, 1, \ldots,$ the magnetic quantum number $m_{\mathrm{e,h}} = -l_{\mathrm{e,h}}, \ldots, l_{\mathrm{e,h}}$, and the principal quantum number $n_{\mathrm{e,h}} = 1, 2, \ldots$. One should form linear combinations of such products, corresponding to the definite values of l and m (the total angular momentum). If $R \gg a_{\mathrm{B}}$ (weak confinement), then the exciton may be considered as a rigid particle moving in the well, the relative motion of the electron and the hole and the center-of-mass motion are effectively separated. For the numerical estimations we use the parameters typical for II-VI semiconductors (e.g., CdSe): $a_{\mathrm{B}} \simeq 50$ Å, $\varepsilon \simeq 6$; for the organic part we need to know only the dielectric constant, and we take it to be $\tilde{\varepsilon} \simeq 4 + 3i$. Having fixed $L_{\mathrm{b}} = 30$ Å, in Fig. 10.10 we plot the transfer times for several strongly confined states, averaged over m as a function of R. We see that for the lowest state the transfer time is quite short – of the order of 10 ps, its dipole moment and transfer rate being [49, 51]

$$Q_1^{\mathrm{eff}} = \frac{3\tilde{\varepsilon}\, d^{\mathrm{vc}}}{\varepsilon + 2\tilde{\varepsilon}}\,, \quad \frac{1}{\tau} = \frac{12\,\mathrm{Im}\,\tilde{\varepsilon}}{|\varepsilon + 2\tilde{\varepsilon}|^2}\,\frac{1}{\hbar}\,\frac{|d^{\mathrm{vc}}|^2}{(R + L_{\mathrm{b}})^3}\,. \tag{10.63}$$

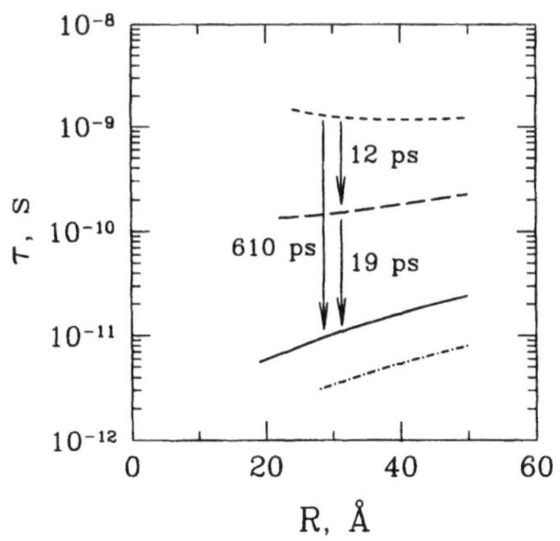

Fig. 10.10. The Förster transfer time for II-VI semiconductor dots in the strong confinement limit from the states $\{l_{\mathrm{e}} = 0, n_{\mathrm{e}} = 1, l_{\mathrm{h}} = 0, n_{\mathrm{h}} = 1\}$ (*solid line*), $\{0, 1, 1, 1\}$ (*long-dashed line*), $\{0, 1, 2, 1\}$ (*short-dashed line*), $\{1, 1, 1, 1\}$ ($l = 0$, *dash-dotted line*) versus the dot radius R ($L_{\mathrm{b}} = 30$ Å). The *vertical arrows* show schematically the intraband transition times (see text for details). (From [51])

The states with larger l have longer times, one may roughly say that every unit of l "costs" about an order of magnitude of τ.

The transfer time for the two lowest states in the weak-confinement limit is shown in Fig. 10.11 ($\{l=0,n=1\}$ and $\{1,1\}$ for 1s-exciton – solid and dashed curves at $R>50$ Å respectively).

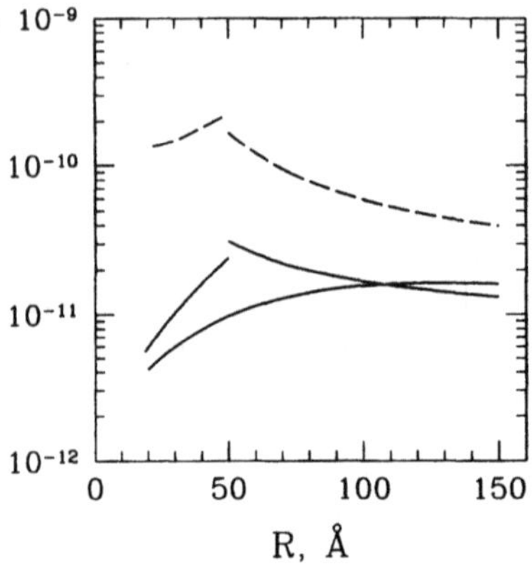

Fig. 10.11. The transfer times from the lowest state of the electron–hole pair (*discontinuous solid line*) and the second excited state (*discontinuous dashed line*) in both limiting cases versus the dot radius for the same parameters as in Fig. 10.10. The *continuous solid line* represents the result of the variational calculation for the lowest state. (From [51])

The times do not depend strongly on R; the difference between the dipole and quadrupole transitions is weaker than for the strong confinement (due to the fact that the ratio $R/(R+L_b)$ is closer to 1). The dipole moment and transfer rate for the lowest state $\{l=0,n=1\}$ are given by [49,51]

$$Q_1^{\text{eff}} = \frac{1}{\pi}\left(\frac{2R}{a_B}\right)^{3/2}\frac{3\tilde{\varepsilon}\,d^{\text{vc}}}{\varepsilon+2\tilde{\varepsilon}} \quad \text{and} \quad \frac{1}{\tau} = \frac{96\,\text{Im}\,\tilde{\varepsilon}}{\pi^2|\varepsilon+2\tilde{\varepsilon}|^2}\frac{1}{\hbar}\frac{|d^{\text{vc}}|^2}{a_B^3}\left(\frac{R}{R+L_b}\right)^3 . \tag{10.64}$$

To investigate the crossover region between the two limiting cases, that is $R\sim a_B$, we perform a simple variational calculation for the lowest excited state [51]. The wavefunction with three variational parameters (a, b_e, b_h) is chosen to reproduce correctly the polarization in the dot in both limiting cases and to pass continuously from one to the other as the ratio R/a_B is varied

$$\psi_{a,b_e,b_h}(\boldsymbol{r}_e,\boldsymbol{r}_h) \propto \frac{\chi_0(r_e)}{\sqrt{\chi_0(r_e)+b_e}}\frac{\chi_0(r_h)}{\sqrt{\chi_0(r_h)+b_h}}\exp\left(-\frac{|\boldsymbol{r}_e-\boldsymbol{r}_h|}{a}\right), \tag{10.65}$$

with

$$\chi_0(r) = \frac{1}{\sqrt{2\pi R}}\frac{\sin(\pi r/R)}{r} . \tag{10.66}$$

In the Hamiltonian we include the kinetic energy, the confining potential, and the Coulomb interaction, screened by the semiconductor background dielectric constant ε_0, neglecting the image charges due to the dielectric discontinuity. Fig. 10.11 shows the transfer times from the lowest state of the electron–hole pair (discontinuous solid line) and the second excited state (discontinuous dashed line) in both limiting cases versus the dot radius for the same parameters as in Fig. 10.10. The continuous solid line represents the result of the variational calculation. We conclude that the estimations using the limiting expressions for the strong and weak confinement describe the situation reasonably well (within a factor of 3).

Above we mentioned Förster energy transfer, accompanied by an interband transition in the semiconductor. But such transfer may also occur via an intraband transition as well, if the surrounding material (in principle, not necessarily organic) has non-zero absorption at the corresponding frequency. In such a situation the presence of an absorbing material may affect the single-carrier relaxation in the dot. A transition between single-particle states $\{l_1, m_1, n_1\} \rightarrow \{l_2, m_2, n_2\}$ also corresponds to 2^l-pole moments, $l = |l_1 - l_2|, \ldots, l_1 + l_2$. Its frequency lies typically in the infrared, where the absorption of organics is usually much weaker than in the visible range. Hence, we set $\tilde{\varepsilon} = 4+0.3\,i$ and evaluate the intraband transition times, averaged over m_1 and summed over m_2, for the transitions shown by arrows in Fig. 10.10: $\{l_1 = 1, n_1 = 1\} \rightarrow \{l_2 = 0, n_2 = 1\}$ (dipole transition – 19 ps), $\{2,1\} \rightarrow \{1,1\}$ (dipole transition – 12 ps), $\{2,1\} \rightarrow \{0,1\}$ (quadrupole transition – 610 ps) for $R = 30$ Å and $L_b = 30$ Å (in our model the times are the same for electrons and holes). As we see, dipole transitions are much faster than higher multipole ones, and the times may be relatively short (a few tens of ps). Thus, if other relaxation processes are quenched, the Förster transfer may play some role. In this connection one may also consider a single-carrier relaxation process, related both to the above considered Förster transfer and to the Auger relaxation (dealt with in [50]). If one considers an array of quantum dots of characteristic size R, separated by distance r from each other on the wetting layer of thickness L_w (no surrounding organics is assumed), then the following process may occur: Two carriers in the neighboring dots interact via Coulomb interaction (the first non-vanishing multipole component is the dipole–dipole one), then, as a result of this interaction, one of the carrier relaxes to a lower state in the dot, while the other is promoted to the wetting-layer continuum. This process is adequately described by the Fermi Golden Rule, and a simple estimate gives the corresponding time of the order of

$$\tau \simeq \frac{\hbar^3 \varepsilon^2}{m e^4} \frac{r^6}{R^5 L_w} . \tag{10.67}$$

Estimating the first fraction as 10–100 fs, one may obtain numbers of the order of 10–100 ps in a favorable geometry.

In conclusion, our calculations for quantum dots show that the Förster energy transfer, corresponding to dipole transitions in the dot is fast enough compared to the carrier recombination time, while for higher-multipole transitions it is slower. However, the radiative recombination of the pair in the corresponding states should also be suppressed, so we may expect that large part of the quantum-dot energy may be transferred to the organic molecules, with possible subsequent emission of light if other relaxation channels are weak. We have also estimated the effect of an organic substance on the carrier intraband relaxation in the case when the organics have a non-zero absorption coefficient at the frequency corresponding to the given intraband transition. Unlike intradot Auger processes [50], this relaxation mechanism will also work if only one carrier in the dot is present.

10.3 Excitons at Donor–Acceptor Interfaces

Interfaces between donor and acceptor molecular materials can have very interesting optoelectronic and photovoltaic properties. The Organic Molecular Beam Deposition (OMBD) technique considerably enlarges the possibilities to grow such structures, and it is therefore timely to discuss a number of phenomena which could take place in them and which could be also of potential interest for applications.

10.3.1 Cold Photoconductivity

At a donor–acceptor (D–A) interface, the molecules on either side are neutral in the ground state, but the lowest energy electronic excitations are charge-transfer excitons (CTEs) accompanied by the displacement of an electron across the interface from the donor to the acceptor side. As CTEs are accompanied by large electric dipole moments which are all oriented in the same direction (perpendicular to the interface plane), a strong dipole–dipole repulsion between excitons appears with energy $V = \mu^2/(\epsilon\, \rho^3)$, where μ is the static CTE dipole moment, ϵ the background dielectric constant and ρ the distance between dipoles along the interface plane. Assuming, for example, $\mu = 20$ Debye, which is not unusual for a CTE, and $\epsilon \simeq 3$, we have $V \simeq 0.7$ eV for a distance between neighbors of $\rho \simeq 5$ Å. For a CTE concentration larger than a critical value (of order 0.1), this repulsion can induce a population of higher-lying levels corresponding to the appearance of free electrons and holes, and therefore to conductivity even at low temperatures (cold photoconductivity) [52]. Such type of non-thermal ionization of the CTEs also sets a limit to the 2D density of CTEs; in all of the following considerations, we will consider excitation densities well below the critical value of this instability.

10.3.2 All-Optical Nonlinearities

In this section, we discuss the influence of the 2D CTE repulsion on the non-linear optical properties of D–A interfaces [53]. For this case the average exciton–exciton distance ρ_o is related to the 2D density of CTEs n by $n \simeq 1/\rho_o^2$. The ensuing repulsion increases the energy of the CTEs and the corresponding energy shift $\Delta\epsilon$ is given by the average interaction of one exciton with all the others. The partly mobile (e.g., by hopping) CTEs will have a tendency to order in such a way as to minimize the energy of repulsion. Assuming for the sake of simplicity that the CTEs are located at the sites of a square lattice, we find

$$\Delta\epsilon \simeq 9\ \mu^2/(\epsilon\rho_o^3) = 9\ V_o\ (n/N)^{(3/2)}\ , \tag{10.68}$$

with $V_o = \mu^2/(\epsilon a^3)$, and the total 2D density of molecules N is given by $N = 1/a^2$, where a is the lattice constant along the interface ($a \simeq 5$ Å). It is clear that such an additional repulsion energy is required to create a CTE at an empty site (i.e., at a vacancy in the CTE lattice). Even if the CTEs are not ordered on a lattice, their mean field additional energy due to the dipole–dipole repulsion will be related to their 2D density n in the same way, and also the numerical prefactor will not be much different (this consideration also applies to the photovoltaic effects discussed in Sect. 10.3.3). We wish to stress here the non-analytic dependence of $\Delta\epsilon$ on n. The D–A interface polarizability due the CTE contribution can be written as

$$\chi(\omega) \simeq \frac{A}{(\epsilon_o + \Delta\epsilon)^2 - \omega^2}\ , \tag{10.69}$$

where ϵ_o is the CTE energy for $n = 0$ and A is a constant proportional to the CTE oscillator strength. Expanding $\chi(\omega)$ in series of $\Delta\epsilon/\epsilon_o$ we find that

$$\chi(\omega) = \chi_o^{(1)}(\omega)(1 - \frac{2\epsilon_o\Delta\epsilon(n)}{\epsilon_o^2 - \omega^2}) = \chi_o^{(1)} + \Delta\chi(n)\ , \tag{10.70}$$

where $\chi_o^{(1)}$ is the polarizability for $n = 0$. Thus, for the non-linear correction to the polarizability which usually contributes to the Kerr non-linearity, we have

$$\Delta\chi/\chi_o^{(1)} = -\frac{2\epsilon_o\Delta\epsilon}{\epsilon_o^2 - \omega^2} \simeq -\frac{\Delta\epsilon}{\epsilon_o - \omega}\ , \tag{10.71}$$

where for resonant pumping $|\epsilon_o - \omega| \simeq \gamma$, γ being the exciton linewidth. Taking a concentration $n/N \simeq 1/100$ and using the previous estimates, we have $\Delta\epsilon \simeq 50\ \mathrm{cm}^{-1}$ which, even for $\gamma \simeq 500\ \mathrm{cm}^{-1}$ gives $\Delta\chi/\chi_o^{(1)} \simeq 1/10$.

Such a big change in polarizability is not due to the saturation-like anharmonicity which is the main non-linear mechanism for WEs in semiconductors [18, 24] (notice also that the phase-space-filling parameter n/N is only 1/100), but is caused by the exciton–exciton interaction which in the present case is particularly large. We note also that a 2D concentration along each interface of 0.01 can hardly be dangerous for organic crystals as, taking into account the thickness of the layer between successive interfaces, it may correspond to a 3D concentration of less than 0.001. We point out also that the optical non-linearity of 2D CTEs here considered is peculiar as it goes beyond the usual perturbation theory à la Bloembergen. As a matter of fact, because in equilibrium conditions under continuous-wave (CW) excitation $n \propto |E(\omega)|^2$, the shift of the CTE energy $\Delta\epsilon(n)$ and the non-linear correction to the polarizability turn out to be proportional to $|E|^3$. Thus, the non-linear part of the polarization is $\Delta \boldsymbol{P} \propto |E(\omega)|^3 \boldsymbol{E}(\omega)$ and this effect cannot be included in the usual expansion of the polarization $\boldsymbol{P}$ in powers of the components of the electric field $\boldsymbol{E}$. This peculiarity stems from the long-range exciton–exciton interaction that shifts the CTE energy in a non-analytic way with respect to n.

In the above discussion, we have only considered the non-linear correction to the polarizability due to the CTE–CTE repulsion, which, for example, contributes to the non-linear absorption by CTEs themselves. A more general mechanism by which the non-linear optical properties of media containing CTEs in the excited state can be enhanced is due to the influence on nearby molecules of the strong static electric field arising in the vicinity of an excited CTE. If, for example, the CTE (or CT complex) electric static dipole moment is 10 Debye, at a distance of 5 Å it creates a field larger than 10^6 V/cm. Such strong electric fields have to be taken into account in the calculation of non-linear polarizabilities, because they change the non-resonant hyperpolarizabilities α, β, γ of all molecules close to the CT complex. For instance, in the presence of this CTE-induced static local field, the microscopic molecular hyperpolarizabilities are modified as follows

$$\alpha_{ij} = \alpha_{ij}^{(0)} + \alpha_{ijl}^{(1)} E_l^{\mathrm{CTE}} + \alpha_{ijlk}^{(2)} E_l^{\mathrm{CTE}} E_k^{\mathrm{CTE}} + \dots , \tag{10.72}$$

$$\beta_{ijl} = \beta_{ijl}^{(0)} + \beta_{ijlk}^{(1)} E_k^{\mathrm{CTE}} + \dots , \quad \text{etc.} \tag{10.73}$$

The changes in the molecular hyperpolarizabilities are reflected in all macroscopic non-linear optical constants of media. As CTE-induced electrical fields increase with the concentration of CTEs, a stronger dependence of the non-linear optical response on the intensity of light is obtained. For example, $\chi^{(3)}$ will include a contribution from $\chi^{(4)}$, and the corresponding polarization will be given by

$$\Delta P_i \simeq \chi_{ijlm}^{(3)} E_j E_l E_m + \chi_{ijlmq}^{(4)} E_j E_l E_m E_q^{CTE} , \tag{10.74}$$

where E is the pump light electric field; therefore, if the second term is not negligible, $|\Delta P|^2$ will depend on the pump intensity $I \propto |E(\omega)|^2$ more strongly than I^3, through the dependence of E^{CTE} on I.

These considerations may also apply to non-crystalline polymeric media with impurities having a CT complex in the excited state. If the polymer structure is such that the CT complexes along the polymer have nearly the same orientation, we have a 1D system of CTEs. Again, as for 2D CTEs, the 1D CTEs can have a strong repulsive interaction, and we expect an enhancement of optical non-linearities and a stronger intensity dependence of the non-linear optical response. Indeed, for a 1D system the correction to the energy is already proportional to n^3 and the non-linear contribution to the polarizability due to the CTE–CTE repulsion is proportional to $|E(\omega)|^6$. In any case, the appearance of CTEs under the influence of an intense laser light drastically changes the intensity dependence of the non-linear optical properties (for the 2D case and, particularly, the 1D case, this dependence becomes larger), and this theoretical prediction can be checked easily by experiments. If the CT complexes in the medium are not preferentially oriented in one direction, these effects after averaging can be suppressed at least to some degree.

10.3.3 Photovoltaic Mechanism

It is also very interesting to consider the possibility of a stack of D–A interfaces in an organic superlattice (based, for instance, on C_{60} as acceptor material and perylene or coronene as donors). To avoid having a center of inversion in order to allow for second harmonic generation, it is possible to employ a structure of the type $...AAA|DDD|NNN|AAA|DDD|NNN...$ where N is any different molecule which should be neither a good donor nor a good acceptor. In such a structure, we also expect striking photovoltaic effects [54]: with all the dipoles pointing in the same direction, the pumping of CTEs will produce a net macroscopic voltage drop proportional to the product of the number of D–A interfaces and the 2D density of CTEs per interface.

Assuming as above that the CTEs are located at the sites of a square lattice of side ρ_o, the potential along a line perpendicular to the interface and passing through the center of a lattice unit cell (i.e., avoiding as much as possible the CTEs themselves) is given by

$$V(z) = \frac{\mu z}{\epsilon \rho_o^3} \sum_{j,l=-\infty}^{+\infty} \left((j+0.5)^2 + (l+0.5)^2 + (\frac{z}{\rho_o})^2 \right)^{-3/2} = \frac{2\pi\mu}{\epsilon \rho_o^2} \phi(\frac{z}{\rho_o}) ; \tag{10.75}$$

the function ϕ is shown in Fig. 10.12.

It is clear that the interface electric field extends over a thickness of order ρ_o and that the total potential drop across an interface is simply $\Delta V = 4\pi\mu/(\epsilon\rho_o^2)$. When the 2D density of CTEs and the distance L between successive D–A interfaces are such that $\rho_o < L$, the potential across

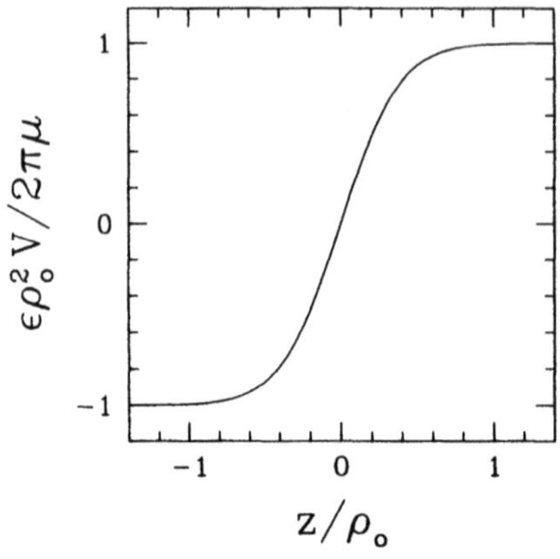

Fig. 10.12. Normalized electrostatic potential across an A–D interface in the presence of a 2D CTE density $n \simeq 1/\rho_o^2$

the superlattice has a staircase profile; otherwise, it has a smooth slope (see Fig. 10.13).

The same conclusions are reached pretending that the CTEs give rise to a uniform static dipole moment per unit area μn perpendicular to the interface; the potential drop across one interface would be $4\pi\mu n/\epsilon$, which, for $\mu \simeq 20$ Debye, $\epsilon \simeq 3$ and $n \simeq 10^{12}\,\mathrm{cm}^{-2}$, is of about 25 mV. This actually corresponds to an interface electric field which is extending over a thickness of about 100 Å and is larger than 10^4 V/cm (i.e., comparable to the electric field in the depletion layer of a typical semiconductor p-n junction). Of course, the uniformity along the superlattice planes will never be perfect in a real

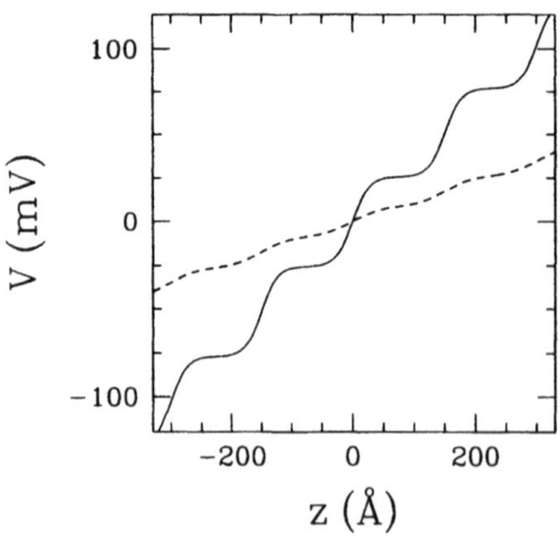

Fig. 10.13. Macroscopic potential along an asymmetric A–D–N superlattice with a period $L \simeq 150$ Å at different 2D CTE densities per interface ($n \simeq 2 \times 10^{12}\,\mathrm{cm}^{-2}$, *solid line*; $n \simeq 7 \times 10^{11}\,\mathrm{cm}^{-2}$, *dashed line*); see text for details

structure and, in general, a rather complicated spatial pattern of electric field force lines (and therefore current filaments) can be expected.

Under steady illumination, a condition of dynamical equilibrium will be reached with a constant 2D density of interface CTEs. Free carriers will also be present, either photogenerated directly or, for instance, as a result of the thermal ionization of the CTEs at room temperature. The electric field above will effectively separate the electrons and holes and drive a current along the superlattice crossing the D–A interfaces from the donor to the acceptor side. We are neglecting the effects of the periodic changes of the band edges due to the superlattice compositional changes (such discontinuities can be minimized by a proper choice of materials) with respect to the additive effect of the potential variations at the D–A interfaces. Of course, these effects would be important in estimating the electron and hole mobilities along the growth axis. In the structures here considered, electrons could be effectively injected into the valence band on the acceptor side through a contact with a large work function metal and extracted from the conduction band on the donor side through a contact with a small work function metal (as done in the reverse direction in LED devices); such a current would deliver power to the external circuit load at the expense of the light absorbed. In the absence of sufficient experimental information on structures of the kind we propose, we are unable to comment on the expected efficiency of photovoltaic conversion; we have only pointed out an interesting new physical mechanism to provide the internal electric field necessary to separate and drive the photogenerated carriers. Finally, we stress that such dc electric field is controlled by the pump intensity determining the a 2D density of CTEs; therefore, it may also lead to large opto-electronic effects.

10.4 Conclusions

The method of OMBD combined with a careful sample characterization has been successful in growing layered systems including organic crystalline materials (e.g., tetracene, perylene, fullerene, PTCDA). There is little doubt that in the future the number of molecular species used will rapidly increase and that a large flexibility in the design of novel type of structures will be achieved. In particular, hybrid multilayer systems containing layers of organic and inorganic materials could be widely used in different optical devices.

Here, we have studied organic–inorganic heterostructures having resonating Frenkel and Wannier–Mott excitons in both the strong- and weak-coupling regimes. In the former case, we have discussed the linear and non-linear optical properties of novel hybrid excitons in organic–inorganic heterostructures. With respect to the usual semiconductor quantum wells, a very strong enhancement of both the linear and non-linear parts of the susceptibility is predicted. These results are rationalized in terms of the large oscillator strength of FEs and the low saturation density of WEs: both characteristics at the same time belong to the hybrid resonances. Moreover, the HEs are a promising system for novel electro-optical and magneto-optical properties as well as for second harmonic generation. We have also considered the new possibilities which may appear in microcavities containing resonating organic and inorganic QWs. In such structures we can expect a drastic shortening of the relaxation time of excitons into states having a large radiative width and a short fluorescence decay time. In the weak-coupling regime, on the other hand, when strongly broadened excitonic resonances are present in the organic material, the resonant dipole–dipole coupling leads to efficient Förster energy transfer from the inorganic to the organic medium. We can also expect that the combination of electrical pumping of excitons in inorganic semiconductors with the fast relaxation and fluorescence of excitons in organics will open up a new scenario of excitonic processes, of interest for both basic science and device applications.

We have finally reported a few studies on charge transfer excitons at donor–acceptor interfaces, which can have very interesting opto-electronic and photovoltaic properties. We have discussed the electronic phase transition to a conducting state in the system of 2D CTEs at high excitation density (cold photoconductivity) and some other properties related to the strong Coulombic dipole–dipole CTE repulsion. This type of strong long-range exciton–exciton interaction can be responsible for the enhancement of non-linear optical properties. In asymmetric superlattices, the electric field associated with CTEs provides a novel photovoltaic mechanism.

We believe that this is a very promising field of research: more detailed theoretical calculations will be needed, but probably the crucial factor will be the technological progress in the growth of such structures. If successfully synthesized, systems of the type here considered would exhibit novel phenomena of potential technological interest.

References

1. S.R. Forrest, Chem. Rev. **97**, 1793 (1997)
2. E.I. Rashba, M.D. Sturge (eds.) *Excitons* (North-Holland, Amsterdam 1982)
3. F. Bassani, G. Pastori Parravicini, In: *Electronic States and Optical Transitions in Solids* (Pergamon Press, Oxford 1975) Chap. 6
4. M. Sakurai, M. Furukawa, K. Mizuno, A. Matsui, J. Phys. Soc. Jpn. **61**, 445 (1992)
5. M. Orrit, J. Bernard, J.M. Turlet, P. Kottis, J. Chem. Phys. **78**, 2847 (1983)
6. V.M. Agranovich, R. Atanasov, F. Bassani, Solid State Comm. **92**, 295 (1994)
7. V.M. Agranovich, D.M. Basko, G.C. La Rocca, F. Bassani, J. Phys.: Condens. Matter **10**, 9369 (1998)
8. B.I. Green, J. Orenstein, S. Schmitt-Rink, Science **247**, 679 (1990)
9. G. Bastard, *Wave Mechanics Applied to Semiconductor Heterostructures* (CNRS, Paris 1988)
10. R. Atanasov, F. Bassani, V.M. Agranovich, Phys. Rev. B **49**, 2658 (1994)

11. R. Cingolani, In: *Semiconductors and Semimetals*, Vol. 44 (Academic Press 1997) pp. 163–226
12. T. Yamada, H. Hoshi, T. Manaka, K. Ishikawa, H. Takezoe, A. Fukuda, Phys. Rev. B **53**, R13314 (1996)
13. A.F. Hebard, R.C. Haddon, R.M. Fleming, A.R. Kortan, Appl. Phys. Lett. **59**, 2109 (1991)
14. D.M. Basko, G.C. La Rocca, F. Bassani, V.M. Agranovich, Eur. Phys. J. B **8**, 353 (1999)
15. M. Kohl, D. Heitmann, P. Grambow, K. Ploog, Phys. Rev. B **42**, 2941 (1990)
16. G.C. La Rocca, Physica Scripta T **66**, 142 (1996)
17. O. Madelung (ed.), Landolt-Börnstein Vol. III/22a (Springer-Verlag, Berlin 1987)
18. S. Schmitt-Rink, D.S. Chemla, D.A.B. Miller, Phys. Rev. B **32**, 6601 (1985); Adv. Phys., **38**, 89 (1989)
19. S. Jaziri, S. Romdhane, H. Bouchriha, R. Bennaceur, Phys. Lett. A **234**, 141 (1997)
20. L. Schulteis, A. Honold, J. Kuhl, K. Köhler, C.W. Tu, Phys. Rev. B **34**, 9027 (1986)
21. A. Vinattieri, J. Shah, T.C. Damen, D.S. Kim, L.N. Pfeiffer, L.J. Sham, Solid State Comm. **88**, 189 (1993)
22. A. D'Andrea, R. Muzi, Solid State Comm. **95**, 493 (1995)
23. G.C. La Rocca, F. Bassani, V.M. Agranovich, In: *Notions and Perspectives of Non-linear Optics*, ed. by O. Keller (World Scientific, Singapore 1996); Nuovo Cimento D **17**, 1555 (1995)
24. H. Haug, S.W. Koch, *Quantum Theory of the Optical and Electronic Properties of Semiconductors*, 3rd ed. (World Scientific, Singapore 1994)
25. G.C. La Rocca, F. Bassani, Phys. Lett. A **247**, 365 (1998)
26. V. Agranovich, H. Benisty, C. Weisbuch, Solid State Comm. **102**, 631 (1997)
27. E. Burstein, C. Weisbuch (eds.) *Confined electrons and photons: New physics and applications* (Plenum, New York 1995)
28. Bei Zhang, Lei Zhuang, Yong Lin, Zhongju Xia, Yong Ma, Xiamin Ding, Shumin Wang, Dejian Zhou, Chunhui Huang, Solid State Comm. **87**, 445 (1996); A. Dodabalapur, L.J. Rothery, T.M. Miller, E.W. Kwoch, Appl. Phys. Lett. **64**, 2486 (1994); N. Takada, T. Tsutsui, S. Saito, Appl. Phys. Lett. **63**, 2032 (1993)
29. G. Vaubel, H. Baessler, Mol. Cryst. Liq. Cryst. **12**, 39 (1970); J.M. Turlet, M.R. Philpott, Jour. Chem. Phys. **62** 4260 (1975)
30. D.G. Lidzey, D.D.C. Bradley, M.S. Skolnick, T. Virgili, S. Walker, D.M. Whittaker, Nature **395**, 53 (1998)
31. F. Tassone, C. Piermarocchi, V. Savona, A. Quattropani, P. Schwendimann, Phys. Rev. B **56**, 7554 (1997)
32. V.I. Yudson, P. Reineker, V.M. Agranovich, Phys. Rev. B **52**, R5543 (1995)
33. A. Engelmann, V.I. Yudson, P. Reineker, Phys. Rev. B **57**, 1784 (1998)
34. J.R. Sheats, Science **277**, 191 (1997).
35. V.G. Kozlov, V. Bulović, P.E. Burrows, S.R. Forrest, Nature **389**, 362 (1997)
36. M. Bergren, A. Dodabalapur, R.E. Slusher, Z. Bao, Nature **389**, 466 (1997)
37. V.M. Agranovich, M.D. Galanin, *Electronic excitation energy transfer in condensed matter* (North Holland, Amsterdam 1982)
38. V.M. Agranovich, G.C. La Rocca, F. Bassani, JETP Lett. **66**, 748 (1997)
39. L.D. Landau, E.M. Lifshitz, *Electrodynamics of continuous media* (Pergamon Press, Oxford 1960)

40. M. Hoffmann, private communication
41. M. Umlauff, J. Hoffmann, H. Kalt, W. Langbein, J.M. Hvam, M. Scholl, J. Söllner, M. Heuken, B. Jobst, D. Hommel, Phys. Rev. B **57**, 1390 (1998)
42. L.C. Andreani, F. Tassone, F. Bassani, Solid State Comm. **77**, 641 (1991); B. Deveaud, F. Clerot, N. Roy, K. Satzke, B. Sermage, D.S. Katzer, Phys. Rev. Lett. **67**, 2355 (1991)
43. S. Permogorov, Phys. Status Solidi (b) **68**, 9 (1975)
44. S. Permogorov, V. Travnikov, Solid State Comm. **29**, 615 (1979)
45. N. Pelekanos, J. Ding, Q. Fu, A.V. Nurmikko, S.M. Durbin, M. Kobayashi, R.L. Gunshor, Phys. Rev. B **43**, 9354 (1991)
46. J. Shah, *Ultrafast Spectroscopy of Semiconductors and Semiconductor Nanostructures* (Springer-Verlag, Heidelberg 1996)
47. R. Zimmerman, Pure and Applied Chemistry **69**, 1179 (1997)
48. U. Woggon, *Optical Properties of Semiconductor Quantum Dots* (Springer-Verlag, Berlin, 1997)
49. V.M. Agranovich, D.M. Basko, JETP Lett. **69**, 250 (1999)
50. R. Ferreira, G. Bastard, Appl. Phys. Lett. **74**, 2818 (1999)
51. D.M. Basko, G.C. La Rocca, F. Bassani, V.M. Agranovich, Eur. Phys. J. B **13**, 653 (2000)
52. V.M. Agranovich, K.N. Ilinski, Phys. Lett. A **191**, 309 (1994); S.A. Kiselev, E. Hartung, Z. Soos, S. Forrest, V.M. Agranovich, Chem. Phys. **283**, 365 (1998)
53. V.M. Agranovich, G.C. La Rocca, F. Bassani, Chem. Phys. Lett. **247**, 355 (1995)
54. V.M. Agranovich, G.C. La Rocca, F. Bassani, JETP Lett. **62**, 418 (1995)

11 Excitons and Energy Transfer in Doped Luminescent Molecular Organic Materials

V. Bulović, M.A. Baldo and S.R. Forrest

11.1 Overview

Excitons are integral to nearly all electronic processes in molecular organic solids. Exciton diffusion is a dominant method of energy transfer; exciton absorption dominates the absorption spectrum; and exciton formation is necessary for luminescence and photoconductivity. Indeed, tracking excitonic processes is the key to understanding the physical behavior of molecular organic semiconductors. In this chapter, we examine excitonic behavior in doped molecular organic thin films, focusing on energy-transfer processes between dissimilar organic molecules.

Fundamental findings described in this chapter have been used in practical devices such as organic light-emitting devices (OLEDs) and organic semiconductor lasers (OSLs). For example, it has been shown that luminescence efficiency of OLEDs can be improved when excitons in a host organic film transfer their energy to a dopant lumophore [1]; similarly, the lasing threshold in optically pumped OSLs was shown to be significantly reduced for doped as compared to neat organic films [2]. However, just as scientific investigations encourage technological development, technological progress spurs new investigations and provides a versatile test bed of scientific principles. Indeed, intense efforts during the past decade to develop commercially viable OLED displays have led to improvements in material purity and growth methods, both essential to the fundamental studies of exciton energy-transfer processes described here. Material purity achieved by train sublimation and vacuum deposition of organic thin films allows for unambiguous identification of physical processes that in earlier studies were distorted by the presence of impurities [3]. We can now precisely locate the exciton generation region in thin films, making possible the investigations of interlayer exciton transfer and determination of diffusion lengths of singlet and triplet excitons [4]. Development of vacuum deposition and monitoring techniques allows for the growth of organic layers to the precision of a molecular monolayer, defining the spatial scale of excitonic processes in these thin films [3]. Finally, new microscopic techniques with atomic scale resolution make possible observations of molecular order in the thin films and at heterointerfaces, with the ultimate goal of revealing the behavior of a single molecule in a molecular assembly [5].

Investigations of excitonic processes presented in this chapter primarily focus on OLED structures consisting of amorphous molecular organic thin films. Such molecular aggregates are typically held together by weak van der Waals forces, and their optical properties are primarily determined by the properties of the constituent molecules. Accordingly, the chapter begins with a review of molecular energy structure (Sect. 11.2.1), and excitation and relaxation processes of molecules (Sect. 11.2.2), and then describes the types of excitons in organic molecular solids (Sect. 11.2.3).

When dissimilar molecules are present in the same organic material, excitons can transfer from one molecular species to the other via a Förster or Dexter process (Sect. 11.2.4). Introducing just a few percent of dopant molecules into a host molecular matrix can significantly alter the luminescence, absorption, and exciton diffusion properties (Sect. 11.2.5) as compared to those of neat films. Changes in the luminescence spectra of OLEDs consisting of doped organic layers are examined in subsequent sections to explore these energy-transfer processes. The physical underpinning of OLED operation is described in Sect. 11.3.

Sect. 11.4 provides examples of Förster energy-transfer in fluorescent organic thin films. We first describe an interlayer energy-transfer process across significant distances to determine the length scale governing Förster transfer. This process is found useful in adjusting the spectral emission of white OLEDs (Sect. 11.4.1). Next, we discuss how polar dopants influence the emission energy of luminescent molecules via the recently identified solid-state solvation effect (Sect. 11.4.2). Examples of Dexter energy transfer to red and green phosphorescent dopants are described in Sects. 11.5.1 and 11.5.2, respectively.

In Sect. 11.6.1 we discuss the ratio of singlet to triplet excitons generated in OLEDs. Under electrical excitation, it is found that three times more triplet than singlet excitons are generated, in accordance with the spin multiplicities of the two states. A combined Förster–Dexter energy transfer can then generate efficient luminescence (Sect. 11.6.2). We show how a phosphorescent dye in a doped OLED structure can be used to gather both singlet and triplet excitons and transfer their energy to a second fluorescent dopant. Via this "sensitized fluorescence" process, the luminescence efficiency of fluorescent dyes in electrically pumped structures can be increased by as much as a factor of four. Through these examples we attempt to present a comprehensive description of energy transfer in doped molecular organic thin films. Practical applications of physical phenomena are emphasized throughout the chapter to illustrate utility of these effects to the performance of organic devices.

11.2 Background

11.2.1 Excited States of a Molecule

The energy states of molecules correspond to electronic, vibrational, rotational, and translational degrees of freedom. The dominant energy transitions, described by the Jablonski diagram in Fig. 11.1, are electronic with energy spacing of a few eV (designated by bold horizontal levels with labels S_n or T_n).

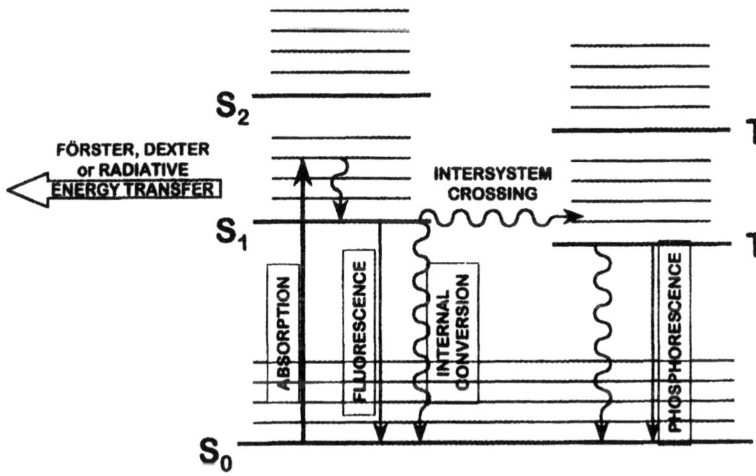

Fig. 11.1. Jablonski diagram of an organic molecule depicting typical energy levels and energy-transfer process. The *straight-line arrows* represent radiative processes, and *wavy-line arrows* non-radiative processes

To each electronic level corresponds a vibrational manifold with typical sublevel spacings on the order of ~0.1 eV, and to each vibrational sublevel corresponds a manifold of rotational and translational molecular energies with ~0.01 eV interlevel spacing. In Fig. 11.1, only vibrational sublevels are indicated by light horizontal lines. Straight vertical arrows indicate allowed or radiative transitions, while non-radiative processes are shown by wavy arrows.

Each electronic state in the Jablonski diagram corresponds to a bonding or anti-bonding molecular orbital. The orbitals associated with a carbon–carbon bond can be either σ or π type, where in a σ-bond electrons are localized between the carbon atoms, while the π-bond establishes delocalized electron density above and below the atomic plane. The corresponding anti-bonding orbitals are designated σ^* and π^* and are of higher energy. There are also valence-shell electrons that do not participate in formation of molecular bonds. Their molecular orbitals are designated as n.

Absorption of a photon can promote an electron from its σ, π, or n ground state to a σ^* or π^* excited state. Such electronically excited states of a molecule contain two unpaired electrons in different orbitals. The spin states of the two electrons can be either even or odd under exchange, giving rise to triplet (T_n) or singlet (S_n) states, respectively. For clarity, the states of different spin multiplicity are separated horizontally in the Jablonski diagram. We note that the ground state of practically all organic molecules is a singlet, since in the lowest energy state molecular orbitals are almost always filled with paired electrons.

The triplet state is typically at a lower energy than the corresponding singlet state. This is explained by considering the spatial distribution of the lowest energy electron orbital Φ_0 and the excited-state orbital Φ_1. The orbital overlap must be non-negligible in order to promote an electron from Φ_0 to Φ_1. However, according to the exclusion principle, two electrons with parallel spins cannot occupy the same eigenstate. Since there is no such restriction on electrons with antiparallel spins these can be found on average closer together. As a result, the average Coulombic repulsive energy is larger in the singlet than in the triplet state, so the former must have a higher energy.

The singlet–triplet (S–T) energy difference varies according to the degree of spatial overlap of Φ_0 and Φ_1. Therefore for (n, π^*) excited states where typically the overlap is small, the S–T splitting is smaller than for the (π, π^*) states where the molecular orbitals have a larger overlap.

11.2.2 Electronic Transitions Between Energy States of a Single Molecule

An electron can be promoted to an excited state by absorbing photons, phonons, by electrostatic interactions between neighboring molecules, or by electrical injection. Once in the excited state the electron tends towards the ground state, losing energy via both radiative and non-radiative processes. Various dissipative pathways (see Fig. 11.1) compete for the energy of the excited state where individual contributions are determined by the relative magnitudes of their rate constants.

Radiative Transitions

Two radiative processes are fluorescence and phosphorescence. *Fluorescence* is a rapid radiative process (with a typical rate constant of $K_\mathrm{f} \sim 10^6$ to $10^9\ \mathrm{s}^{-1}$) in which the spin multiplicities of the initial and the final states are the same. For typical organic molecules this corresponds to $S_1 \rightarrow S_0$, and occasionally $T_n \rightarrow T_m$ transition (where $n > m \geq 1$). *Phosphorescence* is a result of radiative transition between states of different multiplicity, typically $T_1 \rightarrow S_0$. This process is spin forbidden [6] and typically has a much smaller rate constant ($K_\mathrm{p} \sim 10^{-2}$ to $10^6\ \mathrm{s}^{-1}$) than that of fluorescence. In reality, $S \rightarrow T$ and $T \rightarrow S$ transitions may often occur due to spin-orbit coupling.

Even so, the $S_0 \rightarrow T_1$ transitions are, in general, not observed in absorption. When $T_1 \rightarrow S_0$ transitions contribute to radiative emission, the slow rate of this weakly allowed transition results in long excited state lifetimes (on the order of 10^{-6} s to >1 s). If other, faster non-radiative processes are allowed they can quench phosphorescent emission.

Non-radiative Transitions

If vibrational manifolds of two different electronic states on the same molecule overlap in energy, an electron can transfer from one state to the other. Such electron transfer between two isoenergetic energy levels does not change the total energy of the system, and no photon is emitted. When the spin multiplicities of the isoenergetic levels are the same, the process is known as *internal conversion*, while electron transfer between states of different multiplicities is called (Fig. 11.1).

An electron in an excited vibrational level of a molecular electronic state can lose its energy non-radiatively by vibrational relaxation. If the molecule is in its gas phase, vibrational energy is rapidly lost in collisions with other molecules. Equivalently, in the solid state, the vibrational energy of the molecule is lost to its neighbors via phonon modes of the solid (with rate constant $K_{\mathrm{v}} \sim 10^{12}\ \mathrm{s}^{-1}$). In equilibrium, the population of vibrational levels follows a Boltzmann distribution, which implies that at room temperature the ground state is predominantly populated.

An excited molecule can also non-radiatively transfer its energy to a neighboring molecule via dipole–dipole interactions or electron exchange, also known as Förster and Dexter transfer, respectively (see Sect. 11.2.4).

Franck–Condon Shift

The molecular configuration diagram in Fig. 11.2 assumes no change in the nuclear coordinates during electronic transitions (that occur on the time-scale of $\sim 10^{-14}$ s). Absorption or emission of a photon is, therefore, drawn as a vertical transition. The excited state (S_1) is shown offset in the nuclear coordinate, r, relative to the ground state (S_0) due to redistribution of electrons upon excitation.

Absorption of a photon of energy E_{A} can promote an electron from S_0 to S_1. The excited molecule is confined between classical turning points a_2^* and c_2^* and may emit phonons to relax to sublevel a_0^*–c_0^*. Electron transfer from S_1 back to S_0 is exothermic, releasing a photon of energy E_{B}. Finally, the molecule can relax to its lowest sublevel a_0-c_0 by further emission of phonons. The rate of non-radiative, phonon emission is much larger than the radiative transition rate. Therefore, radiative transitions are most likely to occur from the lowest vibrational level of an electronic state. This principle is known as Kasha's rule.

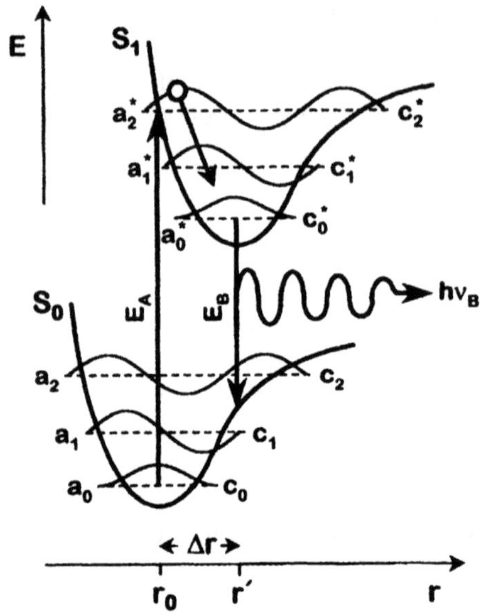

Fig. 11.2. Molecular configuration energy diagram of electronic states S_0 and S_1 for a diatomic molecule. *Dashed lines* indicate the vibrational energy levels on top of which the vibrational wavefunctions are drawn. The equilibrium distance, r_0, and nuclear displacement Δr are indicated

The energy of the emitted photon E_B is lower than that of the absorbed photon E_A. This energy difference is called the Franck–Condon energy, leading to an energetic red-shift, or "Stokes shift". Therefore, the Franck–Condon shift is a Stokes shift due to the nuclear displacement Δr.

Virtually all molecular organic compounds exhibit a Franck–Condon shift, which in some cases is as large as 0.5 eV. As an example of the Franck–Condon shift in amorphous molecular organic thin films, in Fig. 11.3 we plot the absorption and luminescence spectra of aluminum *tris*-(8-hydroxyquinoline) (Alq_3). Alq_3 is extensively used in OLEDs and will be discussed in the context of a number of applications and physical processes described in Sects. 11.3 to 11.6. Comparison of the Alq_3 absorption spectrum to the absorption features of similar compounds allows one to unambiguously attribute the longer- and shorter-wavelength absorption bands to transitions from the ground state to the ^1L_a and ^1B_b excited states, respectively [7]. Luminescence is then due to the radiative transitions from ^1B_b to the ground state. The Franck–Condon shift renders Alq_3 and numerous other molecular species transparent to their own radiation.

Influence of Molecular Dipole Moments on Luminescence

A dipole moment is associated with each bond between inequivalent atoms. Asymmetric molecules have a non-uniform charge distribution, and therefore a net molecular dipole moment. Electronic transitions lead to further redistribution of electrons, leading to a difference in the dipole moment between the ground state and excited state.

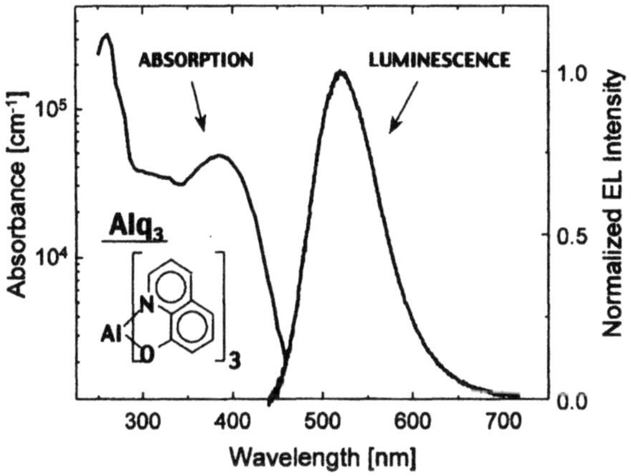

Fig. 11.3. Absorption and luminescence of aluminum *tris*-(8-hydroxyquinoline) (Alq_3). Due to the Franck–Condon shift this material is transparent to its own radiation. Structural formula of Alq_3 is also shown. (From [7])

Spectroscopic studies on molecules in polar solvents have shown that the peak luminescence (λ_{max}) and absorption wavelengths can be affected by dipole moment of the solvent molecules. Both red and blue shifts in λ_{max} have been observed as the solvent dipole moment is increased [6]. This is known as the "solvation effect" or "solvent effect" and has been the subject of numerous theoretical and experimental studies [8]. The theoretical treatments are generally limited by the complexity of the solute–solvent system which involves dipole–dipole, higher-order multipole, and hydrogen-bonding interactions. Their combined effect on the molecular orbitals of the solute are not easily determined. Furthermore, even if all physical interactions were well described, the detailed arrangement of solvent atoms around the solute is not well known, especially when the solute is in a short-lived excited state. Qualitative treatments of the problem, however, give reasonable agreement with experimental measurements. In essence, it is found that the dipolar solvent strongly influences the polar state of the molecule. If an excited state of a molecule has a larger dipole moment than the ground state, an increase in the solvent molecular dipole moments will preferentially stabilize the excited state, lowering its energy and red shifting λ_{max}. This is the case for the organic laser dye DCM2 described in Sect. 11.4.2. Conversely, if the dipole moment of the ground state is larger than that of the excited state, the ground state is preferentially stabilized upon increasing the dipole moment of the solvent molecules, and a blue shift in λ_{max} is observed. These solvent effects are such a prevalent phenomenon that they are routinely used as a tool to probe molecular transitions. For example, a blue shift in λ_{max} with the increase in solvent polarity suggests a $\pi^* \rightarrow n$ transition in the luminescence process, while a $\pi^* \rightarrow \pi$ transition typically results in a bathochromic shift in λ_{max} [6,9].

Solvation effects in the solid state have only recently been documented [10], as discussed further in Sect. 11.4.2.

11.2.3 Excitons

A mobile molecular excited state is referred to as an *exciton* . In solids with strong intermolecular interactions an exciton can be delocalized over a number of molecules. Depending on the degree of delocalization, the excitons are identified as Frenkel, charge-transfer, or Wannier–Mott [11]. The two extreme cases, the extended Wannier–Mott [12,13] and the localized Frenkel [14] excitons (Fig. 11.4), have been intensively investigated, and quantum mechanical descriptions of both have been thoroughly developed.

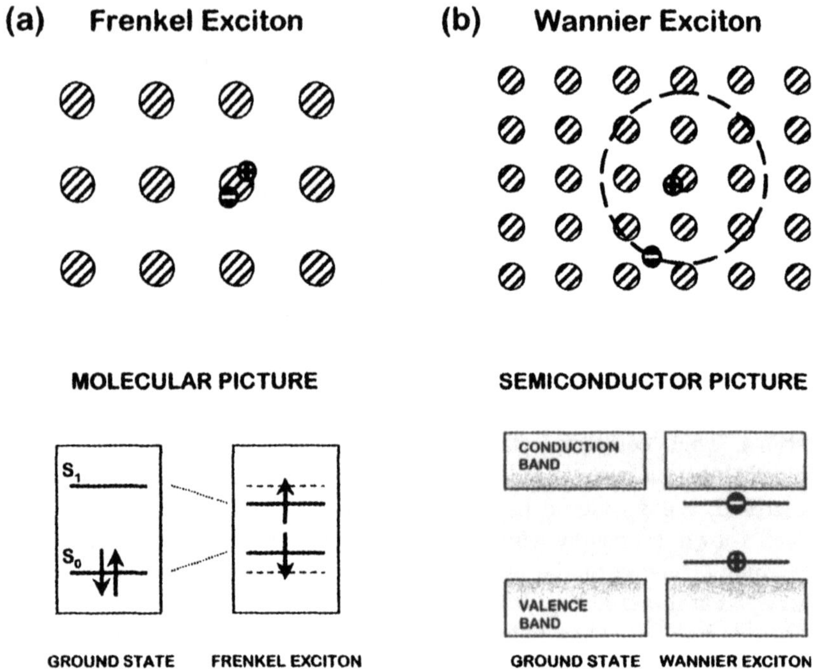

Fig. 11.4. Schematic of a (**a**) Frenkel exciton showing the energy level diagram of discrete molecular states and (**b**) Wannier–Mott (Wannier) exciton with the corresponding semiconductor band diagram

Frenkel Excitons

The Frenkel exciton corresponds to a correlated electron–hole pair localized on a single molecule. Its radius is comparable to the size of the molecule (typically <5 Å) or is smaller than the intermolecular distance. A Frenkel exciton has no internal degrees of freedom and, hence, no internal quantum numbers. It is considered a neutral particle that can diffuse from site to site. The local electric field of the exciton polarizes the surrounding lattice, and its motion therefore generates polaron-type quasiparticles [15]. The single-electron approximation is therefore not sufficient to describe the system and can be replaced by a Hamiltonian [16] that describes the Frenkel exciton in terms of an exciton band number and a wavevector.

Wannier–Mott Excitons

Wannier–Mott (WM) excitons occur in uncorrelated, crystalline materials (e.g., Si, Ge, GaAs and InP), in which overlap between neighboring lattice atoms reduces the Coulombic interaction between the electron and the hole of the exciton, resulting in a large exciton radius ($40 \sim 100$ Å), many times the size of the lattice constant. The WM states can be described by an orthonormal complete set of hydrogen-like wavefunctions. The intervening medium between the electron and hole is then treated as a dielectric continuum. Consequently, energy levels of WM excitons follow a Rydberg progression. These excitons are not found in van-der-Waals-bonded molecular organic solids.

Charge-Transfer Excitons

The charge-transfer (CT) exciton is an intermediate between a Frenkel and a WM state, being neither very extended nor tightly bound to a single molecular site. Its theoretical formulation has proven to be difficult as it requires reconciliation between the localized (Frenkel) and the extended (WM) character of two extreme exciton cases.

A CT exciton is sometimes considered as an unrelaxed polaron pair with the positive and negative polarons located on discrete, identifiable, and nearly adjacent molecules [16]. However, this localized picture is only true if each molecule forms a deep potential well in which charges are confined. Such a localized potential would be expected for a molecular crystal with weak intermolecular interactions and a small overlap between the neighboring orbitals.

An alternative picture of a CT exciton assumes that its wavefunction is a partially uncorrelated, extended state of considerable spatial coherence, with shape and symmetry dependent on the crystalline and molecular structure. This leads to the quantum (extended) description of the exciton, in contrast to the simple electrostatic (localized) models.

Recently, delocalized CT excitons were described by a self-consistent field approximation model [15]. At low exciton densities, the exciton–exciton interaction is negligible; thus, the CT exciton was described as a quasiparticle interacting with an average field of the remaining charges. The one-electron Hamiltonian is then

$$H_0 = T + V_{\text{e-h}} + V_{\text{PSEUDO}} , \tag{11.1}$$

where T is the kinetic energy, $V_{\text{e-h}}$ is the electron–hole binding energy (Coulombic energy) and V_{PSEUDO} is the periodic crystalline pseudopotential. For covalently bonded solids where the interaction between neighboring sites is strong and excited electrons are mostly uncorrelated with the ionic cores, the pseudopotential is small compared to $V_{\text{e-h}}$. In this limit the motion of the electron can be described by a one-electron Schrödinger equation in a periodic lattice. An incipient band structure is formed with a typical bandwidth greater than or on the order of the electron–hole exchange energy. Figure 11.5a shows the schematic of $V_{\text{e-h}}$ (dashed line) and $V_{\text{e-h}} + V_{\text{PSEUDO}}$ (solid line) typical of ionic or covalently bonded semiconductors.

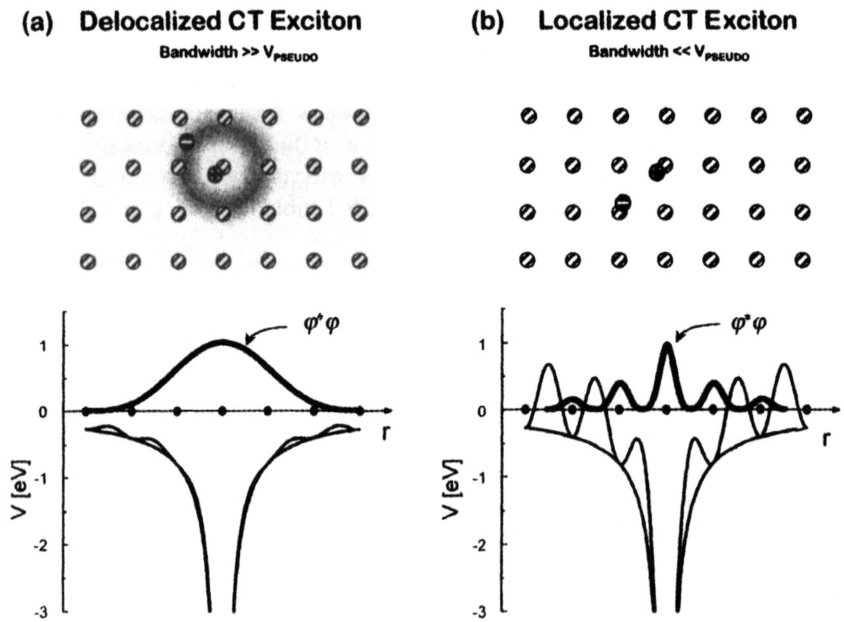

Fig. 11.5. Schematic of (**a**) delocalized and (**b**) localized charge-transfer (CT) excitons. For both cases the Coulombic potential, $V_{\text{e-h}}$ (*dashed line*), and the total potential, $V_{\text{e-h}} + V_{\text{PSEUDO}}$ (*solid line*), are also shown together with the resulting wavefunction density, $\varphi^* \varphi$ (*bold line*)

The resultant excitons are hydrogenic, with the lowest energy electron wavefunction density, $\varphi^* \varphi$ (bold line), showing a characteristic monotonic decay and extended spatial coherence.

In weakly bonded molecular crystals the overlap between the neighboring molecular orbitals is small, and the use of the one-electron approximation and the concept of extended bands becomes questionable. Additionally, the coupling between the charge carriers and lattice becomes strong, further increasing charge localization [15]. In this case $V_{\rm PSEUDO}$ is larger than $V_{\rm e\text{-}h}$ (Fig. 11.5b) and the carriers are localized on individual molecules. For some organic molecular crystals, however, $V_{\rm PSEUDO}$ is smaller or on the order of the bandwidth [15]. In this case the above model may be reasonable. Indeed, this one-electron Hamiltonian was shown to successfully describe both a relatively delocalized CT exciton in non-polar anisotropic organic crystals, and a WM exciton in an inorganic semiconductors such as GaAs [15].

11.2.4 Electronic Energy Transfer

In general terms, during energy transfer between molecules, an excited donor molecule, D^*, transfers its energy to an acceptor, A, which in turn is promoted to an excited state:

$$D^* + A \rightarrow D + A^* \, . \tag{11.2}$$

For example, in some doped organic thin films, the exciton energy can be transferred from the host to the guest molecules, quenching the luminescence of the host while increasing that of the guest. If the donor and acceptor are the same species, the term *energy migration* is used.

The interaction between D^* and A is described by the perturbation Hamiltonian, H'. In this case $(D^* + A)$ is not a stationary state of the total Hamiltonian, but can evolve to other isoenergetic states, such as $(D + A^*)$. Time-dependent perturbation theory then assigns the probability of evolution from $(D^* + A)$, described by $\Psi_{\rm i}$, to $(D + A^*)$, described by $\Psi_{\rm f}$, as:

$$P \propto \rho \, \langle \Psi_{\rm i} | H' | \Psi_{\rm f} \rangle^2 \, , \tag{11.3}$$

where ρ is the number density of coupled isoenergetic donor–acceptor transitions. For organic molecules with typically broad spectral features, ρ can be estimated by calculating the overlap integral of donor luminescence and acceptor absorption. The exciton can then transfer from D to A, and with the same rate, $K_{\rm ET}$, transfer back from A to D. When the coupling between the neighboring molecules is very strong, $K_{\rm ET}$ is so large that we cannot think of the exciton as being localized on any one molecule, as it becomes associated with the whole lattice (see Sect. 11.2.5).

In a typical amorphous organic molecular solid, $K_{\rm ET}$ is slower than the rate of vibrational relaxation. In this case, after the exciton transfers from D to A, it promptly relaxes to the lowest vibrational level of A^*. This destroys the degeneracy between $(D^* + A)$ and $(D + A^*)$, making the exciton transfer unidirectional.

The perturbation, H', contains several terms, of which the most important are the electrostatic (Coulombic or Förster) and electron-exchange (or Dexter) interactions, both independently capable of inducing energy transfer.

Förster Energy Transfer

The electrostatic interaction can be expressed as a sum of multipole–multipole interactions. The dominant, dipole–dipole interaction has been treated by Förster, who found that the rate of energy transfer depends on the distance R between the donor and the acceptor molecules, given by [6,11,17]:

$$K_{\mathrm{ET}}(R) = \left(\frac{1}{\tau}\right)\left(\frac{R_0}{R}\right)^6 . \tag{11.4}$$

Here, R_0 is the Förster radius and τ is the average donor exciton lifetime for recombination in the absence of energy transfer, corresponding to the rate $K_{\mathrm{D}} = 1/\tau$. When $R = R_0$ then $K_{\mathrm{ET}} = K_{\mathrm{D}}$, and the probability that an exciton will recombine at the donor is equal to its transfer probability. The critical distance R_0 is given by the integral over all energies $h\nu$

$$R_0^6 = 1.25 \times 10^{17} \frac{\phi_{\mathrm{E}}}{n^4} \int F_{\mathrm{D}}(\bar{\nu})\alpha_{\mathrm{A}}(\bar{\nu})\frac{\mathrm{d}\bar{\nu}}{\bar{\nu}^4} , \tag{11.5}$$

where h is Planck's constant, ν is the photon frequency, ϕ_{E} is the quantum efficiency of donor emission, n is the refractive index of the host, F_{D} is the normalized emission spectrum of the donor, and α_{A} is the molar extinction coefficient (absorption spectrum) of the acceptor. The dipole–dipole interaction can be significant even at distances as large as 100 Å due to phonon broadened emission and absorption spectra typical of organic molecular species.

In Förster energy transfer, the spin of both D and A is conserved. Therefore, the allowed transitions are

$$^1D^* + {}^1A \rightarrow {}^1D + {}^1A^*$$

and (11.6a)

$$^1D^* + {}^3A(T_n) \rightarrow {}^1D + {}^3A^*(T_m) . \tag{11.6b}$$

The triplet–singlet transition

$$^3D^* + {}^1A \rightarrow {}^1D + {}^1A^* \tag{11.6c}$$

is forbidden, but is sometimes observed since $^3D^*$ has a long lifetime and, although slow, K_{ET} can be larger than the $^3D^* \rightarrow {}^1D$ transition rate [6].

Dexter Energy Transfer

The theory of triplet energy transfer was first described by Dexter [18]. Unlike Förster transfer, the Dexter process does not require allowed transitions on both donor and acceptor molecules. By calculating the exchange interaction, it was found that, given sufficient molecular overlap, an exciton could diffusively hop from one molecule to the next with no change in spin. For transfers between states with allowed transitions, Dexter transfer is typically overwhelmed by long-range Förster processes. But since triplet-to-triplet energy transfer is forbidden by spin conservation in dipole–dipole interactions, Dexter transfer is the only mechanism permitting triplet energy transfer.

The transfer probability from the donor to the acceptor as a function of energy may be expressed as follows [18]:

$$P_{\mathrm{da}} = \frac{2\pi}{\hbar} Z^2 \int f_{\mathrm{d}}(E) F_{\mathrm{a}}(E) \mathrm{d}E \,. \tag{11.7}$$

Here, f_{d} is the normalized emission spectrum of the donor and F_{a} is the normalized absorption spectrum of the acceptor. Now Z is determined by the molecular overlap and is related to the intermolecular spacing R between D^* and A and an effective average Bohr radius L:

$$Z^2 \propto \exp\left(-\frac{2R}{L}\right) . \tag{11.8}$$

The oscillator strength for singlet-to-triplet transitions on the acceptor is likely to be exceedingly small, making determination of F_{a} difficult. However experimentally [19], it is found that the rate of Dexter transfer depends monotonically on the energy difference between donor and acceptor triplet levels, i.e., the transfer rate decreases dramatically as the donor and acceptor levels approach resonance. Since Z^2 and P_{da} are rapidly decreasing functions of intermolecular separation, Dexter transport occurs only over a short distance of typically ~10 Å, and is effectively restricted to neighboring molecules.

Dexter transport may also be thought of as a simultaneous transfer of an electron and a hole between D^* and A. In this model, the mobility of a triplet exciton is related to the product of electron (K_{e}) and hole (K_{h}) transfer rates from the donor to the acceptor, i.e.,

$$K_{\mathrm{ET}} \propto K_{\mathrm{e}} K_{\mathrm{h}} \,. \tag{11.9}$$

Indeed, there is experimental evidence [20] confirming this link. Dexter transfer of triplets should therefore be thermally activated, similar to charge-carrier transport.

In the Dexter transfer process, only the total spin of the D^*A system is conserved. Triplet–triplet energy transfer, which is doubly forbidden in the Förster approximation, is now allowed:

$$^3D^* + {}^1A \rightarrow {}^1D + {}^3A^* \,. \tag{11.10}$$

A representation of both Förster and Dexter processes is shown in Fig. 11.6 for a singlet–singlet energy transfer.

11.2.5 Exciton Transport

Although excitons are initially localized on a single molecule, these quanta of energy can travel through the solid, moving sometimes hundreds of molecules away from their origins. Excitons transport energy without involving

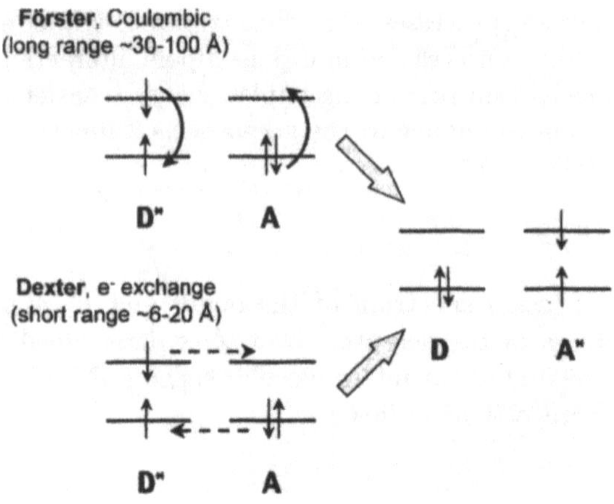

Fig. 11.6. Representation of the Förster and the Dexter processes in singlet–singlet energy transfer

the migration of net electric charge. As well as *radiative* transport in which an exciton emits a photon that is absorbed in another part of the solid to create a second exciton, three other non-radiative transport processes have been identified. These are electromagnetic wave packet transport, hopping transport, and long-range resonant transfer. Regardless of the particular details of the transport process, the exciton distribution can be described by diffusion equations [21].

Electromagnetic wave packet transport is a coherent process, where the exciton binding energy is large compared to the intermolecular hopping potential. Coherence refers to the fact that the wavelength of light which is absorbed to create excitons (for example, $\lambda \sim 5000$ Å) is much larger than the lattice spacing ($a \sim 100$ Å), or inverse absorption length ($\alpha^{-1} \sim 500$ Å), and therefore all the generated excitons initially have a well-defined phase relationship. Because of intermolecular coupling, the excitation can move in a wave-like manner [11]. The energy is transported by a polariton, which is a mixture of a photon and an exciton. For coherent transport, the time required to transfer energy from one molecule to the next is much shorter than the time it takes for the atomic nuclei to rearrange. The exciton motion is therefore not accompanied by lattice distortions, and it can move over several molecular sites before being scattered in time τ_c, defined as the coherence time. For molecular organic materials, τ_c is typically $<10^{-13}$ s because of strong exciton–phonon interactions. At times greater than τ_c the exciton is incoherent and is typically viewed as following hopping-like transport.

Another important exciton transfer process is long-range (<100 Å) resonant transfer via the Förster dipole–dipole interaction. In this process it is not necessary to have a periodic order to transfer energy. Rather, overlap between the donor luminescence and acceptor absorption spectra is sufficient, as described in Sect. 11.2.4.

11.3 Organic Light-Emitting Devices

Organic light-emitting devices (OLEDs) are uniquely suited for examination of many important physical properties of excitons in molecular organic thin-films. Under electrical pumping of the thin-film devices, excitons can form in well-defined regions of the OLED. With such localized exciton generation it is possible to precisely monitor both exciton diffusion and energy-transfer processes.

11.3.1 Operating Principles

The simplest OLED configuration is the single organic layer device consisting of an electron-transporting or hole-transporting organic layer (ETL or HTL, respectively) sandwiched between a cathode and an anode. Besides serving to transport carriers, the organic thin film also acts as the electroluminescent layer (LL). The mobilities of electrons and holes in most organic materials are considerably different, leading to exciton formation nearer one of the two contacts. Since non-radiative exciton recombination, or quenching, is enhanced at the electrode–organic interface, the single organic layer structure typically exhibits a low quantum efficiency. Furthermore, a large percentage of the higher mobility carriers can pass through the organic without forming excitons, also decreasing device efficiency.

The luminescence efficiency is significantly improved in heterojunction OLEDs formed by sequential deposition of two or more organic layers [22]. Figure 11.7 shows the structure of a low-operating-voltage, single-heterojunction OLED [24] where the function of the ETL and LL is performed by Alq_3, and that of HTL by N,N'-dipenyl-N,N'-bis(3-methylphenyl)-1,1'-biphenyl-4,4'-diamine (TPD). In this structure holes are injected from TPD [25] into Alq_3. The small hole mobility in Alq_3 [26] results in a short diffusion length of 30–100 Å [27]; hence, excitons are formed near the ETL–HTL interface. Figure 11.7 also shows the EL spectrum of an Alq_3 TPD OLED and the photoluminescence (PL) spectrum of Alq_3. The similarity between the two indicates their common origin from the radiative recombination of Alq_3 excitons.

The exciton recombination region near the ETL–HTL interface is known as the "recombination zone", whose width is determined by the diffusion length of carriers and excitons in the emitting material. It has been found [7] that singlet exciton diffusion lengths in amorphous thin films such as Alq_3 are approximately 100–300 Å. Hence, in optimal devices, this is on the order of the Alq_3 layer thickness. Similar conclusions about the width and position of the recombination zone have been confirmed by measurements of exciton dynamics in Alq_3 and Alq_3/TPD LEDs [1,28], where most excitons radiatively recombine in the ETL within 400 Å of the heterointerface. In Alq_3/diamine OLEDs, the distribution of excitons within the recombination zone is also weakly dependent on the drive current [29]. These conclusions

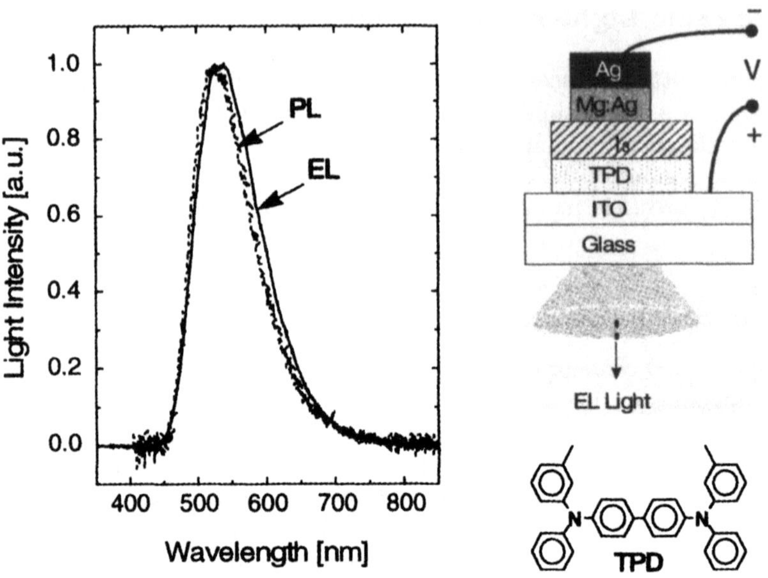

Fig. 11.7. The photoluminescence (PL) spectrum of an Alq_3 thin film and the electroluminescence (EL) spectrum of an Alq_3/TPD organic light-emitting device (OLED) sketched at right. Structural formula of TPD is also shown. (From [23])

can be generalized to almost all small-molecular-weight, vacuum-deposited OLEDs.

11.3.2 Doped OLEDs

Doping the active layers of OLEDs with dyes of high PL quantum efficiencies can significantly improve the device luminescence efficiency [1]. The host exciton can transfer its energy to a luminescent dopant molecule. Similarly, the energy-level structure of the dopant might render it an electron or a hole trap for mobile carriers, which can also lead to exciton formation on the dopant. Even small doping concentrations (from a fraction of a percent to a few percent) in the electrically pumped host layer are sufficient to change the luminescence spectrum, often completely quenching the host luminescence. Low dopant concentration implies a large average distance between these molecules, which reduces exciton quenching from dopant–dopant interactions and yields highly efficient luminescence. The lifetime and efficiency of doped OLEDs have been found to depend on the nature and concentration of dopant molecules in the light-emitting layer [30].

Some examples of doped Alq_3 OLEDs are shown in Fig. 11.8. For comparison, the luminescence spectra of undoped Alq_3 is also shown. Förster energy transfer in Alq_3:(2%) DCM2 ([2-methyl-6-[2-(2,3,6,7-tetrahydro-1H, 5H-

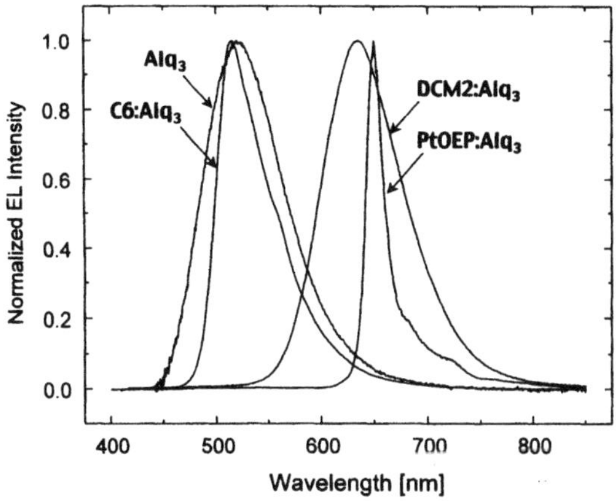

Fig. 11.8. The normalized electroluminescence (EL) spectra of heterojunction OLEDs with an undoped Alq_3, DCM2-doped Alq_3 (DCM2:Alq_3), PtOEP-doped Alq_3 (PtOEP:Alq_3), and Coumarin-6-doped Alq_3 (C6:Alq_3). (From [23])

benzo [ij] quinolizin - 9 - yl) ethenyl] - 4H - pyran - 4 - ylidene] propane - dinitrile) OLEDs leads to orange EL with forward external emission efficiency of $\eta_{\mathrm{ext}}^{(\mathrm{forward})} \sim 0.5\%$ photons per injected carrier [1, 31]. Dexter energy transfer in Alq_3:(6%) PtOEP (2,3,7,8,12,13,17,18-octaethyl-21H,23H-porphine platinum (II)) OLEDs leads to highly efficient, $\eta_{\mathrm{ext}}^{(\mathrm{forward})} \sim 4\%$, saturated red phosphorescence [32]. Also, coumarin 6 doped in Alq_3 improves the color saturation and efficiency ($\eta_{\mathrm{ext}}^{(\mathrm{forward})} \sim 2\%$) of green OLEDs [1], presumably due to carrier trapping.

In doped OLEDs, different molecules perform the functions of electron–hole transport, and of the luminescent center. Such separation of functionality allows for optimization of device performance.

11.4 Examples of Förster Energy Transfer

11.4.1 White OLEDs Based on Interlayer Sequential Energy Transfer

When more than one dopant species is present in an organic thin film, an exciton can transfer its energy in a multistep process from host to a first dopant, and subsequently to a second dopant, etc. This process, known as *cascade energy transfer* [33], simply requires that there is sufficient overlap of energy levels between molecular species at each step in the Förster transfer series such that the integral in (11.5) is non-vanishing. Alternatively, the host exciton can transfer its energy to another molecular species in an adjacent, physically distinct layer, leading to *interlayer energy transfer* [34]. Both of these processes have been used to tailor the luminescence spectrum of organic thin films. In this section we describe how interlayer energy transfer can be used to generate white-light-emitting OLEDs. To obtain white emission, excitation of more than one molecular species is necessary, since the luminescence of a single organic molecule typically does not span the entire visible spectrum.

Incomplete energy transfer from the host to the dopant results in light emission from both species. This technique has been used to obtain white light from small-molecule-based multilayer devices comprising two [35] or three [36] different layers, each emitting one or a combination of the three primary colors. Similarly, doping a polymer host with red, blue and green fluorescent dyes or a combination of several electroluminescent polymer blends has been shown to generate white light emission [37–40].

Alternatively, the emission color of luminescent devices can be optimized by controlling the spatial location of the exciton recombination zone. Sandwiching a layer that preferentially blocks one type of carrier between the hole and electron transporting layers enables exciton formation in both the HTL and ETL. If the layers are themselves luminescent or contain fluorescent dopants, simultaneous emission from the two layers can be obtained to achieve the appropriate color balance.

Both of these techniques have been combined to develop white OLEDs [34, 41]. For example, a balanced white emission (Commission Internationale d'Eclairage (CIE) coordinates (0.33, 0.33) [42]) can be obtained by using Förster energy transfer to balance the emission intensity from blue-emitting 4,4' bis [N-(1-napthyl-N-phenyl-amino)] biphenyl (α-NPD), green-emitting Alq_3, and red/orange-emitting DCM2, all used in the OLED structure shown in the inset of Fig. 11.9.

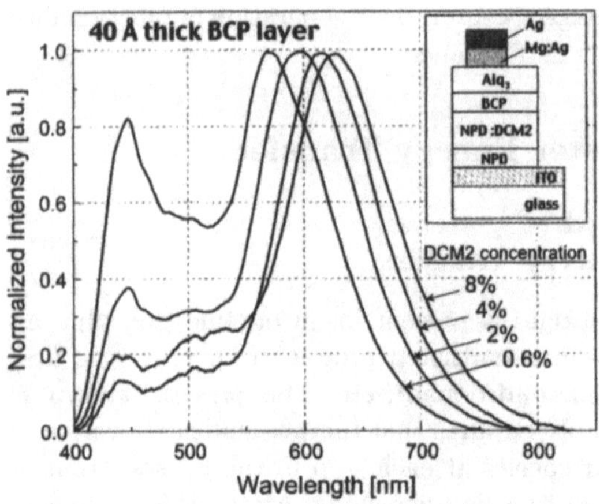

Fig. 11.9. Effect of DCM2 concentration in α-NPD on the spectra of devices with a 40-Å-thick BCP layer. Inset: Structure of the white-emitting organic light-emitting device. (From [34])

The device was grown by high vacuum thermal deposition of organic materials onto the surface of an indium tin oxide (ITO) coated, precleaned [43] glass substrate. The organic materials were deposited in the following sequence: 200 Å of α-NPD was used as a hole-injection layer, followed by a 200-Å-thick α-NPD HTL doped with from 0.6% to 8% (by mass) of the red-emitting DCM2. Next, a hole-blocking layer [44] of 2,9-dimethyl, 4,7-diphenyl, 1,10-phenanthroline, or bathocuproine (BCP), was deposited with a thickness between 20 Å and 120 Å. Finally, a 300-Å-thick layer of Alq_3 was used as the green-emitting ETL. The cathode consisted of a Mg:Ag alloy capped with a thin layer of Ag, as deposited through a shadow mask defining the electrodes.

The spectral emission of these OLEDs was tuned by changing the DCM2 concentration in the α-NPD HTL and by varying the thickness of the BCP layer. Figure 11.9 shows the spectrum for various DCM2 doping concentrations in α-NPD, using a 40-Å-thick BCP layer. Two effects are clearly evident as the DCM2 doping is increased: a bathochromic (red) shift in the DCM2 emission, and a decrease in the ratio of α-NPD to DCM2 emission. The former is a manifestation of the solid-state solvation effect [10, 31] (see Sect. 11.4.2) while the latter results from increased energy transfer from α-NPD to DCM2. For the highest doping (8%), formation of DCM2 aggregates reduces its emission intensity [31], resulting in a relative increase in the α-NPD peak.

The spectra for devices with different BCP thicknesses at a fixed DCM2 doping concentration of 0.6% (by mass) in α-NPD are shown in Fig. 11.10.

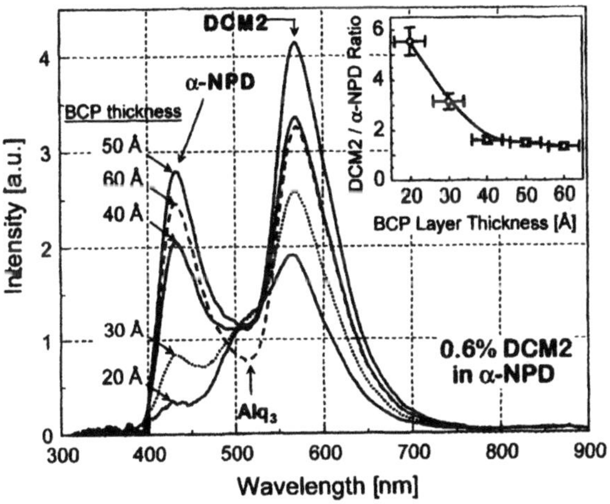

Fig. 11.10. Effect of BCP-layer thickness on the device spectra of 0.6% (by mass) DCM2 doped in α-NPD at 10 mA/cm^2. Inset: Variation of the ratio of DCM2 to α-NPD peak emission intensities as a function of the BCP thickness. (From [34])

Upon reducing the BCP-layer thickness, three effects are observed: a decrease in the α-NPD emission, an increase in the Alq_3 emission, and an increase in the ratio of DCM2 to α-NPD emission. A thin BCP layer facilitates Förster energy transfer from α-NPD to Alq_3, provided that the BCP layer is on the order of a transfer radius, R_0. Also, the probability for transport of holes through the BCP layer increases as its thickness is reduced. Both of these mechanisms can reduce the number of excitons formed on α-NPD, and thereby its blue-emission intensity. Conversely, a greater number of excitons are formed on Alq_3 on thinning the BCP layer, enhancing its green-emission intensity and flattening the visible-emission spectrum.

Neither of these mechanisms, however, explains the observed increase in the DCM2 to α-NPD emission ratio as the BCP-layer thickness is decreased (see Fig. 11.10 inset). Such an effect would arise if the excitons, once on Alq_3, transferred their energy to the DCM2 molecules via a second Förster transfer back across the BCP layer. That is, to explain the relative changes in the emission of α-NPD, Alq_3 and DCM2 with BCP-layer thickness, the excitation must be transferred sequentially from α-NPD to Alq_3 to DCM2, *twice* across the thin BCP layer. This is known as "interlayer sequential energy transfer" (ISET) since energy is transferred in a sequential fashion between species in different layers. Hence, thinning the BCP layer allows for partial depletion of excitons on the blue-emitting α-NPD, and an increase in concentration of the green-emitting Alq_3 excitons. Since the energy transfer from Alq_3 to DCM2 is more probable than the direct transfer from α-NPD to DCM2, due to the improved energetic resonance between the former two species, the thin BCP layer also allows for the return of the exciton back from Alq_3 to DCM2 in the HTL.

Using the spectral overlap integral (11.5), Förster energy transfer radii of $R_0 = 23$ Å, 32 Å, and 39 Å for α-NPD/DCM2, α-NPD/Alq_3 and Alq_3/DCM2, respectively, were calculated. For the latter two cases, R_0 is comparable to or larger than the BCP-layer thicknesses of 20 Å to 40 Å, where ISET is observed. The radii indicate that for low (<1%) DCM2 dopings, the energy transfer from α-NPD to Alq_3 across a thin layer of BCP will be more efficient than that from α-NPD to DCM2 due to the R^{-6} dependence of the rate of Förster transfer. Hence, for a BCP-layer thickness of 30 Å, ISET is favored over direct excitation of a DCM2 molecule by α-NPD excitons.

The spectral emission of the 0.6%-DCM2-doped α-NPD device with a 40-Å-thick BCP layer most closely matches the CIE coordinates of balanced white light (0.33, 0.33). The emission spectrum is largely insensitive to the drive current density: the CIE coordinates change from (0.34, 0.34) at drive current density of 10 mA/cm^2 to (0.33, 0.33) at 160 mA/cm^2, indicating minimal redistribution of the exciton recombination region (Fig. 11.11).

The maximum brightness of 13500 cd/m^2 is achieved at 1100 mA/cm^2 when the voltage is 18.2 V. At a brightness of 100 cd/m^2, $I = 5.2$ mA/cm^2, $V = 11.5$ V, the external quantum efficiency (η_{ext}) is 0.5% and the power efficiency is 0.35 lm/W. For all currents $\eta_{\text{ext}} > 0.4\%$, while the average power efficiency is 0.3 lm/W (see Fig. 11.11 inset).

The advantage of using ISET is its ability to *independently* tune relative red-, green-, and blue-emission intensities by varying the concentration of DCM2 in the HTL, and the thickness of the hole-blocking BCP layer. In this manner emission spectra of OLEDs can be precisely tailored.

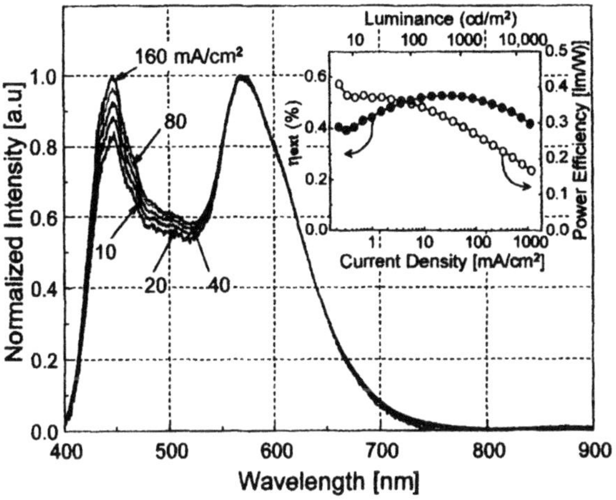

Fig. 11.11. Spectral variation with drive current density of a device with 0.6% DCM2 in α-NPD and a 40-Å-thick BCP layer. Inset: Variation of the external quantum efficiency, luminance and power efficiency with drive current density for the same device. (From [34])

11.4.2 Solid-State Solvation Effect

Absorption and emission spectra of many organic dyes in liquid solutions depend on the dipole moments of the surrounding solvent molecules [9]. This "solvation effect" is a result of intermolecular solute–solvent interaction forces (such as dipole–dipole or dipole induced dipole) which tend to alter the energy difference between the ground and excited states of the solute. It is expected that dye molecules doped into solid matrices should experience a similar effect. Indeed, recent studies show that the polar organic laser dye molecule DCM2 doped into a vacuum-deposited molecular organic thin film of Alq_3 exhibits a fluorescence shift that is dependent on the average local dipole moment of the host [10,31]. Through investigation of Alq_3:DCM2 and other polar fluorescent doped materials, this section describes the interplay of the solvation effect with Förster energy transfer.

Physical Origins of the Solvation Effect

In solutions, solvatochromism is generally assigned to the change in the solute dipole moment upon excitation and its interaction with a polarizable environment [9]. The same explanation is consistent with observations in doped organic thin films; hence, we refer to this condensed phase phenomenon as the "solid-state solvation effect" (SSSE). Theories of solvation effects principally assume that the chemical states of the isolated and solvated chromophore-containing molecules are the same, and processes such as solvent–solute charge transfer, solvent-dependent aggregation, ionization, complexation, or isomerization do not occur. Simply, solvatochromism is considered as a physical perturbation of the relevant molecular states of the chromophore due to the presence of the solvent. Spectroscopic studies on solutions of polar molecules in polar solvents have shown that the peak luminescence

(λ_{max}) and absorption wavelength can be affected by the dipole moment of the solvent molecules. Both bathochromic (red) and hypsochromic (blue) shifts in λ_{max} have been observed as the solvent polarity increased [9]. Theoretical treatments of this phenomenon are generally limited by the complexity of the solute–solvent system which involves dipole–dipole, dipole-induced dipole, higher-order multipole, and hydrogen-bonding interactions. Their combined effect on the molecular orbitals of the solute are not easily determined. Furthermore, even if all physical interactions were well described, the detailed arrangement of solvent atoms around the solute is not well known, especially when the solute is in a short-lived excited state. Qualitative treatments of the problem, however, give reasonable agreement with experimental measurements. In essence, it is found that the polar solvent strongly influences the more polar state of the molecule. Since molecular electronic transitions lead to redistribution of charge, the dipole moments of the ground state (μ_0) and an excited state (μ_1) of a molecule will, in general, be different. If the excited state of the molecule has a larger dipole moment than the ground state, an increase in solvent polarity will more strongly stabilize the excited state (Fig. 11.12a), lowering its energy and red-shifting λ_{max} [9]. Conversely, if $\mu_0 > \mu_1$ (Fig. 11.12b) the ground state is more stabilized upon increasing the solvent polarity, and a blue shift in λ_{max} is observed.

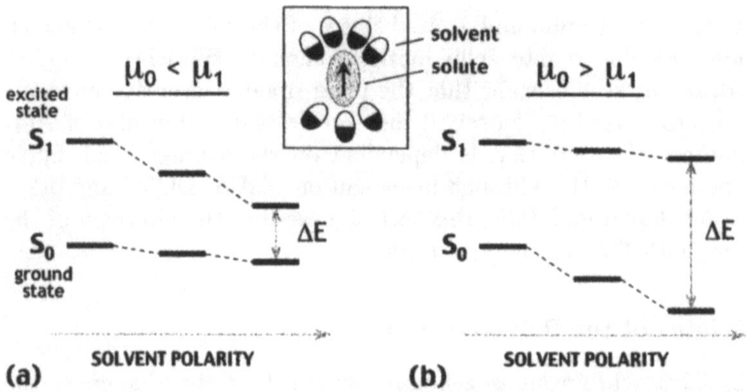

Fig. 11.12. Schematic of solvatochromism. Decrease [increase] in the luminescence energy of molecules with excited state dipole moment (μ_1) larger [smaller] than the ground state moment (μ_0) as a function of solvent molecule dipole moment (polarity) is shown in (**a**) [(**b**)]. Inset: Solvent molecules organize around the polar solute to minimize total system energy

DCM2 Photoluminescence

The solvation effect is pronounced for DCM2 dissolved in solvents of different dielectric susceptibility (Fig. 11.13). The solutions were very dilute (~1/1000 of the saturated solution concentration), ensuring that the spectra observed are due to monomer fluorescence, and that the contribution from aggregates is negligible. There is no significant difference in the quantum yield from DCM2 in any of the solutions used. It is apparent that the peak emission wavelength (λ_{max}) depends strongly on solvent composition, with λ_{max} varying from 565 nm for the nonpolar solvent, benzene, with a low-frequency dielectric constant of $\varepsilon = 2.27$ in solution, to 675 nm for dimethylsulfoxide (DMSO) with a molecular dipole moment of 3.9 Debye and $\varepsilon = 46.7$ (solvent dipole moments are indicated in Fig. 11.13). From these data we conclude that the dipole moment of the excited DCM2 molecule is higher than that of the ground state, consistent with earlier measurements and calculations of the dipole moment of DCM1 [45].

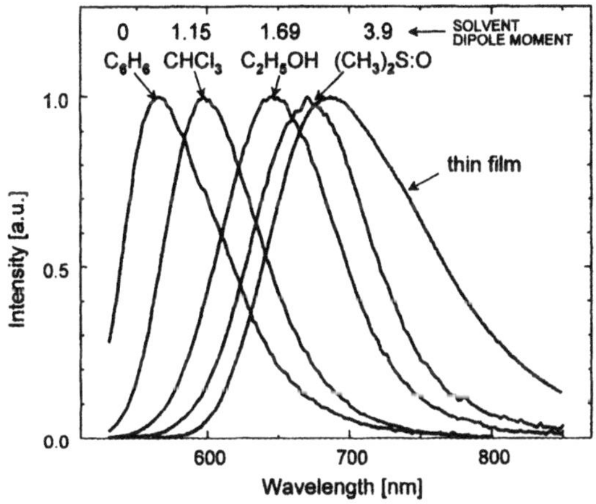

Fig. 11.13. Normalized photoluminescence spectra of dilute DCM2 solutions in benzene (C_6H_6), chloroform ($CHCl_3$), ethanol (C_2H_5OH), and DMSO ($(CH_3)_2S:O$). Solvent dielectric constant and molecular dipole moment are also indicated. Photoluminescence spectrum of DCM2 thin film is also shown. (From [31])

The solution luminescence shift can be compared to the change in the photoluminescence spectrum of DCM2 doped into different vacuum-deposited thin films. The PL of 1% DCM2 in Alq_3 ($\lambda_{max} = 635$ nm) is red-shifted with respect to that of 1% DCM2 in Zrq_4 ($\lambda_{max} = 6$ nm) (Fig. 11.14a,b). This is consistent with the trend found in the solution studies, since the ground state dipole moment of Alq_3 ($\mu_0 = 5.5$ Debye) is larger than that of the more symmetric Zrq_4 ($\mu_0 \sim 0$ Debye). The dipole moments of molecules in organic thin films are obtained from semi-empirical molecular orbital calculations based on an intermediate neglect of differential overlap algorithm [46], parameterized from solution electronic spectra for the optimized molecular geometries.

As the concentration of DCM2 changes in both Alq_3 and Zrq_4, the luminescence red-shifts. It has been suggested that this spectral shift is a result of

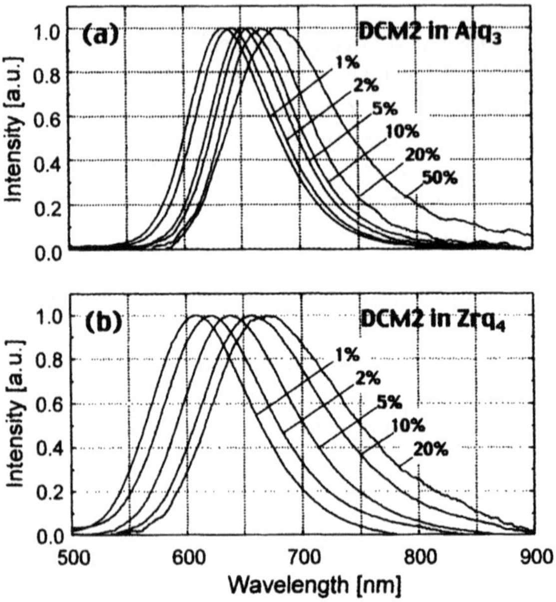

Fig. 11.14. Normalized photoluminescence spectra of 400-Å-thick vacuum-deposited thin films of (**a**) Alq_3:DCM2 and (**b**) Zrq_4:DCM2 as a function of DCM2 concentration

excimer formation [1]. In related studies on solutions of quinacridone in polar solvents [47], such pronounced spectral shifting was attributed to hydrogen bonding in solution. These conclusions are, however, not consistent with the observations in Fig. 11.14: excimer formation would not result in a continuous, rigid shift of the EL spectrum, while hydrogen bonding in solution is not possible for DCM2 molecules.

The similarity of the spectral widths and the magnitudes of the peak shifts in both Figs. 11.13 and 11.14 suggests that DCM2 suspended in solid matrices of Alq_3 and Zrq_4 undergoes an energy shift due to self-polarization. That is, as the concentration of highly polar ($\mu_0 = 11.3$ Debye) DCM2 molecules in the relatively non-polar Alq_3 and Zrq_4 is increased, the distance between nearest-neighbor DCM2 molecules decreases, thereby increasing the local electric field. This field perturbation tends to red-shift the DCM2 emission spectrum. The effect is increased until the concentration is 100% in an undiluted DCM2 film, where $\lambda_{max} = 685$ nm is close to that attained in the polar solvent, DMSO. By comparison, the most dilute films of 1% DCM2 in Alq_3 and 1% DCM2 in Zrq_4 are both red-shifted from DCM2 PL in $CHCl_3$ (with $\mu_0 = 1.15$ Debye and $\varepsilon = 4.81$). This suggests that even at such low concentrations, the DCM2 emission spectrum is influenced by the presence of other DCM2 molecules, given that a smaller shift than that observed would be expected from interactions with only the non-polar Zrq_4.

The self-polarization of DCM2 due to dipole–dipole interactions can thus strongly influence the energy levels of the lumophore. Indeed, for DCM2 concentrations above 10%, λ_{max} is independent of the host material (Fig. 11.15) as the local fields are overwhelmingly determined by the highly polar DCM2 molecules.

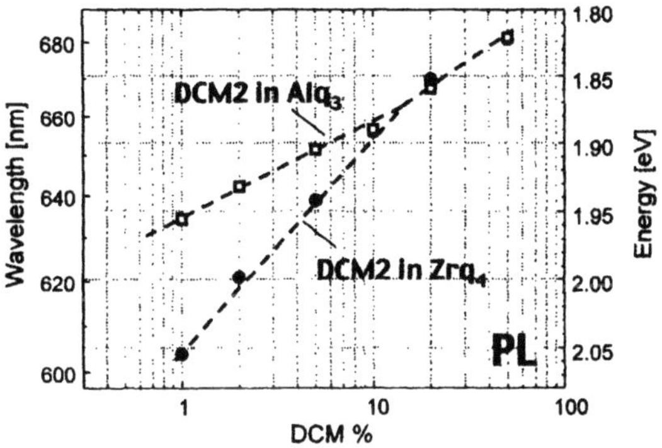

Fig. 11.15. Peak photoluminescence (PL) wavelengths of films from Fig. 11.14 as a function of DCM2 concentration

At concentrations below 10%, emission from Alq_3 is always red-shifted from that of Zrq_4, since the local field is largely determined by the host.

Over a large volume, the net DCM2 dipole moment averages to zero, since the molecules in the solid solution should be randomly distributed. However, considering that the dipole field decreases as $1/r^3$, where r is the distance between dipoles, only those molecules nearest the luminescent center contribute to the local electric field and hence influence the emission spectrum. Local fields due to randomly oriented dopants would contribute equally to blue and red spectral shifts, broadening the overall spectrum, but would not change the peak wavelength. Polar dopants such as DCM2, however, tend to aggregate in oriented domains that minimize the overall energy of the system, causing the spectral red shift. The polar nature of DCM2 can be inferred from the dielectric constant for DCM2 ($\varepsilon = 4$) which is somewhat higher than is typically observed for non-polar molecular solids, while the inhomogeneous distribution of dipoles is reflected in the luminescence efficiency of lightly DCM2-doped Alq_3 films, as shown in the next section.

DCM2-Doped OLEDs

With a well-defined exciton recombination region, OLEDs are ideally suited for studying SSSE. By growing OLED structures with a doped ETL or HTL, we can examine how host molecules with different dipole moments influence the spectral emission of polar molecules such as DCM2. Structures A in Fig. 11.16 corresponds to a conventional OLED in which excitons are generated in the Alq_3 ETL [1].

In structure B excitons are generated in the HTL due to the presence of the BCP hole-blocking layer [48] (see Sect. 11.3.1). The HTL consists of either TPD, α-NPD, or a mixture of TPD and Alq_3. Förster energy transfer

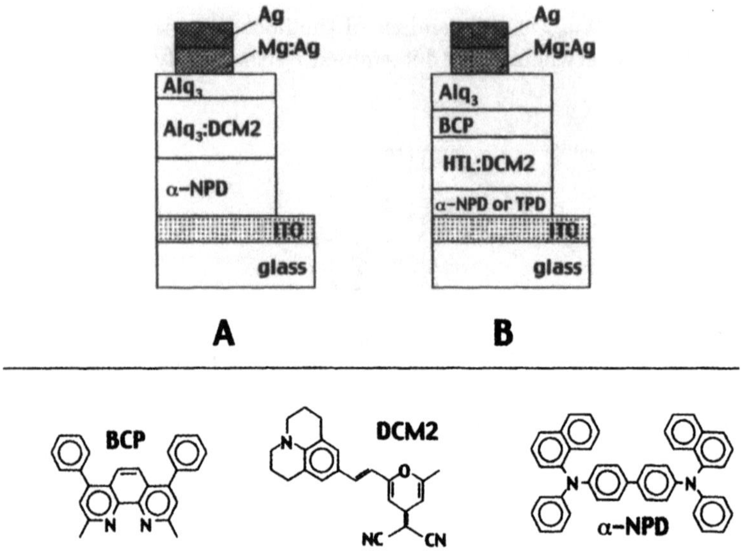

Fig. 11.16. *Structure A*: A conventional OLED with an exciton recombination zone in the DCM2-doped electron transport layer (ETL). *Structure B*: An OLED with an exciton recombination zone in the DCM2-doped hole transport layer (HTL). Structural formulae of molecules used are also shown

from these host layers to DCM2 generates luminescence spectra that vary from yellow to red, depending on the DCM2 concentration.

All devices were grown by high vacuum (10^{-6} to 10^{-8} Torr) thermal evaporation on the surface of ITO-coated glass substrates. The organic materials, shown in Fig. 11.16, were layered as follows: for structure A a 400-Å-thick α-NPD HTL was deposited, followed by a 300-Å-thick Alq_3 ETL doped with from 0.5% to 6% by mass of DCM2, and a 100-Å-thick Alq_3 cap layer. The cathode consisted of a Mg:Ag alloy capped with a Ag layer. For structure B, first a 100-Å-thick HTL of α-NPD or TPD was deposited, followed by a 300-Å-thick layer of DCM2 doped into the HTL, a 100-Å-thick hole blocking layer of BCP, a 300-Å-thick layer of Alq_3, and the same cathode as in A.

Singly Doped Luminescent Layers

A demonstration of the SSSE is apparent for DCM2 doped into a the non-polar host, TPD, as shown in Fig. 11.17b for OLED structure B. For all spectra, two energy peaks are discernible due to incomplete Förster energy transfer from TPD (where the excitons are formed) to DCM2. The lower energy peak corresponds to DCM2 and undergoes a shift from $\lambda_{max} = 570$ nm for 0.9% DCM2 (doping fraction $Q = 0.009$) in TPD to $\lambda_{max} = 645$ nm for $Q = 0.11$, whereas the peak at $\lambda_{max} = 440$ nm corresponds to emission from

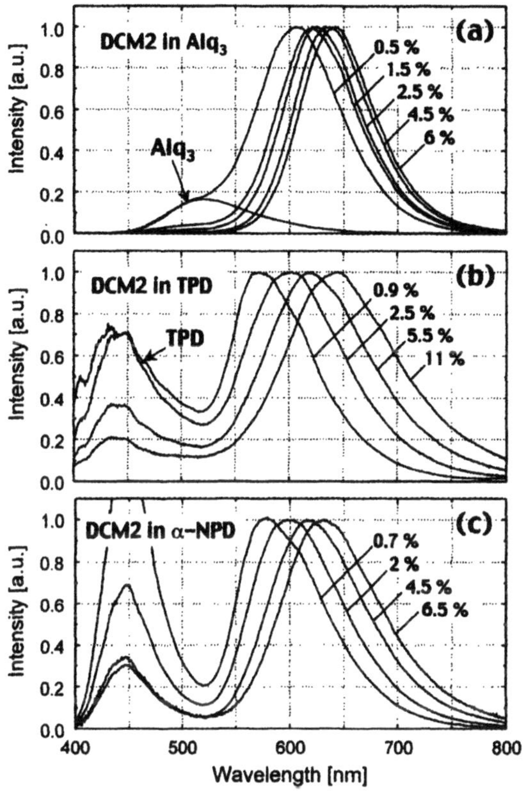

Fig. 11.17. Normalized electroluminescence (EL) spectra of three series of doped devices, where the luminescent layer is (**a**) DCM2 doped in Alq_3, (**b**) DCM2 doped in TPD, and (**c**) DCM2 doped in α-NPD. (From [10] and [31])

TPD. The DCM2 peak shift of $\Delta\lambda_{max} = 75$ nm is considerably larger than for DCM2 in Alq_3, where $\Delta\lambda_{max} = 50$ nm (Fig. 11.17a) [31]. Just as in Zrq_4, the larger shift in the TPD host is due to a small dipole moment [49] of $\mu = 1.5$ Debye associated with its nearly symmetric structure, whereas $\mu = 5.5$ Debye for Alq_3. At the lowest doping, DCM2 molecules are sufficiently far apart such that the average local dipole moment is due primarily to that of the host. In this case, SSSE is smaller in non-polar hosts, resulting in a more blue-shifted spectrum as $Q \to 0$. The shifts in λ_{max} of DCM2 doped in Alq_3 and TPD are plotted together in Fig. 11.18. Similar results to those in TPD are obtained when DCM2 is doped into another non-polar host, α-NPD (Fig. 11.17c).

Förster Transfer From Alq_3 to DCM2

The luminescence of DCM2 molecules in Alq_3 is preceded by Förster transfer of energy from Alq_3. The rate of energy transfer depends on the distance R between the host (Alq_3) and the guest (DCM2) molecules, and is given by (11.4), which describes the interaction of a single host molecule with a single guest molecule. In a doped thin film, however, the exciton on the host

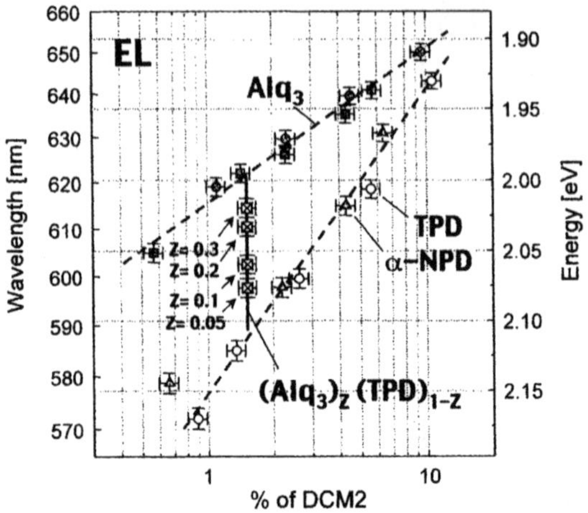

Fig. 11.18. Peak emission wavelengths of OLEDs in Figs. 11.17 and 11.19 as a function of DCM2 concentration. (From [10])

can interact with a *distribution* of guest molecules, and hence the total rate of transfer is obtained by integrating over the film volume:

$$K_{\mathrm{ET}}^{\mathrm{TOT}} = 4\pi \int_{R=2a}^{R=\infty} \left(\frac{1}{\tau}\right)\left(\frac{R_0}{R}\right)^6 \rho_a Q R^2 dR = \left(\frac{1}{\tau}\right)\frac{R_0^6 Q}{8a^6} . \tag{11.11}$$

Here, a is the radius of an $\mathrm{Alq_3}$ molecule, and $\rho_a = 3/(4\pi a^3)$. The probability for exciton transfer from an $\mathrm{Alq_3}$ to a DCM2 molecule is then

$$P_{\mathrm{ET}} = K_{\mathrm{ET}}^{\mathrm{TOT}}/[K_{\mathrm{D}} + K_{\mathrm{ET}}^{\mathrm{TOT}}] , \tag{11.12}$$

where $K_{\mathrm{D}} = 1/\tau$, and τ is the average $\mathrm{Alq_3}$ (donor) exciton lifetime in the absence of energy transfer.

From (11.11) and (11.12) we estimate R_0 by determining the percentage of EL from $\mathrm{Alq_3}$ and DCM2 as a function of doping fraction Q. Using linear fitting of the spectrum of an OLED with $Q = 0.01$, we find that 15% of the EL is due to direct radiative recombination of excitons on $\mathrm{Alq_3}$, resulting in a small shoulder in the spectrum at $\lambda = 520$ nm (Fig. 11.17a). From the internal luminescence quantum efficiency of $\eta_{\mathrm{int}}^{\mathrm{Alq_3}} = 0.32$ for $\mathrm{Alq_3}$ thin films [7], and assuming an efficiency of $\eta_{\mathrm{int}}^{\mathrm{DCM2}} = 1.0$ for DCM2, we infer that ~35% of the excitons recombine on $\mathrm{Alq_3}$ molecules without transferring to DCM2, or equivalently $P_{\mathrm{ET}} = 65\%$, resulting in $R_0 = 16$ Å. However, it is likely that $\eta_{\mathrm{int}}^{\mathrm{DCM2}}$ is significantly less than unity since DCM2 molecules can interact with the $\mathrm{Alq_3}$ host, resulting in non-radiative decay; in which case an even smaller fraction of excitons recombine on the host. For example, with $\eta_{\mathrm{int}}^{\mathrm{DCM2}} = 0.33$, then $P_{\mathrm{ET}} = 85\%$ and $R_0 = 19$ Å. This distance is within a factor of two of $R_0 = 39$ Å, calculated using the $\mathrm{Alq_3}$ PL and DCM2 absorption spectral overlap in the Förster integral.

In considering the differences between these two estimates of R_0, we note that the distribution of DCM2 in $\mathrm{Alq_3}$ is locally inhomogeneous, as already inferred from analysis of the spectral shifts. We note that for devices with $Q = 0.01$, the average separation between neighboring DCM2 monomers is >5 Alq_3 molecules. However, the average spacing between Alq_3 molecules is >9 Å [50], from which we estimate an average DCM2 spacing of $d_{DCM2} = 50$ Å. In fact, some pairs of DCM2 molecules will be farther apart than d_{DCM2}, resulting in a higher probability for recombination at an Alq_3 molecule prior to transfer. The net effect of the inhomogeneous dopant distribution is therefore to increase the host and decrease the guest luminescence, or equivalently P_{ET} (homogeneous) > P_{ET} (inhomogeneous). The inhomogeneous distribution of distances between DCM2 molecules in an Alq_3:DCM2 film with $Q = 0.01$ can be modeled as a Gaussian distribution centered at $d_{DCM2} = 50$ Å, and with a full-width at half-maximum of w. With $R_0 = 39$ Å, calculated via the Förster integral, and $w = 12$ Å to 15 Å, P_{H-G} is 85% to 75%, in agreement with P_{ET} obtained from the EL spectral measurement. Therefore, the local variations in the DCM2 doping due to aggregation and other inhomogeneities resulting from the growth process can account for the observed enhancement in Alq_3 luminescence.

Double-Doped Luminescent Layers

The functions of the luminescence center and the dipole source in SSSE can be separately optimized by assigning them to two separate dopants. This property, for example, can be used to minimize or even eliminate the loss of efficiency due to aggregation quenching at high lumophore concentrations. Separation of functionality was tested using a series of OLEDs containing a TPD:Alq_3:DCM2 luminescent HTL in structure B. Here, DCM2 is retained as the lumophore, while Alq_3 is introduced into the TPD host to generate the local dipole moment. Figure 11.19 shows the EL spectra of a series of these devices with a fixed DCM2 doping of $Q = 0.015$ in a $(Alq_3)_Z(TPD)_{1-Z}$ host. Two spectral peaks are once again observed: the low-energy peak corresponds to DCM2 and the high-energy peak to Alq_3 emission. The DCM2 luminescence shows a shift from $\lambda_{max} = 597$ nm at $Z = 0.05$ to 614 nm at $Z = 0.30$.

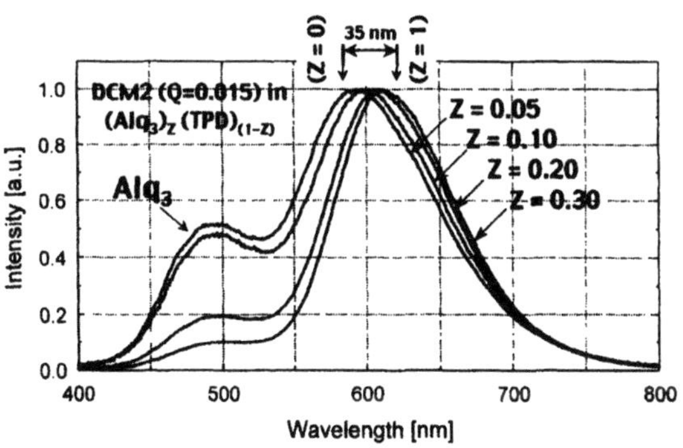

Fig. 11.19. Normalized electroluminescence spectra (EL) of a series of devices of structure B, with the HTL consisting of a mixture of Alq_3 and TPD, and doped with 1.5% DCM2. The HTL fraction consisting of Alq_3, Z, is indicated. (From [10])

The Alq_3 emission intensity decreases with increasing Z due to the efficient cascade energy transfer [33] via TPD $\rightarrow$ Alq_3 $\rightarrow$ DCM2. This is also apparent in Fig. 11.20, where the quantum efficiencies for several different Z and drive currents are plotted. The efficiency of all devices falls slowly with increasing current, typical of the behavior of doped fluorescent devices. For low Z, the average Alq_3-to-DCM2 distance exceeds the Förster radius of 40 Å [31], leading to exciton recombination on Alq_3 molecules prior to their transfer to DCM2. Hence, in this regime (region A in Fig. 11.20b), the efficiency decreases with increasing Alq_3 concentration. The luminescence quantum efficiency has a maximum at $Z = 0.20$ (region B in Fig. 11.20b). Here, the average Alq_3-to-DCM2 distance is comparable to or less than the Förster radius, leading to maximum energy transfer and hence DCM2 emission. At still higher Z, the efficiency declines, presumably due to concentration quenching at Alq_3 clusters (region C in Fig. 11.20b). Therefore, the luminescent efficiency is increased due to a combination of reduction in dopant aggregation quenching.

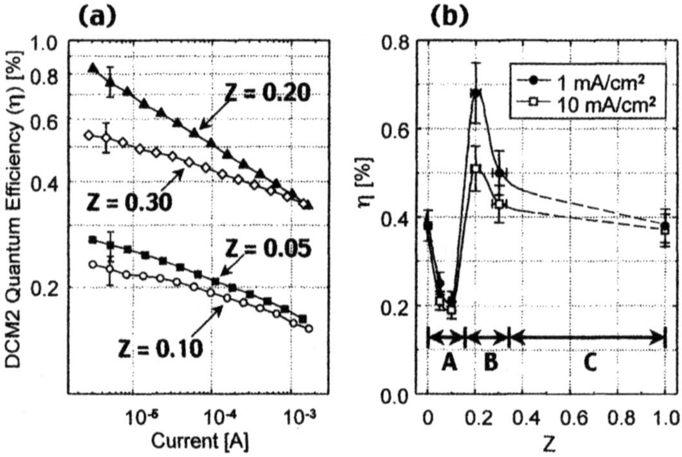

Fig. 11.20. Performance of devices from Fig. 11.19: (**a**) Luminescence quantum efficiency (η) of DCM2 emission versus current density at several values of Z, and (**b**) DCM2 luminescence quantum efficiency versus Z at currents of 1 mA/cm^2 and 10 mA/cm^2. In (**b**) region A corresponds to incomplete energy transfer from Alq_3 to DCM2, resulting in significant Alq_3 emission. In region B, cascade energy transfer from TPD to Alq_3 to DCM2 results in the highest η. In region C, η decreases due to Alq_3 aggregation quenching. (From [10])

The dependence of λ_{max} on Z for DCM2 ($Q = 0.015$) in $(Alq_3)_Z(TPD)_{1-Z}$ is plotted in Fig. 11.18 as a vertical line. The DCM2 spectral emission can be tuned between $\lambda_{max} = 585$ nm and $\lambda_{max} = 620$ nm, corresponding to

$Z = 0$ and $Z = 1$, respectively. Keeping a low DCM2 concentration prevents formation of DCM2 aggregates, thus retaining a high luminescence efficiency while the red spectral shift is independently tuned by adjusting the Alq_3 concentration and cascade energy transfer between host and lumophore.

SSSE in Other Lumophores

The solvation effect also influences the emission of Alq_3 molecules. Note that in Fig. 11.19 the Alq_3 spectral peak is somewhat shifted from $\lambda_{max} \sim 525$ nm observed in conventional OLEDs [24]. As in the case of DCM2 in TPD, this shift is also due to self-polarization-induced SSSE due to the small dipole moment of Alq_3 in the ground state. Figure 11.21 shows that with an increase of Alq_3 concentration in a TPD:Alq_3 film λ_{max} of Alq_3 shifts from 498 nm to 522 nm. This same effect could explain the spectral emission of methylated-Alq_3 ($Almq_3$) thin films [51], which is similar in shape to that of Alq_3, but peaks at $\lambda_{max} \sim 500$ nm. The dipole–dipole interaction between neighboring $Almq_3$ molecules should be smaller since the intermolecular spacing is likely larger due to steric hindrance of CH_3 groups. Hence, the $Almq_3$ emission spectrum is expected to be blue-shifted with respect to Alq_3, as observed. Indeed, the solvation effect applies generally to all polar lumophores.

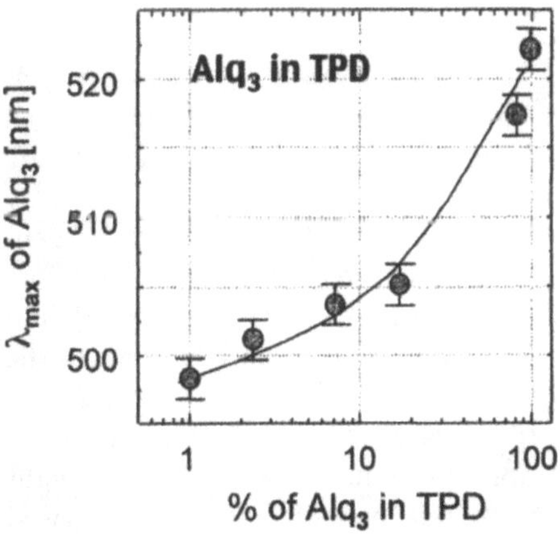

Fig. 11.21. Peak wavelength of photoluminescence spectra of 400-Å-thick Alq_3:TPD thin films for different Alq_3 concentrations. (From [10])

We summarize the color emission of most of the OLEDs from this section by plotting their chromaticity coordinates in Fig. 11.22.

For the OLEDs of Fig. 11.17b, the coordinates of only the DCM2 emission are plotted (open squares), to indicate the color assuming complete energy transfer from a non-polar host, such as TPD, to DCM2. With control of the DCM2 concentration, saturated yellow-to-red emission is achieved, indicating the remarkably broad tuning range accessible to this material system via SSSE. Similarly, Alq_3 emission can be tuned when doped in TPD to span the

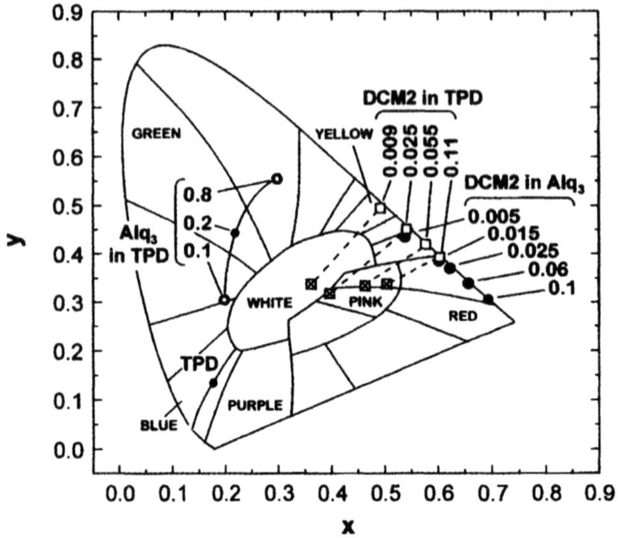

Fig. 11.22. Commission Internationale d'Eclairage (CIE) coordinates of DCM2:Alq$_3$ OLEDs in Fig. 11.17a (*solid circles*), DCM2:TPD OLEDs in Fig. 11.17b (*crossed squares*), DCM2 component of Fig. 11.17b OLEDs (*open squares*), and Alq$_3$:TPD PL spectra of Fig. 11.21 (*open circles*). The numbers associated with data points represent the dopant fraction, Q, in the host material. (From [10])

color spectrum from yellowish-green to green, while undoped TPD emits in the blue. Fine tuning of the fluorescence spectra of luminescent molecules by SSSE can be applied to optimize the emission of OLEDs. That is, a frequently used method for changing the OLED emission color is by designing and synthesizing a new fluorescent molecule. This typically requires significant effort to obtain the desired color along with reasonable fluorescent efficiency and high photochemical stability. Using SSSE, a material which meets all of the other demanding requirements of an OLED used in a display can be *independently* tuned to also match the particular emission criteria of the application. The SSSE is general in nature and is ubiquitous in small-molecular-weight organic thin films. Indeed, similar effects are also expected in polymer-based luminescence.

11.5 Examples of Dexter Energy Transfer

11.5.1 PtOEP Phosphorescence

As discussed in Sect. 11.2.4, Förster transfer from a triplet donor to a triplet acceptor state is doubly forbidden under spin-conservation rules. However,

Dexter transfer occurs between donor and acceptor states of similar spin symmetry. Hence phosphors which can accept triplet states when doped into a host provide an ideal system with which to study Dexter processes. Furthermore, if luminescence can be obtained from triplets as well as singlets, then the maximum efficiency of an electroluminescent device could be quadrupled. To achieve this, it is necessary to find materials with a propensity for radiative decays from their triplet states, and the most effective materials to date [32, 52, 53] are organic complexes with heavy metal atoms such as Pt, Ir, Os, Ru or Au. The heavy metal enhances spin-orbit coupling and mixes singlet and triplet states, thereby reducing the triplet lifetime and enhancing the efficiency of phosphorescence.

Efficient phosphorescence has been observed from thin films of Alq_3 doped with the phosphor PtOEP. Porphyrin complexes are known to possess long-lived triplet states useful in oxygen detection [54]. The addition of platinum to the porphyrin ring reduces the phosphorescence lifetime by increasing spin-orbit coupling; the triplet states gain additional singlet character and vice versa. This also enhances the efficiency of intersystem crossing from the first singlet excited state to the triplet excited state. Transient absorption spectrometry gives a singlet lifetime in PtOEP of ~1 ps, and the fluorescence efficiency is extremely weak [55]. In contrast, the room-temperature phosphorescence efficiency of PtOEP in a polystyrene matrix [56] is 0.5, with an observed lifetime of 91 μs. Thus, both singlet and triplet excitations in PtOEP yield efficient phosphorescence. Consequently, no significant singlet emission at approximately 580 nm [55] is observed, although strong emission from the triplet excited state at 650 nm, with weaker emission at the vibronic harmonic overtones at 623 nm, 687 nm and 720 nm, is apparent.

Conclusive evidence of triplet transfer to PtOEP from Alq_3 is obtained by examining the unnormalized spectra of the two devices in Fig. 11.23. A 100-Å-thick layer of Alq_3 doped with ~1% of the fluorescent dye DCM2 is placed at the heterojunction between Alq_3 and α-NPD. Alq_3 preferentially transports electrons, whereas α-NPD preferentially transports holes; thus, exciton formation is localized at this interface. Since DCM2 exhibits efficient energy transfer from Alq_3 (see Sect. 11.4.2), the combination of 1% DCM2 in Alq_3 effectively removes singlet excitons. The remaining singlets eventually recombine in Alq_3, yielding the small shoulder in the spectrum at ~530 nm. However, in device 2, an additional layer of ~10% PtOEP in Alq_3 is introduced 200 Å away from the exciton formation zone. In this device, emission is seen from PtOEP, without any change in the intensity of emission from either DCM2 or Alq_3. Hence, PtOEP cannot be an efficient charge trap, since carriers removed by PtOEP in this device would result in a decrease in the DCM2 and Alq_3 emission compared with that in device 1, an effect clearly not observed. Since the DCM2 acts as a "filter" that removes singlet Alq_3 excitons, the only possible origin of the PtOEP luminescence is Alq_3 triplet states that have diffused through the DCM2 and intervening Alq_3 layers.

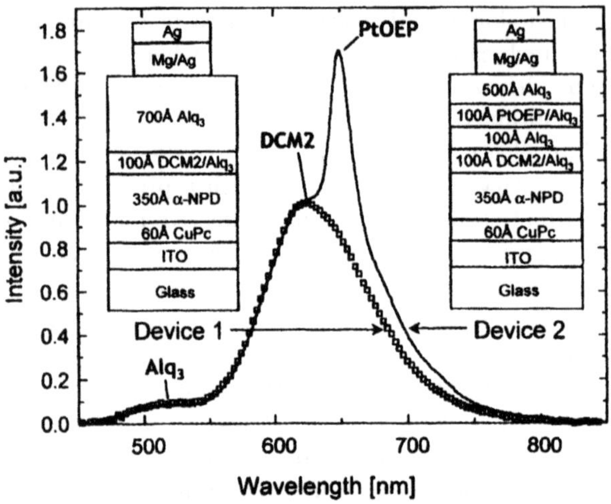

Fig. 11.23. Two electroluminescent devices demonstrating that Alq_3 triplets are transferred to PtOEP. Each device contains a 100-Å-thick layer of ~ 1% DCM2 in Alq_3 at the recombination zone. This layer acts to remove singlet states. Remaining singlets recombine in Alq_3, yielding the shoulder apparent in the spectra at 530 nm. Device 2 contains an additional layer of ~ 10% PtOEP in Alq_3 positioned 200 Å away from the Alq_3/α-NPD interface. Strong emission is seen from the PtOEP without a corresponding decrease in emission from DCM2 or Alq_3. (From [32])

An alternative demonstration of triplet transfer to PtOEP was described by Cleave et al. [57]. There, PtOEP was doped into the polymer host poly[4-(N-4-vinylbenzyloxyethyl,N-methylamino)-N- (2,5-di-tert-butylphenylnapthalimide)] (PNP) and the transient electroluminescence was examined in OLEDs with the general structure ITO/polyvinylcarbazole/PtOEP:PNP/Ca [57]. The observed decay rate is the convolution of the rate of energy transfer from PNP to PtOEP and the PtOEP triplet relaxation rate. Analysis of the electroluminescent decay determined [57] that a fraction of excitons participated in slow (~10 μs) energy transfer. Moreover, such slow energy transfer from PNP to PtOEP was not found in the photoluminescent transient response of PtOEP:PNP. Thus, it was inferred that Dexter transfer of host triplet excitons was responsible for the difference between the photoluminescent and electroluminescent decay transients.

11.5.2 Triplet Diffusion

To examine triplet energy transfer and diffusion, transient studies of phosphorescence may be performed using OLEDs with the structures shown in Fig. 11.24. Four host materials are examined: Alq_3, TPD, α-NPD and CBP (4,4'-N,N'-dicarbazole-biphenyl). Two structures were made dependent on whether the host material preferentially conducts electrons or holes. If the

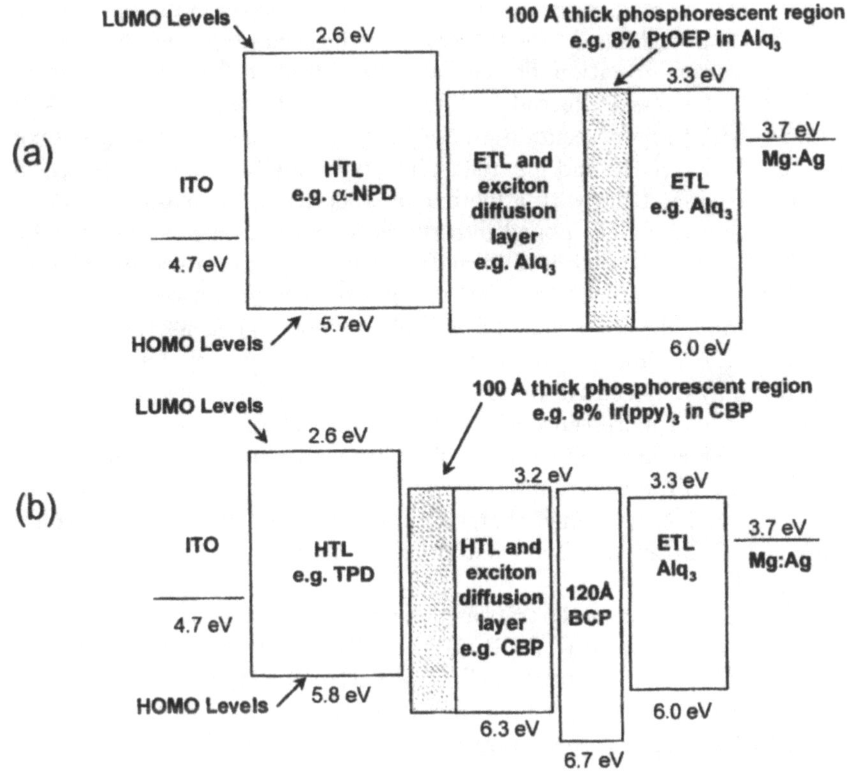

Fig. 11.24. The structure of the electroluminescent devices used in transient response measurements of triplet diffusion in organic host materials. In (**a**) the host preferentially transports electrons, and the exciton formation zone is at the interface between the host and α-NPD; in (**b**) the host preferentially transports holes, and the exciton formation is at the interface between the host and BCP. In both devices triplets diffuse through the host before reaching a phosphorescent region, created by doping a narrow (~100 Å) layer of the host with a phosphorescent dye. Singlets formed during electrical excitation cause fluorescence within the host; thus, triplet dynamics are reflected in the delay between fluorescence and phosphorescence. (From [58])

host acts as an electron transport layer it was employed in the structure of Fig. 11.24a; of the materials studied, only Alq_3 falls in this category. Of the remaining materials, TPD and α-NPD are predominantly hole conductors, and CBP conducts both electrons and holes. Hence, these three materials can be used as hole transport layers in conjunction with BCP to force emission from the HTL (Fig. 11.24b).

Where the host material is Alq_3, the exciton formation zone is located at the interface of the HTL and Alq_3. In the case of HTL host materials, the exciton formation zone occurs at the interface between the HTL and BCP. To study exciton diffusion, undoped layers of the host material were inserted between the exciton formation interface and the phosphorescent zone. As shown in Fig. 11.24, triplets are forced to diffuse through this zone to reach a "sensing layer", where luminescence from the phosphor is observed. Two phosphors were employed: PtOEP and *fac* tris(2-phenylpyridine) iridium ($Ir(ppy)_3$), the latter molecule emitting with a lifetime of $\sim$500 ns at $\lambda \sim 510$ nm.

Since the phosphor is doped into the host material in a narrow band at some distance from the exciton formation zone, triplet excitons must diffuse through the host material to reach the phosphorescent region. Singlets, on the other hand, do not reach the phosphorescent region, as they are constrained to the vicinity of the formation zone by their very short lifetimes; thus, they may only contribute to host fluorescence. The delay between fluorescence and phosphorescence is therefore a function of exciton transfer rate, diffusion distance, and temperature. In addition, it is possible to apply strong negative bias to the electroluminescent device during the period after excitation. This removes trapped charge and may be used to distinguish among energy transfer, trapping, and exciton formation directly on the luminescent dye molecule.

In Fig. 11.25, we show the transient response of four electrophosphorescent systems: (a) PtOEP in Alq_3, (b) PtOEP in CBP, (c) $Ir(ppy)_3$ in CBP, and (d) $Ir(ppy)_3$ in TPD. In the case of the PtOEP-based OLEDs, a 1 μs voltage pulse is applied before returning the bias to either zero or -10 V. For the $Ir(ppy)_3$-based OLEDs, a 200 ns voltage pulse is applied. All systems exhibit delayed phosphorescence in the absence of reverse bias; however, only PtOEP:Alq_3 and $Ir(ppy)_3$:TPD retain delayed phosphorescence in the presence of a strong negative bias. Thus, we conclude that triplet energy transfer from the host to the guest is present in these systems. In PtOEP:CBP and $Ir(ppy)_3$:CBP, luminescence is dominated by charge-trapping and exciton formation directly on the phosphor. Since the trap-based devices all emit from the HTL, it seems that in these host materials the phosphorescence occurs after hole-trapping and electron diffusion to the emitting sites.

11.5.3 $Ir(ppy)_3$ Phosphorescence

Analysis of $Ir(ppy)_3$ doped in TPD reveals some surprising phenomena. If we displace $Ir(ppy)_3$ from the exciton formation region and use TPD as the host material, we find that the phosphorescent peak is delayed by $\sim$100 μs for a diffusion distance of only 200 Å(see Fig. 11.25d). In most host materials $Ir(ppy)_3$ possesses a radiative lifetime of $\sim$1 μs, but in TPD its electroluminescent lifetime is increased to 15 μs. In this case, the rate of energy transfer from host to the guest is much slower than guest phosphorescence. Photoluminescence of $Ir(ppy)_3$:TPD gives somewhat similar results; a 15 μs lifetime is also observed, but in this case an additional fast transient is observed di-

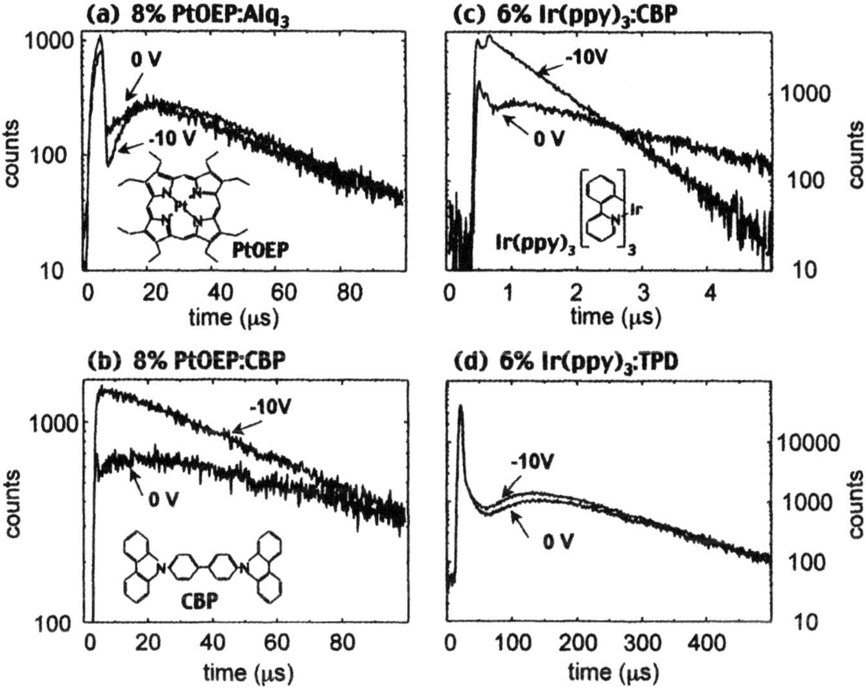

Fig. 11.25. The transient response of four phosphorescent guest–host systems under 1 μs electrical excitation. Between excitations, bias voltage is either zero or −10 V. Triplet energy transfer is demonstrated in (**a**) by 8% PtOEP:Alq_3. These transients exhibit delayed phosphorescence due to triplet diffusion in Alq_3, and also show minimal change when reverse bias is applied, indicating the absence of traps. In (**b**) 8% PtOEP:CBP shows no delayed phosphorescence when reverse-biased, indicating that charge trapping on PtOEP is significant. In (**c**) the transient response of 6% $Ir(ppy)_3$:CBP also shows evidence for charge-trapping on $Ir(ppy)_3$. (**d**) Energy transfer to $Ir(ppy)_3$ is observed in $Ir(ppy)_3$:TPD with an $Ir(ppy)_3$ radiative lifetime of ~15 μs, significantly longer than its natural lifetime of ~1 μs. Taken together with the apparent absence of charge-trapping on $Ir(ppy)_3$ in TPD, this long lifetime indicates that energy transfer from TPD to $Ir(ppy)_3$ might be the rate-limiting step in $Ir(ppy)_3$ phosphorescence. Insets: Molecular structural formulae of $Ir(ppy)_3$, CBP, and PtOEP. (From [58])

rectly after excitation. At 200 K the fast transient is not observed to change significantly from ~1 μs, but the slow transient extends to a lifetime of 80 μs.

In the electroluminescence of $Ir(ppy)_3$:TPD, the initial (fast) transient is less intense, and principally comprises emission from the α-NPD HTL. However, in photoluminescence all emission originates from $Ir(ppy)_3$ and the two transients in the decay are indicative of two processes in this film. The extension in the triplet lifetime in $Ir(ppy)_3$:TPD is possibly caused by triplets residing in TPD, where they presumably have long lifetimes. Ultimately, triplets may transfer to $Ir(ppy)_3$, where they undergo comparatively rapid (1 μs) decay. The difference observed in the photo- and electroluminescent response of $Ir(ppy)_3$:TPD arises because significantly more $Ir(ppy)_3$ molecules are directly excited by optical pumping without requiring energy transfer from TPD as an intermediate step. Since the 15 μs $Ir(ppy)_3$ transient is also observed in photoluminescence, significant populations of triplets must also be generated by photoexcitation. It is most likely that these triplets are formed by intersystem crossing in $Ir(ppy)_3$ and then rapidly transfer to TPD by either a Dexter or Förster process.

11.5.4 Triplet–Triplet Annihilation

Transient analysis is not restricted to the study of energy transfer in organic materials; it may also be used to examine exciton dynamics. An example is the investigation of triplet–triplet (T–T) annihilation, an effect with important implications for phosphorescent efficiency.

T–T annihilation proceeds via the following exothermic reaction [59]:

$$^3M^* + {}^3M^* \rightarrow {}^1M^* + M \, . \tag{11.13}$$

Here, $^3M^*$ represents the triplet excited state of a phosphorescent molecule, $^1M^*$ the singlet excited state, and M the ground state. Since the phosphorescent molecules considered all possess strong spin-orbit coupling, then process (11.13) is followed by

$$^1M^* + M \xrightarrow{\text{ISC}} {}^3M^* + M \, . \tag{11.14}$$

Due to the participation of two identical triplet excitons, the T–T annihilation rate should increase with the square of the triplet excited state concentration, $[^3M^*]$. Hence, to second order, $[^3M^*]$ is determined by the rate of triplet generation (proportional to the current density J), and the rates of mono- and biexcitonic [60] triplet recombination, via:

$$\frac{\mathrm{d}[^3M^*]}{\mathrm{d}t} = -\frac{d\,[^3M^*]}{\tau} - k_\mathrm{q}[^3M^*]^2 + \frac{J}{qd} \, . \tag{11.15}$$

Here, q is the electron charge, d is the thickness of the exciton formation zone, τ is the phosphorescent recombination lifetime, and k_q is the triplet quenching parameter. Following an excitation pulse, the solution to (11.15) is

$$[^3M^*(t)] = \frac{[^3M^*(0)]}{(1 + [^3M^*(0)]\tau k_\mathrm{q})\,\mathrm{e}^{t/\tau} - [^3M^*(0)]\tau k_\mathrm{q}} \, . \tag{11.16}$$

Assuming that the luminescence intensity (L) is linearly proportional to the concentration of excited states, i.e., $L(t) \propto [^3M^*(t)]/\tau$, then the time evolution of phosphorescent emission is derived from (11.16):

$$L(t) = \frac{L(0)}{(1 + K\tau)\mathrm{e}^{t/\tau} - K\tau} \, , \tag{11.17}$$

where K is defined by:

$$k_{\mathrm{q}} = \frac{K}{[{}^3M^*(0)]} \, . \tag{11.18}$$

The quantum efficiency of emission (η) can also be calculated from the steady state solution of (11.15):

$$\frac{\eta}{\eta_0} = \frac{J_{\mathrm{T}}}{4J}\left(\sqrt{1+8\frac{J}{J_{\mathrm{T}}}}-1\right) , \tag{11.19}$$

where η_0 is the quantum efficiency in the absence of T–T annihilation and

$$J_{\mathrm{T}} = \frac{2qd}{k_{\mathrm{q}}\tau^2} \tag{11.20}$$

is the onset current density for $\eta = \eta_0/2$.

To observe T–T annihilation, a film of 4% PtOEP doped in CBP was excited by a 500 ps pulse from a N_2 laser at a wavelength of $\lambda = 337\,\mathrm{nm}$. This photoexcitation generates singlet excitons on both the CBP and PtOEP molecules; the CBP singlets are rapidly transferred to PtOEP, and PtOEP singlets intersystem cross to the triplet state of PtOEP. From the phosphorescent spectra of CBP and PtOEP, it is known that the energy of the CBP triplet is significantly higher than that of PtOEP [58]. Hence, triplets are localized on PtOEP molecules and for times $\geq$10 ns after the optical pulse, the only excited species in the film are PtOEP triplets. As seen in Fig. 11.26,

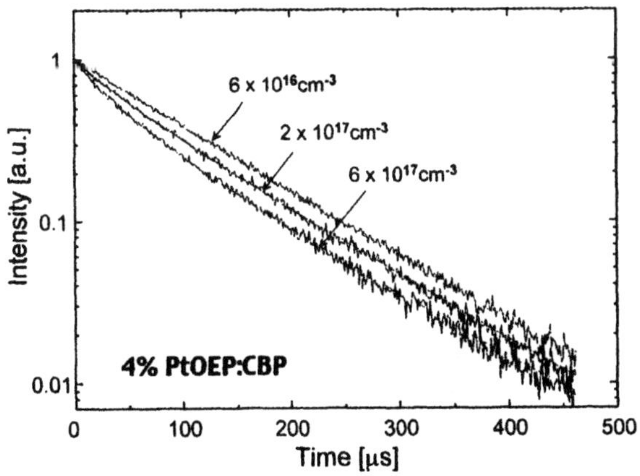

Fig. 11.26. Transient phosphorescent decays observed after the pulsed photoexcitation of a 4% PtOEP:CBP device. The initial concentration of PtOEP triplet excitons absorbed from the pump pulse are indicated. An intensity-dependent nonlinearity is observed as the triplet concentration increases. The curves are normalized for comparison. Fits to the transients (*smooth lines*) using (11.17) are also shown. (From [61])

with increasing pulse intensities, the transient decay of PtOEP deviates from a single exponential and becomes increasingly curved, indicating an intensity-dependent quenching of PtOEP triplets.

Transients were fit to (11.17), as shown in Fig. 11.26, to obtain τ and K. Both electro- and photoluminescent transients were studied, and for the electroluminescent transients $[^3M^*(0)]$ was determined by integrating the current in a 1 μs excitation pulse for the device structure shown in Fig. 11.27. It is assumed that trapped charge does not contribute to the current, and hence all charge carriers are presumed to form excitons. For the photoluminescent studies, $[^3M^*(0)]$ was calculated using an absorption coefficient at $\lambda = 337$ nm of 3×10^4 cm^{-1} for CBP [62].

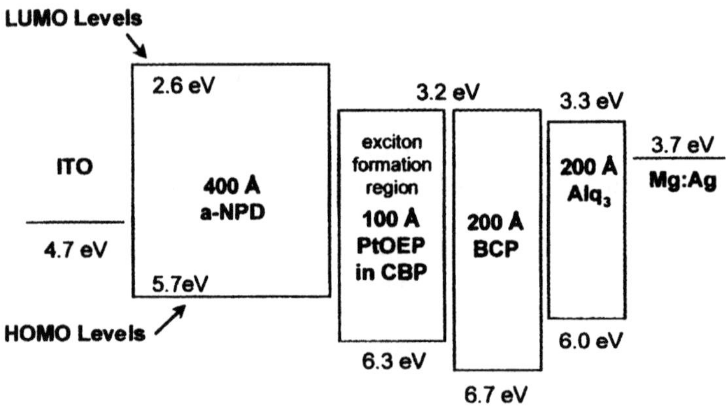

Fig. 11.27. Energy-level diagram of the electrophosphorescent devices. To control the concentration of triplet excitons, the devices contain a narrow recombination and luminescent zone. (From [61])

The dependence of τ and k_q on $[^3M^*(0)]$ in 8% PtOEP:CBP are plotted in Fig. 11.28; photoluminescent data are shown as squares, and electroluminescent data as circles. Within error, both PL and EL results give $k_q = (1.5 \pm 1) \times 10^{-14}$ cm^3 s^{-1}. The lifetime τ, however, varies significantly between PL and EL measurements. Now, τ is related to the radiative and non-radiative lifetimes (τ_r and τ_{nr}, respectively) via

$$\frac{1}{\tau} = \frac{1}{\tau_r} + \frac{1}{\tau_{nr}} . \tag{11.21}$$

The short EL lifetime of PtOEP:CBP is approximately independent of PtOEP concentration and excitation strength; therefore, it is assumed that the decrease in τ relative to the PL lifetimes is due to the electroluminescent cavity. Indeed, given the thickness of the organic layers employed in these OLEDs,

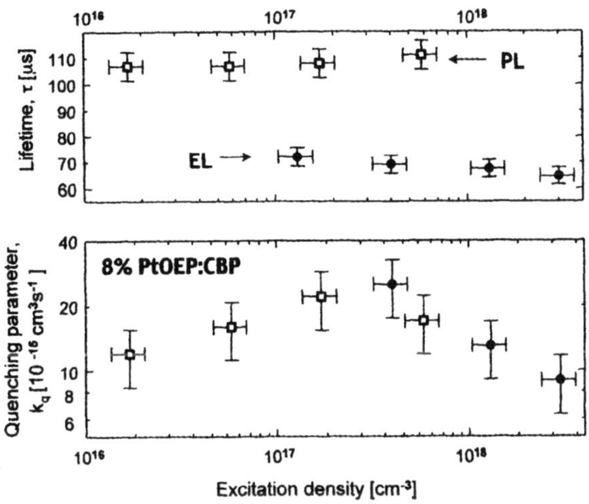

Fig. 11.28. Lifetime (τ) and biexcitonic quenching rate (k_q) as a function of the initial triplet concentration in 8% PtOEP:CBP. Data points marked by *squares* and *circles* are calculated from photoluminescent (PL) and electroluminescent (EL) decays, respectively. The difference between the PL and EL lifetimes is due to microcavity-induced absorption by the cathode in the EL structure. (From [61])

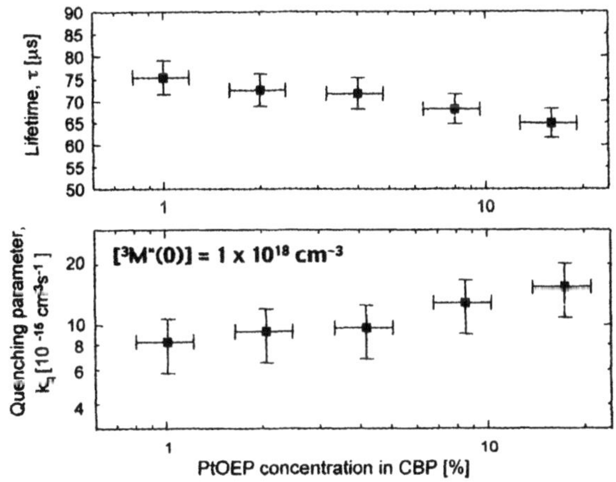

Fig. 11.29. Lifetime (τ) and biexcitonic quenching rate (k_q) as a function of the concentration of PtOEP in CBP. The data are from an electroluminescent device with an initial triplet concentration of $[M^*(0)] \sim 1\times10^{18}$ cm^{-3}. (From [61])

calculations confirm a ~25% reduction in the EL lifetime, as compared to the PL lifetime in the absence of a cavity [29].

The variation of k_q and τ with PtOEP concentration in a PtOEP:CBP electroluminescent device is shown in Fig. 11.29. The measurements were recorded at a triplet density of $[^3M^*(0)] \sim 1\times10^{18}$ cm^{-3}. As the concentration of PtOEP increases, τ decreases slightly, possibly due to an increase in the number of non-radiative pathways available to triplets in PtOEP aggregates – an example of concentration quenching frequently found in doped organic systems [1]. A very weak increase in k_q is also observed as the con-

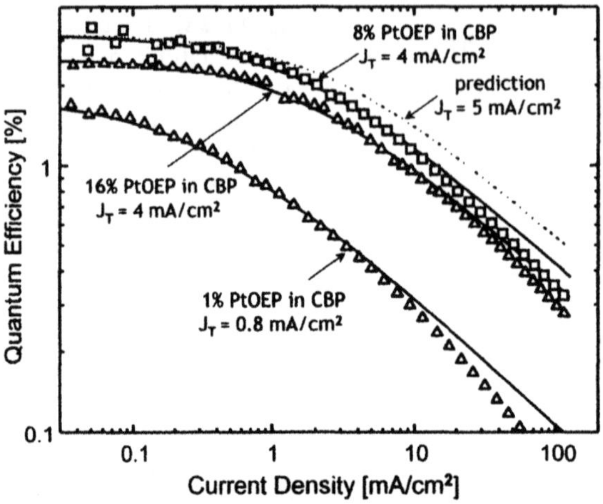

Fig. 11.30. External quantum efficiency of PtOEP:CBP devices. The curves are fit to (11.19) and demonstrate good agreement with the behavior expected for biexcitonic quenching. The only deviation occurs at high current densities ($\sim$100 mA/cm^2) where saturation of PtOEP is first apparent. Parameters extracted from Fig. 11.29 were used to predict an onset current for the 8% PtOEP:CBP device of $J_T = 5$ mA/cm^2. The best fit to the data yielded $J_T = 4$ mA/cm^2. (From [61])

centration of PtOEP increases. If triplets are assumed to localize on PtOEP molecules, triplet interactions and T–T annihilation should increase as percolation between adjacent or clustered PtOEP molecules is enhanced.

The external quantum efficiency of PtOEP:CBP electroluminescent devices is shown in Fig. 11.30. The characteristics of 1%, 8% and 16% PtOEP doped into CBP devices were fit to (11.19) to determine J_T. The characteristic of the 8% device was compared to a prediction made from the transient measurements of Figs. 11.28 and 11.29. Both the fits to theory and the prediction taken directly from the transient decay characteristics are accurate reproductions of the quantum efficiency trend observed in the data. The deviation from theory that occurs at high current densities ($\sim$100 mA/cm^2) is most probably due to the onset of saturation of PtOEP sites.

11.6 Combined Förster and Dexter Energy Transfer

11.6.1 Singlet-to-Triplet Exciton Ratio in Electrically Excited Thin Films

It is possible [4] to determine the ratio of singlet-to-triplet excitons formed during electrical excitation of Alq_3 by using a combination of Förster and Dexter energy transfer. It has already been established here that triplet excitons can be extracted from a semiconducting host material using a phosphorescent dye dopant [32, 44, 52, 57], and it is well known that singlets can be similarly extracted using a fluorescent dye [1]. Thus, it follows that the singlet fraction of excitons within a host (χ_s) can be determined from the ratio of fluorescence to phosphorescence if efficient energy transfer is possible to both types of dye from a single host molecule, such as Alq_3. In previous sections we described efficient phosphorescence of PtOEP:Alq_3 films [32], and efficient fluorescence of DCM2:Alq_3 films [1]. Furthermore, direct emission from Alq_3 originates from ligand-localized fluorescence [63] (i.e., no triplet emission), providing a second, independent measurement of the singlet fraction without the complication of energy transfer.

Fluorescent dyes such as DCM2 possess no phosphorescent properties and luminescence arises solely from singlet excitons. In a system characterized by strong Förster coupling such as DCM2:Alq_3, the singlet transfer rate from Alq_3 to DCM2 is approximately fifty times faster [4] than non-radiative losses in Alq_3. Hence, DCM2 emission accurately reflects the population of singlets formed in Alq_3. Similarly, if BCP is used as a barrier to triplet diffusion, it has been observed that the triplet transfer efficiency from Alq_3 to PtOEP is approximately 100% [64]. While Dexter transfer of triplets is most probably slower than Förster transfer of singlets, the longer lifetime of triplets increases the efficiency of the transfer. To quantify energy transfer from Alq_3 to DCM2 or PtOEP, the external EL quantum efficiency is written as [65,66]:

$$\Phi_{el} = \left[\chi_s \Phi_{fl} \eta_s + \chi_t \Phi_{ph} \eta_t\right] \eta_r \eta_e \,. \tag{11.22}$$

Here, $\chi_t = (1 - \chi_s)$ is the triplet fraction of excitons. Also, Φ_{fl} and Φ_{ph} are the photoluminescent efficiencies of fluorescence and phosphorescence of the acceptor, and are the transfer efficiencies of singlet and triplet excitons from D to A, η_r is the fraction of injected charge carriers that form excitons on the donor, and η_e is the fraction of emitted photons that are coupled out of the device. By quantitatively comparing the electroluminescent efficiencies ($\Phi_{el}^{(fl)}$ and $\Phi_{el}^{(ph)}$) of separate devices employing either a fluorescent or phosphorescent dye [4], it has been shown that (11.22) can be used to determine χ_s and χ_t, provided that η_r and η_e are identical in both devices. Furthermore, if all singlets are transferred in both cases (i.e., $\eta_s^{(fl)} = \eta_s^{(ph)} \sim 1$), then χ_s can be obtained from the ratio of fluorescent and phosphorescent efficiencies as follows:

$$\chi_s = \frac{\eta_t^{(ph)}}{\left(\dfrac{\Phi_{el}^{(ph)}}{\Phi_{el}^{(fl)}} \cdot \dfrac{\Phi_{pl}^{(fl)}}{\Phi_{pl}^{(ph)}}\right) - \left(1 - \eta_t^{(ph)}\right)} \,. \tag{11.23}$$

Since the ratios reflect relative measurements, the only absolute measurement required is the triplet-state transfer efficiency from the host to the phosphor (i.e., $\eta_t^{(ph)}$); and as discussed above, it is found that $\eta_t^{(ph)} \sim 1$.

To determine the ratio of the electroluminescent efficiency of PtOEP and DCM2 in Alq_3 (i.e., $\Phi_{el}^{(ph)}/\Phi_{el}^{(fl)}$), a series of devices similar to that in Fig. 11.27

(except with Alq_3 as the host in the exciton formation region) were made using either of these two materials doped into the Alq_3 host.

At current densities of 10^{-5} A/cm^2, the quantum efficiencies of devices in Fig. 11.31 are nearly constant and there is a low density of excited states.

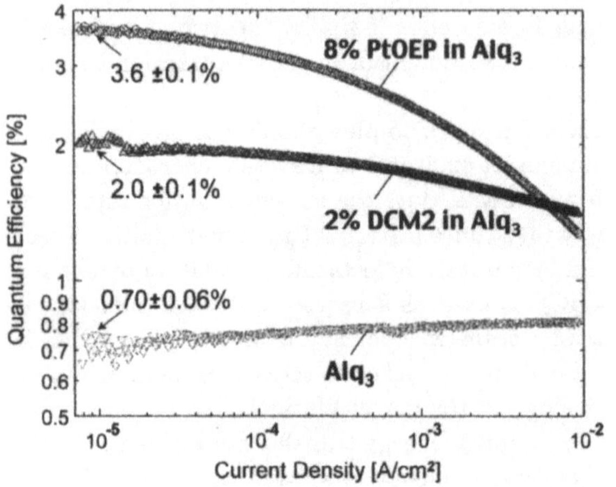

Fig. 11.31. Electroluminescent quantum efficiency of the DCM2-, PtOEP- and Alq_3-only devices. All curves are nearly constant in the low-current regime, where multi-exciton interactions leading to quenching are negligible. All devices contain an exciton- and hole-blocking layer consisting of BCP, as in Fig. 11.27. (From [4])

In this regime, we get $(\Phi_{\mathrm{el}}^{(\mathrm{DCM2})}/\Phi_{\mathrm{el}}^{(\mathrm{PtOEP})}) = 0.56 \pm 0.06$ and $(\Phi_{\mathrm{el}}^{(\mathrm{Alq_3})}/\Phi_{\mathrm{el}}^{(\mathrm{PtOEP})}) = 0.20 \pm 0.02$. The values of $(\Phi_{\mathrm{pl}}^{(\mathrm{PtOEP})}/\Phi_{\mathrm{pl}}^{(\mathrm{DCM2})})$, $(\Phi_{\mathrm{pl}}^{(\mathrm{PtOEP})}/\Phi_{\mathrm{pl}}^{(\mathrm{Alq_3})})$, $\eta_{\mathrm{s}}^{(\mathrm{DCM2})}$ and $\eta_{\mathrm{s}}^{(\mathrm{PtOEP})}$ were obtained by optically pumping doped films near the absorption peak of Alq_3 at a wavelength of $\lambda = 400$ nm [7]. As in the case of electrical pumping, low optical pump intensities of 0.2 mW/cm^2 were used to minimize bimolecular quenching due to dye-exciton interactions [32]. Equating the number of photons absorbed with the number of carriers injected in the EL devices, and assuming $\eta_{\mathrm{r}} = 1$, yields an equivalent current density of $\sim 10^{-5}$ A cm^{-2} for these optical pump intensities. The integrated PL efficiency ratios were $(\Phi_{\mathrm{pl}}^{(\mathrm{PtOEP})}/\Phi_{\mathrm{pl}}^{(\mathrm{DCM2})}) = 0.37 \pm 0.03$ and $(\Phi_{\mathrm{pl}}^{(\mathrm{PtOEP})}/\Phi_{\mathrm{pl}}^{(\mathrm{Alq_3})}) = 1.2 \pm 0.2$. The relative PL quantum efficiencies as a function of pump intensity are shown in Fig. 11.32.

The trends in the relative photoluminescence data of Fig. 11.32 match those observed in the electroluminescence data of Fig. 11.31, indicating that similar quenching phenomena occur in both cases. The rapid decrease in PtOEP:Alq_3 EL efficiency as the current density increases is not observed in PL, indicating that triplet annihilation in the host may be significant. These discrepancies are minimized by taking measurements at very low excitation densities.

No Alq_3 emission was observed in the doped films; hence, we conclude that $\eta_{\mathrm{s}}^{(\mathrm{fl})} = \eta_{\mathrm{s}}^{(\mathrm{ph})} = 1$. Given the previous results for the relative EL and PL quantum efficiencies of the devices containing BCP and a triplet energy trans-

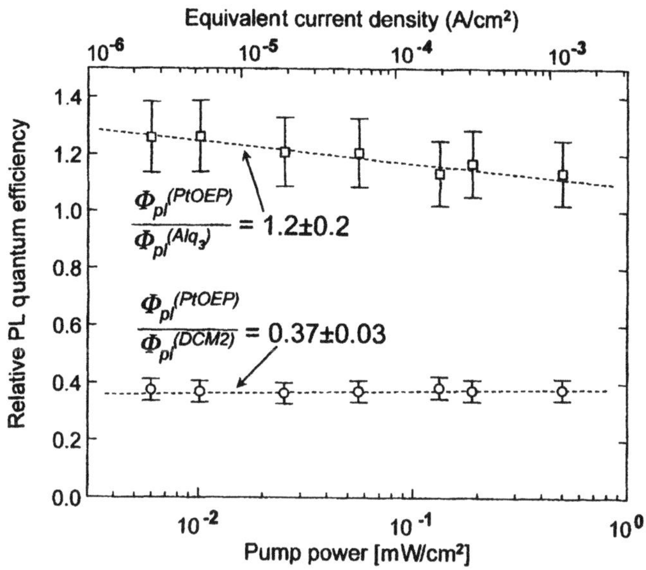

Fig. 11.32. Photoluminescent efficiencies of Alq_3 and DCM2:Alq_3 relative to PtOEP:Alq_3. The efficiencies are plotted versus pump power (at $\lambda = 400$ nm) and also, for comparison with the electroluminescent data, versus an estimation of the current density equivalent to the pump power. (From [4])

fer efficiency of $\eta_t^{(ph)} \sim 1$, (11.23) is used to obtain the spin multiplet fractions of $\chi_s = (21 \pm 3)\%$ for the DCM2/PtOEP system, and $\chi_s = (24 \pm 4)\%$ for the Alq_3/PtOEP system. Taking the weighted average of these results, we obtain an overall value for the singlet fraction in Alq_3 of $\chi_s = (22 \pm 3)\%$. This is in accordance with exciton statistics determined by the multiplicities of singlet and triplet states.

11.6.2 Phosphor-Sensitized Fluorescence

Since dipole–dipole energy transfer is possible if the donor and acceptor transitions are allowed, an interesting implication of Förster's theory is the possibility of long-range energy transfer from a phosphorescent material [67]. Such transitions are advantageous to the efficiency of OLEDs, but to understand why this is so, it is necessary to examine phosphorescence in greater detail.

In organic materials, the rate of all radiative transitions is at least partly determined by the requirement of spin conservation; given that the ground state of organic materials is typically spin anti-symmetric with total spin $S = 0$, only the decay of $S = 0$ excitons conserves spin, leading to fluorescence. Phosphorescence originates in the decay of $S = 1$ excitons and is typically extremely inefficient at room temperature. However, as noted in Sect. 11.6.1,

$S = 1$ excitons are three times as likely to be generated by electrical pumping than $S = 0$ excitons [4]. If luminescence can be obtained from triplets as well as singlets, then the maximum efficiency of an EL device could be quadrupled.

Singlet–triplet mixing using heavy metal atoms increases the oscillator strength of the triplet decay, although it is typically still several orders of magnitude below that of a spin-conserving fluorescent transition. However,

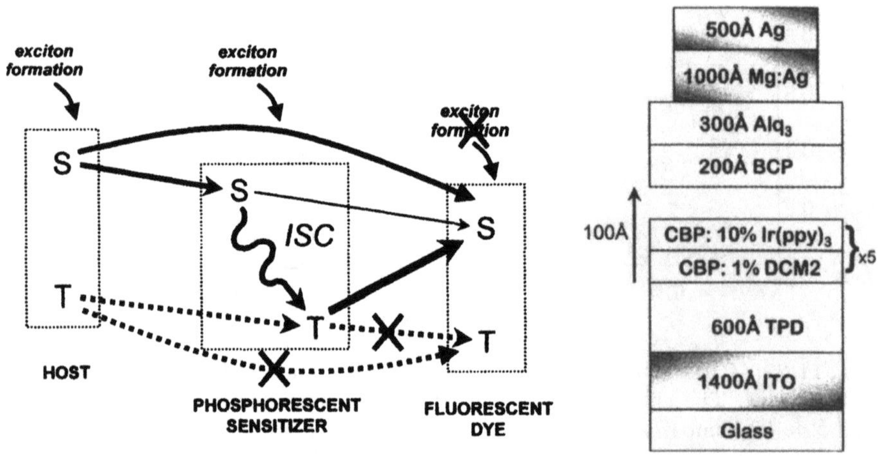

Fig. 11.33. Proposed energy-transfer mechanisms in a fluorescent guest–host system with a phosphorescent sensitizer. Ideally, all excitons are transferred to the singlet state of the fluorescent dye, as triplets in the dye non-radiatively recombine. Förster transfers are represented by *solid lines* and Dexter transfers by *dotted lines*. Electron–hole recombination creates singlet (S) and triplet (T) excitons in the host material. These excitons are then transferred to the phosphorescent sensitizer. There is also a lower probability of direct transfer into the singlet state of the fluorescent dye by a Förster process, or Dexter transfer into the triplet state. This latter mechanism is a source of loss and is indicated in the figure by a *cross*. Singlet excitons in the phosphor are then subject to intersystem crossing (ISC) and transfer to the triplet state. From this state, the triplets may either dipole–dipole couple with the singlet state of the fluorescent dye or, in another loss mechanism, they may Dexter transfer to the triplet state. Note that electron–hole recombination is also possible on the phosphor and fluorescent dye, and direct formation of triplets on the fluorescent dye is an additional path to loss. In fact, it is believed that trapping on $Ir(ppy)_3$ and DCM2 is the primary source of exciton formation in CBP:$Ir(ppy)_3$:DCM2. Inset: The structure of electroluminescent devices used to study phosphor-sensitized fluorescence. To demonstrate the multiple-stage transfer, CBP is used as the host material [44], the green-emitting phosphor [52] $Ir(ppy)_3$ as the sensitizer, and the red-fluorescent dye [1] DCM2 as the acceptor. An alternating series of five 10-Å-thick layers of 10% $Ir(ppy)_3$:CBP followed by 1% DCM2:CBP was used to encourage energy transfer. Excitons were confined within the luminescent region by a layer of BCP. (From [68])

since triplet states have extended lifetimes, energy transfer may still be very efficient, as can be seen by analyzing the efficiency of energy transfer (η_{ET}):

$$\eta_{\mathrm{ET}} = \frac{K_{\mathrm{ET}}}{K_{\mathrm{ET}} + K_{\mathrm{R}} + K_{\mathrm{NR}}} . \tag{11.24}$$

Here, K_{ET} is the rate of Förster energy transfer from D to A, and K_{R} and K_{NR} are the radiative and non-radiative rates on the donor, respectively. In Förster's theory, K_{ET} is proportional to the oscillator strength of the donor transition [59], as is K_{R}. Thus, η_{ET} is approximately independent of oscillator strength if $K_{\mathrm{R}} \gg K_{\mathrm{NR}}$, i.e., *if the donor is efficiently phosphorescent then it is possible to obtain triplet–singlet energy transfer via the Förster mechanism.* Long-range triplet–singlet transfers were predicted by Förster [67] and confirmed by Ermolaev and Sveshnikova [64], using a range of phosphorescent donors and fluorescent acceptors in rigid media. Large transfer distances were observed, for example, with triphenylamine as the donor and chrysoidine as the acceptor, the interaction range is 52Å.

Triplet–singlet transfers via the Förster mechanism have also been used to generate high-efficiency fluorescence from an OLED [68]. For this application, both the phosphorescent donor and the fluorescent acceptor were doped into a charge-transporting host material. The fluorescent dye was DCM2, the phosphor was $\mathrm{Ir(ppy)_3}$ and the host material was CBP. To ensure energy transfer between the phosphor and the fluorescent dye, the materials are chosen such that the emission spectrum of the phosphor overlaps the absorption spectrum of the fluorescent dye. If excitons are formed in the host material, then the phosphor may act as a "sensitizer" for triplet transfer from the host to the singlet state of the fluorescent dye, as shown in Fig. 11.33.

Note, however, that in the case of CBP:$\mathrm{Ir(ppy)_3}$:DCM2, excitons are formed predominantly after charge-trapping [58] on $\mathrm{Ir(ppy)_3}$ and DCM2. To

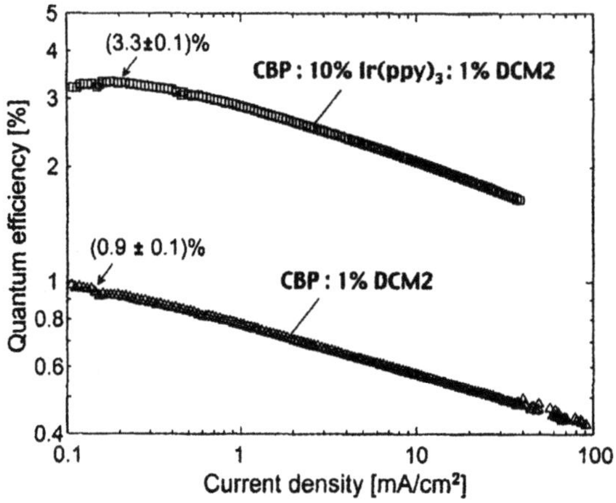

Fig. 11.34. The external quantum efficiencies of DCM2 emission with and without $\mathrm{Ir(ppy)_3}$. The sensitizing action of $\mathrm{Ir(ppy)_3}$ improves the efficiency by more than a factor of three. (From [68])

avoid loss in efficiency, direct triplet exciton formation on DCM2 should be minimized; thus, the doping fraction of $Ir(ppy)_3$ is significantly higher (8% by mass) than the concentration of DCM2 (1%). A low fraction of DCM2 also assists in the prevention of concentration quenching [1] of its emission. In the inset of Fig. 11.33, the structure of an OLED exhibiting triplet-to-singlet energy transfer is shown. The external quantum efficiency of this device with and without $Ir(ppy)_3$ is shown in Fig. 11.34.

Since the emission spectrum is comprised of >80% DCM2 in both cases, it is evident that the presence of the phosphor increases the efficiency of fluorescence in the device by more than three times.

Conclusive evidence of energy transfer from the triplet state of $Ir(ppy)_3$ to DCM2 was obtained by transient analysis of the emission spectrum of an OLED containing $Ir(ppy)_3$ and DCM2 in CBP. In Fig. 11.35, it can be observed that after some initial DCM2 fluorescence, DCM2 emission decays at the same rate as a small amount of residual $Ir(ppy)_3$ phosphorescence. Moreover, the decay rate of $Ir(ppy)_3$ is considerably increased in the presence of DCM2 (from ~500 ns to ~100 ns), indicating quenching of $Ir(ppy)_3$ triplets. This technique has the potential to quadruple the efficiency of fluorescence by employing the energy of both triplets and singlets. It also exhibits the advantages of Förster transfer, notably large interaction lengths, high transfer rates and high efficiencies.

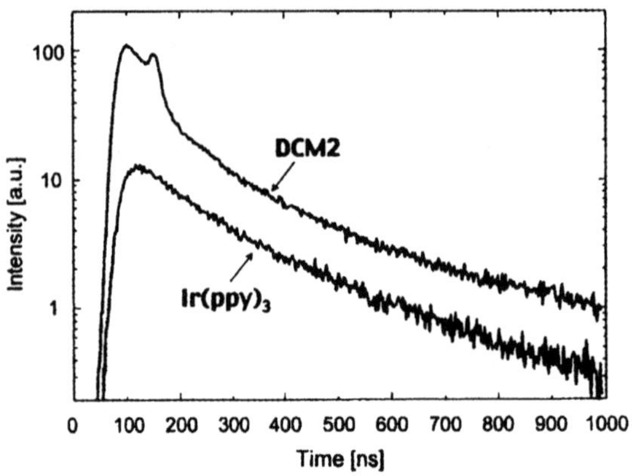

Fig. 11.35. The transient response of the DCM2 and $Ir(ppy)_3$ spectral components in the CBP:10% $Ir(ppy)_3$:1% DCM2 device. The transient lifetime of DCM2 is ~1 ns; thus, in the case of energy transfer from $Ir(ppy)_3$, the response of DCM2 should be governed by the transient lifetime of $Ir(ppy)_3$. After the initial 100-ns-wide electrical excitation pulse, this is clearly the case, demonstrating that energy is transferred from the triplet state in $Ir(ppy)_3$ to the singlet state in DCM2. During the excitation pulse, singlet transfer to DCM2 is observed, resulting in the ripples in the transient response. These ripples are due to fluctuations in the current density and the discharge of traps at the falling edge of the pulse. (From [68])

11.7 Concluding Remarks

The demonstrations of energy-transfer processes described in this chapter would not be possible without the ability to grow molecular organic structures containing molecular-scale layers of high purity. In fact, the increasing understanding of the behavior and properties of OLEDs in particular has turned these devices into valuable test vehicles for the examination of optical and electronic processes in organic materials. Reflecting this codependency, we have provided several examples of effects isolated in OLEDs with broad implications for the general understanding of molecular materials. Principal among these is SSSE; given its ubiquity and technological potential, it promises to play an important role in future research on organic solid-state light emitters. Indeed, the combination of SSSE and high-resolution imaging techniques such as near field scanning optical microscopy (NSOM) promises new insights into the molecular-scale behavior of organic materials. Whereas SSSE allows the examination of microscopic properties of organic materials such as local field effects, phosphorescence has revealed many of the previously obscure properties of triplet excitons. This too has significant technological ramifications, not the least of which is the potential for quadrupling the efficiencies of fluorescent OLEDs. However, a thorough understanding of triplet energy transfer and diffusion is now also possible, and no doubt this will in turn alter the design of OLEDs. Thus, with materials sensitive to microscopic phenomena and the ability to precisely fabricate molecular-scale features, we have gathered the tools to facilitate greater understanding of organic materials.

Acknowledgments. We would like to acknowledge our colleagues M. E. Thompson, R. Deshpande, D. O'Brien, Z. Shen, and P. E. Burrows for participation in the studies discussed in this chapter. We are also grateful to the National Science Foundation Materials Research Science and Engineering Center, Defense Advanced Research Projects Agency, the Air Force Office of Scientific Research, and Universal Display Corporation for their generous support of this research.

References

1. C.W. Tang, S.A. VanSlyke, C.H. Chen, J. Appl. Phys. **65**, 3610 (1989)
2. V.G. Kozlov, V. Bulović, P.E. Burrows, M. Baldo, V.B. Khalfin, G. Parthasarathy, S.R. Forrest, J. Appl. Phys. **84**, 4096 (1998)
3. S.R. Forrest, Chem. Rev. **97**, 1793 (1997)
4. M.A. Baldo, D.F. O'Brien, M.E. Thompson, S.R. Forrest, Phys. Rev. B **60**, 14422 (1999)
5. P.F. Barbara, D.M. Adams, D.B. O'Connor, Annu. Rev. Mat. Sci. **29**, 433 (1999)

6. J.A. Barltrop, J.D. Coyle, *Principles of Photochemistry* (John Wiley & Sons, New York 1978)
7. D.Z. Garbuzov, V. Bulović, P.E. Burrows, S.R. Forrest, Chem. Phys. Lett. **249**, 433 (1996)
8. B. Pullman (ed.) *Environmental Effects on Molecular Structure and Properties, Proceedings of the Eight Jerusalem Symposium on Quantum Chemistry and Biochemistry* (D. Reidel Publishing Company, Boston 1976)
9. C. Reichardt, *Solvents and Solvent Effects in Organic Chemistry* (VCH Verlagsgesellschaft mbH, Weinheim 1988)
10. V. Bulović, R. S. Deshpande, S. R. Forrest, Chem. Phys. Lett. **308**, 317 (1999)
11. M. Pope, C.E. Swenberg, *Electronic Processes in Organic Crystals* (Oxford University Press, Oxford 1982)
12. G.H. Wannier, Phys. Rev. **52**, 191 (1937)
13. N.F. Mott, Trans. Faraday. Soc. **34**, 500 (1938)
14. J. Frenkel, Phys. Rev. **37**, 17 (1931); **37**, 1276 (1931)
15. Z. Shen, S.R. Forrest, Phys. Rev. B **55**, 10578 (1997)
16. E.A. Silinsh, V. Čápek, *Organic Molecular Crystals* (American Institute of Physics, New York 1994)
17. T. Förster, In: *Modern Quantum Chemistry*, Part 2, ed. by O. Sinanoglu (Academic Press, 1965)
18. D.L. Dexter, J. Chem. Phys. **21**, 838 (1953)
19. A.A. Lamola, Photochem. Photobiol. **8**, 601 (1968)
20. G.L. Closs, M.D. Johnson, J.R. Miller, P. Piotrowiak, J. Am. Chem. Soc. **111**, 3751 (1989)
21. K.C. Kao, W. Hwang, *Electrical Transport in Solids* (Pergamon Press, Oxford 1981)
22. C. Adachi, T. Tsutsui, S. Saito, Optoelect. Devices Technol. **6**, 25 (1991)
23. V. Bulović, PhD thesis, Princeton University (1998)
24. C.W. Tang, S.A VanSlyke, Appl. Phys. Lett. **51**, 913 (1987)
25. M. Stolka, J.F. Janus, D.M. Pai, J. Phys. Chem. **88**, 4707 (1984)
26. R.G. Kepler, P.M. Beeson, S.J. Jacobs, R.A. Anderson, M.B. Sinclair, V.S. Valencia, P.A. Cahill, Appl. Phys. Lett. **66**, 3618 (1995)
27. L.-B. Lin, S.A. Jenekhe, R.H. Young, P.M. Borsenberger, Appl. Phys. Lett. **70**, 2052 (1997)
28. J. Kalinowski, G. Giro, P.D. Marco, N. Camaioni, V. Fattori, Chem. Phys. Lett. **265**, 607 (1997)
29. V. Bulović, V.B. Khalfin, G. Gu, P.E. Burrows, S.R. Forrest, Phys. Rev. B **58**, 3730 (1998)
30. C.W. Tang, *Society for Information Display, International Symposium*, Digest Tech. Pap. **27**, 181 (1996)
31. V. Bulović, A. Shoustikov, M.A. Baldo, E. Bose, V.G. Kozlov, M.E. Thompson, S.R. Forrest, Chem. Phys. Lett. **287**, 455 (1998)
32. M.A. Baldo, D.F. O'Brien, Y. You, A. Shoustikov, S. Sibley, M.E. Thompson, S.R. Forrest, Nature **395**, 151 (1998)
33. M. Berggren, A. Dodabalapur, R.E. Slusher, Z. Bao, Nature **389**, 466 (1997)
34. R.S. Deshpande, V. Bulović, S.R. Forrest, Appl. Phys. Lett. **75**, 888 (1999)
35. R.H. Jordan, A. Dodabalapur, M. Strukelj, T.M. Miller, Appl. Phys. Lett. **68**, 1192 (1996)
36. J. Kido, W. Ikeda, M. Kimura, K. Nagai, Jpn. J. Appl. Phys. Part 2 **35**, L394 (1996)

37. M. Granström, O. Inganas, Appl. Phys. Lett. **68**, 147 (1996)
38. S. Tasch, E. J. W. List, O. Ekström, W. Graupner, G. Leising, P. Schlichting, U. Rohr, Y. Geerts, U. Scherf, K. Müllen, Appl. Phys. Lett. **71**, 2883 (1997)
39. J. Kido, H. Shionoya, K. Nagai, Appl. Phys. Lett. **67**, 2281 (1995)
40. T. Christ, A. Greiner, R. Sander, V. Stümpflen, J. Wendorff, Adv. Mat. **9**, 219 (1997)
41. J. Kido, M. Kimura, K. Nagai, Science **267**, 1332 (1995)
42. CIE standard colorimetric illuminants, Document Number: ISO/CIE 10526 (ANSI, New York 1991)
43. P.E. Burrows, S.R. Forrest, Appl. Phys. Lett. **64**, 2285 (1993)
44. D.F. O'Brien, M.A. Baldo, M.E. Thompson, S.R. Forrest, Appl. Phys. Lett. **74**, 442 (1999)
45. M.M. Martin, P. Plaza, Y.H. Meyer, Chem. Phys. **192**, 367 (1995)
46. W.P. Anderson, T.R. Cundari, R.S. Drago, M.C. Zerner, Inorg. Chem. **29**, 1 (1990)
47. J. Kalinowski, W. Stampor, P. Di Marco, V. Fattori, Chem. Phys. **182**, 341 (1994)
48. Y. Kijima, N. Asai, S. Tamura (eds.) *Proc. of the Mat. Res. Soc., Spring '98 Meeting*, paper G2.1, San Francisco, California (April 13–17, 1998)
49. P.M. Borsenberger, J.J. Fitzgerald, J. Phys. Chem. **97**, 4815 (1993)
50. H. Schmidbaur, J. Lettenbauer, D.L. Wilkinson, G. Muller, O. Kumberger, Z. Naturforsch. B **46**, 901 (1991)
51. J. Kido, Y. Iizumi, Appl. Phys. Lett. **73**, 2721 (1998)
52. M.A. Baldo, S. Lamansky, P.E. Burrows, M.E. Thompson, S.R. Forrest, Appl. Phys. Lett. **75**, 4 (1999)
53. M.A. Baldo, M.E. Thompson, S.R. Forrest, Pure Appl. Chem. (2000)
54. A. Mills, A. Lepre, Anal. Chem. **69**, 4653 (1997)
55. G. Ponterini, N. Serpone, M.A. Bergkamp, T.L. Netzel, J. Am. Chem. Soc. **105**, 4639 (1983)
56. D.B. Papkovski, Sensors and Actuators B **29**, 213 (1995)
57. V. Cleave, G. Yahioglu, P. Le Barny, R.H. Friend, N. Tessler, Adv. Mat. **11**, 285 (1999)
58. M.A. Baldo, S.R. Forrest, Phys. Rev. B **62**, 10958 (2000)
59. M. Klessinger, J. Michl, *Excited states and photochemistry of organic molecules* (VCH Publishers, New York 1995)
60. N.J. Turro, *Modern Molecular Photochemistry* (University Science Books, Mill Valley 1991)
61. M.A. Baldo, C. Adachi, S.R. Forrest, Phys. Rev. B **62**, 10967 (2000)
62. C. Adachi, M.A. Baldo, M.E. Thompson, S.R. Forrest, unpublished
63. R. Ballardini, G. Varani, M.T. Indelli, F. Scandola, Inorg. Chem. **25**, 3858 (1986)
64. V.L. Ermolaev, E.B. Sveshnikova, Dokl. Akad. Nauk. SSSR **149**, 1295 (1963)
65. T. Tsutsui, S. Saito, In: *Intrinsically Conducting Polymers: An Emerging Technology*, ed. by M. Aldissi (Kluwer Academic, Dordrecht 1993) pp. 123
66. L.J. Rothberg, A.J. Lovinger, J. Mat. Res. **11**, 3174 (1996)
67. T. Förster, Discuss. Faraday Soc. **27**, 7 (1959)
68. M.A. Baldo, M.E. Thompson, S.R. Forrest, Nature **403**, 750 (2000)

Index

Printing and Binding: AZ Druck und Datentechnik GmbH, Kempten

Springer Series in

MATERIALS SCIENCE

Editors: R. Hull R. M. Osgood, Jr. H. Sakaki A. Zunger

1 **Chemical Processing with Lasers***
By D. Bäuerle

2 **Laser-Beam Interactions with Materials**
Physical Principles and Applications
By M. von Allmen and A. Blatter
2nd Edition

3 **Laser Processing of Thin Films and Microstructures**
Oxidation, Deposition and Etching of Insulators
By. I. W. Boyd

4 **Microclusters**
Editors: S. Sugano, Y. Nishina, and S. Ohnishi

5 **Graphite Fibers and Filaments**
By M. S. Dresselhaus, G. Dresselhaus, K. Sugihara, I. L. Spain, and H. A. Goldberg

6 **Elemental and Molecular Clusters**
Editors: G. Benedek, T. P. Martin, and G. Pacchioni

7 **Molecular Beam Epitaxy**
Fundamentals and Current Status
By M. A. Herman and H. Sitter 2nd Edition

8 **Physical Chemistry of, in and on Silicon**
By G. F. Cerofolini and L. Meda

9 **Tritium and Helium-3 in Metals**
By R. Lässer

10 **Computer Simulation of Ion-Solid Interactions**
By W. Eckstein

11 **Mechanisms of High Temperature Superconductivity**
Editors: H. Kamimura and A. Oshiyama

12 **Dislocation Dynamics and Plasticity**
By T. Suzuki, S. Takeuchi, and H. Yoshinaga

13 **Semiconductor Silicon**
Materials Science and Technology
Editors: G. Harbeke and M. J. Schulz

14 **Graphite Intercalation Compounds I**
Structure and Dynamics
Editors: H. Zabel and S. A. Solin

15 **Crystal Chemistry of High-T_c Superconducting Copper Oxides**
By B. Raveau, C. Michel, M. Hervieu, and D. Groult

16 **Hydrogen in Semiconductors**
By S. J. Pearton, M. Stavola, and J. W. Corbett

17 **Ordering at Surfaces and Interfaces**
Editors: A. Yoshimori, T. Shinjo, and H. Watanabe

18 **Graphite Intercalation Compounds II**
Editors: S. A. Solin and H. Zabel

19 **Laser-Assisted Microtechnology**
By S. M. Metev and V. P. Veiko
2nd Edition

20 **Microcluster Physics**
By S. Sugano and H. Koizumi
2nd Edition

21 **The Metal-Hydrogen System**
By Y. Fukai

22 **Ion Implantation in Diamond, Graphite and Related Materials**
By M. S. Dresselhaus and R. Kalish

23 **The Real Structure of High-T_c Superconductors**
Editor: V. Sh. Shekhtman

24 **Metal Impurities in Silicon-Device Fabrication**
By K. Graff 2nd Edition

25 **Optical Properties of Metal Clusters**
By U. Kreibig and M. Vollmer

26 **Gas Source Molecular Beam Epitaxy**
Growth and Properties of Phosphorus Containing III–V Heterostructures
By M. B. Panish and H. Temkin

* The 2nd edition is available as a textbook with the title: *Laser Processing and Chemistry*

MIX
Papier aus verantwortungsvollen Quellen
Paper from responsible sources
FSC® C105338

If you have any concerns about our products,
you can contact us on
ProductSafety@springernature.com

In case Publisher is established outside the EU,
the EU authorized representative is:
Springer Nature Customer Service Center GmbH
Europaplatz 3, 69115 Heidelberg, Germany

Printed by Libri Plureos GmbH
in Hamburg, Germany